Lecture Notes in Electrical Engineering 237

For further volumes:
http://www.springer.com/series/7818

Wego Wang
Editor

Mechatronics and Automatic Control Systems

Proceedings of the 2013 International
Conference on Mechatronics
and Automatic Control Systems (ICMS2013)

Volume 2

Editor
Wego Wang
University of Massachusetts Lowell
Wellesley, MA, USA

ISSN 1876-1100 ISSN 1876-1119 (electronic)
ISBN 978-3-319-01272-8 ISBN 978-3-319-01273-5 (eBook)
DOI 10.1007/978-3-319-01273-5
Springer Cham Heidelberg New York Dordrecht London

Library of Congress Control Number: 2013950379

© Springer International Publishing Switzerland 2014
This work is subject to copyright. All rights are reserved by the Publisher, whether the whole or part of the material is concerned, specifically the rights of translation, reprinting, reuse of illustrations, recitation, broadcasting, reproduction on microfilms or in any other physical way, and transmission or information storage and retrieval, electronic adaptation, computer software, or by similar or dissimilar methodology now known or hereafter developed. Exempted from this legal reservation are brief excerpts in connection with reviews or scholarly analysis or material supplied specifically for the purpose of being entered and executed on a computer system, for exclusive use by the purchaser of the work. Duplication of this publication or parts thereof is permitted only under the provisions of the Copyright Law of the Publisher's location, in its current version, and permission for use must always be obtained from Springer. Permissions for use may be obtained through RightsLink at the Copyright Clearance Center. Violations are liable to prosecution under the respective Copyright Law.
The use of general descriptive names, registered names, trademarks, service marks, etc. in this publication does not imply, even in the absence of a specific statement, that such names are exempt from the relevant protective laws and regulations and therefore free for general use.
While the advice and information in this book are believed to be true and accurate at the date of publication, neither the authors nor the editors nor the publisher can accept any legal responsibility for any errors or omissions that may be made. The publisher makes no warranty, express or implied, with respect to the material contained herein.

Printed on acid-free paper

Springer is part of Springer Science+Business Media (www.springer.com)

Preface

This book contains papers accepted by the 2013 International Conference on Mechatronics and Automatic Control Systems (ICMS), which was held on 10 and 11 August, 2013, in Hangzhou, China.

It covers emerging technologies in a timely manner for two important subjects: mechatronics and automatic control systems. In particular, it allows researchers and students to become familiar with state-of-the-art developments in the field, and it also helps engineers and practitioners enhance their productivity with the latest technology.

The book is composed of two volumes. The first volume has 64 papers focusing on Advanced Mechanics and Mechatronics (Part I), Intelligent Systems and Algorithms (Part II), and Manufacturing Engineering and Engineering Systems (Part III); the second includes 56 papers in the areas of Control Theory and Application (Part IV), Data Mining and Application (Part V), Sensing Control Theory (Part VI), and Signal Processing and Control (Part VII). Both volumes will serve as references for practitioners, faculty, and students to help them quickly incorporate into their work the most recent advancements in these fields and state-of-the-practice of mechanics, electronics, and computing, to efficiently and effectively improve the reliability and feasibility of high-traffic and mission-critical systems.

We extend our gratitude not only to the authors for contributing their ideas, works, and experiences but also to the reviewers for guiding the evaluation and selection of high-quality papers – this book would not have been possible without them. Special thanks also to Mr. Brett Kurzman and Miss Rebecca Hytowitz at Springer US for their valuable assistance with logistical issues throughout the publication process.

Wellesley, MA, USA   Wego Wang

ICMS2013 Committee

Advisory Chair
Andrew Lu   U.S. Jiangsu Economy, Trade and Culture Association, USA
Aniruddha Bhattacharjya   CSE Department, Amrita School of Engineering, Amrita University, Bangalore, India
Jiaxi Hu   Corporate-Social-Responsibility Zhuzhou Institute Co. Ltd, China
Rakesh Kumar Tripathi   Devi Ahilya university Indore, India

General Chair
Wego Wang   University of Massachusetts Lowell
Yudong Zhao   Columbia University, USA
Zhanming Wei   Fortinet Technologies Inc, Sunnyvale, USA

Program Chair
PakKin Wong   Macau University, China
Shu-lung Wang   Taoyuan Innovation Institute of Technology, Taiwan, China
Tianhong Yan   China Jiliang University, China

Program Committee
Abdalhossein Rezai   Academic Center for Education, Culture and Research and Semnan University, Iran
Aihua Mao   South China University of Technology, China
Amitabha Chanda   University Collage of Science, Technology & Agriculture, University of Calcutta, India
Andres Ivan Oliva Arias   Cinvestav IPN Unidad Mérida, Spain
Ashwini Kumar   Institute for Education by Radio-Television Allahabad, India
Baiqiang You   Xiamen University, China
Caihua Kang   National Chiao Tung University, Taiwan, China
Chadwick Carreto Arellano   National Polytechnic Institute Interpenetrating Polymer Network, México
Denis V. Bogdanovich   Moscow State Technical University, Russia
Dongliang Wang   Army Aviation Institute, China
Fuli Zhao   Sun Yat-Sen University, China
Guojie Qin   Beijing Institute of Technology, China
Hong Hu   Harbin Institute of Technology, China

José RAGOT	Université de Lorraine, France
Kleanthis Thramboulidis	University of Patras, Greece
Kumar Dookhitram	University of Technology, Mauritius
Kuswara Setiawan	UPH Surabaya, Indonesia
Long Zhang	University of Science & Technology Beijing, China
Marco Ivan Ramirez Sosa Moran	Instituto Tecnologico de Nuevo Leon, México
Mohamed Chtourou	National School of Engineering, Sfax-Tunisia
Muhammad Naufal Bin Mansor	University Malaysia Perlis
Rajiv Kumar Chechi	Haryana College of Technology and Management Technical campus, India
Reggie C. Gustilo	De La Salle University Manila, Philippines
Sabina Jeschke	Rheinisch-Westfaelische Technische Hochschule Aachen University, Germany
Taiwo	Obafemi Awolowo University, Nigeria
Tao Li	Institute of Electrical Engineering Chinese Academy of Sciences, China
Tiecheng Xia	Shanghai University, China
Tse-Chen Yeh	Academia Sinica, Taiwan, China
Wenbang Sun	The Airforce and Aeronautical University of the People's Liberation Army, China
Wenyong Gong	Jilin University, China
Wenzheng Lai	Ming Chi University Of Technology, Taiwan, China
Xianqiang Yan	University of Science & Technology, Beijing, China
Xiaobing Sun	Southeast University, China
Xiaoshan Chen	Naval Engineering University of the People's Liberation Army, China
Xingfei Yuan	Zhejiang University, China
Yanbo Zhang	Henan University, China
Yang Zhang	Emhiser Research, Inc., USA
Yi Li	Hefei Nova Institute of Technology, China
Yudong Gao	National University of Defense Technology, China
Yu-ying Shi	North China Electric Power University, China
Yuxia Sun	Jinan University, China
Zielinska Teresa	Warsaw University of Technology, Poland

Contents

Part IV Control Theory and Application

Monte Carlo Based Predictive Method for Determining Work
Envelope of Spacesuit in EVA Operation.................................. 583
Huaiji Si, Qianfang Liao, and Wanxin Zhang

Reduced Thrust Take-Off of Large Passenger Aircraft Based
on Derate Method... 591
Xinmin Wang, Haitao Yin, Yi Zheng, and Rong Xie

Permanent Magnet Synchronous Motor Feedback
Linearization Vector Control.. 601
Hehua Wang and Xiaohe Liu

Design and Simulation of Image Compensation
Control System.. 609
Chan Tan and Lei Ding

µ-Method for Robust Stability of Active Aeroelastic Wing
with Multiple Control Surfaces... 617
Fu-hu Liu, Xiao-ping Ma, and Zi-jian Zhang

Nonlinear Flight Controller Design Using Combined
Hierarchy-Structured Dynamic Inversion and Constrained
Model Predictive Control.. 629
Chao Wang, Shengxiu Zhang, and Chao Zhang

Cooperative Multitasking Software Design for Gas Pressure
Control Based on Embedded Microcontroller System................. 641
Pubin Wang and Longxiang Lou

Robust H_∞ Filtering for Uncertain Switched Systems
Under Asynchronous.. 649
Guihua Li and Jun Cheng

A Kind of Adaptive Backstepping Sliding Model Controller Design for Hypersonic Reentry Vehicle 657
Congchao Yao, Xinmin Wang, Yao Huang, and Yuyan Cao

Design of Smith Auto Disturbance Rejection Controller for Aero-engine 665
Fang-Zheng Luo, Shi-Ying Zhang, Min Chen, and Yu Hu

Asynchronous Motor Vector Control System Based on Space Vector Pulse Width Modulation 675
YingZhan Hu and SuNa Guo

Design of Direct Current Subsynchronous Damping Controller (SSDC) 683
Shiwu Xiao, Xiaojuan Kang, Jianhui Liu, and Xianglong Chen

Three Current-Mode Wien-Bridge Oscillators Using Single Modified Current Controlled Current Differencing Transconductance Amplifier 693
Yongan Li

Improved Phase-Locked Loop Based on the Load Peak Current Comparison Frequency Tracking Technology 701
Bingxin Qi, Hui Zhu, Yonglong Peng, and Yabin Li

Optimization of Low-Thrust Orbit Transfer 711
Zhaohua Qin, Min Xu, and Xiaomin An

Self-Healing Control Method Based on Hybrid Control Theory 719
Qiang Zhao, Meng Zhou, and XinQian Wu

PMSM Sensorless Vector Control System Based on Single Shunt Current Sensing 727
Hongyan Ma

Wind Power System Simulation of Switch Control 737
Yuehua Huang, Guangxu Li, and Huanhuan Li

Simulation of Variable-Depth Motion Control for the High-Speed Underwater Vehicle 745
Tao Bai and Yuntao Han

A Low Power Received Signal Strength Indicator for Short Distance Receiver 755
Qianqian Lei, Erhu Zhao, Min Lin, and Yin Shi

Ship Shaft Generator Control Based on Dynamic Recurrent Neural Network Self-Tuning PID 765
Ming Sun

**Control System Design and Simulation
of Microelectromechanical Hybrid Gyroscope**............ 773
Haiyan Xue, Bo Yang, and Shourong Wang

Part V Data Mining and Application

**New Detection Technology Based on the Theory of Eddy
Current Loss**.. 783
Yafei Si and Jianxin Chen

**Pedestrian Detection Based on Road Surface Extraction
in Pedestrian Protection System**........................ 793
Hao Heng and Huilin Xiong

**Methods on Reliability Analysis of Friction Coefficient Test
Instrument**.. 801
Tianrong Zhu, Xizhu Tao, Xinsheng Xu, and Tianhong Yan

**Design and Implementation of a New Power Transducer
of Switch Machine**..................................... 811
Yanli Wang and Bin Li

**Data Model Research of Subdivision Cell Template Based
on EMD Model**.. 817
Delan Xiong and Qingzhou Xu

**A Spatial Architecture Model of Internet of Things Based
on Triangular Pyramid**................................. 825
Weidong Fang, Lianhai Shan, Zhidong Shi, Guoqing Jia,
and Xin Wang

Design and Realization of Fault Monitor Module........ 833
Dongmei Zhao, Huanhuan Dong, and Xiang Xue

**Dynamical System Identification of Complex Nonlinear System
Based on Phase Space Topological Features**............. 843
Lei Nie, Zhaocheng Yang, Jin Yang, and Weidong Jiang

**Field Programmable Gate Array Configuration Monitoring
Technology for Space-Based Systems**.................... 851
Longxu Jin, Jin Li, and Yinan Wu

**HVS-Inspired, Parallel, and Hierarchical Traffic
Scene Classification Framework**........................ 859
Wengang Feng

**Analysis and Application on Reactive Power Compensation
Online Monitoring System of 10 kV Power
Distribution Network**.................................. 867
Yan-jun Shen and Jin-liang Jiang

Histogram Modification Data Hiding Using Chaotic Sequence 875
Xiaobo Li and Quan Zhou

A Family of Functions for Generating Colorful Patterns with Mixed Symmetries from Dynamical Systems 883
Jian Lu, Yuru Zou, Guangyi Tu, and Haiyan Wu

Automatic Calibration Research of the Modulation Parameters for the Digital Communications 891
Kai Wang, Zhi Li, Ming-zhao Li, Chong-quan Zhu, Qiang Ye, and Juan Lu

Adaptive Matching Interface Technology Based on Field: Programmable Gate Array 899
Xuejiao Zhao, Xiao Yan, and Kaiyu Qin

Power Grid Fundamental Signal Detection Based on an Adaptive Notch Filter 909
Zhi-xia Zhang, Xin-yu Zhang, Chang-liang Liu, and Tsuyoshi Funaki

The Precise Electric Energy Measurement Method Based on Modified Compound Simpson Integration 917
Min Zhang, Huayong Wei, and Weimin Feng

A Frequency Reconfigurable Microstrip Patch Antenna 925
Yong Cheng, ZhenYa Wang, XuWen Liu, and HongBo Zhu

Electricity Consumption Prediction Based on Non-stationary Time Series GM (1, 1) Model and Its Application in Power Engineering 933
Xiaojia Wang

Electricity Load Forecasting in Smart Grid Based on Residual GM (1, 1) Model 941
Jianxin Shen, Haijiang Wang, and Shanlin Yang

Electricity Consumption Forecasting Based on a Class of New GM (1, 1) Model 947
Mei Yao and Xiaojia Wang

Spectral Visibility of High-Altitude Balloon by the Ground-Based Detection 955
Xiaoping Du, Yu Zhang, and Dexian Zeng

Part VI Sensing Control Theory

A Novel Demodulation Method for Fiber Optic Interferometer Sensor Using 3 × 3 Coupler 967
Haiyan Xu and Zhongde Qiao

High Resolution Radar Target Recognition Based on Distributed Glint 977
Baoguo Li, Zongfeng Qi, Ying Zhou, and Jing Lei

A Differential Capacitive Viscometric Sensor for Continuous Glucose Monitoring 985
Zhijun Yang, Meng Wang, Youdun Bai, and Xin Chen

An Improved Secure Routing Protocol Based on Clustering for Wireless Sensor Networks 995
Lin Chen and Long Chen

Spatiotemporal Dynamics of Normalized Difference Vegetation Index in China Based on Remote Sensing Images 1003
Yaping Zhang and Xu Chen

Part VII Signal Processing and Control

The Application of Digital Filtering in Fault Diagnosis System for Large Blower 1013
Changfei Sun, Yong Han, Zhishan Duan, and Yingge Xu

Performance Analysis on ST-ASLC with Wide-Band Interference 1019
Xingcheng Li and Shouguo Yang

Velocity Error Analysis of INS-Aided Satellite Receiver Third-Order Loop Based on Discrete Model 1029
Dong-feng Song, Bing Luo, Xiao-ping Hu, An-cheng Wang, Pu-hua Wang, and Kang-hua Tang

Data-Processing for Ultrasonic Phased Array of Austenitic Stainless Steel Based on Wavelet Transform 1041
Xiaoling Liao, Qiang Wang, and Tianhong Yan

The Non-stationary Signal of Time-Frequency Analysis Based on Fractional Fourier Transform and Wigner-Hough Transform 1047
Jun Han, Qian Wang, and Kaiyu Qin

Weakness in a Serverless Authentication Protocol for Radio Frequency Identification 1055
Miaolei Deng, Weidong Yang, and Weijun Zhu

The Radar Images Smooth Rolling 1063
Peng Gao

Index of Volume 1 1073

Index of Volume 2 1077

Part IV
Control Theory and Application

Monte Carlo Based Predictive Method for Determining Work Envelope of Spacesuit in EVA Operation

Huaiji Si, Qianfang Liao, and Wanxin Zhang

Abstract Astronauts worn spacesuit manipulate objects in the extravehicular environment is strikingly different from that on the ground. A critical issue addressed in planning for Extra-Vehicular Activity (EVA) and evaluating EVA worksites is whether the astronaut can reach and comfortably work in the designed worksite or not. In this paper, a 9-DOF arm model of spacesuit is established, Monte Carlo based computer simulation and predictive method are researched and the arm workspace is predicted by the limitations of spacesuit joint angles. The prediction result is verified by a Articulated Arm Coordinate Measuring Machines (AACMM). This method provides a basis for planning EVA tasks of space suit and further study of mobility of space suit.

Keywords Monte Carlo • Work envelope • Spacesuit • Mobility • Extravehicular activity

1 Introduction

A spacesuit for EVA is a small space craft with complicated life support systems, which can to keep astronauts alive in the harsh environment (vacuum and extreme temperature) of outer space [1]. A critical issue addressed in planning for EVA and evaluating EVA worksites is whether the astronaut can reach and comfortably work in the designed worksite. Reaching envelope is the boundary of the space that a person can reach and the work envelope is a subset of the reach envelope, in which a person can work comfortably [2]. These two types of envelopes depend not only on the size but also the flexibility of the individual. Work envelope analysis that

H. Si (✉) • Q. Liao • W. Zhang
National Key Laboratory of Human Factors Engineering, Astronaut Center of China, Beijing 100094, China
e-mail: sihuaiji@126.com

incorporates the mechanics of the space suit and ergonomics of the astronauts is a useful method for both assessing potential worksite locations and evaluating the functional significance of modifications to space suit mobility and visibility [3].

The microgravity environment and spacesuits constrain astronauts' body motions in significant and complicated ways make EVA operations strikingly different from those performed on the ground [4]. Since the cost of experimenting in microgravity is high and underwater training lead to motions inappropriate for microgravity. Traditional computational methods used inverse kinematic method with simplified model to avoid the complicated calculation, which were adopted by researchers to simulate in the ground [5]. In this article, a prediction method based on Monte Carlo using a 9 DOF kinematic arm model was present and the result was agreed well with the experiment.

2 Monte Carlo Predictive Model

2.1 Human Upper Limb and Spacesuit Arm

When an astronaut wearing a spacesuit to move his body to manipulate objects, he must do extra work every time as bending the joint and maintain a force to keep the joint bent, which is strikingly different from operates on the ground. All movable upper limbs joints of human wearing spacesuit are as shown in Fig. 1, in which three glenohumeral joints (abduction-adduction, flexion-extension, and rotation), elbow extension joint, elbow abduction joint, wrist extension and wrist abduction are included. All joints are sequenced by the spacesuit joints. Generally, a 7-DOF arm model without acromioclavicular joints is adopted to construct upper-limb kinematics model. Acromioclavicular joints have rather small motion limits, their existence greatly enlarge motion limits of upper limbs. In this paper, a 9-DOF model including acromioclavicular joints is applied, so as to make workspace of mobility better approximate that of real human body wearing spacesuit. As many researchers have done, the Denavit-Hartenberg convention is used to compute segment parameters of the upper limbs of spacesuit. Parameters of transform matrix between different coordinate systems with DH parameters are shown in Table 1.

$$_{i}^{i-1}T = \begin{bmatrix} c\theta_i & -s\theta_i & 0 & a_{i-1} \\ s\theta_i c\alpha_{i-1} & c\theta_i c\alpha_{i-1} & -s\alpha_{i-1} & -s\alpha_{i-1}d_i \\ s\theta_i s\alpha_{i-1} & c\theta_i s\alpha_{i-1} & c\alpha_{i-1} & c\alpha_{i-1}d_i \\ 0 & 0 & 0 & 1 \end{bmatrix} \quad (1)$$

By replacing parameters in Table 1 with the transform matrix in Eq. 1, kinematics model of spacesuit arm can be obtained.

Fig. 1 Kinematic model of human upper left limb by the sequence of spacesuit joints

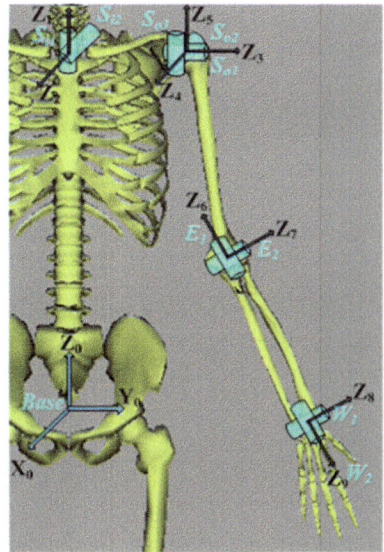

Table 1 DH Parameters for spacesuit-human arm

Num	Joint	α_{i-1}	a_{i-1}	d_i	θ_i
1	S_{i1}	0°	0	L_1	$\theta_1(90°)$
2	S_{i2}	90°	0	0	$\theta_2(-90°)$
3	S_{o1}	−90°	0	L_2	$\theta_3(0°)$
4	S_{o2}	90°	0	0	$\theta_4(90°)$
5	S_{o3}	−90°	0	0	$\theta_5(90°)$
6	E_1	0°	0	$-L_3$	$\theta_6(-180°)$
7	E_2	−90°	0	0	$\theta_7(90°)$
8	W_1	0°	L_4	0	$\theta_8(180°)$
9	W_2	−90°	0	0	$\theta_9(-90°)$
End	Palm	−90°	0	$-L_5$	−90°

$$^0_{end}T = {}^0_1T\,{}^1_2T\,{}^2_3T \ldots {}^8_9T\,{}^9_{end}T = \begin{bmatrix} n_x & o_x & a_x & p_x \\ n_y & o_y & a_y & p_y \\ n_z & o_z & a_z & p_z \\ 0 & 0 & 0 & 1 \end{bmatrix} \qquad (2)$$

Obviously, the model is rather complex, and it is impossible to accomplish by inverse kinematics analysis method. In this paper, forward kinematics analysis method based on Monte Carlo is applied.

To supply enough oxygen for respiration, a spacesuit using pure oxygen must have a pressure of about 32.4 kPa (240 Tor, 4.7 psi), which means an astronaut move his body to operate objects need extra effort to bend the limbs, resisting a soft pressure garment's natural tendency to stiffen against the vacuum.

Compared with human arm joints angular limits, EVA spacesuit has smaller angular limits due to manufacturing process and limitation of joints. Thus, it will be unreliable to determine workspace of spacesuit by mobility of human arm. Some originally achievable attitudes can no longer be achieved, and some originally reachable areas can no longer be reached, since astronauts are limited and affected by spacesuit when he bends joints.

2.2 Monte Carlo Predictive Method

Reachable workspace is defined as W_R, that is the maximum mobility limits of arms' ends, or volume that reachable points of arms' ends take. According to kinematics definition, W_R can be regarded as mapping from joints space variables to reachable workspace, and can be expressed as:

$$W_R = \{p(q) : q \in Q\} \subset \mathbf{R}^3 \tag{3}$$

In Eq. 3, where in $p(q) : Q \to \mathbf{R}^3$ represents a position component of kinematics mapping positive solution, q represents generalized joints variables, Q represents joints workspace, W_R represents reachable workspace and $\mathbf{R}^3$ the whole three-dimensional space.

According to mapping relationship of forward kinematics, a certain number of random quantities in accord with demand of joints' changes are assigned to joints' variables by uniform distribution, in order to obtain a graph made up by random points of reachable workspace, which forms Monte Carlo workspace. Decision method of Monte Carlo workspace has following steps:

Step1. According to space-coordinate point cloud acquired by kinematics equations, firstly identical numbers of random quantities are assigned to nine joints' variables (θ_i, $i = 1$ to 9) within their value ranges on the basis of the mapping relationship of each joint's variable and arm's reachable workspace. Then the nine joints' variables are mapped to workspace by kinematics equation to form three-dimensional "nephogram". But this nephogram cannot be observed or analyzed, and needs further processing in the following steps.

Step2. Divide series of strata of the nephogram according to the height, extract boundaries on different strata, and connect extracted boundaries by sequence to form boundary lines. The Monte Carlo workspace and Curve shown in Fig. 2 is acquired by kinematics model of spacesuit and spacesuit angular limits.

Step3. Boundary lines acquired in Step2 is reachable workspace envelop of spacesuit, that is maximum limits actually reached by spacesuit arm. The final workspace is obtained by combining reachable workspace with visible workspace. Visibility of points in reachable workspace can be obtained on the basis of mesh generation method and successive judgment of each point's visibility. Visible points' reachable points are the workspace of spacesuit. Visibility analysis of reachable workspace envelop is shown in Fig. 3 and the final maximum workspace envelop is shown in Fig. 4.

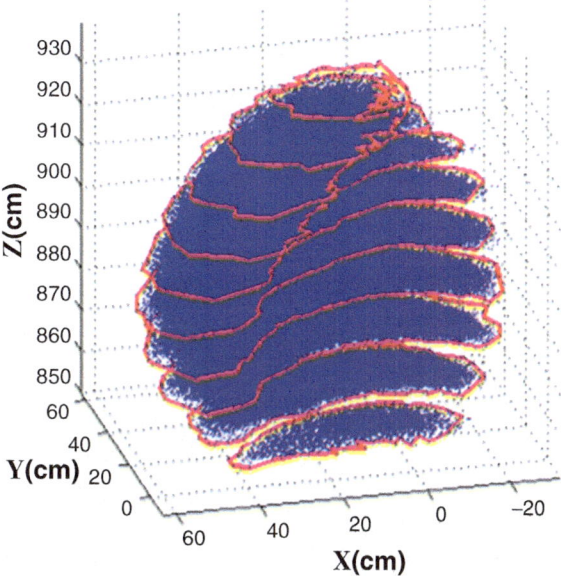

Fig. 2 Reach envelope with Monte Carlo predictive method

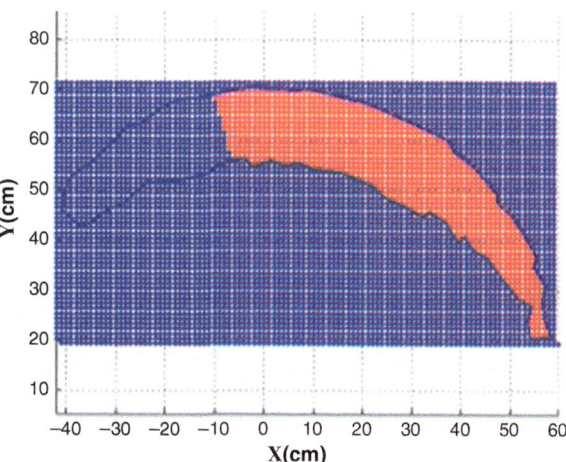

Fig. 3 Dividing reach envelope by visible (*red*) or not (*blue*)

Work envelope and reach envelope was distinguished by the visibility of points, which were shown in Fig. 3.

This model can clearly report the region of left hand, right hand and both hands. And the volumes of the operation regions are accurately calculated. The measurement provides a reliable support in the spacesuit engineering design. The shape and size of work envelop calculate by this method was shown in Fig. 4.

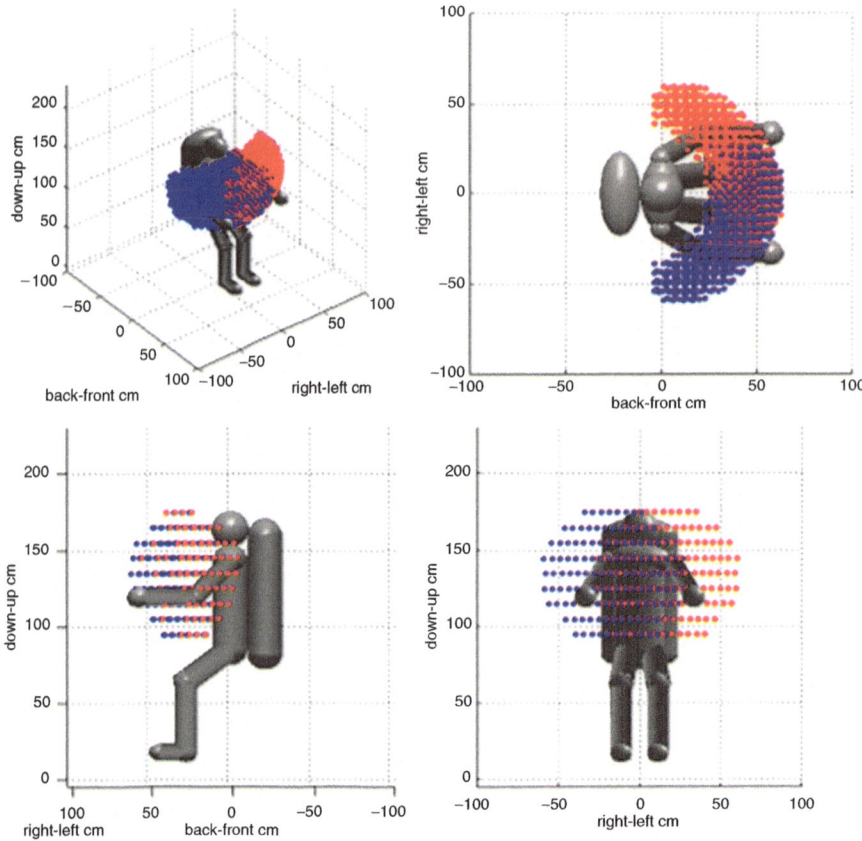

Fig. 4 Simulation of work envelope of spacesuit (*blue* for *right hand* and *red* for *left hand*)

3 Experimental

For the verification of the Predictive Model, a mechanical manipulator has been developed to measure the mobility and work envelope of astronaut wearing spacesuit. The measuring arm was designed combining the advantages of perpendicular ACMM which includes four rotation mechanical arms in horizontal plate and elevation device. Interface of test routine is shown in Fig. 5.

During reach envelop testing process, two subjects (male youth) wearing spacesuit respectively to operate the test handle moving in a horizontal plane. The motion path was recorded by testing routine, which is shown in Fig. 6. When a horizontal plane test was finished, height of test handle can be adjusted by lift to another horizontal plane. By testing motion path in each horizontal plane, the reach envelope can be calculated with reverse engineering method.

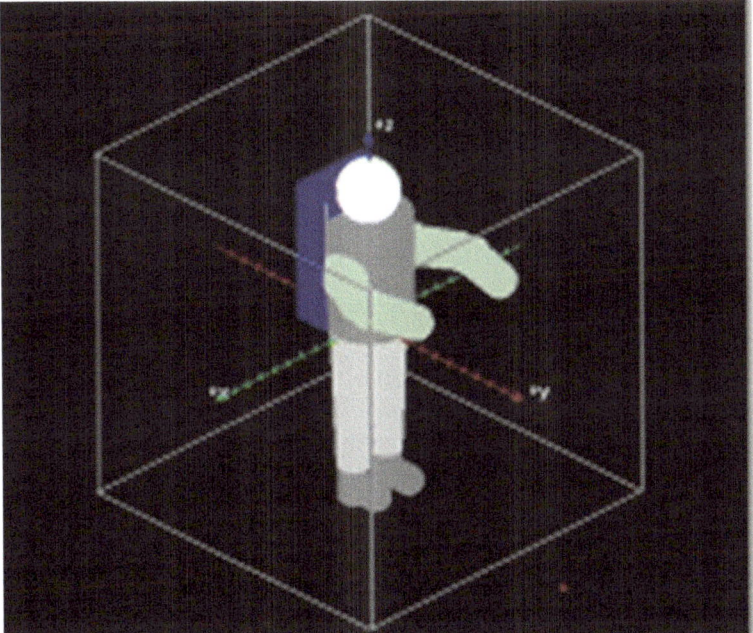

Fig. 5 Interface of spacesuit work envelope testing system

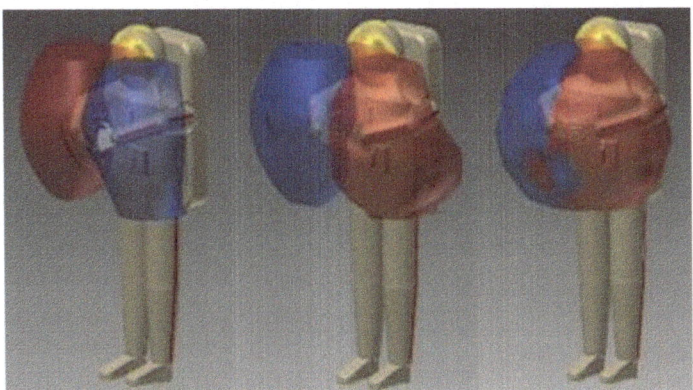

Fig. 6 Compared with experimental work envelope (*left* and *mid*) and simulation work envelope (*right*)

Experiments were performed to verify the Monte Carlo method simulation result, which were shown in Fig. 6. Compared with experiment result, the simulation result was similarity of shape and volume differ within 5 %, which prove the method meet the project need.

4 Conclusion

In this paper, a predictive method is used to determine work envelope of astronaut wearing spacesuit during EVA operating which considered the influence of spacesuit and human upper limbs. Monte Carlo based work envelope predictive model was established with kinematic model of human upper limbs by considering the sequence of spacesuit joints. A mechanical testing system was developed to verify the model. Different individuals wearing different spacesuit operated the manipulator to test the reach and visible envelope. Testing results and computational results were compared to verify the model.

The results of the project serve as a stepping stone for further research in this area, the availability of spacesuit-arm model, predictive model, and testing system will allow us to implement different modify and control methods to analyze the effect on the spacesuit's mobility performance.

References

1. Schmidt PB (2001) An investigation of space suit mobility with applications to EVA operations. Doctoral dissertation, Massachusetts Institute of Technology
2. Stirling L, Willcox K, Newman D (2010) Development of a computational model for astronaut reorientation. J Biomech 43:2309–2314
3. Judnick D, Newman D, Hoffman J (2007) Modeling and testing of a mechanical counterpressure bio-suit system. In: Proceedings of the international conference on environmental systems, Massachusetts Institute of Technology, Chicago, pp 87–112
4. Xu WF, Li LT (2007) Workspace analysis of space 3R robot. J Astronaut 28(5):1389–1394
5. Frazer A (2003) Modeling human-spacesuit interactions. Master's thesis, Massachusetts Institute of Technology, Cambridge, pp 43–51

Reduced Thrust Take-Off of Large Passenger Aircraft Based on Derate Method

Xinmin Wang, Haitao Yin, Yi Zheng, and Rong Xie

Abstract As the actual take-off weight of the aircraft is less than the maximum take-off weight, the technology which use thrust less than maximum take-off thrust for take-off is called reduced thrust take-off. Derate Method is one of the methods. In this paper, the principle of reduced thrust take-off is analyzed, and process and limitation are studied, and principle and application conditions of Derate Method are analyzed. At last, the vertical control law of Derate Method Reduced Thrust Take-off is designed and simulated to a large commercial aircraft. The results show that, under the safe flight condition, Derate Method reduced thrust take-off reduces fuel consumption effectually and is very meaningful to improve economic.

Keywords Reduced thrust takeoff • Derate method • Aircraft engine

1 Introduction

During the flight of the large commercial aircraft, the take-off stage is one of the most important stage in the whole flight. It not only directly affects flight safety, but also has significant influence on the engine performance. Research further reducing the operating costs of airlines on the basis of meeting safety performance of take-off is an important issue faced by Civil Aviation [1–3].

Civil Aviation large transport aircraft mostly use the Reduced Thrust Take-off technology. Reduced Thrust Take-off is that aircraft use thrust less than normal engine thrust to take off on the premise to ensure the flight safety [4, 5].

X. Wang • H. Yin • Y. Zheng • R. Xie (✉)
School of Automation, Northwestern Polytechnical University, Xi'an, China
e-mail: Yinhaitao198691@163.com

This paper first describes the principle of Reduced Thrust Take-off, and then introduces the principle of Derate Method, and the example is simulated at last. Necessity to implement Reduced Thrust Take-off in large aircraft is analyzed with quantitative data.

2 Principle of Reduced Thrust Take-Off

Reduced Thrust Take-off (known as flexible thrust take-off) is that aircraft use thrust less than normal engine thrust to take off on the premise to ensure the flight safety (meet to the requirements of the appropriate regulations).

2.1 Process of Reduced Thrust Take-Off

The take-off process of the aircraft is shown as Fig. 1. In the process of take-off, ground acceleration stage is from loosing brake to lifting the front wheel; take-off decision speed V_1 is reached at some point of that stage; take-off cannot be interrupted when V_1 is reached. The speed when lifting the front wheel refers to as V_R, pilots should keep the nose about $8°$ in this stage until aircraft leave the ground. Then aircraft needs to accelerate to take-off safety speed. After climbing, the aircraft usually keeps speed $(V_2 + 10kn)$ before it reaches enough height [6].

When the weight of the aircraft is less than weight limited by airport length, V_1, V_2 and V_R will be reached in advance. Thus the aircraft will take-off without the whole runway and Reduced Thrust Take-off can be used by excess runway. And it is possible to implement Reduced Thrust Take-off.

2.2 Restriction of Reduced Thrust Take-Off

The lighter weight of the aircraft is, the larger climb gradient is, and the higher flight altitude is. The greater thrust is, the larger climb gradient is [7]. The aircraft cannot

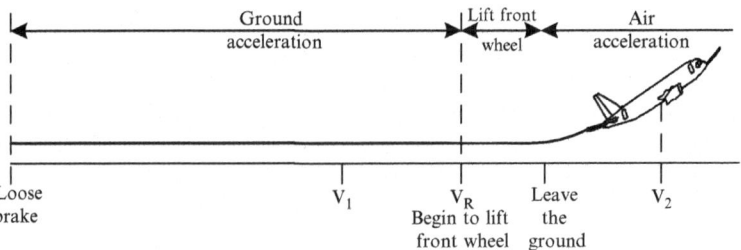

Fig. 1 Longitudinal sectional view of take-off

take off safely when the thrust of the engine cannot satisfy the minimum climb gradient.

Now we analyze the relationship of the take-off weight and thrust by the climb gradient [7]. As angle of attack α and angle of incidence of engine φ_T is very small, we can get,

$$F - D - mg \sin\theta - m\frac{dV}{dt} = 0 \quad (1)$$

By Eq. 1 and definition of rate of climb (r/c) we can obtain,

$$r/c = V \sin\theta = \frac{(F-D)V}{mg} - \frac{V}{g}\frac{dV}{dt}$$

By the definition of the climb gradient (C.G), we gen formula of C.G,

$$C.G = \frac{r/c}{V} = \frac{\frac{F}{mg} - \frac{C_D}{C_L}}{1 + \frac{V}{g}\frac{dV}{dh}} \quad (2)$$

In formula (2), F is thrust of engine, D is air resistance, m is take-off weight, V is take-off speed, θ is pitch angle, C_L is lift coefficient, C_D is drag coefficient. With given conditions of take-off, the greater the take-off weight is, the smaller the climb gradient is. The minimum available climb gradient provided by American Airlines Management Regulations (FAR-25) limits the maximum take-off weight. Equation 2 shows that, when lift coefficient and drag coefficient are dug out and take-off speed is determined, we can determine the relationship of thrust and take-off weight.

3 Derate Method

The essence of Derate Method is engine is regarded as a smaller power engine. The thrust of take-off must not exceed the maximum thrust of the virtual small power engine.

Figure 2 is variation of engine thrust dependent on temperature which aircraft take off with whole thrust and different levels of Derate Method. In general engine set two derating thrust level, TO1 and TO2. Each level has its restriction. The maximum thrust will decrease when derate level is selected in take-off. As Derate Method determines the take-off performance, take-off performance chart corresponded to power must be used. Derate Method have no operating limitation. It can be used under any circumstances provided aircraft performance is allowed.

Fig. 2 The principle of Derate method

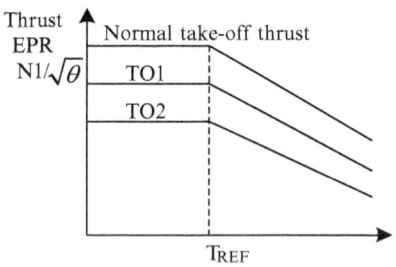

Different derate level usually corresponds to fixed reduced thrust, and specific data reduced by each airline may also be different (TO1 is 10 % and TO2 is 20 % in China Southern Airlines) [6]. And not all type can use Derate Method. Currently, all of types of Boeing can use Derate Method, and only can A319, A321, A330 and A340 use Derate Method in Airbus. There are six levels in A330 and A340. And thrust reduced is 4 %, 8 %, 12 %, 16 %, 20 % and 24 %. It cannot be used on other types.

4 Establishment of Fight and Propulsion Comprehensive Model

Establishing fight and propulsion comprehensive system requires that thrust provided by engine and thrust needed by plane is balance. So we can get general expression of state-space equation of the fight and propulsion comprehensive system [8],

$$\begin{cases} \dot{X} = AX + BU \\ Y = CX + DU \end{cases}. \tag{3}$$

In Eq. 3,

$$X = \left[X_p, X_e\right]^T = \left[V, \alpha, \theta, q, n_l, n_h\right]^T, \ Y = \left[Y_p, Y_e\right]^T = \left[V, \alpha, \theta, q, n_l, n_h, F\right]^T,$$
$$U = \left[U_p, U_e\right]^T = \left[\delta_e, m_f, A_e\right]^T$$

$$A = \begin{bmatrix} -0.0069 & 6.5191 & -9.8000 & 0 & 0.0007 & 0 & 0 \\ -0.0003 & -0.6105 & 0 & 1 & 0 & 0 & 0 \\ 0 & 0 & 0 & 1 & 0 & 0 & 0 \\ 0.0027 & -2.3540 & 0 & -0.4970 & 0 & 0 & 0 \\ 0 & 240 & 240 & 0 & 0 & 0 & 0 \\ 0 & 0 & 0 & 0 & -2.3410 & -3.5730 & 0.6720 \\ -0.0318 & 0 & 0 & 0 & -3.2410 & 0.3370 & -3.4920 \end{bmatrix},$$

$$B = \begin{bmatrix} 0.3504 & 0.0003 & 0 \\ -0.0251 & 0 & 0 \\ 0 & 0 & 0 \\ -1.8867 & 0 & 0 \\ 0 & 0 & 0 \\ 0 & 0.4960 & 0.6690 \\ 0 & 0.6360 & 3.6040 \end{bmatrix} \quad C = \begin{bmatrix} 1 & 0 & 0 & 0 & 0 & 0 & 0 \\ 0 & 1 & 0 & 0 & 0 & 0 & 0 \\ 0 & 0 & 1 & 0 & 0 & 0 & 0 \\ 0 & 0 & 0 & 1 & 0 & 0 & 0 \\ 0 & 0 & 0 & 0 & 1 & 0 & 0 \\ 0 & 0 & 0 & 0 & 0 & 1 & 0 \\ 0 & 0 & 0 & 0 & 0 & 0 & 1 \\ 0.3950 & 0 & 0 & 0 & 67.3260 & 3.8357 & 1.3573 \end{bmatrix},$$

$$D = \begin{bmatrix} 0 & 0 & 0 \\ 0 & 0 & 0 \\ 0 & 0 & 0 \\ 0 & 0 & 0 \\ 0 & 0 & 0 \\ 0 & 0 & 0 \\ 0 & 0 & 0 \\ 0 & 29.8606 & -0.5174 \end{bmatrix}$$

5 Simulation Example

5.1 Structure and Processes of Simulation

Vertical Derate Method control law is taken to design and simulate to a large passenger aircraft. The given aircraft is equipped with four turbofan engines; the maximum thrust of single engine is 84.48 kN; the maximum thrust of aircraft is

Fig. 3 Simulation process of take-off

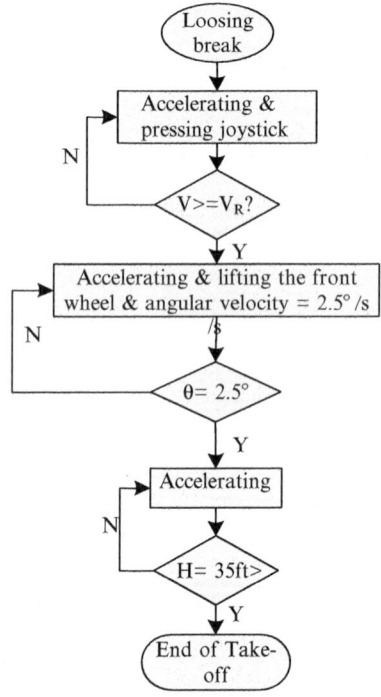

320.8 kN. Speed of lifting the front wheel is read as $V_R = 69.72 m/s$, and Take-off Safety Speed is $V_2 = 72.5 m/s$ [7]. Simulation process is shown as Fig. 3.

The aircraft takes off from loosing brake. First, press the joystick; when speed reaches to the speed of lifting the front wheel V_R, aircraft lifts the front wheel with the pitch angular rate of $2.5°/s$ and prepares to take off; pitch angle maintains to $7.5°$ at last and continues to climb. The take-off stage ends when the altitude reaches to 35 ft (10.7 m).

Now we make simulation with Derate Method take-off. Derate Method is set as two levels by the rules of a airline. TO1 is set as 10 %, and TO2 is set as 20 %. The structure of longitudinal take-off combining of aircraft and engine is shown as Fig. 4, which aircraft model is controlled with PID and engine model is controlled with Optimal Servo.

5.2 Simulation Results

The model is simulated by MATLAB 7.8. Responses of speed, altitude, distance rolling and fuel quantity of the full thrust and Derate Method is show as Figs. 5, 6, 7, and 8.

Fig. 4 Control structure of take-off

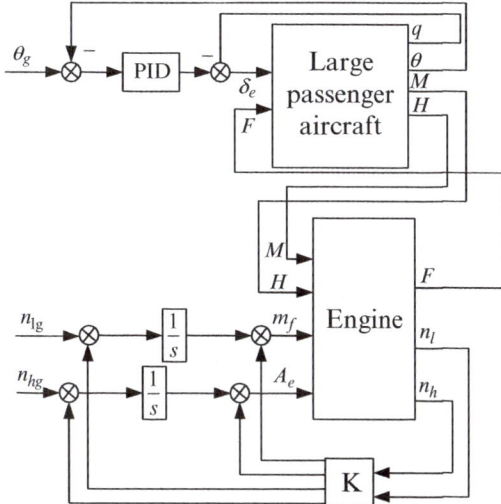

Fig. 5 Speed response

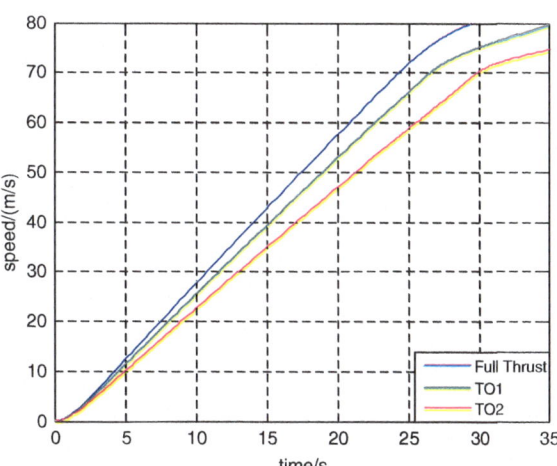

As shown from Figs. 5, 6, 7, and 8, time of TO1 and TO2 are respectively 28.77 and 31.95 s; speed of them are respectively 73.81 and 72.6 m/s; distance of them are 928 and 1052 m; fuel quantity of them are 265 and 262 kg. Compared Derate Method with full thrust, take-off time of TO1 is prolonged 4.05 %, and distance is increased 5.82 %, and fuel is reduced 4.35 %, take-off speed is $V = 73.81 > V_2$. Take-off time of TO2 is prolonged 15.55 %, and distance is increased 19.95 %, and fuel is reduced 5.07 %, take-off speed is $V = 72.6 > V_2$. Simulation results show that, as length of runway allowing, using Derate Method can ensure the security of take-off, and reduce fuel consumption greatly, and has a significant role in improving economic efficiency.

Fig. 6 Altitude response

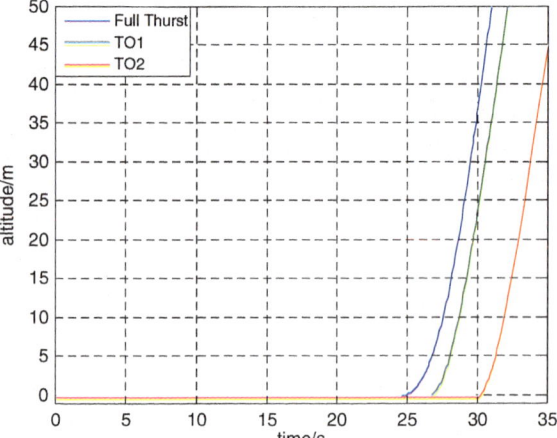

Fig. 7 Distance response

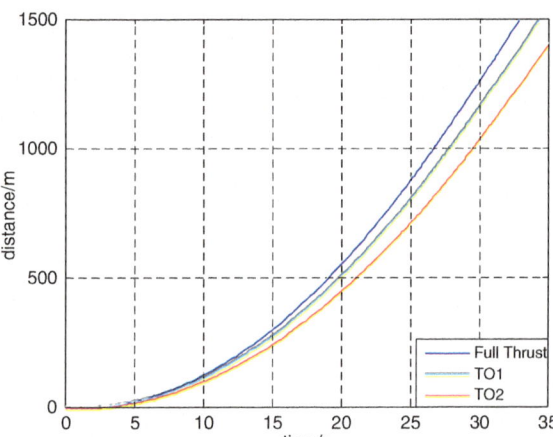

Fig. 8 Fuel quantity response

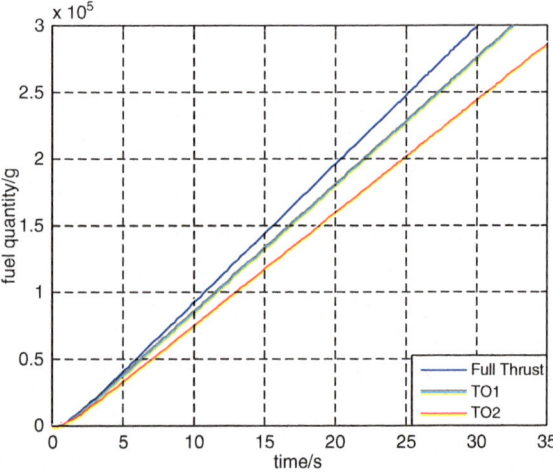

6 Conclusion

In this paper, the simulation is proved that Derate Method can reduce fuel consumption and transportation costs. Derate Method can also reduce in-flight shutdown rate and unscheduled engine removal rate, and improve safety standard of airlines. So it is necessary to study and spread Reduced Thrust Take-off.

References

1. Tingyu Zhao (2005) Necessity analysis of reduced-thrust taking-off of turbofan engines. Tianjin J Civil Aviat Univ China 23(3):6–8 (In Chinese)
2. Yong Wu (2005) An analysis of CFM56-5B Derated takeoff & flex takeoff. Beijing China Civil Aviat 1(49):73 (In Chinese)
3. Tingyu Zhao (2005) Principle of assumed temperature reduced thrust takeoff. Chengdu J Southwest Jiaotong Univ 40(5):677–679 (In Chinese)
4. Chmiela B, Sozanska M, Cwajna J (2012) Identification and evaluation of freckles in directionally solidified casting made of PWA 1426 nickel-based superalloy. USA Arch Metall Mater 57(2):559–564
5. Ao Liangzhong (2012) Thrust lever angle signal processing of an aircraft engine. In: Proceedings – 2012 international conference on computer science and electronics engineering, ICCSEE 2012, Los Alamitos, CA, USA, pp 613–616
6. Xiaoming Liu, Tingyu Zhao, Xiaohang Wen et al (2009) Reduced-thrust takeoff technique applied to passenger airplanes. Xi'an Flight Dynam 27(3):83–85 (In Chinese)
7. Tongjun Tian (2002) The integrated flight/propulsion H_∞ controller design[D]. Northwestern Polytechnical University, Xi'an, pp 19–22 (In Chinese)
8. Zhihuai Chen, Runping Gu, Junjie Liu (2006) Aircraft performance engineering[M]. Ordnance Industry Press, Beijing, pp 49–55 (In Chinese)

Permanent Magnet Synchronous Motor Feedback Linearization Vector Control

Hehua Wang and Xiaohe Liu

Abstract In order to solve the control problem of a class multiple-input multiple-output nonlinear system, the feedback linearization method is introduced. By calculating the output variables of Lie derivative, the appropriate coordinate transform and nonlinear state feedback are obtained, then through the coordinate transformation and state feedback, the input-output linearization is realized and the system decoupling is achieved. According to the system's linear model, the actual control rate is designed. For illustration, a multiple-input multiple-output nonlinear system example is utilized to show the feasibility of the feedback linearization in solving the permanent magnet synchronous motor, and then combines with the vector control method. Using MatLab7.6/Simulink to build modular and simulation verifies the effectiveness of the algorithm. Empirical results show that the feedback linearization is a better method to handle nonlinear system. From the simulation results we can be obtained that the feedback linearization vector control method has a good control effect.

Keywords Feedback linearization · Vector control · Effectiveness

1 Introduction

In recent years, with the high performance permanent magnet material technology, power electronics technology and microelectronics technology growing fast. Make permanent magnet synchronous motor be characterized by small volume, high efficiency, and the advantages of small losses. The PMSM plays an increasingly important role in small power motion control system. With the deepening of vector control theory and automatic control principle, the permanent magnet synchronous

---

H. Wang (✉) · X. Liu
Beijing Information Science & Technology University, Beijing, China
e-mail: wanghehua1985@126.com; liuxiaohe551026@163.com

motor control system has developed rapidly. As a nonlinear system, the precise control methods have been studied by many scholars, among which the one based on differential geometry feedback linearization has achieved a big development. People have successfully solved the many problems about motor control with it, and got a good control effect. Such as the PMSM control systems based on SVPWM [1], the designs of PMSM in the vector control system, which has a good control effect for PMSM [2], and research the application feedback linearization in PMSM [3].

2 Description of Problem

For the following n-order multi-variable nonlinear system, using state space form to describe as follow types of equations:

$$\dot{x} = f(x) + \sum_{i=1}^{m} g_i(x)u_i$$
$$y_1 = h_1(x)$$
$$\cdots$$
$$y_m = h_m(x) \tag{1}$$

Here $f(x), g_1(x), \cdots, g_m(x)$ is n-dimensional smooth vector function; $h_1(x), \cdots, h_m(x)$ is a scalar function. These equations can be more compact form:

$$\dot{x} = f(x) + g(x)u$$
$$y = h(x) \tag{2}$$

Here $g(x) = (g_1(x), \cdots, g_m(x))$ is $n \times m$ order matrix; $h(x) = (h_1(x), \cdots, h_m(x))^T$ is m-dimensional column vector. Given the definition of the relation degree:

1. $L_{g_i} L_f^k h_i(x) = 0$;
2. The m-dimensional square $A(x)$ is singular at x_0.

$$A(x) = \begin{pmatrix} L_{g_1} L_f^{r_1-1} h_1(x) & \cdots & L_{g_m} L_f^{r_1-1} h_1(x) \\ L_{g_1} L_f^{r_2-1} h_2(x) & \cdots & L_{g_m} L_f^{r_2-1} h_2(x) \\ L_{g_1} L_f^{r_m-1} h_m(x) & \cdots & L_{g_m} L_f^{r_m-1} h_m(x) \end{pmatrix}$$

Here $L_{g_i} L_f^k h_i(x) = \frac{\partial L_f^k h_i(x)}{\partial x} g_i(x)$, $L_f^k h_i(x) = \frac{\partial L_f^{k-1}}{\partial x} f(x)$, $L_f h(x) = \frac{\partial h(x)}{\partial x} f(x)$, the $L_f h(x)$ can also be written in the form of $(x), f(x) >$ that is Lie bracket, they $L_f h(x)$ can also be defined as the derivative of $h(x)$ along the vector field $f(x)$. The formula $L_g L_f h(x) = \frac{\partial L_f h(x)}{\partial x} g(x)$ is the derivative of $h(x)$ along the vector field $f(x)$, and then along the vector field $g(x)$.

The relation degree vector is $\{r_1, \cdots, r_m\}$, the relation degree is $r = r_1 + \cdots r_m$. Now we discuss system feedback linearization problem.

1. If $r = r_1 + \cdots + r_m \leq n$, for any $1 \leq i \leq m$ meets:

$$\begin{aligned} \phi_1(x) &= h_1(x) \\ \phi_2(x) &= L_f h_2(x) \\ &\cdots \\ \phi_i(x) &= L_f^i h_i(x) \end{aligned} \quad (3)$$

Here $L_f^i h_i(x) = \frac{\partial L_f^{i-1} h_{i-1}(x)}{\partial x} f(x)$, $L_f h_1(x) = \frac{\partial h_1(x)}{\partial x} f(x)$.

When $r = r_1 + \cdots + r_m$ is strictly less than n, so can find $n - r$ functions $\phi_{r+1}(x), \cdots, \phi_n(x)$, which make mapping $\Phi(x) = (\phi_1^1(x), \cdots, \phi_{r_1}^1(x), \cdots, \phi_1^m(x), \cdots, \phi_{r_m}^m(x), \phi_{r+1}(x), \cdots, \phi_n(x))^T$ have a nonsingular Jacobian matrix at x_0, so the mapping is equivalent to a local coordinate transformation at a neighborhood of x_0.

The values of these additional functions can be arbitrary chose according to the following condition:

$L_{g_j}\phi_i(x) = \frac{\partial \phi_i(x)}{\partial x} g_j(x) = 0$, For any x within a field of x_0, $r + 1 \leq i \leq n$ and $1 \leq j \leq m$.

2. When $r = r_1 + r_2 + \cdots + r_m$ is strictly equal to n, so can find out a set of functions: $\phi_k^i(x) = L_f^{k-1} h_i(x)$, $1 \leq k \leq r_i$, $1 \leq i \leq m$. We can define $\phi_k^i(x)$ as a local coordinate transformation in a neighborhood of x_0. By the local coordinate transformation, the system can use this form of m-group equations described under the new coordinate, making $z_i = \phi_i(x) = L_f^{i-1} h(x)$.

$$\begin{cases} \dot{z}_1 = z_2 \\ \dot{z}_2 = z_3 \\ \vdots \\ \dot{z}_{r-1} = z_r \\ \dot{z}_r = L_f^r h(x) \end{cases} \quad (4)$$

By the above-mentioned transforms, z is selected as a new variable, the system becomes a linear system.

3 Mathematical Model of AC PMSM and Simulation

3.1 Mathematical Model of AC PMSM

The stators of Ac permanent magnet synchronous motor (PMSM) and general electric excitation three-phase synchronous motor are similar, if the permanent magnet produces the induced electromotive force and the excitation coil generates induced electromotive force are the same, also to be the sine wave. The mathematical models of PMSM and electric excitation synchronous motor are the same. Some assumptions are made as follows:

1. Magnetic saturation, eddy current loss and hysteresis loss are neglected
2. Oversight space harmonic, three-phase winding symmetry, the air gap magnetic field is sine distribution
3. Three-phase power supply voltage balance

Based on the above assumptions, the mathematical model of PMSM under d-q coordinate system is [4]:

$$\begin{bmatrix} \dot{i}_d \\ \dot{i}_q \\ \dot{\omega} \end{bmatrix} = \begin{bmatrix} -\frac{R}{L} & P_n\omega & 0 \\ -P_n\omega & -\frac{R}{L} & -\frac{P_n\varphi_f}{L} \\ 0 & \frac{3P_n\varphi_f}{2L} & -\frac{B}{J} \end{bmatrix} \begin{bmatrix} i_d \\ i_q \\ \omega \end{bmatrix} + \begin{bmatrix} \frac{u_d}{L} \\ \frac{u_q}{L} \\ -\frac{T_L}{J} \end{bmatrix} \quad (5)$$

Here i_d, i_q denote d-q reference current, u_d, u_q denote d-q equivalent voltage, R is stator resistance, P_n denotes the number of pole pairs, L is equivalent inductance, ω is rotor speed, T_L represents load torque, J is the moment of inertia, B denotes the friction coefficient, φ_f represents the magnetic flux. Then Eq. 5 can be written in the form of affine nonlinear systems standards:

$$\dot{x} = f(x) + g_1(x)u_d + g_2(x)u_q \quad (6)$$

Here $x = [i_d, i_q, \omega]^T$, $g_1(x) = [\frac{1}{L}, 0, 0]^T$, $g_2(x) = [0, \frac{1}{L}, 0]^T$, $f(x) = \begin{bmatrix} -\frac{R}{L}i_d + P_n\omega i_q \\ -\frac{R}{L}i_q - P_n\omega i_d - \frac{P_n\varphi_f}{L}\omega \\ \frac{3P_n\varphi_f}{2J}i_q - \frac{B}{J}\omega - \frac{T_L}{J} \end{bmatrix}$. In order to make the system into linear system can make the following transformation, choosing ω, i_d as the output of the system, defining new output variables:

$$\begin{cases} z_1 = h_1(x) = \omega \\ z_2 = L_f h_1(x) \\ z_3 = h_2(x) = i_d \end{cases} \quad (7)$$

State equation of the system under new coordinate system as follows:

$$\begin{cases} \dot{z}_1 = z_2 \\ \dot{z}_2 = L_f^2 h_1(x) + L_{g1}L_f h_1(x)u_d + L_{g2}L_f h_1(x)u_q \\ \dot{z}_3 = L_f h_2(x) + L_{g1}h_2(x)u_d + L_{g2}h_2(x)u_q \end{cases} \quad (8)$$

The system can be feedback linearized because it contains three inputs, three outputs and its relation degree vector is $\{1, 1, 1\}$. That means the sum of vectors is equal to the rank of the system. So the system can be described as exact feedback linearization and won't appear zero dynamics problems, selecting the new variables:

$$\begin{cases} \dot{z}_1 = z_2 \\ \dot{z}_2 = v_1 \\ \dot{z}_3 = v_2 \end{cases} \quad (9)$$

This system has become a linear system. We can select v_1 and v_2 as the control variables of the system. According to linear system poles configuration theory and state feedback, we can achieve the linear design of the system [5].

$$\begin{aligned} v_1 &= -k_1(y_1 - y_{1ref}) - k_2\dot{y}_1 = -k_1(\omega - \omega_{ref}) - k_2\dot{\omega} \\ v_2 &= -k_3(y_3 - y_{3ref}) = -k_3(i_d - i_{dref}) \end{aligned} \quad (10)$$

Let v_1 and v_2 be the control variables of system, so the actual control variables of the system are u_d u_q.

$$\begin{aligned} u_q &= \tfrac{2JL}{3P\varphi_f}\left[\tfrac{3P\varphi_f}{2JL}\left(\tfrac{R}{L}i_q + P\omega i_d + \tfrac{P\varphi_f}{L}\omega\right) + \tfrac{B}{J}\left(\tfrac{3P\varphi_f}{2J}i_q - \tfrac{B}{J}\omega - \tfrac{T_L}{J}\right) + v_1\right] \\ u_d &= L\left(\tfrac{R}{L}i_d - P\omega i_q + v_2\right) \end{aligned} \quad (11)$$

3.2 Simulation Module Introduction

According to the established mathematical model and feedback linearization of knowledge, and the knowledge of the motor control [4], we can get the system control diagram as follows the Fig. 1. Depending on the Eq. 10, we can know that the P controller can be used in the comparison between i_d and i_{dref}, the comparison between ω and ω_{ref} can use the *PI* controller.

In the MATLAB7.6/Simulink environment, set up its simulated modules based on the system control diagram. The following will describe a few of the more important in the simulated model.

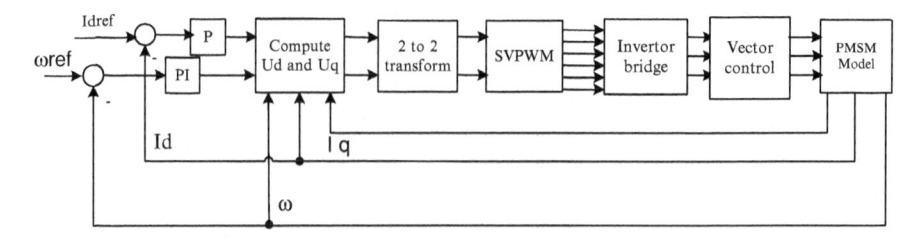

Fig. 1 Block diagram of the control system

Fig. 2 Abc2alfa-beta module

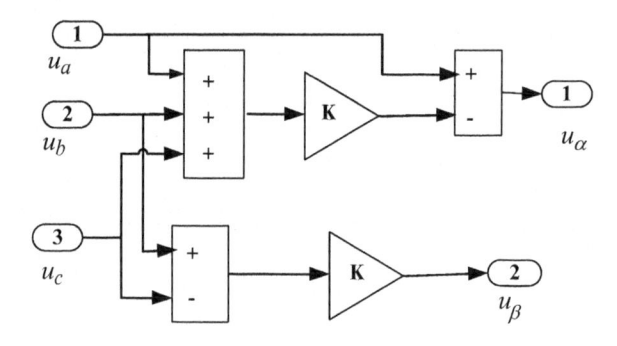

Fig. 3 Alfa-beta2dq module

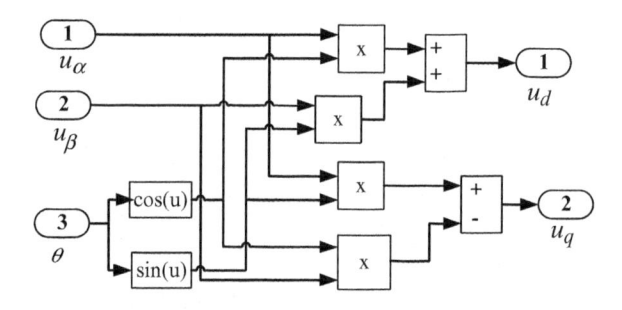

1. Vector control module

    The basic thought of vector control is that through three to two transformation make two phases stationary coordinate system of the alternating current i_α, i_β equivalent to three-phase static coordinate system of the stator alternating current i_a, i_b, i_c. Then after rotor field orientation rotation transformation, letting i_α, i_β equivalent into two phases rotating coordinate system of the current i_d, i_q, i_d equivalents to DC motor of the excitation current, i_q equivalents to DC motor armature current. Realize vector control need two modules [6]: abc2*alfa − beta* and *alfa − beta*2*dq* transformation modules. As shown Figs. 2 and 3:

Fig. 4 Speed variation

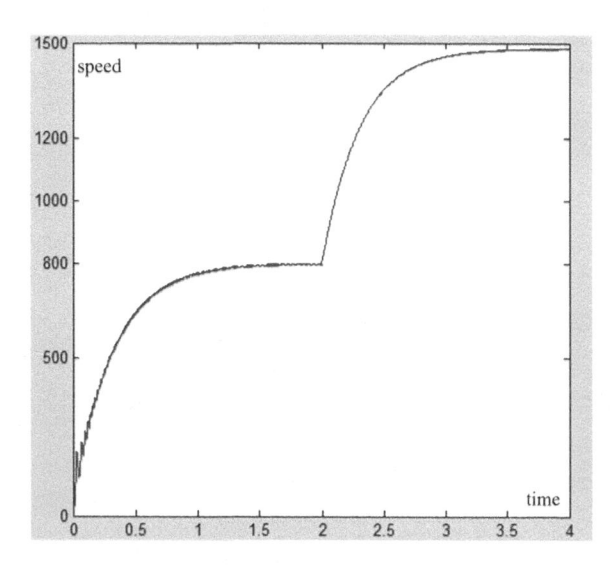

3.3 Simulation Results Analysis

According to the above each simulation module, In the MATLAB7.6/Simulink environment, the PMSM parameter is set as follows: $R = 4.2\ \Omega$, $L = 0.0153$ H, $\varphi_f = 0.175\ Wb$, $J = 0.0021$ kg $*$ m^2, $P_n = 3$, $B = 0.006$ Nm.Sec/rad. At first, given the reference ω_{ref} is 800 rad/s, after two seconds the reference ω_{ref} is 1,500 rad/s, the simulated results as follows. The speed variation is Fig. 4:

From the Fig. 4 we can clearly see that the system will soon be able to track the reference ω_{ref}, and no overshoot and steady state error. It can be known from the simulated result that the system responds rapidly and has good robust stability.

At the same time we can get phase current i_q change curve Fig. 5. We can clearly see that the phase current i_q soon enters a stable state and the change of the reference ω_{ref} will cause the current change of i_q. We can know that different sizes of ω_{ref} disturbance to i_q is not the same, but the phase current i_q can be able to quickly run in the stable state.

From the above the simulated result we can know that feedback linearization vector control method has a good control effect. The system responds rapidly and has good robust stability. We also know that the system can quickly run in the stable state, the response is fast, almost no steady-state error.

Fig. 5 Phase current Iq variation

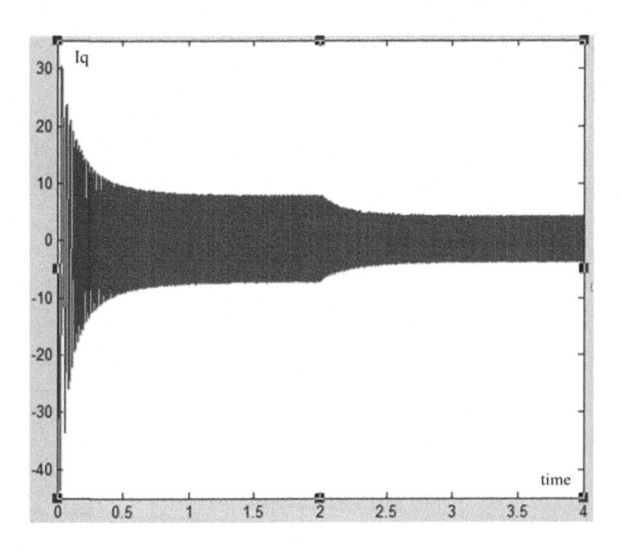

4 Conclusion

This paper described the Feedback linearization vector control of PMSM, using the MATLAB7.6/Simulink to construct the system of the simulated model. The simulated result shows that the system runs smoothly, with good dynamic performance; the feedback linearization vector control method has a good control effect; the designed system responds rapidly and has good robust stability, which will provide useful reference to the PMSM control system design and analysis later.

References

1. Dong SY, Sun SH (2010) Modeling and simulation of PMSM control system based on SVPWM. Mod Electron Tech 18(329):188–191
2. Wu B (2008) Modeling and simulation of PMSM vector control system based on MATLAB/Simulink & SimPower systems. Electro-Mech Eng 24(3):57–59
3. Liu DL, Zhao GY (2006) Speed tracking control of PMSM based on direct feedback linearization. Electr Drive Autom Control 28(2):8–10
4. Kou BQ, Cheng SK (2008) Ac servo motor and control. China Machine Press, Beijing, pp 112–115
5. Wang CY (2009) Modern control technology for electric machines. China Machine Press, Beijing, pp 56–70
6. Zhang XF, Wang J, Chen XY, Li YJ (2011) MATLAB mechanical and electrical control system technology and application. Tsinghua University Press, Beijing, pp 98–105

Design and Simulation of Image Compensation Control System

Chan Tan and Lei Ding

Abstract Image compensation is a key to high resolution imaging. This paper designs and simulates an image compensation control system based on swing pointing mirror. The principle of pointing mirror compensation is presented and the mathematical model of pointing mirror's driving motor is established. Based on this model, a high accurate controller which consists of current loop and velocity loop is designed and simulated. Considering that there is disadvantage of traditional speed PID controller in dynamic performance, which is speed overshoot, the speed loop adopts pseudo derivative feed-forward (PDFF) controller. Theoretical analysis and simulation results show that the designed image compensation control system has good characteristic in following performance, dynamic response and noise resistance.

Keywords Image compensation control • Swing pointing mirror • PDFF controller

1 Introduction

If there is relative motion between camera and aim during exposure time, the target image recorded in detectors is moved and blur. So it is necessary to compensate image shift due to the limitation of MTF (Modulation Transfer Function). In order to resolve the image motion, many methods have been proposed, including mechanical compensation [1], electronic compensation such as TDI (Time Delay Integration) CCD and whole frame transfer CCD [2], optical compensation such as optical joint correlators [3, 4] and swing mirror, and software compensation [5] etc.

C. Tan (✉) • L. Ding
Key Laboratory of Infrared System Detection and Imaging Technology, Shanghai Institute of Technical Physics, Chinese Academy of Sciences, Shanghai, China
e-mail: helen367@126.com

However, mechanical compensation is only applied in film camera, and electronic compensation based on TDI CCD merely achieves one-dimension compensation, and optical joint correlator is strict to the environment, while compensation based on swing mirror could not only compensate forward image shift but also pitching and yawing image shift.

To achieve the compensation, the swing pointing mirror has to swing at certain speed during satellite's forward flight, and due to its extreme low compensation speed, the servo system of pointing mirror must be stable and high accuracy. Because PMSM have excellent speed adjustment performance and high efficiency which is fit for high-powered close-loop control system and small torque ripple. So this paper chooses PMSM as swing pointing mirror's driving motor.

Firstly, this paper gives the principle and calculation of swing pointing mirror compensation, and then establishes mathematical model of PMSM. Based on this model, a high accurate controller is designed including current loop and speed loop. Besides, in order to reduce speed overshoot, the speed loop adopts PDFF controller. Theoretical analysis and simulation results show that the designed image compensation control system has good characteristic in following performance, dynamic response and noise resistance.

2 Swing Mirror-Based Image Compensation Control System

2.1 Principle of Swing Pointing Mirror Compensation

The principle of pointing mirror compensation is shown in Fig. 1 [6].

It can be seen that before compensation, with the forward flight of the camera, the aim's image is moved and blur on CCD plane. After compensation, when the camera is flying forward, the mirror is swing at certain speed, by whose lead the aim

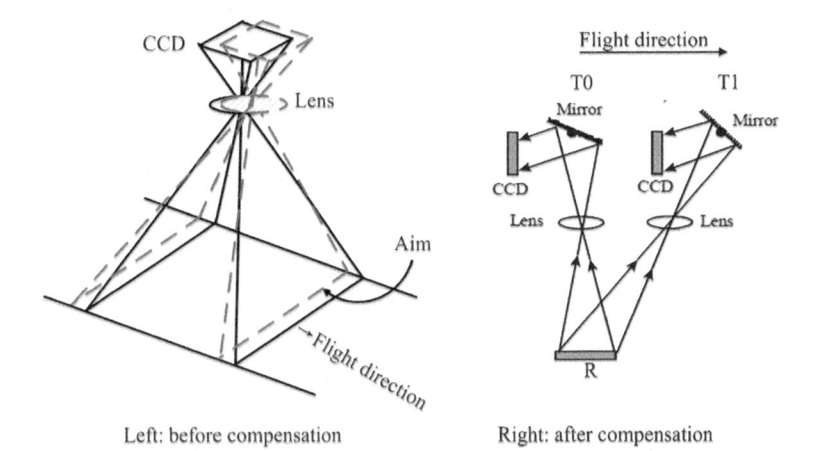

Fig. 1 Pointing mirror compensation principle chart

images on CCD, in other words, CCD camera "sees" the same aim at time T0 and T1, thus the image's residence time increase. In a word, the purpose of pointing mirror compensation is to make sure CCD camera imaging the same aim for several continuous times with the pointing mirror swing on its axis at certain speed, thus increase the resident time of CCD camera.

2.2 Calculation of Swing Speed of Compensation Pointing Mirror

All detail analysis and calculation of motion compensation speed could be seen in reference [7], here just gives the final calculation formula shown as follows.

$$\omega_m = \frac{K-1}{2K} \frac{H\sqrt{R^2 - (R+H)^2 \sin^2 2\theta_m} \cos 2\theta_m}{[(R+H)\cos 2\theta_m - \sqrt{R^2 - (R+H)^2 \sin^2 2\theta_m}]^2} \Omega \quad (1)$$

Where ωm is pointing mirror's angle speed, θm is pointing mirror's pointing angle, H is orbit height, R is the earth radius, Ω is satellite's angle speed and K is multiple of CCD camera's residence time. The relationship between ωm and θm in different orbit height H and different K is shown in Fig. 2.

It can be seen that with the increase of orbit height, the compensation speed becomes lower. Even when the orbit height is 100 Km, the largest compensation speed is 2° per second, which needs the servo control system to be highly stable and accurate. This paper chooses PMSM as compensation mirror's driving motor due to its high accuracy in closed-loop control system.

3 Mathematical Model of PMSM

The mathematical model of PMSM in d-q synchronous rotating coordinate is shown as follows [8]:

$$\begin{aligned}
\frac{d}{dt}i_d &= \frac{1}{L_d}U_d - \frac{R}{L_d}i_d + \frac{L_q}{L_d}P_n\omega_r i_q \\
\frac{d}{dt}i_q &= \frac{1}{L_q}U_q - \frac{R}{L_q}i_q - \frac{L_d}{L_q}P_n\omega_r i_d - \frac{\psi_f}{L_q}P_n\omega_r \\
\psi_q &= L_q i_q \\
\psi_d &= L_d i_d + \psi_f \\
T_e &= 1.5P_n[\psi_f i_q + (L_q - L_d)i_d i_q] \\
J\frac{d\omega_r}{dt} &= T_e - B\omega_r - T_l
\end{aligned} \quad (2)$$

Fig. 2 Relationship of pointing mirror's speed to its angle

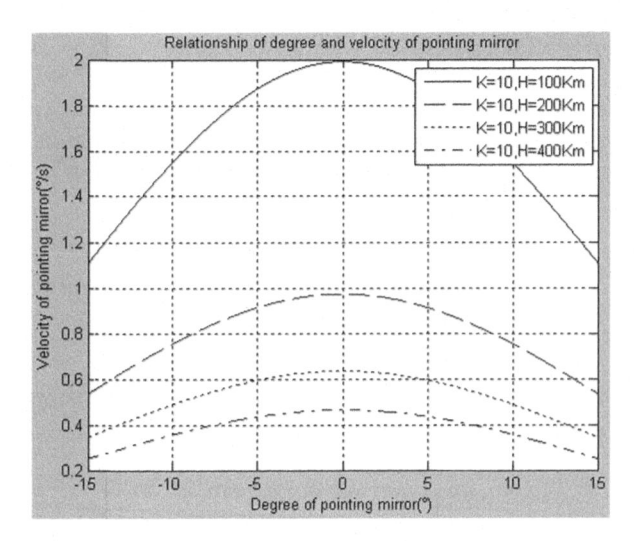

Fig. 3 Control diagram of PMSM

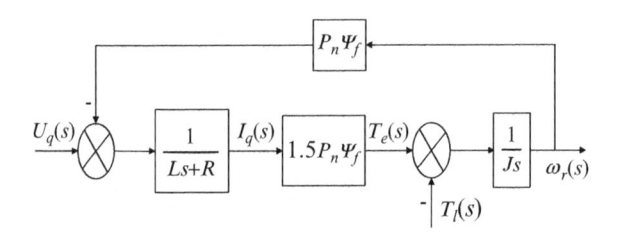

Where Ud and Uq is d-axis and q-axis stator voltage respectively, id and iq is d-axis and q-axis stator current respectively, ψd and ψq is d-axis and q-axis stator magnet linkage respectively, ψf is magnetomotive, R is stator resistance, ωr is rotator angle speed, Te is magnet torque, Tl is load torque, J is rotary inertia and B is friction coefficient. The control diagram of PMSM, which is shown in Fig. 3, could be achieved by transforming the mathematical model of PMSM into Laplace form.

There are many control methods of PMSM and the most widely used control strategy is field orientated control. When id is equal to zero, the motor could gain the largest torque, which is space vector control method [9]. Thus the control structure of PMSM is shown in Fig. 4, where GASR and GACR is speed and current PID adjustor respectively. Especially, when the pointing mirror performs the function of image compensation, the input is compensation speed calculated in second part, so the control structure of PMSM is just speed loop and current loop. In next part, this paper will discuss how to design speed and current PID adjustor.

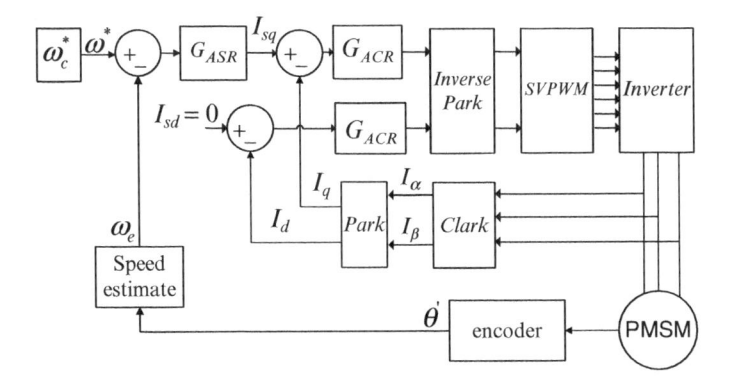

Fig. 4 Control structure of PMSM

4 Design of PID Controller

To design multiple loop system, outer loop sees inner loop as a whole element, and according to practice experience, when the bandwidth of inner loop is over five times of outer loop's, the inner loop could be seen as first-order inertial element.

4.1 Current Loop Controller

As the inner most loop, current loop must have the characteristic of fast following to the change of input voltage and excellent noise resistance, no matter when the pointing mirror performs orientation or compensation. According to the model of PMSM, the motor armature is equivalent to a first-order inertial loop with time constant Tm = L/R, and the PWM inverter is also a first-order inertial loop with time constant Ti, so the control object of current loop is two first-order inertial elements, whose transfer function is as follows:

$$G_{iobj} = \frac{1}{(Ls+R)(T_i s+1)} \tag{3}$$

According to the design method in engineering, the system should be adjusted into I-system, and the controller has an integrating element and a proportional element [10], thus the controller's transfer function is shown in Eq. 4,

$$G_{ACR}(s) = K_{ip}\frac{\tau_i s+1}{\tau_i s} \tag{4}$$

Where Kip is the proportional coefficient and τi is integrating time constant. Making τi = L/R and KI = Kip/Rτi, the current loop's open-loop transfer function is

$$G_i(s) = K_I \frac{1}{s(T_i s + 1)} \tag{5}$$

To consider dynamic response and overshoot, make KI * Ti = 0.5. Supposing the motor's inductance L = 8.5 mH, and resistance R = 6.42 Ω, using MATLAB/SIMULINK to simulate the current loop, and the result shows that the −3 dB bandwidth is 3.37 KHz and the gain in low frequency band is smooth, which approaches the first-order inertial element, besides, the step response is stable at 1.8 ms and no overshoot. Moreover, the amplitude of noise response is 2.8 % of input noise, and decreases to zero in 10 ms. So the designed current loop satisfies the characteristic in fast response, no overshoot and strong noise resistance.

4.2 Speed Loop Controller

From above discussion, the current loop is similar to first-order inertial element, and it could be seen as an element of speed loop. Besides, according to Fig. 3, the current loop and the integrated element Kt/Js constitute the control object of speed loop. In general, the system should be adjusted into II-system, however, this kind of system always has large overshoot, which is not tolerable in image compensation because as long as the mirror's speed deviates the compensation speed shown in Fig. 2, the compensation loses its function. To resolve this problem, speed differential negative feedback is introduced; nevertheless, speed differential would introduce noise, even though some of the noise could be eliminated by filter, while the filter would lead to phase delay, limiting system's performance improvement. Thus, this paper adopts a pseudo derivative feed-forward controller, which has a gain adjustable feed-forward channel, and changing this gain could improve dynamic response and decrease overshoot. Use MATLAB/SIMULINK to build the simulation model as Fig. 5, where Kvfr is gain constant of feed-forward channel, τs is integration time constant and KS is proportional coefficient, and GACR is the current loop above designed.

So the speed loop's close-loop transfer function is shown in Eq. 6.

$$H_s = \frac{K_{vfr}K_t K_s K_I \tau_s s + K_t K_s K_I}{J\tau_s s^3 + J\tau_s K_I s^2 + K_t K_s K_I \tau_s s + K_t K_s K_I} \tag{6}$$

When Kvfr = 0.6, τs = 0.025 and KS = 14, use MATLAB to analyze the transfer function, and the result shows that −3 dB bandwidth of speed loop is

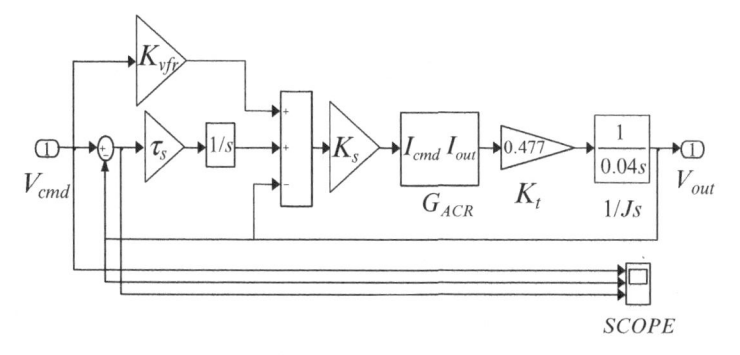

Fig. 5 Simulation model of speed loop

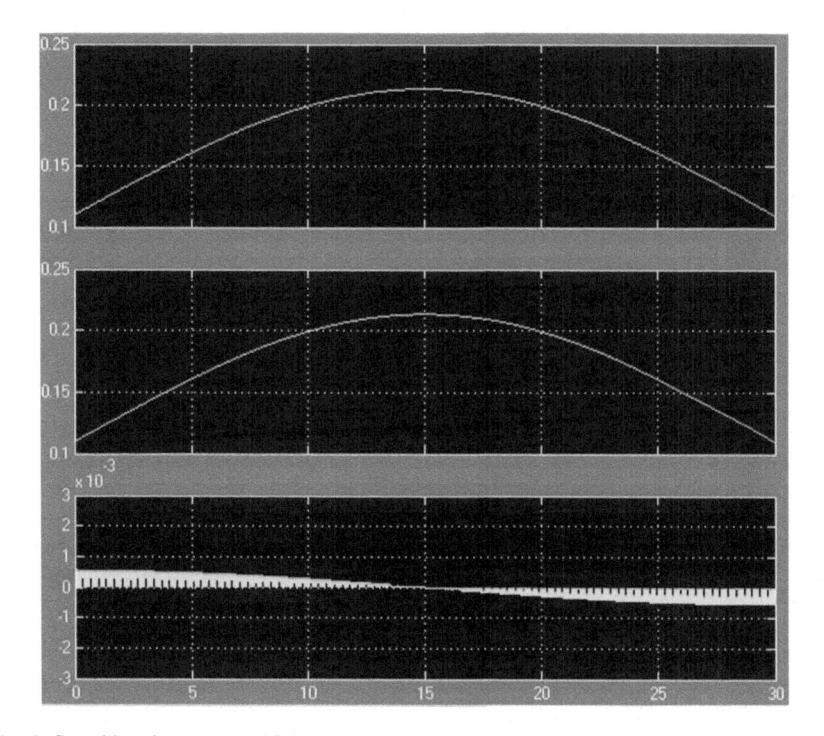

Fig. 6 Speed loop's response to input compensation speed (from up to down: input, output, error)

101 Hz and the step response is stable at 0.06 s and no overshoot. Furthermore, the speed loop's step noise response is only 5 % of input and decreases to zero in 0.1 s.

Besides, use SIMULINK to see how the speed loop's following performance is and the result is shown in Fig. 6. It can be seen that the speed loop follow the input compensation speed very well and the follow error is less than 0.001, thus the designed speed loop satisfy the requirements.

5 Conclusion

This paper designs and simulates an image compensation control system based on pointing mirror driven by PMSM. The principle and calculation of pointing mirror compensation is given firstly and then the mathematical model and control structure of PMSM is established. Finally, based on this model, a high accurate controller including current loop and speed loop is designed and simulated, the simulation results show that the designed controller has good performance in fast response, no overshoot and noise resistance.

References

1. Gu Song, Yan Yong, Xu Kai et al (2011) Design of motion compensation mechanism of satellite remote sensing camera. Proc SPIE 8196:81961Z-1–81961Z-8
2. Wang De-jiang, Kuang Hai-peng, Cai Xi-chang et al (2008) Digital implementation of forward motion compensation in TDI-CCD panoramic aerial camera. Opt Precis Eng 16:2465–2471 (In Chinese)
3. Tchernykh V, Dyblenko S, Janschek K, Seifart K, Harnisch B (2004) Airborne test results for a smart Pushbroom imaging system with optoelectronic image correction. Proc SPIE 5234:550–559
4. Janschek K, Tchernykh V, Dyblenko S, Flandin G, Harnisch B (2004) Compensation of focal plane image motion perturbations with optical correlator in feedback loop. Proc SPIE 5570:280–288
5. Liu Ming (2005) Research on detection and compensation technology of forward image motion in aerial photography based on image restoration. Ph.D. thesis, Graduate School of Chinese Academy of Sciences (In Chinese)
6. Lv Peng, Tang Yuanhe et al (2007) Study CCD image motion for remote sensing detection. Proc SPIE 6279:62795S-1–62795S-7
7. Xie Ren-Biao (2009) The research of aerospace low-speed high accuracy scanning-controlling technology. Ph.D. thesis, Graduate School of Chinese Academy of Sciences (In Chinese)
8. Zhang Yu (2010) The research of image compensation and control of space staring imaging. Ph.D. thesis, Graduate School of Chinese Academy of Sciences (In Chinese)
9. Chen Rong (2004) The research of PMSM servo system. Ph.D. thesis, Nanjing University of Aeronautics and Asnautics (In Chinese)
10. Chen Boshi (2003) Electrical towage automatic control system. China Machine Press, Beijing, pp 59–80 (In Chinese)

μ-Method for Robust Stability of Active Aeroelastic Wing with Multiple Control Surfaces

Fu-hu Liu, Xiao-ping Ma, and Zi-jian Zhang

Abstract μ-method for robust stability of an active aeroelastic wing section with leading and tailing edge control surface is developed. Robust system is constructed to account for the uncertainty parameters associated with the variable structural damping and the nonlinear structural stiffness. The nominal and robust stability margins, critical flutter airspeeds and frequencies are computed to analyze the aeroelastic and aeroservoelastic robust stability in the μ-framework. The analysis process shows μ method for robust stability analysis of aeroservoelastic system with uncertainties is effective. The simulation results indicated that uncertain perturbation reduces stability margin of system. The aeroservoelastic system increases flutter speed and critical dynamic pressure to the aeroelastic (openloop) system, specifically increases in flutter speed is 12 % when leading edge flap activated and 32 % when both leading and trailing edge flap activated. The system tends to stabilize more quickly and trailing edge flap deflects smaller by using both the leading and trailing edge control surfaces simultaneously.

Keywords Robust stability • Aeroservoelastic • Uncertain perturbation

1 Introduction

For the active aeroelastic wing [1], researchers have focused much attention on active flutter suppression (AFS). Great progress has been made since active control technology has been further developing. However, few literatures have been written about AFS on an active aeroelastic wing with multiple control surfaces. Platanitis

---

F.-h. Liu (✉)
Northwestern Polytechnical University, Xi'an, China
e-mail: liufuhu2008@163.com

X.-p. Ma • Z.-j. Zhang
UAV Research Institute, Northwestern Polytechnical University, Xi'an, China

and Strganac [2] used feedback linearization and adaptive control method for the suppression of limit-cycle oscillations (LCO) of a typical wing section with leading and trailing edge control surfaces. It is compared with the study of the wing with only trailing edge control surface, which investigated by Jeffry and Thomas [3] using full-state feedback control law. The result shows that globally stabilizing control may be achieved by using two control surfaces.

Uncertainty is an important issue to its stability for the modern control system. Therefore, robust stability has been put forward. It means the systems with uncertainty can keep stable. Aeroservoelastic (ASE) system considers the interaction of aeroelastic (AE) system and servoactuators. Aeroservoelastic stability has been a necessary condition for safety of air vehicles flight. In the classical control systems analysis of aeroservoelastic stability, the precision magnitude and phase stability margin can be reduced from Nyquist method, which is for single-input-single-output (SISO) system. For multi-input and multi-output (MIMO) system stability analysis, it is a well-known and practically effective technique to use minimum single value method. However, it cannot analyze robust stability for the system with parametric uncertainties. The result of minimum single value method is deficient because it only considering unparametric uncertainties, such as additive perturbation and multiplicative perturbation. Therefore, so many researchers examined μ-method in their analysis to solve this problem. Livne indicated that the uncertainties are one of the important effects in future aeroelasticity research [4]. Lind and Brenner at the NASA Dryden Flight Research Center combining μ-method and flight test data for the estimation of robust flutter and aeroservoelastic margins of F/A-18 research aircraft, and which improve safety of flight test [5, 6]. Lind suggested a match point solution method for a robust flutter prediction. He analyzed the variation of the air vehicles flutter speed in terms of the altitude [7]. Borglund suggested a μ-k method for a robust aeroelastic stability analysis [8]. It proves that μ-method for robust flutter and analysis is more practicable.

In this paper, a typical wing section with leading and trailing edge control surfaces is concerned with the structural and dynamic pressure uncertainty. We construct uncertain AE and ASE system. Furthermore, robust stability for uncertain system is analyzes with μ-method. Herein, robust stability of active aeroelastic wing with multiple control surfaces is studied.

2 Structured Singular Value μ-Method

Standard $P - \Delta$ configuration shown in Fig. 1, where P is the plant of a system, and Δ is structured matrix of perturbations. For $P \in C^{n \times n}$ and a known structure of Δ,

$$\Delta = [\text{diag}(\delta_1 I_{r1}, \ldots \delta_s I_{rs}, \Delta_1, \ldots \Delta_f) : \delta_i \in C, \Delta_j \in C^{m_j \times m_j}] \quad (1)$$

Fig. 1 $P - \Delta$ block diagram for robust stability analysis

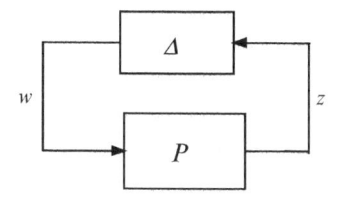

Note the structured perturbation Δ is a block diagonal matrix, where $\sum_{i=1}^{s} r_i + \sum_{j=1}^{f} m_j = n$ with n is the dimension of the block Δ.

The structure singular value μ is defined by

$$\mu_\Delta(P) = (\min\{\sigma_{\max}(\Delta) : \Delta \in \underline{\Delta}, \det(I - P\Delta) = 0\})^{-1} \quad (2)$$

and unless $I - P\Delta$ is not singular for any $\Delta \in \underline{\Delta}$, then $\mu_\Delta(P) = 0$ [9].

It is obvious that the structured singular value μ denotes a measure of the smallest destabilizing perturbation. The robust stability result with regard to structured uncertainty is given as follow.

Robust stability theorem [10]: Let the nominal feedback system be stable and let $\gamma > 0$ be an uncertainty bound, i.e. $\|\Delta\|_\infty \leq \gamma$. The perturbed system of Fig. 1 is robustly stable, with respect to Δ, if and only if $\mu_\Delta(P) < 1/\gamma$.

Generally we assume the set of Δ is bounded to unity $\|\Delta\|_\infty \leq 1$, the robust stability condition is then $\mu_\Delta(P) < 1$.

3 System Model

3.1 A Typical Wing Section

A typical wing section with both a leading and trailing edge control surface is shown in Fig. 2, where b is semichord of the wing, a is nondimensional distance from the midchord to the elastic axis, h is plunge displacement, α is pitch angle, β and γ are trailing and leading edge flap deflection, x_α is nondimensional distance between the elastic axis and the center of mass, K_h is structural stiffness in plunge, K_α is structural stiffness in pitch.

The wing has four degrees of freedom, and the frequency dynamics of the control surfaces are far higher the dynamics of the primary system. Thus the motion of the system, with two degrees of freedom pitch and plunge, may be described by

$$\begin{bmatrix} m_t & m_w x_\alpha b \\ m_w x_\alpha b & I_\alpha \end{bmatrix} \begin{bmatrix} \ddot{h} \\ \ddot{\alpha} \end{bmatrix} + \begin{bmatrix} C_h & 0 \\ 0 & C_\alpha \end{bmatrix} \begin{bmatrix} \dot{h} \\ \dot{\alpha} \end{bmatrix} + \begin{bmatrix} K_h & 0 \\ 0 & K_\alpha \end{bmatrix} \begin{bmatrix} h \\ \alpha \end{bmatrix} = \begin{bmatrix} -L \\ M \end{bmatrix} \quad (3)$$

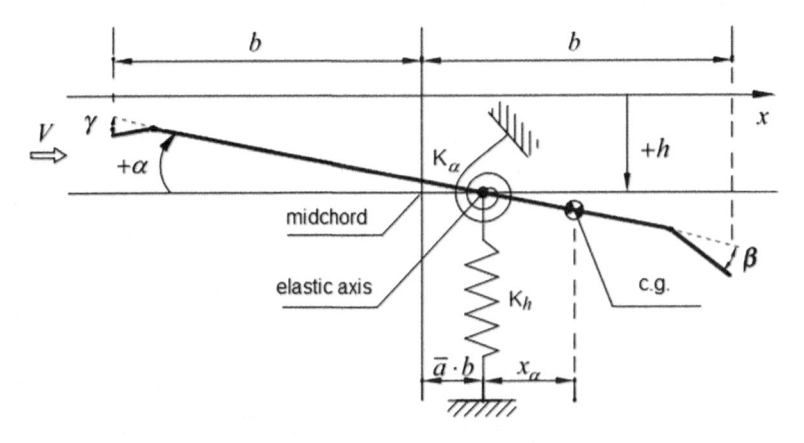

Fig. 2 The wing section with two degrees of freedom with multiple control surfaces

where m_t is total mass of the pitch-plunge system, m_w is total mass of the wing plus mount mass, I_α is total mass moment of inertia about the elastic axis. C_h and C_α are structural damping coefficients in plunge and in pitch. The lift L and moment M are quasi-steady aerodynamic forces and moments with leading and trailing edge have been based on Theodorsen's theory and Fung's theory [11]

$$L = q\left(2bs_p C_{L\alpha}[\alpha + (\dot{h}/V) + (0.5 - a)b(\dot{\alpha}/V)] + 2bs_p C_{L\beta}\beta + 2bs_p C_{L\gamma}\gamma\right) \quad (4)$$

$$M = q\left(2b^2 s_p C_{m\alpha}[\alpha + (\dot{h}/V) + (0.5 - a)b(\dot{\alpha}/V)] + 2b^2 s_p C_{m\beta}\beta + 2b^2 s_p C_{m\gamma}\gamma\right) \quad (5)$$

where s_p is wing span, q is dynamic pressure, $q = \frac{1}{2}\rho V^2$, ρ is air density, V is velocity, $C_{L\alpha}, C_{L\beta}$ and $C_{L\gamma}$ are lift coefficient per angle of attack, trailing edge flap deflection and leading edge flap deflection, $C_{m\alpha}, C_{m\beta}$ and $C_{m\gamma}$ are moment coefficient per angle of attack, trailing edge flap deflection and leading edge flap deflection.

3.2 Nominal AE Systems

To compute stability margin, the unsteady dynamic pressure can be modeled as the nominal dynamic and perturbation associate with unsteady dynamic pressure

$$q = q_{nom} + \delta_q \quad (6)$$

where δ_q is the perturbation of unsteady dynamic pressure, $\delta_q \in R$.

In order to construct μ framework, we let

Fig. 3 Nominal AE system block diagram

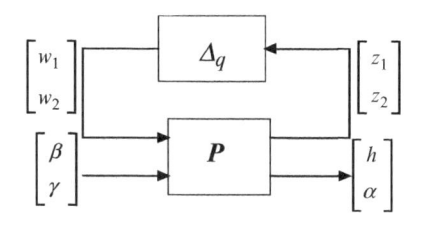

$$z_1 = \left(2bs_p C_{L\alpha}[\alpha + (\dot{h}/V) + (0.5 - a)b(\dot{\alpha}/V)] + 2bs_p C_{L\beta}\beta + 2bs_p C_{L\gamma}\gamma\right) \quad (7)$$

$$z_2 = \left(2b^2 s_p C_{m\alpha}[\alpha + (\dot{h}/V) + (0.5 - a)b(\dot{\alpha}/V)] + 2b^2 s_p C_{m\beta}\beta + 2b^2 s_p C_{m\gamma}\gamma\right) \quad (8)$$

$$w_1 = \delta_1 z_1 \quad (9)$$

$$w_2 = \delta_2 z_2 \quad (10)$$

where $\delta_1 = \delta_2 = \delta_q$.

Combining Eqs. 4, 5, 6, 7, 8, 9, and 10, the aerodynamic loads are reformulated

$$L = q_{nom}(2bs_p C_{L\alpha}[\alpha + (\dot{h}/V) + (0.5 - a)b(\dot{\alpha}/V)] \\ + 2bs_p C_{L\beta}\beta + 2bs_p C_{L\gamma}\gamma) + w_1 \quad (11)$$

$$M = q_{nom}(2b^2 s_p C_{m\alpha}[\alpha + (\dot{h}/V) + (0.5 - a)b(\dot{\alpha}/V)] \\ + 2b^2 s_p C_{m\beta}\beta + 2b^2 s_p C_{m\gamma}\gamma) + w_2 \quad (12)$$

Combining Eqs. 3, 4, 5, 6, 7, 8, 9, and 10 we obtain nominal system is shown in Fig. 3, where $\Delta_q = diag(\delta_1, \delta_2)$.

3.3 Robust AE and ASE Systems

Robust system is the nominal system considers structure perturbation. The structural damping coefficient due to variable damping can be defined as

$$C_h = C_{hnom} + W_3\delta_3 = C_{hnom}(1 + e_3\delta_3) \quad (13)$$

$$C_\alpha = C_{\alpha nom} + W_4\delta_4 = C_{\alpha nom}(1 + e_4\delta_4) \quad (14)$$

where δ_3 and δ_4 are perturbation to the structural damping coefficient in plunge C_{hnom} and in pitch $C_{\alpha nom}$, $\delta_3, \delta_4 \in R$ and $|\delta_3| \leq 1$, $|\delta_4| \leq 1$; W_3 and W_4 are the weighting on perturbation to the structural damping coefficient in plunge C_{hnom} and in pitch $C_{\alpha nom}$, $W_3, W_4 \in R$; e_3 and e_4 are weighted modeling error on the structural damping coefficient in plunge C_{hnom} and in pitch $C_{\alpha nom}$, and $|e_3| \leq 1, |e_4| \leq 1$;

Similar the structural stiffness coefficient with perturbation can be defined as

$$K_h = K_{hnom} + W_5\delta_6 = K_{hnom}(1 + e_5\delta_5) \tag{15}$$

$$K_\alpha = K_{\alpha nom} + W_6\delta_6 = K_{\alpha nom}(1 + e_6\delta_6) \tag{16}$$

where δ_5 and δ_6 are perturbation to the structural stiffness coefficient in plunge K_{hnom} and in pitch $K_{\alpha nom}$, $\delta_5, \delta_6 \in R$ and $|\delta_5| \leq 1, |\delta_6| \leq 1$; W_5 and W_6 are the weighting on perturbation to the structural stiffness coefficient in plunge K_{hnom} and in pitch $K_{\alpha nom}$, $W_5, W_6 \in R$; e_5 and e_6 are weighted modeling error on the structural damping coefficient in plunge K_{hnom} and in pitch $K_{\alpha nom}$, and $|e_5| \leq 1, |e_6| \leq 1$;

In order to construct μ framework, we let

$$\begin{bmatrix} z_3 \\ z_4 \\ z_5 \\ z_6 \end{bmatrix} = \begin{bmatrix} W_3 & & & \\ & W_4 & & \\ & & W_5 & \\ & & & W_6 \end{bmatrix} \begin{bmatrix} \dot{h} \\ \dot{\alpha} \\ h \\ \alpha \end{bmatrix} \tag{17}$$

$$\begin{bmatrix} w_3 \\ w_4 \\ w_5 \\ w_6 \end{bmatrix} = \begin{bmatrix} \delta_3 & & & \\ & \delta_4 & & \\ & & \delta_5 & \\ & & & \delta_6 \end{bmatrix} \begin{bmatrix} z_3 \\ z_4 \\ z_5 \\ z_6 \end{bmatrix} = \Delta_p \begin{bmatrix} z_3 \\ z_4 \\ z_5 \\ z_6 \end{bmatrix} \tag{18}$$

where Δ_p is the perturbation block, $\Delta_p = \text{diag}(\delta_3, \delta_4, \delta_5, \delta_6)$ and $\|\Delta_p\|_\infty \leq 1$.

Combining Eqs. 2, 17, and 18, the motion of the system are reformulated

$$\begin{cases} m_t\ddot{h} + S_\alpha\ddot{\alpha} + C_{hnom}\dot{h} + K_{hnom}h + w_3 + w_5 = -L \\ S_\alpha\ddot{h} + I_\alpha\ddot{\alpha} + C_{\alpha nom}\dot{\alpha} + K_{\alpha nom}\alpha + w_4 + w_6 = M \end{cases}. \tag{19}$$

The robust AE system block diagram can be constructed as shown in Fig. 4, where $[w_i] = diag(w_3, w_4, w_5, w_6)$, $[z_i] = diag(z_3, z_4, z_5, z_6)$.

The robust ASE system block diagram is constructed by adding two servocontrols to the robust AE system. The transfer functions for trailing edge flap K_β and leading edge flap K_γ can be defined as

$$K = \frac{\delta(s)}{\delta_c(s)} = \frac{k}{(1/\omega^2)s^2 + (2\zeta/\omega)s + 1} \tag{20}$$

Fig. 4 Robust AE system block diagram

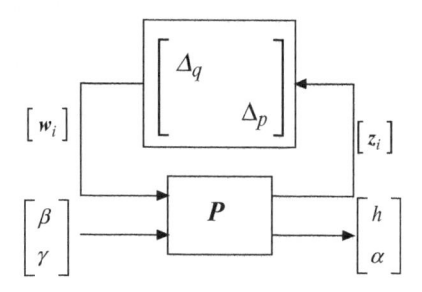

Fig. 5 Robust ASE system block diagram

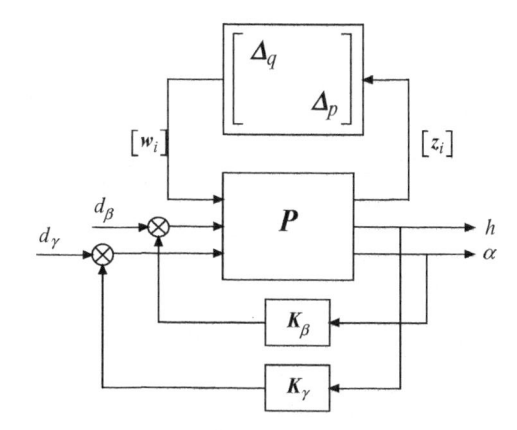

where δ is the deflection of the control surface, δ_c is the input deflection command of the control surface, k is the gain of servoactuators, ω is frequency of servoactuators, ζ is damp ratio of servoactuators.

In this paper, based on characteristic of the trailing edge flap and leading edge flap, servoactuators parameters are $\omega_\beta = \omega_\gamma = 50Hz$, $\xi_\beta = \xi_\gamma = 0.5$, $k_\beta = -1.02$, $k_\gamma = 23.4$. The robust ASE system block diagram shown in Fig. 5, where d_β and d_γ are additional inputs of trailing and leading edge flap deflections.

3.4 State-Space Model

For the robust stability analysis, the uncertainty system should be constructed as state-space model form. Let $X_s = [h \quad \alpha]^T$ and combining Eq. 19 and Fig. 4, the motion of the system are reformulated in matrix form

$$\begin{cases} M_s \ddot{X}_s = (C_a - C_s)\dot{X}_s + (K_a - K_s)X_s + B_a u \\ [z_i] = C_z [X_s \quad \dot{X}_s]^T + D_z [\beta \quad \gamma]^T \end{cases}. \qquad (21)$$

where $u = [w_1 \; w_2 \; w_3 \; w_4 \; w_5 \; w_6 \; \beta \; \gamma]$,

$$M_s = \begin{bmatrix} m_t & m_w x_\alpha b \\ m_w x_\alpha b & I_\alpha \end{bmatrix}, \quad C_s = \begin{bmatrix} C_{hnom} & 0 \\ 0 & C_{\alpha nom} \end{bmatrix}, \quad K_s = \begin{bmatrix} K_{hnom} & 0 \\ 0 & K_{\alpha nom} \end{bmatrix}$$

$$C_a = q_{nom} \begin{bmatrix} -2bs_p C_{L\alpha}(1/V) & -2b^2 s_p C_{L\alpha}(0.5-a)(1/V) \\ 2b^2 s_p C_{m\alpha}(1/V) & 2b^3 s_p C_{m\alpha}(0.5-a)(1/V) \end{bmatrix},$$

$$K_a = \begin{bmatrix} 0 & -q_{nom} 2bs_p C_{L\alpha} \\ 0 & q_{nom} 2b^2 s_p C_{m\alpha} \end{bmatrix},$$

$$B_a = \begin{bmatrix} -1 & 0 & -1 & 0 & -1 & 0 & -q_{nom} 2bs_p C_{L\beta} & -q_{nom} 2bs_p C_{L\gamma} \\ 0 & 1 & 0 & -1 & 0 & -1 & q_{nom} 2b^2 s_p C_{m\beta} & q_{nom} 2b^2 s_p C_{m\gamma} \end{bmatrix},$$

$$C_z = \begin{bmatrix} 0 & 2bs_p C_{L\alpha} & 2bs_p C_{L\alpha}(1/V) & 2b^2 s_p C_{L\alpha}(0.5-a)(1/V) \\ 0 & 2b^2 s_p C_{m\alpha} & 2b^2 s_p C_{m\alpha}(1/V) & 2b^3 s_p C_{m\alpha}(0.5-a)(1/V) \\ 0 & 0 & W_3 & 0 \\ 0 & 0 & 0 & W_4 \\ W_5 & 0 & 0 & 0 \\ 0 & W_6 & 0 & 0 \end{bmatrix},$$

$$D_z = \begin{bmatrix} 2bs_p C_{L\beta} & 2bs_p C_{L\gamma} \\ 2b^2 s_p C_{m\beta} & 2b^2 s_p C_{m\gamma} \end{bmatrix}.$$

Then defining the state variables $x = [h \; \alpha \; \dot{h} \; \dot{\alpha}]^T$, one obtains the state-space form of the openloop AE system, which represent P in Fig. 5

$$\begin{cases} \dot{x} = Ax + Bu \\ y = Cx + Du \end{cases} \quad (22)$$

where output $y = [z_1 \; z_2 \; z_3 \; z_4 \; z_5 \; z_6 \; h \; \alpha]^T$, input $u = [w_1 \; w_2 \; w_3 \; w_4 \; w_5 \; w_6 \; \beta \; \gamma]$,

$$A = \begin{bmatrix} 0_{2\times 2} & I_{2\times 2} \\ M_s^{-1} K & M_s^{-1} C \end{bmatrix}, \quad B = \begin{bmatrix} 0_{2\times 8} \\ M_s^{-1} B_a \end{bmatrix}, \quad C = \begin{bmatrix} C_z \\ I_{2\times 2} & 0_{2\times 2} \end{bmatrix}, \quad D = \begin{bmatrix} 0_{2\times 6} & D_z \\ 0_{6\times 6} & 0_{6\times 2} \end{bmatrix},$$

$K = K_a - K_s$, $C = C_a - C_s$.

4 Simulation Results and Discussion

Platanitis and Strganac give a wing section with training and leading edge flaps, and we use the parameters of this wing [2]. The wing in experiments occur flutter at 13 m/s for openloop AE system [2]. Therefore, in this research, the systems are simulated at nominal speed 13 m/s. Let the system initial state variables are x = [0.01 0.1 0 0]^T and perturbation parameters are $e_3 = e_4 = 0.4$, $e_5 = e_6 = 0.05$. Both system modeling of analysis and the computation are code with MATLAB μ-Analysis and Synthesis Toolbox [12]. The iteration flow chart of stability margin is shown in Fig. 6, which include AE system, ASE system, nominal system and robust system.

The concerned parameters in this study are flutter speed V_F, flutter frequency f_F, critical dynamic pressures q_{cr}, and stability margins $\delta_{q\,max}$. $\delta_{q\,max}$ is the maximum of dynamic pressure perturbation stabilize system, and $q_{cr} = q_{nom} + \delta_{q\,max}$. Figure 7 shows the iteration process of μ value for robust AE system. The iteration is not interrupt until μ < 1. When system reach critical stable structured singular value μ curves are shown in Fig. 7, where ASE (one flap) represents the ASE system with the leading edge flap, and ASE (two flaps) denotes the ASE system with both the leading and trailing edge flap. The computation results are presented as in Table 1.

By above figure, frequency versus structured singular value μ = 1 means flutter frequency. Robust AE system f_F =9.69 rad/s, robust ASE system with leading edge flap f_F =13.80 rad/s, robust ASE system with both leading and trailing edge flap f_F =13.01 rad/s (Fig. 8).

From table above, we observe that stability margin of system. Flutter frequency is identical to Fig. 7. Uncertain perturbation reduces stability margin of system. The decrease in stability margin is 69 % for AE system, 64 % for ASE system (one flap)

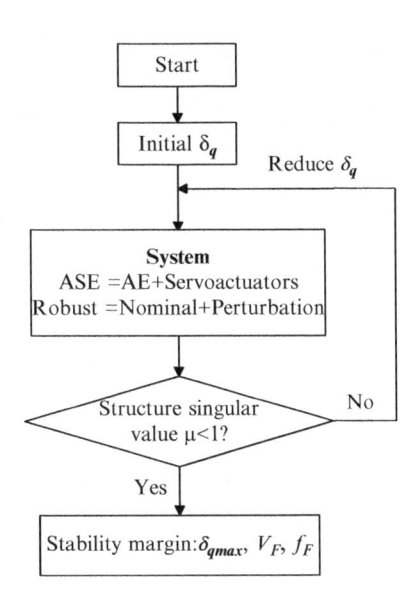

Fig. 6 Iteration flow chart of stability margin

Fig. 7 Iteration process of μ value for robust AE system

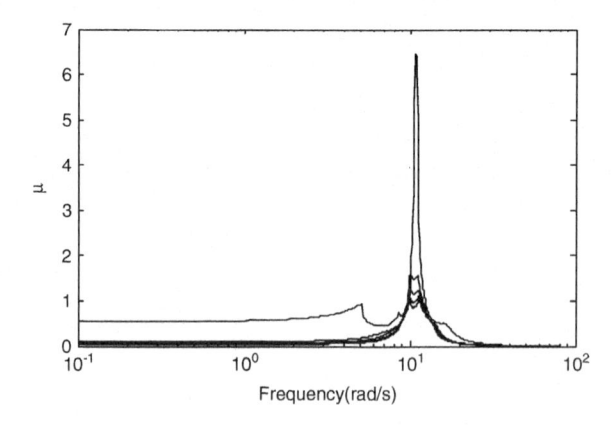

Table 1 Computation results of system at V = 13 m/s

System model	Type	V_F (m/s)	f_F (Hz)	$\delta_{q\max}$ (N/m^2)
AE	Nominal	13.9752	1.7264	16.1133
	Robust	13.3062	1.5425	4.93345
ASE (one flap)	Nominal	17.9044	2.1274	92.8345
	Robust	14.9529	2.1971	33.4359
ASE (two flaps)	Nominal	22.5324	1.9675	207.458
	Robust	17.6148	2.0724	86.5338

Fig. 8 AE system structured singular value μcurves

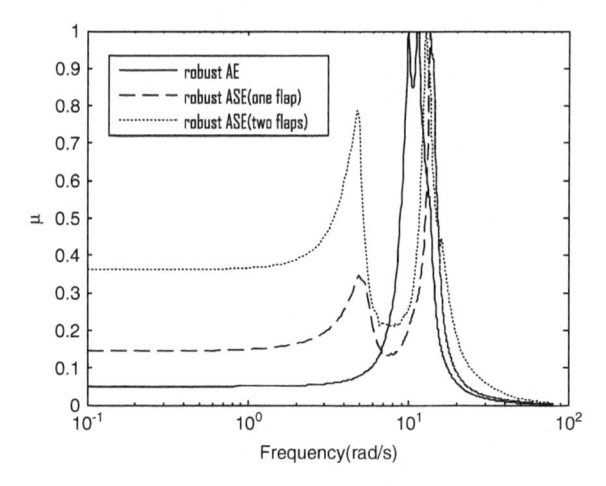

and 58 % for ASE system (two flaps). The ASE system increases flutter speed and critical dynamic pressure to the AE (openloop) system, specifically increases in flutter speed is 12 % when leading edge flap activated and 32 % when both leading and trailing edge flap activated. By using both the leading and trailing edge control surfaces are more effective on flutter suppression.

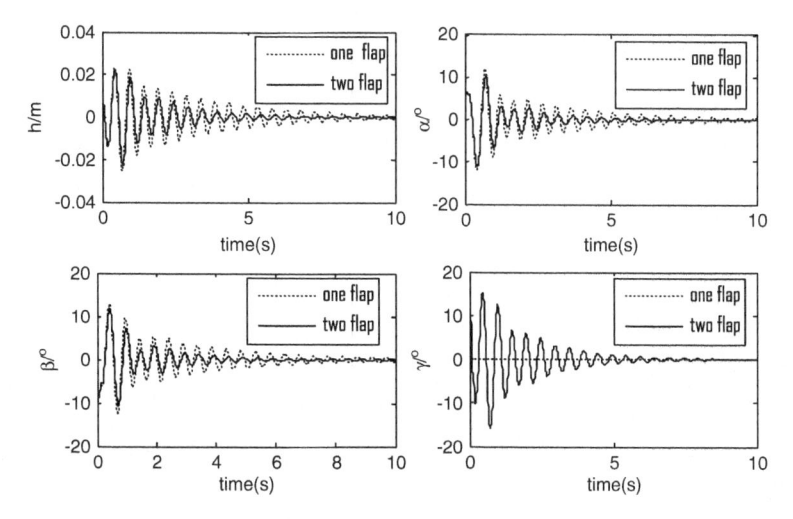

Fig. 9 Time response of robust ASE system with single control surface and multiple control surfaces at V = 14.5 m/s

Figure 9 shows time response of ASE system with only leading edge flap (dotted line) and ASE system with both leading and trailing edge flaps (solid lines). Examining time response process of plunge, pitch and control surface, one notes that the system tended to stabilized more quickly and trailing edge flap deflects smaller by using both the leading and trailing edge control surfaces simultaneously.

5 Conclusion

μ-method for robust stability of an active aeroelastic wing section with leading and tailing edge control surface is developed. Nominal AE system with dynamic pressure perturbation is established. Robust AE system is constructed to account for the uncertainty parameters associated with the variable structural damping and the nonlinear structural stiffness. Robust ASE system is constructed by adding two servocontrols to the robust AE system. The nominal and robust stability margins, critical flutter airspeeds and frequencies are computed. We investigate the aeroelastic and aeroservoelastic robust stability in the μ-framework.

The analysis process shows μ method for robust stability analysis of aeroservoelastic system with uncertainties is effective. The simulation results indicated that uncertain perturbation reduces stability margin of system. The ASE system increase flutter speed and critical dynamic pressure to the AE (openloop) system. The system tends to stabilized more quickly and trailing edge flap deflects smaller by using both the leading and trailing edge control surfaces simultaneously.

References

1. Perry B III, Cole SR, Miller GD (1995) Summary of an active flexible wing program. J Aircr 32(1):10–15
2. Platanitis G, Strganac TW (2004) Control of a nonlinear wing section using leading- and trailing-edge surfaces. J Guid Control Dynam 27(1):52–58
3. Block JJ, Strganac TW (1998) Applied active control for a nonlinear aeroelastic structure. J Guid Control Dynam 21(6):838–845
4. Livne E (2003) Future of airplane aeroelasticity. J Aircr 40(6):1066–1092
5. Lind R, Brenner M (1997) Robust flutter margins of an F/A-18 aircraft from aeroelastic flight data. J Guid Control Dynam 20(3):597–604
6. Lind R, Brenner M (1998) Incorporating flighting data into a robust aeroelastic model. J Aircr 35(3):470–477
7. Lind R (2002) Match-point solutions for robust flutter analysis. J Aircr 39(1):91–99
8. Borglund D (2004) The μ-k method for robust flutter solutions. J Aircr 41(5):1209–1216
9. Gu DW, Hr. Petkov P, Konstantinov MM (2005) Robust control design with MATLAB. Springer-Verlag London Ltd, London, pp 71–73
10. Doyle JC (1982) Analysis of feedback systems with structured uncertainties. IEEE Proc-Part D 129:242–250
11. Buttrill CS, Bacon BJ, Heeg J, Houck JA, Wood DV (1996) Aeroservoelastic simulation of an active flexible wing tunnel model. NASA technical paper 3510, pp 11–13
12. Balas GJ, Doyle JC, Glover K, Packard A, Smith R (2004) μ-analysis and synthesis toolbox user's guide. The Math Works, Natick, MA, USA, pp 299–316

Nonlinear Flight Controller Design Using Combined Hierarchy-Structured Dynamic Inversion and Constrained Model Predictive Control

Chao Wang, Shengxiu Zhang, and Chao Zhang

Abstract In order to account for hard limits on state and actuator range, a nonlinear flight control method combined hierarchy-structured dynamic inversion (HSDI) and constrained model predictive control (MPC) was proposed for a researched unmanned aircraft (RUA). First, the HSDI control law linearizes the nonlinear dynamic model of the aircraft. Then, constrained MPC is designed in accordance with Pseudo-linear system to guarantees HSDI and determines online constrained optimal inputs. Finally, the robust performance of the combined controller is improved by employing on-line identification for uncertainties based on adaptive estimation approaches. The nonlinear model was updated with adaptive parameter T-S fuzzy model to improve the fidelity of the model used for controller synthesis. Simulation results show satisfactory performance of the presented controller for attitude command tracking control, and the robustness to parameters variations and the disturbance rejection are successfully accomplished.

Keywords Flight control • Hierarchy-structured dynamic inversion • Constrained model predictive control • T-S fuzzy model • Adaptive parameter approximation

1 Introduction

New nonlinear guidance and control laws will command and execute agile flight with rapid manoeuvring capability, large thrust, and closer approach to stall boundaries for fast deceleration and rapid turning than previously possible. Such performance improvements expand the operational regime of UAVs. Nevertheless, because of dynamic constraints such as limited manoeuvrability, minimum turn radius for the specified airspeed, and actuator saturation, the physical

C. Wang • S. Zhang (✉) • C. Zhang
Precise Guidance and Simulation Lab, Xi'an Research Institute of Hi-Tech, Xi'an, China
e-mail: Dieche1218@sina.com

impossibility of applying unlimited control signals makes actuator saturation a ubiquitous problem in control systems. A known approach to this task is the treatment of the flight control as a nonlinear model predictive control (NMPC) problem [1–3]. However, this is computationally demanding due to the nonlinearities that have to be considered. Whenever the computational power of the onboard computer is limited, a straight forward NMPC implementation becomes very difficult, or even impossible. One alternative approach is to linearize the nonlinear system on-line using Feedback Linearization (FBL) which can be combined with linear, discrete-time MPC. Van Soest has done this, for example, including only input constraints [4]. Guemghar has also taken state constraints into account, and there the principle of Time Scale Separation (TSS) is used [5]. Nevertheless, uncertainties and disturbances have not been discussed in above literatures.

In this paper, the combination of hierarchy-structured dynamic inversion (HSDI) control and constrained model predictive control (MPC) is applied to track the reference attitude angles for a RUA, which is modeled to be nonlinear with hard limits on state and actuator range. Combining HSDI with MPC gives the best of both worlds: HSDI allows application of the simple linear discrete model predictive control concept; MPC provides explicit constraint handling as part of the optimization process. In order to deal with uncertainties and disturbances, this paper therefore improves the nominal combined controller by employing on-line identification for uncertainties using T-S fuzzy model. The nonlinear model developed based on first principle theory was updated with adaptive parameter T-S fuzzy model to improve the fidelity of the model used for controller synthesis. An online adaptation algorithm for the parameter matrixes of T-S fuzzy model is designed.

The paper is organized as follows: In the next section, the RUA nonlinear dynamic is given. On-line estimator for uncertainties based on adaptive parameter T-S fuzzy model is introduced in Sect. 3. Sect. 4 designs the combined controller and the constraint mapping algorithms. The simulation results are shown in Sect. 5. Finally, Sect. 6 draws the conclusions.

2 Research Unmanned Aircraft Model

Let $x_1 = [\alpha \ \beta \ \mu]^T$, $x_2 = [p \ q \ r]^T$ and control input $u = [\delta_a \ \delta_e \ \delta_r]^T$. Where α, β and μ respectively denote angle of attack, sideslip angle and conical rotation angle; p, q and r respectively denote roll, pitch, and yaw rates about the body axes; δ_a, δ_e and δ_r denote deflections of aileron, elevator and rudder, respectively. Based on the assumption of the flat Earth and constant mass properties [6], the general nonlinear 6DOF dynamic of the RUA with uncertainties and disturbances can be written as [7]:

$$\dot{x}_1 = f_{1n}(x_1) + g_{1n}(x_1)x_2 + h_{1n}(x_1)u + \Delta_1$$
$$\dot{x}_2 = f_{2n}(x_1, x_2) + g_{2n}(x_1)u + \Delta_2 \qquad (1)$$

Where subscript n is added to indicate the nominal situation; The functions $f_{1n}, f_{2n}, g_{1n}, g_{2n}$ and h_{1n} for each controller design are given by Cao [7]; $\Delta_1 = [\Delta_\alpha \ \Delta_\beta \ \Delta_\mu]^T$ and $\Delta_2 = [\Delta_p \ \Delta_q \ \Delta_r]^T$ are lumped disturbances. With assumptions 2 in Cao [7] (which implies that $h_{1n}u \approx 0$) and given a specific sampling time T_s, we have a discrete-time MIMO affine nonlinear system:

$$x_1(k+1) = f_{1n}(k) + g_{1n}(k)x_2(k) + \Delta_1(k)$$
$$x_2(k+1) = f_{2n}(k) + g_{2n}(k)u(k) + \Delta_2(k) \qquad (2)$$

3 On-line Identification for Uncertainties

In the system (2), there are some unknown uncertainties $\Delta f_i(\cdot)$ and $\Delta g_i(\cdot)$ and disturbances d_i. They are combined to form an unknown nonlinear function Δ_i as follows:

$$\Delta_1(k) = \Delta f_1(k) + \Delta g_1(k)x_2(k) + d_1(k)$$
$$\Delta_2(k) = \Delta f_2(k) + \Delta g_2(k)u(k) + d_2(k) \qquad (3)$$

To identify $\Delta_i(k)$ $i = 1, 2$, T-S fuzzy model is applied. The j-th fuzzy rule can be described as:

$$R^j : \text{IF } x_1(k) \text{ is } F_1^j \text{ and } \cdots x_n(k) \text{ is } F_n^j \text{ and } u_1(k) \text{ is } F_{n+1}^j \text{ and } \cdots u_n(k) \text{ is } F_{2n}^j$$
$$\text{THEN } \hat{\Delta}_i(k) = \hat{A}^j(k)x(k) + \hat{B}^j(k)u(k) \qquad (4)$$

Where $x(k) = [x_1(k), x_2(k), \cdots, x_n(k)]^T \in \Re^n$ is the state vector, $u(k) = [u_1(k), u_2(k), \cdots, u_n(k)]^T \in \Re^n$ is control input.

Applying some commonly used defuzzification strategies, system (3) becomes:

$$\hat{\Delta}_i(k) = \sum_{j=1}^r w^j [\hat{A}^j(k)x(k) + \hat{B}^j(k)u(k)] \qquad (5)$$

Where $w^j = \prod_{l=1}^{2n} \mu_{F_l^j}(z_l) \Big/ \sum_{j=1}^r \prod_{l=1}^{2n} \mu_{F_l^j}(z_l)., j = 1, 2, \cdots r$, $\mu_{F_l^j}(z_l)$ is the value of the membership function, $z_l = [x(k) \ u(k)]^T$.

By choosing the following parameter adaptation law:

$$\hat{A}^j(k) = \hat{A}^j(k-1) - \gamma_a w^j [A_s e(k)]^T P x(k)$$
$$\hat{B}^j(k) = \hat{B}^j(k-1) - \gamma_b w^j [A_s e(k)]^T P u(k) \quad (6)$$

And according to a standard Lyapunov theorem extension [8], the tracking error vector for the state variables $e(k)$ is bounded above by ε defined in Eq. 7. Therefore, the adaptive law Eq. 6 permits the adaptation of T-S fuzzy model parameters.

$$\left\{ e(k) : \|e(k)\| < \varepsilon = \rho \sqrt{\lambda_{\max}(P)/\kappa} \right\} \quad (7)$$

Where γ_a, γ_b is a gain in adaptive law; $\lambda_{\max}(P)$ is the largest singular value of P; ρ is a positive constant.

The T-S fuzzy model of the $\Delta_i(k)$, $i = 1, 2$ is identified and validated, as explained above. Using Eq. 5, system (2) can be rewritten as:

$$\begin{aligned} x_1(k+1) &= f_1(k) + g_1(k) x_2(k) \\ x_2(k+1) &= f_2(k) + g_2(k) u(k) \end{aligned} \quad (8)$$

Where $f_1(k) = f_{1n}(k) + \sum_{j=1}^{r} w_1^j \hat{A}_1^j(k) x_1(k)$, $g_1(k) = g_{1n}(k) + \sum_{j=1}^{r} w_1^j \hat{B}_1^j(k)$; $f_2(k) = f_{2n}(k) + \sum_{j=1}^{r} w_2^j \hat{A}_2^j(k) x_2(k)$, $g_2(k) = g_{2n}(k) + \sum_{j=1}^{r} w_2^j \hat{B}_2^j(k)$.

4 The Controller Architecture

This section provides the theory of both HSDI and MPC and discusses in-depth the interconnection issues that arise from this combination. Figure 1 provides an overview of how MPC and NDI are to be combined in each subsystem.

4.1 Hierarchy-Structured Dynamic Inversion

An HSDI-based flight control law is developed in such a way that a general fixed-wing aircraft system is decomposed into subsystems according to the time scales inherent in the dynamics and nonlinear dynamic inversion (NDI) is applied to each subsystem. In each subsystem, the slow variables are controlled by taking the fast variables as fictitious control input. HSDI therefore features a simple nested structure of the first order NDI controllers. Thus, considering the two time scales nonlinear system (8), inner-loop and outer-loop controllers are designed by NDI. Control law in each loop of the system is given by:

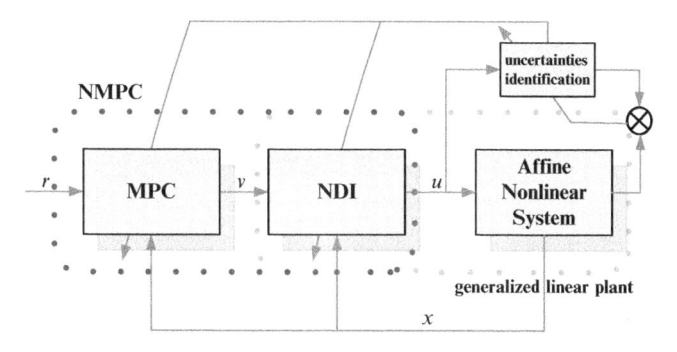

Fig. 1 MPC+NDI control concept. The input to the MPC controller is the reference and current system state, the output is a virtual control input which is mapped by the NDI control law to the input signal of the system

$$x_{2c}(k) = g_1^{-1}(k)[-f_1(k) + v_1] \tag{9}$$

$$u(k) = g_2^{-1}(k)[-f_2(k) + v_2] \tag{10}$$

Nonlinear control of the "inner loop" (i.e., rate loops) and the "outer loop" (i.e., attitude loops) is accomplished via HSDI. This yields a discrete-time linear system with inputs that are subject to state-dependent constraints. Instead of employing a simple HSDI design which has the possibility of generating extremely large control inputs, the feedback-linearized system is regulated with a linear model predictive controller with explicit constraint handling capability.

4.2 Constrained Model Predictive Control

The closed-loop systems with HSDI controllers (9) and (10) are defined as:

$$x_1(k+1) = v_1(k) \tag{11}$$

$$x_2(k+1) = v_2(k) \tag{12}$$

Equations 11 and 12 represent the basic models for the MPC design in this section. The designs of outer loop's MPC_1 and inner loop's MPC_2 are identical, only the design of MPC_2 is discussed here. The basis of MPC controller is an optimization problem. Therefore, the task of the controller has to be translated into a Performance Index (PI). Optimization of this PI leads to an optimal input sequence. The goal of the inner loop controller is to track the reference trajectory commanded by the outer loop controller as good as possible and since this trajectory is defined in terms of angular rates, the PI is based on the error between the actual measured angular rates and the reference angular rates. The PI needs to be minimized is defined as:

$$J = \sum_{i=1}^{N_p} e(k+i|k)^T Q e(k+i|k) \tag{13}$$

Where $e(k+i|k) = \hat{x}_2(k+i|k) - x_{2c}(k+i|k)$, $Q \geq 0$ is the weighting matrix of tracking error, and $\hat{x}_2(k+i|k)$ is the predicted value of $x_2(k+i)$ at time k.

Using relationship (12), the control variable u is replaced by the intermediate variable v_2, the PI (13) of the standard MPC is transformed as follows:

$$J = (\bar{x}_2 - \bar{x}_{2c})^T \bar{Q}(\bar{x}_2 - \bar{x}_{2c}) = (\bar{v}_2 - \bar{x}_{2c})^T \bar{Q}(\bar{v}_2 - \bar{x}_{2c}) = \bar{v}_2^T \bar{Q} \bar{v}_2 - 2\bar{x}_{2c}^T \bar{Q} \bar{v}_2 \tag{14}$$

Where $\bar{x}_2 = [x_2(k+1|k), \cdots, x_2(k+N_p|k)]^T$; $\bar{x}_{2c} = [x_{2c}(k+1|k), \cdots, x_{2c}(k+N_p|k)]^T$; $\bar{u} = [u(k|k), \cdots, u(k+N_p-1|k)]^T$; $\bar{v}_2 = [v_2(k|k), \cdots, v_2(k+N_p-1|k)]^T$; $\bar{Q} = I_{N_p} \otimes Q$; I_{N_p} is an identity matrix of size N_p, and the operator $\otimes$ indicates the Kronecker product of two matrices.

The PI to be minimized on every sampling instant k is a quadratic criterion on $\bar{v}_2$. In this case, fast and reliable quadratic program (QP) optimization routines can be used to find the solution of $\bar{v}_2$. The QP can take any kind of constraint into account, provided it is linear. Due to the NDI controller the actual inputs and states are not directly 'visible' for the MPC controller and the constraints on the actual inputs and states have to be converted by the NDI control law, which is called the constraint mapping. Below first the input constraint mapping will be explained followed by the state constraint mapping.

4.3 Constraint Mapping

Both input and state constraints are considered. The input constraints are defined as aerodynamic control surface deflections, all the control surface deflections are limited to ±25 deg. State constraints are defined in terms of maximum and minimum allowed deviations of the body angular rates p, q and r, and the aerodynamic angles α, β and μ, the magnitudes of the state constraints are given in Table 1.

4.3.1 Input Constraint Mapping

Consider the vectors of upper and lower bounds on u:

Table 1 State constraints

Attitude angle	Amplitude range/°	Angular rate	Amplitude range/°
α	$-10 \sim +15$	p	$-80 \sim +80$
β	$-5 \sim +5$	q	$-30 \sim +30$
μ	$-60 \sim +60$	r	$-15 \sim +15$

$$\underbrace{\begin{bmatrix} -I_n & I_n \end{bmatrix}^T}_{M} u \leq \underbrace{\begin{bmatrix} -u_{\min} & u_{\max} \end{bmatrix}^T}_{b_u} \tag{15}$$

The implementation of the NDI scheme maps the input vector of u into the predictive controller's output v_2 through the inverse of the NDI law (16), thus the original set of linear inequality constraints u in Eq. 15 is transformed into a new set of non-linear inequality constraints in v_2.

$$v_2(k) = f_2(k) + g_2(k)u(k) \tag{16}$$

Expanding Eqs. 15 and 16 over the control horizon as follows:

$$\underbrace{(I_{N_p} \otimes M)}_{\bar{M}} \bar{u}(k) \leq \underbrace{[11 \cdots 1]^T \otimes b_u}_{\bar{b}_u} \tag{17}$$

$$\bar{v}_2 = \underbrace{\begin{bmatrix} f_2(k) \\ \vdots \\ f_2(k+N_p-1) \end{bmatrix}}_{\Phi} + \underbrace{\begin{bmatrix} g_2(k) & 0 & \cdots & 0 \\ \vdots & \vdots & \ddots & \vdots \\ 0 & 0 & \cdots & g_2(k+N_p-1) \end{bmatrix}}_{\Theta} \bar{u} \tag{18}$$

Equation 17 can be substituted for u in Eq. 18. This results in the following expression:

$$\bar{M}_1 \bar{v}_2 \leq \bar{b}_1 \tag{19}$$

Where $\bar{M}_1 = \bar{M}\Theta^{-1}, \bar{b}_1 = \bar{b}_u - \bar{M}\Theta^{-1}\Phi$.

Expression Eq. 19 clearly shows the problem of the constraint mapping, the inequality constraint is not only depending on the state $x(k)$, but also on the future state $x(k+i)$. However, the future state $x(k+i)$ is not known until the MPC problem is solved. To address this problem, we present constant constraint mapping. The inequality constraint for time k of Eq. 19 is valid for the entire control horizon; that is, $\hat{x}(k+i) = x(k)$. Thus, matrix $\Theta \approx I_{N_p} \otimes g(k)$, $\Phi \approx [11 \cdots 1]^T f(k)$.

The reason for using constant constraint mapping can be explained as follows: the constraints on $v_2(k)$ depend on the actual measured state, only the constraints on $v_2(k+i)$ for $i = 1 : N_p - 1$ are approximations. Therefore, the implemented inputs

$v_2(k)$ are feasible to satisfy the original input constraints at time k. The actual implemented input is not affected by the approximation.

4.3.2 State Constraint Mapping

With the state bound defined as:

$$\underbrace{\begin{bmatrix} -I_n & I_n \end{bmatrix}^T}_{M} u \leq \underbrace{\begin{bmatrix} -x_{2\min} & x_{2\max} \end{bmatrix}^T}_{b_s} \tag{20}$$

And the state space model as Eq. 12. For each $i = 1 : N_p - 1$, the state constraints are transformed into input constraints, resulting in the inequality constraint

$$\bar{M}_2 \bar{v}_2 \leq \bar{b}_2 \tag{21}$$

Where $\bar{M}_2 = \bar{M}$, $\bar{b}_2 = [11 \cdots 1]^T \otimes b_s$.

Together with Eqs. 19 and 21 forms the final inequality constraint used in the optimization of $v_2(k)$.

$$\bar{M}_v \bar{v}_2 \leq \bar{b}_v \tag{22}$$

Where $\bar{M}_v = [\bar{M}_1 \quad \bar{M}_2]^T$, $\bar{b}_v = [\bar{b}_1 \quad \bar{b}_2]^T$.

Now, using the NDI and the constraint mapping, the constrained control problem in the original nonlinear system is transformed into the intermediate variables. With this formulation, the nonlinear predictive control problem is solved using a standard quadratic programming (QP). It is much less computationally intensive to solve a MPC problem of the following form:

$$\min_v J = \bar{v}_2^T \bar{Q} \bar{v}_2 - 2\bar{x}_{2c}^T \bar{Q} \bar{v}_2$$
$$\text{s.t.} \quad \bar{M}_v \bar{v}_2 \leq \bar{b}_v \tag{23}$$

The MPC$_2$ controller calculates the optimal inputs $v_2(k)$ based on the reference trajectory in terms p_c, q_c, r_c and $x_2(k)$. The continuous time input $v_2(k)$ is defined by a zero-order hold function on $v_2(k)$. Based on $v_2(k)$ and $x_2(k)$ and aerodynamic data, the combined controller calculates the reference tracking input $u(t)$, which is the actual input to the RUA.

5 Simulation Results

The proposed controller was tested on the GNC simulation platform. The guidance commands had been calculated to get the vehicle follows the trajectory. The controller is evaluated on tracking performance and robustness. The parameters considered for the MPC are T_s, N_p, N_c and Q. The sample time T_s is set at 0.1 s resulted in a satisfying tradeoff between computation speed and tracking performance. The prediction horizon is set at 3 s ($N_p = 30$), with a corresponding control horizon of 2 s ($N_c = 20$). Two simulation scenarios have been carried out to verify the proposed controller in different way, one of which presented in this section is aims at pointing out the explicit constraint handling capabilities enabled by the constrained MPC+HSDI approach. The other aims at demonstrating the improved robustness characteristics of the constrained MPC+HSDI with adaptive parameter T-S fuzzy model.

5.1 MPC+HSDI Versus HSDI

To demonstrate the effective constraint handling capability of the combined formulation, The constrained MPC+HSDI control concept is numerically evaluated and compared with HSDI control design by flown a relatively aggressive attitude reference command. The results of both controllers are starting with the case without uncertainties. See Fig. 2. The blue curve in Fig. 2 shows that the performance of the constrained MPC+HSDI controller is able to keep the vehicle stable and behaved well by tracking the desired references effectively. The inputs and states all change in the given constraint domains. The red curve in Fig. 2 displays the control effect of the HSDI controller. The results show that the closed-loop system is not stable during the conical rotation maneuver. As demonstrated in Fig. 2c, the saturation of ailerons and rudders plays an important role. As a result of saturation, HSDI cannot be accomplished, resulting in large tracking errors.

5.2 Robust MPC+HSDI Versus Nominal MPC+HSDI

To demonstrate the robustness of the combined formulation with adaptive parameter T-S fuzzy model, uncertainty is introduced into model parameters. The simulation presented here assumes that there are +30 % and −30 % uncertainties in the aerodynamic coefficients and aerodynamic moment coefficients respectively. Besides, the disturbance moments upon the RUA are both selected as $\sin(\pi t)[0.05 \quad 0.01 \quad 0.05]^T$. In approximating Δ_1 and Δ_2, the inner and outer loop adaptive parameter T-S fuzzy models are applied. For outer-loop, five fuzzy sets are defined for x_1 and three fuzzy sets for x_{2c}; for inner-loop, three fuzzy sets are

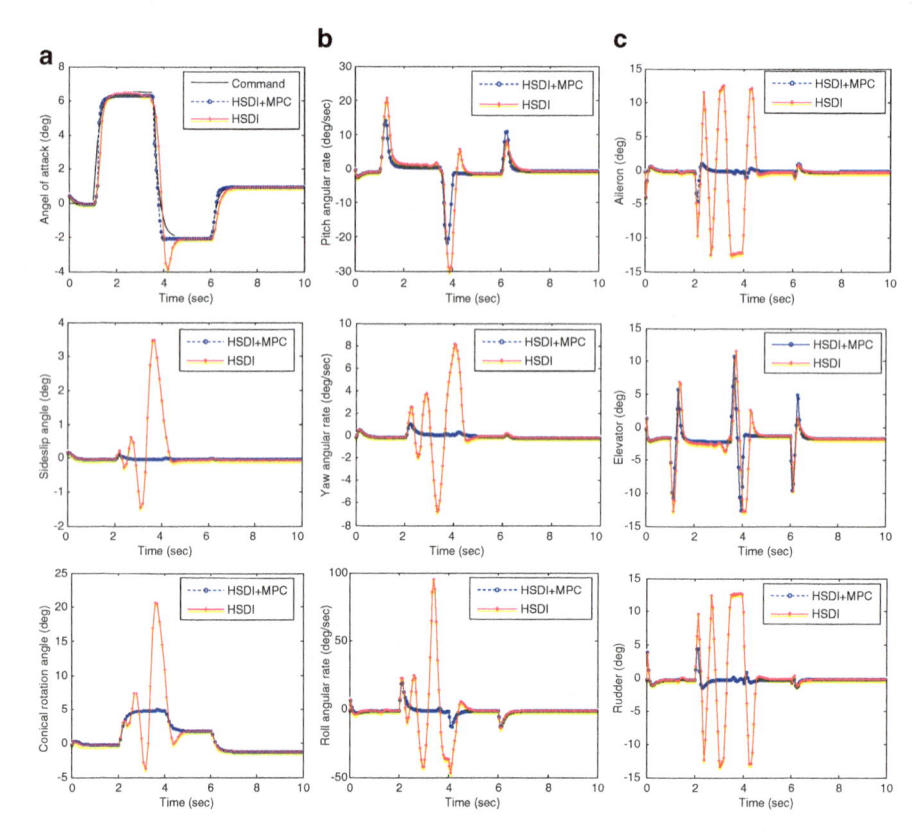

Fig. 2 Tracking performance comparison between the constrained MPC+HSDI and HSDI, (**a**) response of angle commands tracking, (**b**) time histories of body angular rates, (**c**) time histories of control surface deflections

defined for x_2 and five fuzzy sets for u. All adjustable parameters are initialized to 0.01 to represent the initial absence of knowledge about the uncertainties, where the gain of adaptive law $\gamma_a = 1$ and $\gamma_b = 0.5$ are used.

In the first case, we assume that no parameter adaptation law is applied. The blue curve in Fig. 3a shows the tracking performance when uncertainties of aerodynamic parameter and disturbances are considered by adopting the nominal constrained MPC +HSDI method, it can be seen that the overshoot is large and has side-effects like the steady state error. The control results may be worse if uncertainties became much bigger.

In order to improve the control effects in control system, Eq. 6 is applied to update the estimated parameters in the second case. The red curve in Fig. 3a show that the angle of attack, sideslip angle and conical rotation angle commands tracking is quite good for the constrained MPC+HSDI controller with adaptive parameter T-S fuzzy model despite the unknown nonlinear function, the addition of adaptive parameter T-S fuzzy model compensating controllers makes a significant improvement for tracking performance in each channel. The approximation performance for Δ_β and Δ_q are shown in Fig. 4.

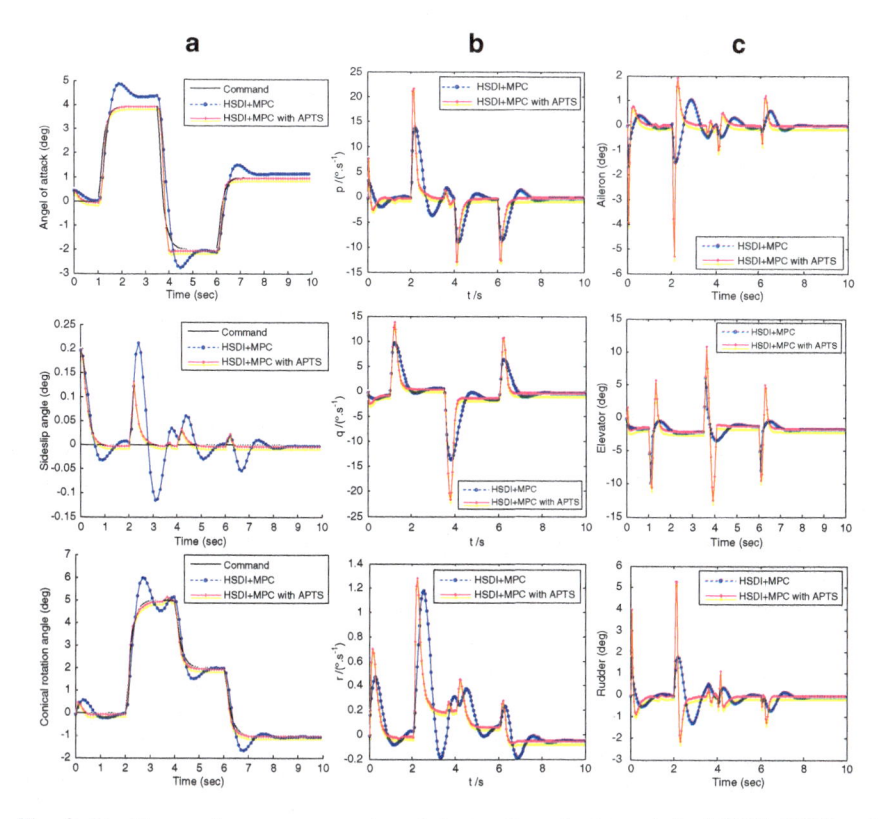

Fig. 3 Tracking performance comparison between the robust constrained MPC+HSDI and nominal constrained MPC+HSDI, (**a**) response of angle commands tracking, (**b**) time histories of body angular rates, (**c**) time histories of control surface deflections

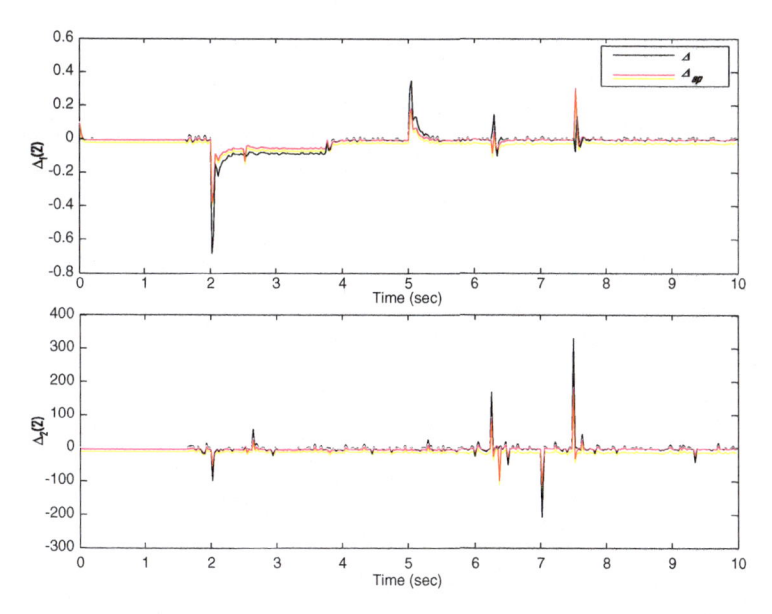

Fig. 4 Approximation performance of the T-S fuzzy model for Δ_β and Δ_q, *dash line* Δ is the actual value and *solid line* Δ_{ap} is the approximation

6 Conclusion

The nonlinear fight control algorithm based on hierarchy-structured dynamic inversion and constraint model predictive control presented in this paper provides a real-time feasible solution to attitude command tracking of a RUA with constrained dynamical behavior. Better results are obtained when compared with the HSDI. Simulation show that constrained MPC+HSDI is superior to HSDI, with respect to both tracking performance and control behavior. Furthermore, when the performance of the constrained MPC+HSDI control law deteriorates much under strong uncertainties, on-line identification for uncertainties using adaptive parameter T-S fuzzy model is combined with the controller design for its capability of canceling out the negative effects of uncertainties and disturbances. The simulation results show that the robust controller indeed has improved robustness characteristics compared to a nominal constrained MPC+HSDI controller, while constraints on the input and states are satisfied.

References

1. Breger L, How JP (2006) Nonlinear model predictive control technique for unmaned air vehicles. J Guid, Control Dyn 29(5):1179–1188
2. Bhattacharya R, Balas GJ, Kaya A, Packard A (2002) Nonlinear receding horizon control of F-16 aircraft. J Guid, Control, Dyn 25(5):924–931
3. WANG XG, Liang M (2011) Simulation and analysis of nonlinear model predictive control for guided projectile attitude. J Nanjing Univ Sci Technol 01:66–71 (In Chinese)
4. van Soest WR, Chu QP, Mulder JA (2006) Combined feedback linearization and constrained model predictive control for entry flight. J Guid Control Dyn 29(2):427–434
5. Guemghar K, Srinivasan B, Mullhaupt P (2005) Analysis of cascade structure with predictive control and feedback linearisation. J Control Theory Appl 152(3):317–324
6. Stevens BL, Lewis FL (2003) Aircraft control and simulation [M], 2nd edn. Wiley, New York, pp 88–106
7. Cao L, Zhang S, Yan S, Li X (2011) Flight controller design using adaptive parameter approximation block backstepping. Acta Aeronauti Astronaut Sin 32(12):2256–2267 (In Chinese)
8. Qi R, Brdys MA (2008) Stable indirect adaptive control based on discrete-time t-s fuzzy model. J Fuzzy Sets Syst 159(8):900–925

Cooperative Multitasking Software Design for Gas Pressure Control Based on Embedded Microcontroller System

Pubin Wang and Longxiang Lou

Abstract In order to control the pressure of coal gas fed to thermal treatment furnaces, a multitasking design method base on the RTX51 Tiny is used in this paper. The control system's hardware was mainly made up of microcontroller, analog, digital and serial interfaces. According to the features of control process and the hardware's performance, the system's software was divided into control task, manual task, display task, alarm task and command task. By running in cooperative multitasking manner, the five tasks handle various functions of the system. For illustration, each task's code structure is provided, and the task scheduling is explained. It is shown by application that the system satisfies the control requirements. With the advantages of multitasking software framework, simple hardware architecture, low cost and reliability, the design method we mentioned is valuable for developing embedded microcontroller control systems.

Keywords Cooperative multitasking · Gas pressure control · Embedded microcontroller · RTX51 Tiny

1 Introduction

One mechanical equipment manufacturing company has several thermal treatment furnaces, which use coal gas as fuel to heating iron and steel parts. The heat treatment processes for many parts often maintain above 10 h. During so long

---

P. Wang (✉)
School of Mechanical Engineering & Automation, University of Science and Technology Liaoning, Anshan, China
e-mail: wangpubin@126.com

L. Lou
Shengzhou Airflow Pipeline Co. Ltd, Shengzhou, China
e-mail: lou.longxiang@163.com

heating durations, the temperature control systems of those furnaces need the coal gas pressure steady or fluctuation less. But the pressure of the coal gas which was supplied by another company are rise or fall every now and then, sometimes the variable ranges were even too high. Operators had to adjust the coal gas main pipe valve manually to decrease the outlet pressure. In order to make the quality of thermal treatment products even better, coal gas pressure control system was required. For this purpose, a coal gas pressure control system was designed. In that system, a microcontroller was used as the core, and the real time operation system RTX51 Tiny was used as software kernel.

2 Hardware Composition

The main framework of the control system is shown in Fig. 1. A buttery valve was fixed on the coal gas main pipe. The inlet and outlet differential pressure sensors were fitted both sides of the valve. When coal gas moves through the main pipe, two differential pressure transducers convert the input and output gas pressures to normal voltages. Two TLC549 A/D converter chips change those two analog voltages to digital values for the microcontroller reading. After averaging the digital values, the AT89S52 microcontroller executes control routine and puts the result to TLC5615 D/A converter chip, which converts the digital value to analog voltage and sends it to an electric actuator connected to the buttery valve. Finally, the actuator drives the buttery valve rotate between 0° and 90° so as to adjust the coal gas outlet pressure within a preset range.

The hardware also includes some parallel interfaces for manual operations. In order to reduce misoperations, only a few buttons and switches were used in the system, keyboard was not assigned. Several LEDs and seven segment displays (7-SEGs) are used to display system running states and the real time pressures. In addition, by RS485 serial interface, the microcontroller can communicate with the factory's host computer, which performs the works of sending commands and receiving history data.

3 Software Design

There are several actions and operations in the control system. Some of them are periodic, such as sampling, real time control computing, reporting data to the host computer with serial communication interface and displaying messages. Manual operations and receiving command information from the host computer are stochastic. So it is reasonable to use real time operation system in control software design [1]. We selected RTX51 Tiny as the software development platform.

Real time applications are composed of one or more tasks that perform specific operations [2]. We can take some sequence operations to one task. As we

```
                ┌──────────┐        ┌──────────┐        ┌──────────┐
                │  LEDs,   │        │ Buttons, │        │          │
                │  7-SEGs, │        │ switches │        │  RS485   │
                │ speaker  │        │          │        │          │
                └────▲─────┘        └────▲─────┘        └────▲─────┘
                     │                   │                   │
                     ▼                   ▼                   ▼
                ┌─────────────────────────────────────────────────┐
                │              Microcontroller                    │
                └─────────────────────────────────────────────────┘
                     │                   │                   │
                     ▼                   ▼                   ▼
                ┌──────────┐        ┌──────────┐        ┌──────────┐
                │ A/D      │        │ D/A      │        │ A/D      │
                │ converter│        │ converter│        │ converter│
                └────┬─────┘        └────┬─────┘        └────┬─────┘
                     ▼                   ▼                   ▼
                ┌──────────┐        ┌──────────┐        ┌──────────┐
                │Transducer│        │ Actuator │        │Transducer│
                └────┬─────┘        └────┬─────┘        └────┬─────┘
                     ▼                   ▼                   ▼
                ┌──────────┐        ┌──────────┐        ┌──────────────┐
                │  Inlet   │        │ Buttery  │        │    Outlet    │
                │ pressure │        │  valve   │        │  pressure    │
                │  sensor  │        │          │        │   sensor     │
                └──────────┘        └──────────┘        └──────────────┘
```

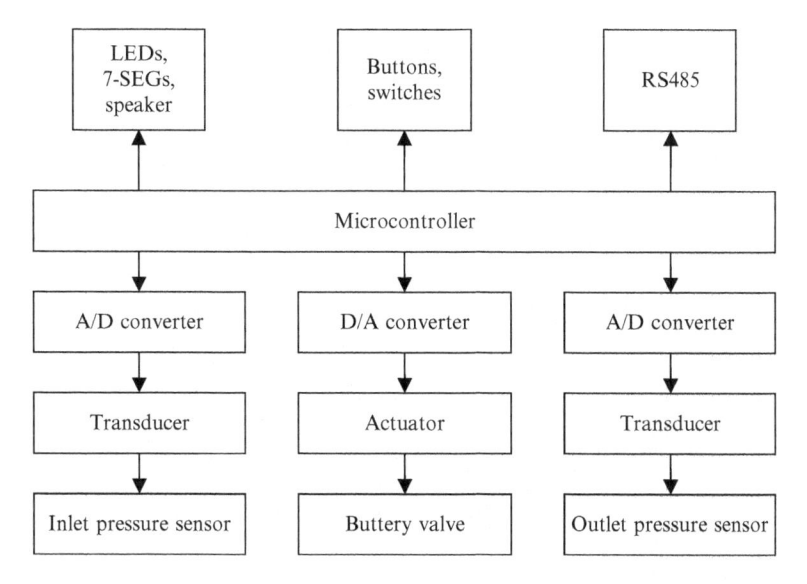

Fig. 1 Framework of system structure

mentioned above, from reading A/D converter values to driving the buttery valve waggling, the procedure is ordinal, thus all operations of that procedure can be assigned in one task. Another manner of planning task is according to hardware functions. For example, manual task is related to buttons and switches digital input interfaces, display task is related to LEDs and 7-SEGs digital output interfaces, command task is depended on the RS485 serial interface. In the end, the whole software includes five tasks, shown in Fig. 2.

3.1 Manual Task, Display Task and Alarm Task

The manual task scans manual inputs and deals with them. A few buttons were assigned to perform manual operations. One reset button was used to clear alarm state and make the system return to initial state. A shift button was used to display preset pressure or set point value. Two buttons were assigned to turn the preset pressure value increasing and decreasing manually.

The control system has two running modes: automatic mode and manual mode. In automatic mode, the buttery valve's swing is adjusted by control program automatically. While in manual mode, the buttery valve's movement is driven directly by pushing manual operation buttons on the control panel. A mode select switch was set to choose one of the two modes.

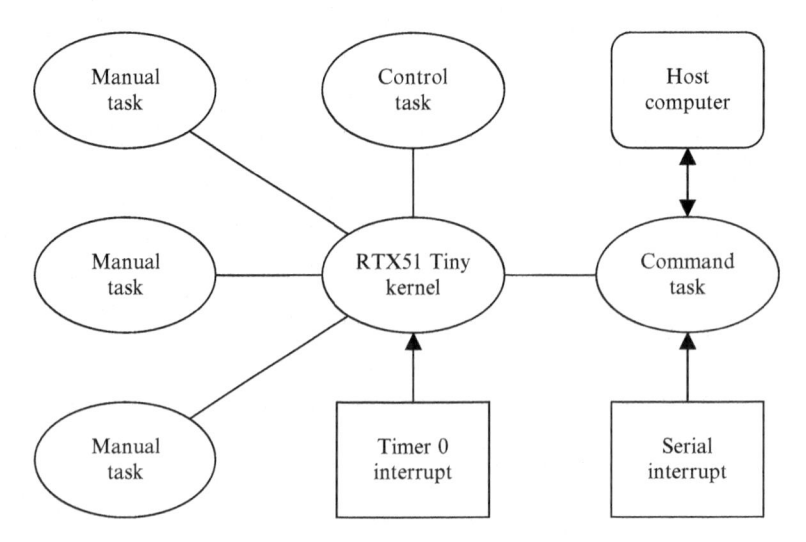

Fig. 2 Software structure

The code structure of manual task is shown as follows:

```
void manual_task (void) _task_ 3 {
while (1) {
if(shiftKEY==HIT && incKEY==HIT)setVAL++;
if(shiftKEY==HIT && decKEY==HIT)setVAL-;
if(RESET_BUTTON==HIT)reload_args( );
run_mode=(MODE_SLCT==ON) ?
AUTOMATIC_MODE : MANUAL_MODE;
os_wait (K_TMO, MANUAL_TICKS, 0); }
}
```

Several large scale 7-SEG chips were used to display real time control variables for the convenience of the operator's observation. The inlet and outlet pressures are displayed commonly. When the shift button is pushed, it can display current preset pressure.

The code structure of display task is shown as follows:

```
void display (void) _task_ 1 {
while (1){
if (shiftKEY==HIT) display_normal( );else display_set
point( );
os_wait (K_TMO, DISPLAY_TICKS, 0); }
}
```

The main function of alarm task is to check if the inlet and/or outlet pressures are out of their preset limits. If those situations occur, it should be noticed and the operator notified. The alarm task is responsible for this.

The code structure of alarm task is shown as follows:

```
void alarm_task (void) _task_ 2 {
while (1) {
alarmLED = realVAL > alarmVAL ? ON:OFF;
alarmBEEP = realVAL > alarmVAL ? ON:OFF;
os_wait (K_TMO, LED_BLINK_TICKS, 0);alarmLED = OFF;
os_wait (K_TMO, LED_BLINK_TICKS, 0); }
}
```

3.2 Control Task

This task runs the gas pressure control loop periodically. The process includes a series of steps as follows: getting the gas inlet and outlet pressure digital values from A/D converters, and several samples are required for filtering typically (sampling); in order to attenuate noise and external disturbance effects, using averaging those digital values or other calculating and filtering methods to get an estimative result (filtering) [3]; calling control computing routine to work out the digital output value (computing); if the system is running in automatic mode, sending the result to D/A converter to convert it to an analog voltage so as to driving the butterfly valve swing (output). Otherwise, the program goes to next loop directly. The code structure of control task is shown as follows:

```
void control (void) _task_ 0 {
init_serial_port ( );
os_create_task (1);os_create_task (2);os_create_task (3);
os_create_task (4);
while (1) {
for(unsigned char i = 0;i < NUMBER_OF_SAMPLING;i++){
get_inlet_pressure( );get_outlet_pressure( );
os_wait (K_TMO,SAMPLING_TICKS , 0);}
filtering( );
control_computing( );
if(Automatic_mode==1)dac_output( );
os_wait (K_TMO,CONTROL_TICKS , 0); }
}
```

3.3 Command Task

This task receives the host computer's commands with the microcontroller serial port and executes them. The commands that sent form host computer include setting control arguments and pressure ranges, reading them, and telling microcontroller report real time data, such as inlet, outlet and preset pressures. The serial interface data receiving is performed by serial interrupt routine, which is independent of RTX51 Tiny kernel.

The code structure of command task is shown as follows:

```
void serial (void) interrupt 4 using 2 {
unsigned char c = SBUF;RI = 0;
if(SM2==1){if(c==STATION_ID){SM2=0; rcvinx=0;}}
else{rcvbuf[rcvinx] = c;
    if(rcvinx<BUF_LENGTH-1)rcvinx++;
if(c=='\r')
isr_send_signal(COMMAND);}
}
void command (void) _task_ 4 {
while (1) {
   os_wait (K_SIG, 0, 0);
unsigned char cmd = rcvbuf[0];
SM2=1;
os_clear_signal(COMMAND);
switch(cmd){
case 'A': set_alarmVAL( ); break;
case 'P': set_presetVAL( ); break;
case 'R': report_data( ); break;
/*other commands......*/
default:break;}
}
}
```

4 Task Scheduling

RTX51 Tiny supports cooperative multitasking, which was used in our software design. By setting the TIMESHARING variable to 0 in the 'Conf_tny.A51' file, task switches are only performed when the running task voluntarily gives up control of the processor [4]. For example, when the function of os_wait (K_TMO, n, 0) is called, the running task will gives up CPU an interval of n ticks at least, in which task switches are performed.

RTX51 Tiny maintains each task in exactly one state (Running, Ready, Waiting, Deleted, or Time-Out). Only one task at a time may be in the Running State. Figure 3 shows the chart of the five tasks' switching sequences. The state '1' indicates a task is in the Running state, while state '0' indicates a task is in one of the other states, or we can say it is idle. In Fig. 3a, the grid interval is 5 s. Due to the execution period of a task is far less than the grid interval, it appears a vertical line. In Fig. 3b, the grid interval is 1 s, so the running states of control task and command task appear two pulses. The control task cycle was set as 30 s, and each cycle

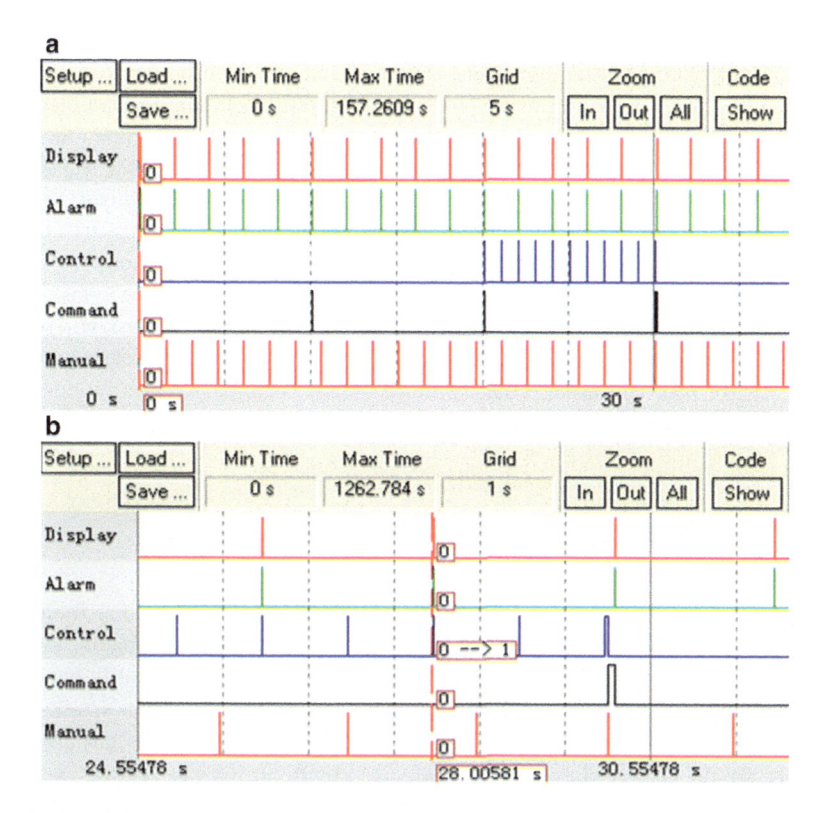

Fig. 3 Chart of task switching

includes 11 times sampling. The command task cycle was set as 10 s for sending data to the host computer. The cycles of display task and alarm task were set as 2 s, and the cycle of manual task was set as 150 milliseconds.

Limited to the electric actuator's action frequency, 10 times per minute maximally, the control task execution cycle should be set not less than 30 s. Furthermore, the on-the-spot tests shown that even that cycle was set as 60 s, the control system could meet the requirement of the factory's actual production process effectively. This means that rapid response would not be needed for the gas pressure control system. Therefore, the processor has enough time to run other tasks in so long control period. Figure 3 shows that other tasks were inserted ceaselessly during one control task execution cycle of 0–30 s, and all tasks were executed seemingly at the same time.

If tasks work cooperatively, the os_wait function or the os_switch_task function must be called somewhere in each task. These functions signal RTX51 Tiny to switch to another task. If one task consumes much CPU time, it will deprive other task's execution. For example, if the display task is always in the Running state, then other tasks will have no opportunity to run. In fact, the cycle of a typical periodic task consists of action time and idle time. Why a running task state shown in Fig. 3 looks like a vertical line? The reason is that its action time is very smaller than its idle time. By calling the os_weit function during its idle time, a task will gives up CPU an interval. Thus, if every task has enough idle time, the task scheduling would be performed effectively. Such a situation was shown in Fig. 3a. With RTX51 Tiny, task-level response is still non-deterministic [5]. As we mentioned above, responsiveness is not important for the gas pressure control system. Thus, cooperative multitasking kernel is a reasonable selection.

A cooperative multitasking kernel does not interrupt a task unless the task itself calls a certain procedure. Thus, a function or a program in a running task will not be broken by other tasks. This feature ensures that the function of control_computing in the control task can be executed simultaneously. Compared with round-robin system, the stability of control system has been improved.

5 Conclusion

Based on the RTX51 Tiny and microcontroller hardware, a real time multitask control system has been build. It is indicated by practical application that the system has good stability and reliability. The thermal treatment product quality has been improved. With cooperative multitasking kernel, though the system's task-level response is non-deterministic, but it can meet the requirement for the gas pressure control system. RTX51 Tiny is a very small system, so the gas pressure control software has the feathers of low cost and small code size. In the control software, many fixed interval operations were arranged in their related task. That makes the control program simple and realistic. By assigning several operations in one task, the system's task number was reduced and the task switching efficiency was improved.

References

1. Wayne Wolf (2008) Computers as components: principles of embedded computing system design, 2nd edn. Elsevier, Burlington, pp 290–302
2. Daniel D Gajski, Samar Abdi, Andreas Gerstlauer, Gunar Schirner (2009) Embedded system design modeling, synthesis and verification. Springer Science+Business Media, New York, USA, pp 173–176
3. Yongwei LU (2007) System of temperature control based on RTX51 Tiny. Mod Electron Tech 15:133–135 (In Chinese)
4. Kai Qian, David den Haring, Li Cao (2009) Embedded software development with C. Springer Science+Business Media, New York, USA, pp 145–146
5. Jean J Labrosse (2000) Embedded system building blocks, complete and ready-to-use modules in C, 2nd edn. Miller Freeman, Lawrence, Kansas, USA, pp 67–68

Robust H$_\infty$ Filtering for Uncertain Switched Systems Under Asynchronous

Guihua Li and Jun Cheng

Abstract This paper investigated the problem of robust H$_\infty$ filtering for uncertain switched systems with asynchronous switching and average dwell time. The asynchronous switching concerned in this paper represents that the switching instants of the presented controllers are lag behind those of the considered subsystems. Based on the Lyapunov function approach and average dwell time technique, sufficient conditions of the robust H$_\infty$ filter are derived. Finally, an example is also given to illustrate the advantages and effectiveness of the theoretical results.

Keywords Switched neutral systems • Asynchronous switching • Exponential stability • Linear matrix inequality • Average dwell time

1 Introduction

The H$_\infty$ filter problem has been a wide range of discussed over the past a few decades and its applications in all kinds of areas, for example signal estimation, signal processing, pattern recognition and communications, control application as well as some other practical control systems. During the last few years, the H$_\infty$ filtering technique was first proposed in [1], which has been received much attentions and it is very important to investigate for many others [2, 3]. A great number of useful results have showed up for analyzing the H$_\infty$ filtering conditions

---

G. Li
School of Computer Science, Civil Aviation Flight University of China, Guanghan, China
e-mail: lighua4519@163.com

J. Cheng (✉)
School of Automation Engineering, University of Electronic Science
and Technology of China, Chengdu, China
e-mail: jcheng6819@126.com

with delay-dependent [4, 5]. However, time delays is very important which always encountered in many practical systems such as electronics, hydraulic or networked control systems and so on [5, 6]. Therefore, it is important to investigate H_∞ filtering issues with the time-varying delays.

It has nothing to do that the aforementioned results are based on the average assumption that, in practice, the filter switching instants is simultaneous with those of the system. The controller switching signal is not match the proposed system switching signal precisely and inevitably exists the asynchronous switching. In real world, during the filter and the system in actual operation, there exists asynchronous switching, which means that, the real switching instants of the filter lag behind or exceed those of the proposed system, which may lead to performance of the systems deteriorated. In fact, during many chemical and mechanical systems, it has been shown the necessity of taking the asynchronous switching into consideration in efficient controller design. There are extensively results about asynchronous switching presented on control synthesis have been proposed [7–9]. However, to the best of authors' knowledge, the problem of asynchronous switching filter design for switched systems has not been fully considered, this motivated us to do this study.

In this paper, the problem of robust H_∞ filtering for uncertain switched systems with asynchronous switching has been proposed. The dwell time approach is utilized for the stability analysis and controller design, then the design asynchronous switching of the filters for switched systems is also developed. At last, a numerical example is provided to illustrate the effectiveness of proposed design approach.

2 Preliminaries

In this paper, we consider the following

$$\begin{cases} \dot{x}(t) = A_{\sigma(t)}(t)x(t) + A_{h\sigma(t)}(t)x(t-\tau_1(t)) + A_{\tau\sigma(t)}(t)x(t-\tau_2(t)) + B_{\sigma(t)}\omega(t), \\ y(t) = C_{\sigma(t)}(t)x(t) + C_{d\sigma(t)}(t)x(t-\tau_1(t)) + D_{\sigma(t)}\omega(t), \\ z(t) = L_{\sigma(t)}(t)x(t), \\ x_{t_0} = x(t_0+\theta) = \varphi(t), \dot{x}_{t_0} = \dot{x}(t_0+\theta) = \phi(t), \theta \in [-\max\{\tau_1,\tau_2\},0]. \end{cases}$$
(1)

Where $x(t) \in R^n$ is the state of the system, $\omega(t) \in R^p$ is the noise input, $y(t) \in R^m$ is the measurement vector, $z(t) \in R^q$ is the signal to be estimated. Switching signal $\sigma(t)$ is a piecewise constant function of time t, and takes values in a finite set $P = \{1, 2, \ldots r\}, r > 1$ denotes the number of subsystems. $\sigma(t) = i \in P$ means the i th subsystem is active. $A_i(t), A_{hi}(t), A_{\tau i}(t), C_i(t)$, and $C_{di}(t)$ are uncertain real-valued appropriate dimensions matrices. $B_i(t), D_i(t)$ are known real constant matrices.

Without loss of generality, we assume that the uncertain real-valued matrices satisfying the following relations:

$$A_i(t) = A_i + \Delta A_i, A_{hi}(t) = A_{hi} + \Delta A_{hi}, A_{\tau i}(t) = A_{\tau i} + \Delta A_{\tau i}, C_i(t) = C_i + \Delta C_i,$$

$$C_{di}(t) = C_{di} + \Delta C_{di}, F_i^T(t)F_i(t) \leq I, [\Delta A_i, \Delta A_{hi}, \Delta A_{\tau i}, \Delta C_i, \Delta C_{di}]$$
$$= M_i F_i(t)[N_{1i}, N_{2i}, N_{3i}, N_{4i}, N_{5i}],$$

Where $A_i, A_{hi}, A_{\tau i}, C_i, C_{di}, N_{1i}, N_{2i}, N_{3i}, N_{4i}, N_{5i}$, are known real constant matrices with appropriate dimensions and $F_i(t)$ is unknown time-varying matrix. $\tau_1(t)$ and $\tau_2(t)$ denote the time-varying delay which satisfy

$$0 \leq \tau_1(t) \leq \tau_1, 0 \leq \tau_2(t) \leq \tau_2, \dot{\tau}_1(t) \leq d_1 \leq \infty, \dot{\tau}_2(t) \leq d_2 \leq \infty,$$

For known constants τ_1, τ_2, d_1, d_2.

In this paper, the switching signal $\sigma(t)$ considered is time-dependent, that is $\sigma(t)$: $\{(t_0, \sigma(t_0)), (t_1, \sigma(t_1)), \ldots, (t_k, \sigma(t_k))\}$, where t_0 is the initial instant and in this paper we denote t_k represent the kth switching instant. For simplicity, $\sigma_f(t)$ is given to denote the filter switching signal, which can be written as

$$\sigma_f(t) : \{(t_0, \sigma_f(t_0)), (t_1 + \Delta_1, \sigma_f(t_1 + \Delta_1)), \ldots, (t_k + \Delta_k, \sigma_f(t_k + \Delta_k))\}$$

where $\sigma_f(t_0) = \sigma(t_0), \sigma_f(t_k + \Delta_k) = \sigma(t_k), \Delta_k > 0$ or $\Delta_k < 0$ denotes the period that the filter switching instants exceed or lag behind those of the system. In the whole paper we consider the case of $\Delta_k > 0$.

It is always assumed that, during the switching instant t_{k-1}, the ith subsystem is activated and at the switching instant t_k the jth subsystem is activated. Thanks to asynchronous switching, the filter switching instant reacting to jth subsystem is $t_k + \Delta_k$, there must exists at least a matched period during the time interval $[t_{k-1} + \Delta_{k-1}, t_k)$, then we can obtain a mismatched period correspond at time interval $[t_k, t_k + \Delta_k)$.

In this paper, we assume the filtering problem considered in this paper is to design a filter with the following structure

$$\begin{cases} \dot{x}_f(t) = A_{f\sigma_f(t)}(t)x_f(t) + B_{f\sigma(t)}y(t), \\ z_f(t) = L_{f\sigma_f(t)}(t)x_f(t), \end{cases} \quad (2)$$

Where $A_{f\sigma_f(t)}, B_{f\sigma_f(t)}, L_{f\sigma_f(t)}$ are filtering matrices which needed to be determined later, $x_f(t) \in R^n$ is the filter state, $z_f(t) \in R^q$ is the output of the filter. Denote

$$\eta(t) = [x^T(t), x_f^T(t)]^T, \bar{z} = z(t) - z_f(t), A_{ei}(t) = \begin{bmatrix} A_i(t) & 0 \\ B_{fi}C_i(t) & A_{fi} \end{bmatrix},$$

$$B_{ei}(t) = \begin{bmatrix} A_{hi}(t) \\ B_{fi}C_{di}(t) \end{bmatrix}$$

$$C_{ei}(t) = \begin{bmatrix} A_{\tau i}(t) \\ 0 \end{bmatrix}, D_{ei}(t) = \begin{bmatrix} B_i(t) \\ B_{fi}D_i \end{bmatrix}, L_{ei} = [L_i \quad -L_{fi}], A_{eij}(t) = \begin{bmatrix} A_f(t) & 0 \\ B_{fi}C_j(t) & A_{fi} \end{bmatrix},$$

$$B_{eij}(t) = \begin{bmatrix} A_{hj}(t) \\ B_{fi}C_{dj}(t) \end{bmatrix}, C_{eij}(t) = \begin{bmatrix} A_{\tau j}(t) \\ 0 \end{bmatrix}, D_{eij}(t) = \begin{bmatrix} B_j \\ B_{fi}D_j \end{bmatrix}, L_{eij} = [L_j \quad -L_{fi}]$$

When $t \in [t_0, t_1) \cup [t_{k-1} + \Delta_{k-1}, t_k), k = 2, 3, \ldots$, that is to say when the proposed switched system runs at the matched period, then we have the filtering error system as follows

$$\Sigma \begin{cases} \dot{\eta}(t) = A_{ei}\eta(t) + B_{ei}\eta(t - \tau_1(t)) + C_{ei}\dot{\eta}(t - \tau_2(t)) + D_{ei}\omega(t), \\ \bar{z}(t) = L_{ei}\eta(t), \end{cases} \quad (3)$$

When $t \in [t_k, t_k + \Delta_k), k = 1, 2, \ldots$, that is to say when the proposed switched system runs at the mismatched period, then we have the filtering error system as follows

$$\Sigma^1 \begin{cases} \dot{\eta}(t) = A_{eij}\eta(t) + B_{eij}\eta(t - \tau_1(t)) + C_{eij}\dot{\eta}(t - \tau_2(t)) + D_{eij}\omega(t), \\ \bar{z}(t) = L_{eij}\eta(t) \end{cases} \quad (4)$$

In this paper, during $[t_0, t]$, let $T^+(t_0, t)$ represents the total matched period and $T^-(t_0, t)$ represents the total mismatched period, respectively.

Definition 1.[7] For given a switching signal $\sigma(t)$, the trajectory of the given filtering error system (3) and (4) satisfies $\|x(t)\| \leq \alpha \|x(t_0)\| e^{-\beta(t-t_0)}$, then filtering error system (3) and (4) are said to be exponential stability, where $\alpha > 0, \beta > 0$ and $t \geq t_0$.

Definition 2.[9] For given scalars $T_2 > T_1 \geq 0$, let $N_\sigma(T_1, T_2)$ denote the switching number of $\sigma(t)$ during (T_1, T_2). If $N_\sigma(T_1, T_2) \leq N_0 + \frac{(T_2-T_1)}{T_a}$ holds for $N_0 \geq 0$ and $T_a > 0$, then N_0 and T_a are called chattering bound and average dwell time, respectively. Moreover, assume $N_0 = 0$ for simplicity as extensively in the literature.

Lemma 1. Suppose $0 < h_m < h_M$, and $x(t) \in R^n$, for any positive matrix $Q \in R^{n \times n}$, then

$$-(h_M - h_m)\int_{t-h_M}^{t-h_m} \dot{x}^T(s)Q\dot{x}(s)ds \leq \begin{bmatrix} x(t-h_m) \\ x(t-h_M) \end{bmatrix}^T \begin{bmatrix} -Q & Q \\ * & -Q \end{bmatrix} \begin{bmatrix} x(t-h_m) \\ x(t-h_M) \end{bmatrix}.$$

3 Main Results

Theorem 1 *For given scalars $\alpha > 0, \beta > 0$, then the system (3) and (4) is exponentially stable, if there exists positive-definite symmetric matrices $P_i, P_{ij}, Q_i, Q_{ij}, R_i, R_{ij}$, such that the following LMIs hold:*

$$\Xi_{1i} = \begin{bmatrix} \Xi^1_{1,1} & P_iB_{ei}+H_ie^{-\alpha\tau_1} & P_iC_{ei} & P_iD_{ei} & A^T_{ei}R_i & \tau_2 A^T_{ei}H_i \\ * & \Xi^1_{2,2} & 0 & 0 & B^T_{ei}R_i & \tau_2 B^T_{ei}H_i \\ * & * & -(1-d_2)e^{-\alpha\tau_2}R_i & 0 & C^T_{ei}R_i & \tau_2 C^T_{ei}H_i \\ * & * & * & -\gamma^2 I & D^T_{ei}R_i & \tau_2 D^T_{ei}H_i \\ * & * & * & * & -R_i & 0 \\ * & * & * & * & * & -H_i \end{bmatrix} < 0 \tag{5}$$

$$\Xi_{2i} = \begin{bmatrix} \Xi^2_{1,1} & P_{ij}B_{eij}+H_{ij}e^{-\alpha\tau_1} & P_{ij}C_{eij} & P_{ij}D_{eij} & A^T_{eij}R_{ij} & \tau_2 A^T_{eij}H_{ij} \\ * & \Xi^2_{2,2} & 0 & 0 & B^T_{eij}R_{ij} & \tau_2 B^T_{eij}H_{ij} \\ * & * & -(1-d_2)e^{-\alpha\tau_2}R_{ij} & 0 & C^T_{eij}R_{ij} & \tau_2 C^T_{eij}H_{ij} \\ * & * & * & -\gamma^2 I & D^T_{eij}R_{ij} & \tau_2 D^T_{eij}H_{ij} \\ * & * & * & * & -R_{ij} & 0 \\ * & * & * & * & * & -H_{ij} \end{bmatrix} < 0 \tag{6}$$

Where $\Xi^1_{1,1} = P_iA_{ei} + A^T_{ei}P_i + \alpha P_i + Q_i + L^T_{ei}L_{ei} - H_ie^{-\alpha\tau_1}, \Xi^1_{2,2} = -(1-d_1)e^{-\alpha\tau_1}Q_i - H_ie^{-\alpha\tau_1}$ $\Xi^2_{1,1} = P_{ij}A_{eij} + A^T_{eij}P_{ij} + \alpha P_{ij} + Q_{ij} + L^T_{eij}L_{eij} - H_{ij}e^{-\alpha\tau_1}, \Xi^2_{2,2} = -(1-d_1)e^{-\alpha\tau_1}Q_{ij} - H_{ij}e^{-\alpha\tau_1}$

In this situation, the average dwell time of switching signal satisfies

$$T^*_{aa} = \frac{\ln \mu_2 \mu_1}{\kappa}, \frac{T^+(t_0,t)}{T^-(t_0,t)} \geq \frac{\beta+\kappa}{\alpha-\kappa}, 0 \leq \kappa < \alpha$$

system (3) and (4) is exponential stability with an H_∞ weighted level γ, where $\mu \geq 1$ satisfying that

$$P_j \leq \mu_1 P_{ij}, P_{ij} \leq \mu_2 P_j, Q_j \leq \mu_1 Q_{ij}, Q_{ij} \leq \mu_2 Q_j, R_j \leq \mu_1 R_{ij}, R_{ij} \leq \mu_2 R_j,$$
$$H_j \leq \mu_1 H_{ij}, H_{ij} \leq \mu_2 H_j, \forall i \neq j, i, j \in P$$

Theorem 2 *For given scalars* $\alpha > 0, \beta > 0, P_i = \begin{bmatrix} P_{1i} & 0 \\ 0 & P_{2i} \end{bmatrix}, P_{ij} = \begin{bmatrix} P_{1ij} & 0 \\ 0 & P_{2ij} \end{bmatrix},$
then the system (3) and (4) is exponential stability with a performance H_∞ level γ for any average dwell time of switching signal and (7) and (8), such that the following LMIs hold:

$$\begin{bmatrix} \Pi_{1i}^1 & \Pi_{2i}^1 \\ * & \Pi_{3i}^1 \end{bmatrix} < 0, \tag{7}$$

$$\begin{bmatrix} \Pi_{1i}^2 & \Pi_{2i}^2 \\ * & \Pi_{3i}^2 \end{bmatrix} < 0, \tag{8}$$

Where

$$\Pi_{1i}^1 = \begin{bmatrix} \Pi_{1,1}^1 & \Pi_{1,2}^1 & \Pi_{1,3}^1 & P_{1i}A_{\tau i} & P_{1i}B_i & A_i^T R_i & \tau_2 A_i^T H_i \\ * & \Pi_{2,2}^1 & X_{2i}C_{di} & 0 & X_{2i}D_i & 0 & 0 \\ * & * & \Pi_{3,3}^1 & 0 & 0 & A_{hi}^T R_i & \tau_2 A_{hi}^T H_i \\ * & * & * & \Pi_{4,4}^1 & 0 & A_{\tau i}^T R_i & \tau_2 A_{\tau i}^T H_i \\ * & * & * & * & -\gamma^2 I & B_i^T R_i & \tau_2 B_i^T H_i \\ * & * & * & * & * & -R_i & 0 \\ * & * & * & * & * & * & -H_i \end{bmatrix},$$

$$\Pi_{2i}^1 = \begin{bmatrix} \varepsilon_{1i} P_{1i} M_i & N_{1i}^T & \varepsilon_{2i} N_{1i}^T & 0 & 0 & 0 & \varepsilon_{4i} N_{4ii}^T & 0 \\ 0 & 0 & 0 & 0 & \varepsilon_{3i} X_{2i} M_i & 0 & 0 & X_{2i} M_i \\ 0 & N_{2i}^T & \varepsilon_{2i} N_{2i}^T & 0 & 0 & N_{5i}^T & 0 & 0 \\ 0 & N_{3i}^T & \varepsilon_{2i} N_{3i}^T & 0 & 0 & 0 & 0 & 0 \\ 0 & 0 & 0 & 0 & 0 & 0 & 0 & 0 \\ 0 & 0 & 0 & R_i M_i & 0 & 0 & 0 & 0 \\ 0 & 0 & 0 & \tau_1 H_i M_i & 0 & 0 & 0 & 0 \end{bmatrix},$$

$$\Pi_{3i}^1 = \begin{bmatrix} -\varepsilon_{1i}I & 0 & 0 & 0 & 0 & 0 & 0 & 0 \\ * & -\varepsilon_{1i}I & 0 & 0 & 0 & 0 & 0 & 0 \\ * & * & -\varepsilon_{2i}I & 0 & 0 & 0 & 0 & 0 \\ * & * & * & -\varepsilon_{2i}I & 0 & 0 & 0 & 0 \\ * & * & * & * & -\varepsilon_{3i}I & 0 & 0 & 0 \\ * & * & * & * & * & -\varepsilon_{3i}I & 0 & 0 \\ * & * & * & * & * & * & -\varepsilon_{4i}I & 0 \\ * & * & * & * & * & * & * & -\varepsilon_{4i}I \end{bmatrix},$$

$$\Pi_{1i}^2 = \begin{bmatrix} \Pi_{1,1}^2 & \Pi_{1,2}^2 & \Pi_{1,3}^2 & P_{1ij}A_{\tau ij} & P_{1ij}B_j & A_j^T R_{ij} & \tau_2 A_j^T H_{ij} \\ * & \Pi_{2,2}^2 & \Pi_{2,3}^2 & 0 & \Pi_{2,5}^2 & 0 & 0 \\ * & * & \Pi_{3,3}^2 & 0 & 0 & A_{hj}^T R_{ij} & \tau_2 A_{hj}^T H_{ij} \\ * & * & * & \Pi_{4,4}^2 & 0 & A_{\tau j}^T R_{ij} & \tau_2 A_{\tau j}^T H_{ij} \\ * & * & * & * & -\gamma^2 I & B_j^T R_{ij} & \tau_2 B_j^T H_{ij} \\ * & * & * & * & * & -R_{ij} & 0 \\ * & * & * & * & * & * & -H_{ij} \end{bmatrix},$$

$$\Pi_{2i}^2 = \begin{bmatrix} \varepsilon_{1ij}P_{1ij}M_j & N_{1j}^T & \varepsilon_{2ij}N_{1j}^T & 0 & 0 & 0 & \varepsilon_{4ij}N_{4j}^T & 0 \\ 0 & 0 & 0 & 0 & \varepsilon_{3ij}P_{2ij}P_{2i}^{-1}X_{2i}M_j & 0 & 0 & P_{2ij}P_{2i}^{-1}X_{2i}M_j \\ 0 & N_{2j}^T & \varepsilon_{2ij}N_{2j}^T & 0 & 0 & N_{5j}^T & 0 & 0 \\ 0 & N_{3j}^T & \varepsilon_{2ij}N_{3j}^T & 0 & 0 & 0 & 0 & 0 \\ 0 & 0 & 0 & 0 & 0 & 0 & 0 & 0 \\ 0 & 0 & 0 & R_{ij}M_j & 0 & 0 & 0 & 0 \\ 0 & 0 & 0 & \tau_1 H_{ij}M_j & 0 & 0 & 0 & 0 \end{bmatrix},$$

$$\Pi_{3i}^2 = \begin{bmatrix} -\varepsilon_{1ij}I & 0 & 0 & 0 & 0 & 0 & 0 & 0 \\ * & -\varepsilon_{1ij}I & 0 & 0 & 0 & 0 & 0 & 0 \\ * & * & -\varepsilon_{2ij}I & 0 & 0 & 0 & 0 & 0 \\ * & * & * & -\varepsilon_{2ij}I & 0 & 0 & 0 & 0 \\ * & * & * & * & -\varepsilon_{3ij}I & 0 & 0 & 0 \\ * & * & * & * & * & -\varepsilon_{3ij}I & 0 & 0 \\ * & * & * & * & * & * & -\varepsilon_{4ij}I & 0 \\ * & * & * & * & * & * & * & -\varepsilon_{4ij}I \end{bmatrix}$$

With

$$\Pi_{1,1}^1 = P_{1i}A_i + A_i^T P_{1i} + \alpha P_{1i} + Q_i + L_i^T L_i - H_i e^{-\alpha\tau_2}, \Pi_{1,2}^1 = C_i^T X_{2i}^T - L_i^T X_{3i},$$

$$\Pi_{1,3}^1 = P_{1i}A_{hi} + H_i e^{-\alpha\tau_2}, \Pi_{2,2}^1 = X_{1i}^T + X_{1i} + \alpha P_{2i} - X_{3i}^T X_{3i},$$

$$\Pi_{3,3}^1 = -(1-d_1)e^{-\alpha\tau_2}Q_i - H_i e^{-\alpha\tau_2}, \Pi_{4,4}^1 = -(1-d_2)e^{-\alpha\tau_2}R_i,$$

$$\Pi_{1,1}^2 = P_{1ij}A_j + A_j^T P_{1ij} - \beta P_{1ij} + Q_{ij} + L_j^T L_j - H_{ij}e^{\beta\tau_1}, \Pi_{1,2}^2 \\ = C_j^T X_{2i}^T P_{2i}^{-1} P_{2ij} - L_j^T X_{3i},$$

$$\Pi_{1,3}^2 = P_{1ij}A_{hj} + H_{ij}e^{\beta\tau_2}, \Pi_{2,3}^2 = P_{2ij}P_{2i}^{-1}X_{2i}C_{dj}, \Pi_{2,5}^2 = P_{2ij}P_{2i}^{-1}X_{2i}D_j,$$

$$\Pi_{2,2}^2 = P_{2ij}P_{2i}^{-1}X_{1i} + X_{1i}^T P_{2i}^{-1} P_{2ij} - \beta P_{2ij} - X_{3i}^T X_{3i},$$

$$\Pi_{3,3}^2 = -(1-d_1)e^{\beta\tau_1}Q_{ij} - H_{ij}e^{\beta\tau_2}, \Pi_{4,4}^2 = -(1-d_2)e^{\beta\tau_2}R_{ij}$$

What's more, a design filter of the system (2) is

$$A_{fi} = P_{2i}^{-1}X_{1i}, B_{fi} = P_{2i}^{-1}X_{2i}, L_{fi} = X_{3i}.$$

Remark 1 It is pointed that the linear matrix inequalities (5), (6), (7) and (8) are dependent on each other. Moreover, we can obtain the feasible solutions through fixed parameter appropriately.

4 Conclusion

This paper is investigated with the problem of robust H_∞ filtering for uncertain switched systems with asynchronous switching, based on the average dwell time technology, the obtained sufficient conditions of a linear filter to ensure the filtering error system is exponential stability.

Acknowledgements The authors would like to give thankful to referees for their valuable suggestions and comments. This work was supported by the Fundamental Research Funds for the Scientific Research Fund for Ph.D. Talents Introduction of CAFUC (Grant No. J2010-46).

References

1. Elsayed A, Grimble M (1989) A new approach to the H_∞ design of optimal digital linear filters. IMA J Math Control Inf 6(2):233–251
2. Karimi H, Zapateiro M, Luo N (2010) A linear matrix inequality approach to robust fault detection filter design of linear systems with mixed time-varying delays and nonlinear perturbations. J Frankl Inst 347(6):957–973
3. Liu J, Hu S, Tian E (2011) A novel method of H_∞ filter design for time-varying delay systems. Int J Innov Comput, Inf Control 7(3):1299–1310
4. Xu S, Chen T (2004) An LMI approach to the H_∞ filter design for uncertain systems with distributed delays. IEEE Trans Circuit Syst 51(4):195–201
5. Yue D, Han Q, Peng C (2004) State feedback controller design of networked control systems. IEEE Trans Circuit Syst 51(11):640–644
6. Yue D, Han Q, Lam J (2005) Network-based robust H_∞ control of systems with uncertainty. Automatica 41(6):999–10007
7. Xiang Z, Wang R (2009) Robust stabilization of switched nonlinear systems with time-varying delays under asynchronous switching. Proc Inst Mech Eng, Part I: J Syst Control Eng 223:1111–1128
8. Zhang L, Shi P (2009) Stability, L2-gain and asynchronous control of discrete-time switched systems with average dwell time. IEEE Trans Autom Control 54:2193–2200
9. Zhang L, Gao H (2010) Asynchronously switched control of switched linear systems with average dwell time. Automatica 46(5):953–958

A Kind of Adaptive Backstepping Sliding Model Controller Design for Hypersonic Reentry Vehicle

Congchao Yao, Xinmin Wang, Yao Huang, and Yuyan Cao

Abstract In order to solve a class of hypersonic vehicle nonlinear multi-input and multi-output (MIMO) systems control problem, a new kind of adaptive controller is designed in this paper. Backstepping is used to build the control law. To ensure the stability of each uncertain subsystem in each step, a virtual zero-order controller is given in sliding surface design, which can guarantee the robustness. A second-order filter is utilized to replace virtual input to avoid the huge computation complexity in multi-step derivatives. The effectiveness of the control law is verified by simulation even under deep uncertainty. The results suggest the feasibility of this method.

Keywords Hypersonic vehicle · Sliding model control · Backstepping control · Adaptive control

1 Introduction

Hypersonic vehicle (HSV) is one kind of aircraft with large envelope. Their special high-speed and complex environments require much more challenges on design of flight control system [1, 2].

Backstepping is considered as an effective method to design flight control system [3, 4]. To achieve global regulation or tracking, the stabilizing controller is designed step by step. Adaptive control methods such as robust, neural network are used based on it [5–7]. Dynamic surface control (DSC) method was first proposed by Swaroop [8, 9]. Calculating expansion can be caused by multi-step derivatives. To avoid this problem, a first-order low-pass filter is introduced to estimate the virtual control law [10].

C. Yao (✉) · X. Wang · Y. Huang · Y. Cao
School of Automation, Northwestern Polytechnical University, Xi'an, China
e-mail: yaocongchao@126.com

Based on the results before, this paper includes the following works: (1) A new kind of adaptive backstepping controller is designed based on dynamic surface technology; (2) A second-order filter is proposed to avoid the "calculating expansion" by instead of virtual control input, which can estimate the measurement noise and time derivative effectively.

2 Problem Statement

To solve the control problem of a general hypersonic vehicle, which is a strict-feedback affine nonlinear uncertain MIMO system, the vehicle can be described as follows:

$$\begin{cases} \dot{x}_1 = f_1(x_1) + g_1(x_1)x_2 + \Delta_1 \\ \dot{x}_2 = f_2(x_2) + g_2(x_2)u + \Delta_2 \\ y = x_1 \end{cases} \quad (1)$$

where $x_1 = [\alpha, \beta, \mu]^T$ are angle of attack, sideslip angle and roll angle, respectively. $x_2 = [p, q, r]^T$ are roll rate, pitch rate and yaw rate, respectively. $u = [L, M, N]^T$ are roll, pitch, yaw control torque caused by aerodynamic control surfaces. Define Δ_1 and Δ_2 as follows:

$$\begin{cases} \Delta_1 = \Delta f_1(x_1) + \Delta g_1(x_1)x_2 + d_1 \\ \Delta_2 = \Delta f_2(x_2) + \Delta g_2(x_2)u + d_2 \end{cases} \quad (2)$$

where d_1 and d_2 are external disturbances. Δ_1 and Δ_2 are parameters and modeling uncertainties.

Because of the existence of the uncertainties, we should design the effective control law to make sure that the system outputs can track the desired trajectory y_d.

3 Design Adaptive Backstepping Sliding Model Control Law

3.1 Definition and Assumption

In order to design the control law, first of all, one definition and three assumptions are given as follows:

Definition 1 *The* Frobenius *norm for a matrix A is defined as* $\|A\| = \sqrt{tr(A^T A)}$.

Assumption 1 The desired trajectories y_d are bounded and continuously differentiable. And so are their first derivatives.

Assumption 2 The inverses of $g_1(x_1)$ and $g_2(x_2)$ are existent.

Assumption 3 The uncertainty terms Δ_1 and Δ_2 are bounded.

3.2 Design Control Law

According to Eq. 1, the virtual feedback errors z_1 and z_2 are defined as bellow respectively:

$$\begin{cases} z_1 = y_1 - y_{1d} = x_1 - x_{1d} \\ z_2 = x_2 - \alpha_1 \end{cases} \tag{3}$$

First Step: according to Eqs. 1 and 3, the differential of z_1 is expressed as

$$\dot{z}_1 = f_1(x_1) + g_1(x_1)z_2 + g_1(x_1)\alpha_1 + \Delta_s \tag{4}$$

Define the second-order sliding mode filter as following:

$$\begin{aligned} \dot{q}_1 &= -\frac{1}{l_1}(q_1 - \alpha_{1d}) - g_1^T(x_1)z_1 - \lambda_1 \frac{2}{\pi}\arctan(q_1 - \alpha_{1d}) \\ \dot{q}_2 &= -\frac{1}{l_2}(q_2 - \dot{q}_1) - z_2 - \lambda_2 \frac{2}{\pi}\arctan(q_2 - \dot{q}_1) \end{aligned} \tag{5}$$

There α_{1d} is the expected virtual control input. And $q_1 = \alpha_1$.

Lyapunov function V_1 and the virtual control input α_{1d} are defined as

$$V_1 = \frac{1}{2}z_1^T z_1 + \frac{1}{2}(\alpha_1 - \alpha_{1d})^T(\alpha_1 - \alpha_{1d}) \tag{6}$$

$$\alpha_{1d} = g_1^{-1}(x_1)[-f_1(x_1) - k_1 z_1 + \dot{y}_d - \gamma_1 \frac{2}{\pi}\arctan(z_1)] \tag{7}$$

Omitting the negative definite term of $\dot{V}_1$, which is the differentiation of V_1, the follow express can be gotten:

$$\begin{aligned} \dot{V}_1 \leq & -\|z_1^T\|\|g_1(x_1)\|\|z_2\| - \Psi_1\|z_1^T\|\left(\eta_1 \frac{2}{\pi}\arctan(\|z_1\|) - 1\right) \\ & -\|(\alpha_1 - \alpha_{1d})^T\|\varpi_1\left(\rho_1 \frac{2}{\pi}\arctan(\|\alpha_1 - \alpha_{1d}\|) - 1\right) \end{aligned} \tag{8}$$

where $\varpi_1 = \dot{\alpha}_{1d\max}$ and $\Psi_1 = \varepsilon_{1\max}$. Select switch gains as:

$$\gamma_1 = \eta_1 \Psi_1, \quad \lambda_1 = \rho_1 \varpi_1 \quad (\eta_1, \rho_1 > 1) \tag{9}$$

The following conditions which are used to guarantee all factors on the right side of Eq. 8 negative, should be satisfied to make $\dot{V}_1 < 0$.

$$\|z_1\| > \tan\left(\frac{\pi}{2\eta_1}\right) \tag{10}$$

$$\|q_2 - \dot{q}_1\| > \tan\left(\frac{\pi}{2\rho_1}\right) \tag{11}$$

Also, if the conditions bellow can be guaranteed, the boundedness can be ensured:

$$\|z_1\| \le c_1 \tan\left(\frac{\pi}{2\eta_1}\right), c_1 > 1 \tag{12}$$

$$\|q_2 - \dot{q}_1\| \le \tau_1 \tan\left(\frac{\pi}{2\rho_1}\right), \tau_1 > 1 \tag{13}$$

Second Step: Lyapunov function V_2 and actual control input u are defined as

$$V_2 = V_1 + \frac{1}{2}z_2^T z_2 + \frac{1}{2}(q_2 - \dot{q}_1)^T (q_2 - \dot{q}_1) \tag{14}$$

$$u = g_2^{-1}(x_2)(-f_2(x_2) - k_2 z_2 - g_1^T(x_1)z_1 + q_2 - \gamma_2 \frac{2}{\pi}\arctan(z_2)) \tag{15}$$

Omitting the negative definite term of $\dot{V}_1$, which is the differentiation of V_1, the follow express can be gotten:

$$\dot{V}_2 \le -\Psi_2 \|z_2^T\| \left(\eta_2 \frac{2}{\pi}\arctan(\|z_2\|) - 1\right) - \|(q_2 - \dot{q}_1)^T\| \varpi_2 \left(\rho_2 \frac{2}{\pi}\arctan(\|q_2 - \dot{q}_1\|) - 1\right) \tag{16}$$

where $\varpi_2 = \ddot{q}_{1\max}$ and $\Psi_2 = \varepsilon_{2\max}$. Select switch gains as:

$$\gamma_2 = \eta_2 \Psi_2, \quad \lambda_2 = \rho_2 \varpi_2 \quad (\eta_1, \rho_1 > 1) \tag{17}$$

Also we need to satisfy the following conditions to keep all the right side factors of Eq. 16 negative, so that $\dot{V}_2 < 0$ and the stability of the system can be ensured:

$$\|z_2\| > \tan\left(\frac{\pi}{2\eta_2}\right) \tag{18}$$

$$\|q_2 - \dot{q}_1\| > \tan\left(\frac{\pi}{2\rho_2}\right) \tag{19}$$

Also, if the conditions bellow can be guaranteed, the boundedness can be ensured:

$$\|z_2\| \leq c_2 \tan\left(\frac{\pi}{2\eta_2}\right), c_2 > 1 \tag{20}$$

$$\|\mathbf{q}_2 - \dot{\mathbf{q}}_t\| \leq \tau_2 \tan\left(\frac{\pi}{2\rho_2}\right), \tau_2 > 1 \tag{21}$$

4 Simulation

4.1 Equation of a Hypersonic Vehicle

In order to demonstrate the effectiveness of the control law, simulation is studied. The hypersonic vehicle six degree-of-freedom (DOF) equations are based on the Winged-Cone model provided by NASA Langley Research Center [11, 12]. Based on singular perturbation theory, affine nonlinear model can be obtained as follows:

$$\begin{cases} \dot{\mathbf{\Omega}} = \mathbf{f}_s + \mathbf{g}_{s1}\boldsymbol{\omega} + \mathbf{g}_{s2}\boldsymbol{\delta}_C \\ \dot{\boldsymbol{\omega}} = \mathbf{f}_f + \mathbf{g}_f \mathbf{M}_C \end{cases} \tag{22}$$

where $\mathbf{\Omega} = [\alpha, \beta, \mu]^T$ are angle of attack, sideslip angle and roll angle, respectively, and $\boldsymbol{\omega} = [p, q, r]^T$ are roll rate, pitch rate and yaw rate, respectively. $\mathbf{M}_C = [l_{ctrl}, m_{ctrl}, n_{ctrl}]^T$ are roll control torque, pitch control torque and yaw control torque, respectively. They are generated by both aerodynamic control surfaces and thrusts. And $\mathbf{f}_s = [f_\alpha, f_\beta, f_\mu]^T, \mathbf{f}_f = [f_p, f_q, f_r]^T$.

$$\begin{aligned}
f_\alpha &= \frac{1}{MV\cos\beta}\left[-\hat{q}SC_{L,\alpha} + Mg\cos\gamma\cos\mu - T_x\sin\alpha\right] \\
f_\beta &= \frac{1}{MV}\left[\hat{q}SC_{Y,\beta}\beta\cos\beta + Mg\cos\gamma\sin\mu - T_x\sin\beta\cos\alpha\right] \\
f_\mu &= -\frac{g}{V}\cos\gamma\cos\mu\tan\beta + \frac{1}{MV}\hat{q}SC_{Y,\beta}\beta\tan\gamma\cos\mu\cos\beta \\
&\quad + \frac{T_x}{MV}\left[\sin\alpha(\tan\gamma\sin\mu + \tan\beta) - \cos\alpha\tan\gamma\cos\mu\sin\beta\right] \\
&\quad + \frac{1}{MV}\hat{q}SC_{L,\alpha}(\tan\gamma\sin\mu + \tan\beta)
\end{aligned} \tag{23}$$

$$\mathbf{g}_{s1} = \begin{bmatrix} -\tan\beta\cos\alpha & 1 & -\tan\beta\sin\alpha \\ \sin\alpha & 0 & -\cos\alpha \\ \sec\beta\cos\alpha & 0 & \sec\beta\sin\alpha \end{bmatrix} \tag{24}$$

$$\begin{aligned}
f_p &= I_{xx}^{-1}\left[l_{aero} - qr(I_{zz} - I_{yy})\right] \\
f_q &= I_{yy}^{-1}\left[m_{aero} - rp(I_{zz} - I_{yy})\right] \\
f_r &= I_{zz}^{-1}\left[n_{aero} - pq(I_{zz} - I_{yy})\right]
\end{aligned} \tag{25}$$

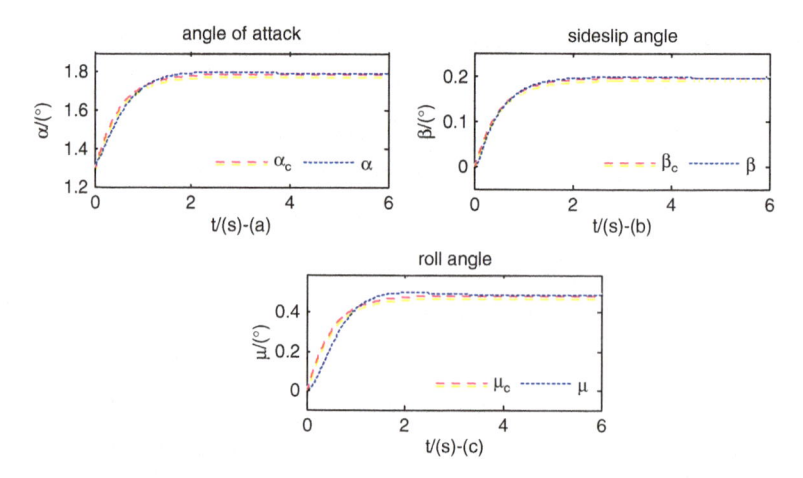

Fig. 1 Responses of attitudes with no uncertainty

$$g_f = diag\{ I_{xx}^{-1} \quad I_{yy}^{-1} \quad I_{zz}^{-1} \} \qquad (26)$$

Where, L is lift force and Y is side force; m is the quality; V is aircraft speed; γ is the flight path angle; T_x, T_y, T_z are the three thrusts. $l_{aero}, m_{aero}, n_{aero}$ are the rolling, pitching and yawing moment, respectively, when deflections of control surfaces are zero.

4.2 Numerical Simulation

The simulation is done under the condition of 15 Mach and 33.5 km. m is 65,530 kg. The I_{xx}, I_{yy}, I_{zz} are 9.15 × 105 kg · m2. 9.49 × 106, 9.49 × 106 kg · m2, respectively. Transfer function of the actuator is assumed as $50/(s+50)$. Deflections of control surfaces are limited in the range of $\pm 30°$, while the thrust vectors are $\pm 15°$. The initial values of attitudes are $\alpha_0 = 1.3°, \beta_0 = \mu_0 = 0°$, and the desired trajectories are $\alpha_c = 1.8°, \beta_c = 0.2°, \mu_c = 0.5°$. Choose the parameters $\eta = \rho = 1.4$, $l = 0.001$, and $k_1 = 4, k_2 = 4.7$. By Simulating, performances of the attitudes α, β, μ are shown in Figs. 1 and 2.

From Fig. 1 it's easy to see that the desired trajectories can be tracked in nearly 2 s. The tracking trajectory is smooth with few overshoots.

It can also be gotten From Fig. 2 that desired ideal trajectories can be tracked in less than four seconds when model parameters have uncertainty of 40 %. The controller also has better performance. Although a few overshoots exist, which are caused by the uncertainty, the overshoots are limited in 5 %. And the purposes designed are achieved.

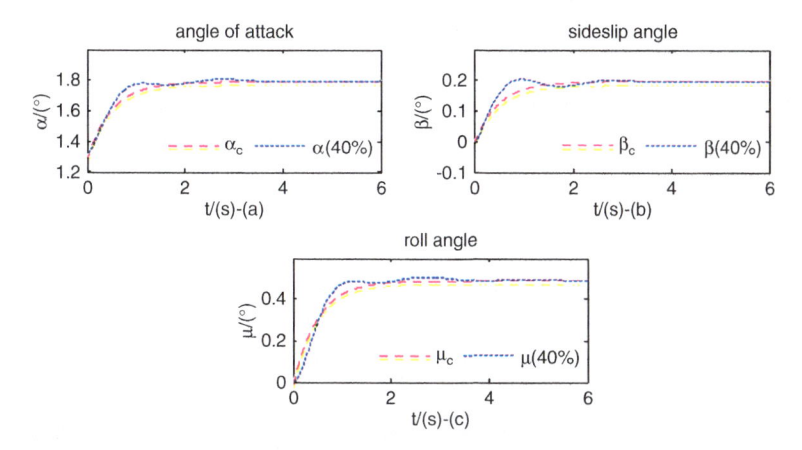

Fig. 2 Responses of attitudes with 40 % uncertainty existing

5 Conclusion

High level of technology is required in HSV reentry flight control system design. The problems of adaptive backstepping control of MIMO system have been discussed in this article. Based on dynamic surface and backstepping technology, Lyapunov function is defined to guarantee system stability. In high supersonic condition, simulation of trajectory tracking is done. The uncertainty is also discussed. The results display strong robustness of the controller.

Acknowledgements This research is financially supported by Foundation of the New Teachers from Northwestern Polytechnical University (No. 11GH0322). The authors are also wish to give thanks to Dr. Wang and Dr. Huang who give great ideas during the HSV FCS design.

References

1. Calise AJ, Buschek H (1992) Research in robust control for hypersonic vehicles. Progress report No.1 to NASA Langley Research Center. NAG-1-1451
2. Hall CE, Gallaher MW, Hendrix ND (1998) X-33 attitude control system design for ascent, transition, and entry flight regimes. NASA Marshall Space Flight Center. AIAA-98-4411
3. Kokotovic PV (2002) The joy of feedback: nonlinear and adaptive. IEEE Control Syst Mag 12 (7):266–285
4. Lian BH, Bang H, Hurtado JE (2004) Adaptive backstepping control based autopilot design for reentry vehicle. In: Proceedings of AIAA guidance, navigation, and control conference and exhibit, AIAA, pp 2004–5328
5. Jagannathan S, Lewis FL (2000) Robust backstepping control of a class of nonlinear systems using fuzzy logic. Inf Sci 133(3/4):223–240

6. Kim KS, Kim Y (2003) Robust backstepping control for slew maneuver using nonlinear tracking function. IEEE Trans Control Syst Technol 11(6):822–829
7. Zhang T, Ge SS, Hang CC (2000) Adaptive neural network control for strict-feedback nonlinear systems using backstepping design. Automatica 36(12):1835–1846
8. Swaroop D, Gerdes JC, Yip PP et al (1997) Dynamic surface control of nonlinear systems. In: Proceedings of the American control conference. IEEE, vol 5, Los Alamitos, CA, pp 3028–3034
9. Swaroop D, Hedrick JK, Yip PP et al (2000) Dynamic surface control for a class of nonlinear systems. IEEE Trans Autom Control 45(10):1893–1899
10. Wang D, Huang J (2005) Neural network-based adaptive dynamic surface control for a class of uncertain nonlinear systems in strict-feedback form. IEEE Trans Neural Netw 16(1):195–202
11. Shaughnessy JD, Pinckney SZ, McMinn JD (1991) Hypersonic vehicle simulation model: winged-cone configuration, NASA TM2102610, pp 1–15
12. Keshmiri S, Colgren R (2005) Development of an aerodynamic database for a generic hypersonic air vehicle. In: Proceedings of AIAA guidance, navigation, and control conference and exhibit, AIAA, San Francisco, CA, USA, pp 2005–6257

Design of Smith Auto Disturbance Rejection Controller for Aero-engine

Fang-Zheng Luo, Shi-Ying Zhang, Min Chen, and Yu Hu

Abstract In order to suppress many strong disturbances while aero-engine is working at steady-state, Auto Disturbance Rejection Control (ADRC) for aero-engine rotor speed was studied in this paper. By analyzing characteristics of the controlled object and making full use of Smith predictor, the two-order Smith-ADRC Controller instead of directly using three-order controller was designed which could avoid the shortcoming of tuning a large number of parameters and effectively compensate time delay of the controlled object. The simulation results show that the designed two-order Smith-ADRC controller can effectively suppress instantaneous and random disturbance, and can be suitable for aero-engine control.

Keywords Aero-engine • Speed control • Smith-ADRC controller • Time delay • Disturbance

1 Introduction

In most time of working process, aero-engine works in cruising state, therefore, the cruising state control called steady-state control is the emphasis of engine control. At this stage, the control law keeping the rotor speed constant is generally used. In practical engineering application, PID control algorithm, which is simple and easily tuned, is usually selected. However, PID control algorithm is only a linear adjustment method based on error feedback signal, which is difficultly used to control the nonlinear objects. Advantages and disadvantages of PID control algorithm were deeply analyzed [1, 2], then ADRC controller was further proposed to hopefully replace PID control algorithm. ADRC controller with strong robustness is a controller that basically not relied on mathematical model of the object and can

---

F.-Z. Luo (✉) • S.-Y. Zhang • M. Chen • Y. Hu
Xi'an Research Institute of Hi-Tech, Xi'an, China
e-mail: 857962178@qq.com

effectively exploit the non-linear function. The controller has been widely exploited since it was proposed [3], while the algorithm of the controller is so complex that research and application was seldom developed in the aero-engine field. The problem of turbo-shaft engine torque disturbed was solved by using ADRC controller [4, 5], and the design of decoupling engine multivariable by using ADRC controller was achieved [6]. ADRC controller was used to control the two-order system of turbofan engine, which obtaining better result than using PID controller [7]. However, at this stage, there is no literature about ADRC controller applied in this situation, in which aero-engine model considering the model of actuating mechanism and delay characteristic of engine working process. To solve the problem, the model of actuating mechanism and the influence of delay characteristic in engine working process were considered. In this paper, taking the high-pressure rotor speed as the control variable, ADRC control of engine steady-state process was studied.

2 Selections of Mathematical Model and Control Scheme

With the change of flight altitude and speed, the model of aero-engine is time-varying. In the field of aero-engine control, the basic mathematical model is non-linear model, which is established based on the structure and thermodynamics characteristics of turbofan engine. However, the solving process of nonlinear model in aero-engine is so complex that it cannot satisfy real-time control. In engineering application, the simplified model can be gotten by using the linearization method. Based on operating characteristics of a certain type of turbofan engine, engine high-pressure speed-fuel control system was regard as the controlled object. Near the steady-state point ($H = 0, Ma = 0, n\% = 85\%$), the simplified engine mathematical model is

$$G_p(s) = \frac{s + 5.38}{s^2 + 8.46s + 16.72} \quad (1)$$

The mathematical model of actuating mechanism can be approximately seen as [7]:

$$G_v(s) = \frac{1}{0.3s + 1} \quad (2)$$

The mathematical model of system delay characteristic (delay time is 0.1 s):

$$G_\tau = e^{-0.1s} \quad (3)$$

The controlled object transfer function could be assembled by these three-part series. Actually, it is a three-order delay model.

$$G(s) = \frac{s + 5.38}{0.3s^3 + 3.538s^2 + 13.48s + 16.72} e^{-0.1s} \quad (4)$$

Generally speaking, while selecting the control scheme, the order of the controlled object transfer function should be same to that of the controller [8]. However, if selecting three-order controller to control three-order system, there are a large number of parameters needing to be tuned. Therefore, it is can be considered that using the two-order controller which has been maturely studied at this stage. Through simulation examples, it was shown that three-order object could be well controlled by using two-order ADRC controller [9]. The method that using two-order ADRC controller to control three-order delay object is also certainly described [8]. However, for the object studying in this paper, there is no specific description to design the controller. In this paper, the control scheme is that engine three-order delay object was controlled by the two-order ADRC controller. By adjusting the parameters, making full use of the strong self-adaption of ADRC controller, the three-order control object could be regard as a two-order object to control. System delay characteristics could be compensated by Smith-predictor.

3 Design of Controller

The principle of ADRC controller is described [1]. Figure 1 shows the basic structure of ADRC controller. Where r is input signal, r_1 is transient process, r_2 is differential signal, u is control variable, ω is disturbance, y is system output.

In Fig. 1, ADRC controller is mainly composed by three parts: tracking-differentiator (TD), nonlinear state error feedback control law (NLSEF) and extended state observer (ESO). TD is used to track input signal and produce an approximate differential signal for original input signal. ESO is designed to estimate extended state of system. NLSEF is introduced to compose margin of error in a nonlinear way and gain control variable. For two-order system, input–output relationship is directly manifested as two-order differential equations. While $u(t)$ is defined as system control variable, $y(t)$ as output variable, control algorithm of

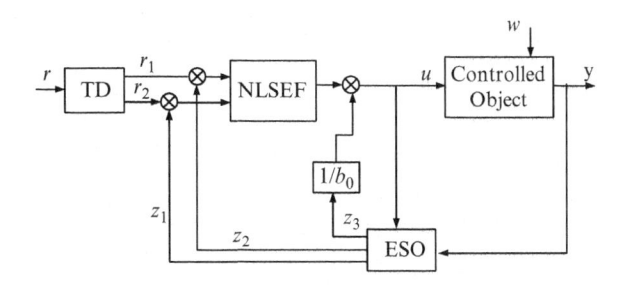

Fig. 1 The basic structure of ADRC controller

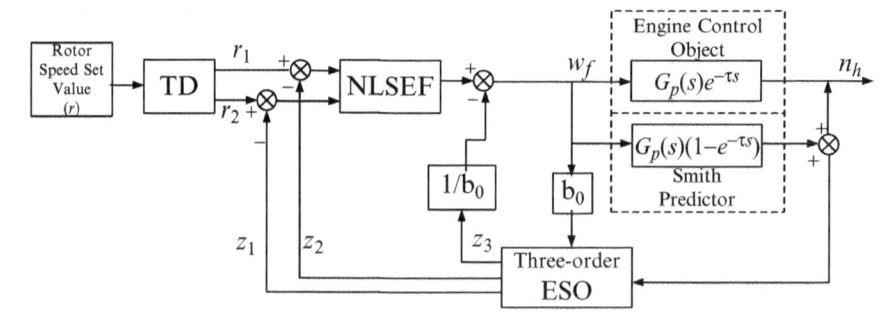

Fig. 2 Engine two-order Smith-ADRC control system

this system is expressed by Eq. 5. In Eq. 5, $fst(\bullet)$ and $fal(\bullet)$ are nonlinear functions, $r, h, \beta_1, \beta_2, \beta_3, \alpha_1, \alpha_2, \alpha_3, \alpha_4, \alpha_5, \delta_1, \delta_2, k_1, k_2$ are all undetermined coefficients, b is input variable magnification coefficient of the controlled object transfer function. Due to space limitation, specific meanings of related functions and parameters can been seen in other references [7, 8].

$$\begin{cases} TD: \dot{r}_1 = r_2 \\ \dot{r}_2 = fst(r_1, r_2, r(t), r, h) \end{cases} \begin{cases} ESO: \varepsilon = z_1(t) - y(t) \\ \dot{z}_1 = z_2 - \beta_1 \cdot fal(\varepsilon, \alpha_1, \delta_1) \\ \dot{z}_2 = z_3 - \beta_2 \cdot fal(\varepsilon, \alpha_2, \delta_1) + bu(t) \\ \dot{z}_3 = -\beta_3 \cdot fal(\varepsilon, \alpha_3, \delta_1) \end{cases}$$

$$\begin{cases} NLSEF: e_1 = r_1 - z_1, e_2 = r_2 - z_2 \\ u_0 = k_1 \cdot fal(e_1, \alpha_4, \delta_2) + k_2 \cdot fal(e_2, \alpha_5, \delta_2) \\ u = u_0 - \dfrac{z_3}{b} \end{cases} \quad (5)$$

Combining the parameter tuning methods and practical experience [1, 7, 8], the parameters can be tuned as follows:

TD: system transition time($t_0 = 0.1s$), amplitude of tracking signal($d = 1$), speed factor($r = 4$); filter factor(h) is equal to emulating time of system, $h = 0.001$.
ESO: according experience, $\alpha_1 = 0.5$, $\alpha_2 = 0.25$, $\alpha_3 = 0.1$; band width ($\omega_0 = 15$ rad/s), $\beta_1 = 45$, $\beta_1 = 675$, $\beta_1 = 3375$.
NLSEF: according experience, $\alpha_4 = 0.5$, $\alpha_5 = 1.5$; the controlled model, $b_0 = 3.3$, $\xi = 1$; according to system band width, $k_1 = 225$, $k_2 = 30$.

To compensate the delay characteristics of the system, Smith predictor could be added to the ADRC controller. The basic principle of Smith Predictor is that import a suitable feedback element to the controller. The delayed controlled variable can be reflected in the controller in advance, which can counteract the influence of time delay [10].

Based on the analysis above, the two-order Smith-ADRC system could be designed in Fig. 2.

4 Results and Analysis

Setting a square wave as the speed tracking signal and applying a Gaussian white noise signal ($\mu = 0$, $\sigma^2 = 1$) to the controlled object, the mentioned above tuned parameters was use to simulate. Simulation results could be seen in Figs. 3 and 4.

In Fig. 3, under initial tuning parameters, system tracking response is stable in square wave band, settling time was equal to 0.937 s ($ts = 0.937s$). The applied noise could be well suppressed by the controller, which has a good tracking

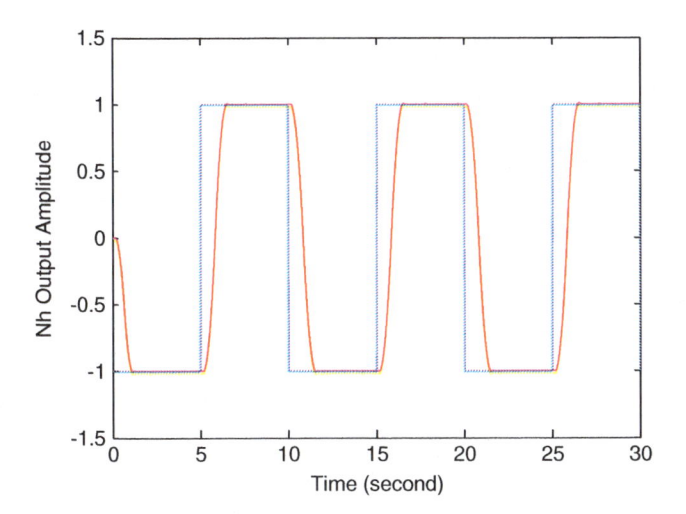

Fig. 3 $r = 4$, system tracking response curve

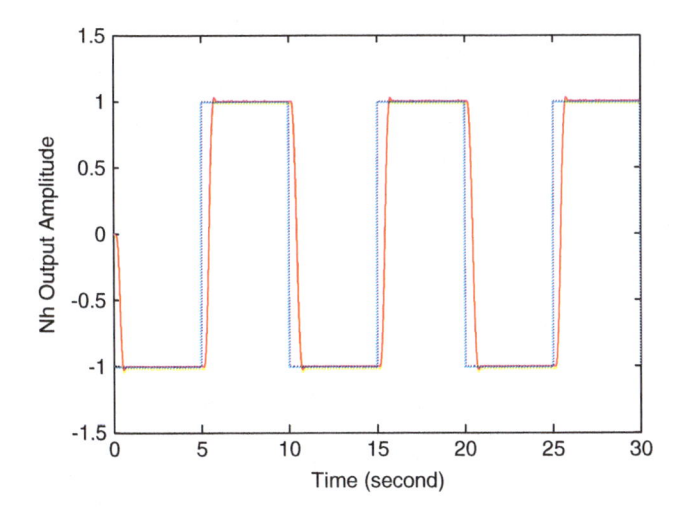

Fig. 4 $r = 20$, system tracking response curve

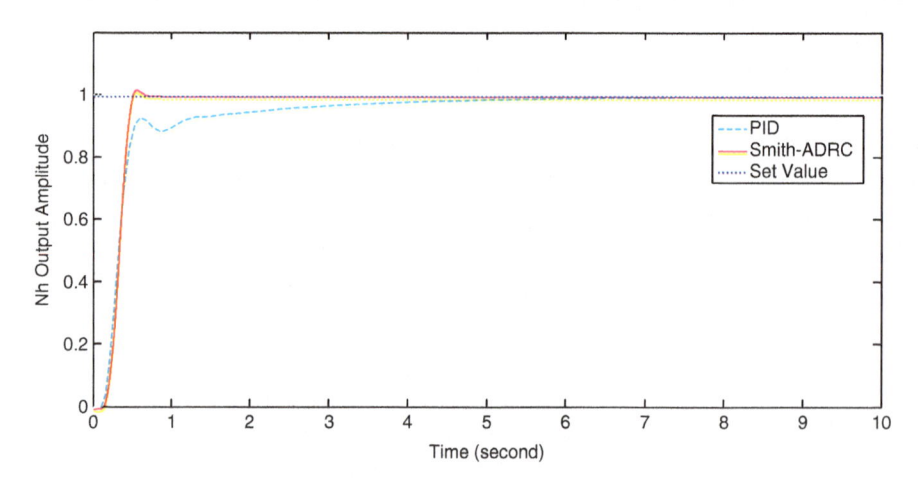

Fig. 5 The comparison of two control methods step response

performance. While adjusting speed factor (r), the shorter setting time could be obtained (Fig. 4). In Fig. 4, increasing r to 20 ($r = 20$), settling time reduced to 0.47 s ($ts = 0.47s$). However, if there was no Smith predictor, oscillation could emerge in system simulation curve. The simulation result shows that Smith predictor can effectively compensate the system time delay.

For engine speed PID and two-order Smith-ADRC control circuits, respectively applying a step signal, without adding disturbance, step responses of control system were displayed in Fig. 5.

Where the adoptive parameters of PID controller were $k_p = 10.22$, $k_i = 5.8$, $k_d = 1.41$. The as seen in Fig. 5, respectively using PID and ADRC controllers to control the system, the obtained response characteristics were similar. However, better tracking performance could be obtained by ADRC controller. Using ADRC controller, it just took 0.4678 s for system to obtain permissible steady-state error, and overshoot was just 2 %.

Simulation results shown that although the system did not use the above tuning parameters or given controlled object, the designed controller all had good control performance. From simulation results, it is also clear that the designed controller with strong robustness and adaptability is not just suitable for the specific object.

4.1 Robustness of Controller

For a controller, the ability, that adjusting the controller's parameters in a certain range has no influence on tracking the set value, is called robustness. For the designed controller, keeping other parameters unchanged, adjusting the speed factor (r) within a wide range, the simulation results shown that while r was less than 10,000, overshoot could be controlled within ±5 %.

Design Aero-Engine Smith-ADRC Controller

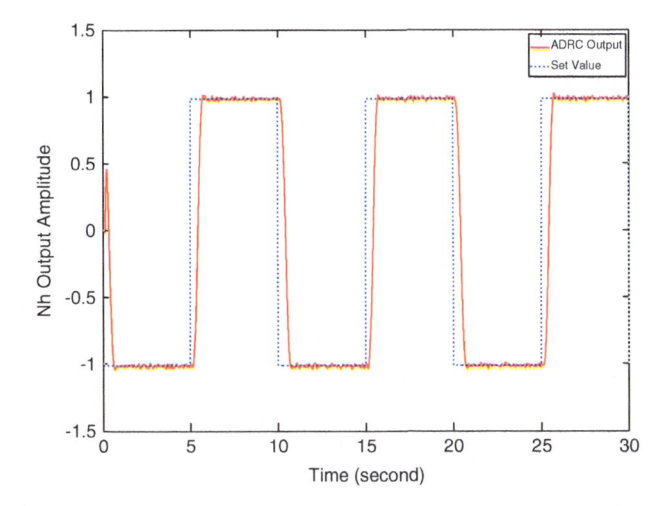

Fig. 6 Applying Gaussian white noise ($\mu = 100$, $\sigma^2 = 200$), speed tracking response curve

While adjusting these parameters (β_1, β_2, β_3, k_1, k_2), except the contracted parameters in most of literatures, good curves could be obtained. It is proved that the controller has strong robustness. Compared with PID controller, although the ADRC controller has more parameters needing to be tuned, these parameters can be easily tuned within a wide range.

4.2 Anti-interference Performance of Controller

Increasing intensity of Gaussian white noise signal applied to system, average value was adjusted to 100 ($\mu = 100$), variance was adjusted to 200 ($\sigma^2 = 200$), system tracking curve was displayed in Fig. 6. If applying a white noise with power of 10 ($P = 10$), system simulation curve was displayed in Fig. 7.

Figure 6 shows that Gaussian white noise could be well suppressed by the system. In Fig. 7, while applying a white noise with power of 10, oscillation of high-pressure rotor was within ± 0.2. Therefore, the white noise of high-power could be also well suppressed.

For chronological disturbance, the designed controller also has strong anti-interference capability. After system working into steady-state, at the 6 s of simulation time, applying a tunable pulse signal with amplitude of A to system, the experiment is used to obtain the responses while the two systems undergoing instantaneous shock. Duration time of the pulse signal was 0.001 s, system response curve could be seen in Figs. 8 and 9.

In Fig. 8, for PID control circuit, while undergoing instantaneous shock which was 180 times amplitude of input signal, the response of high-pressure rotor output

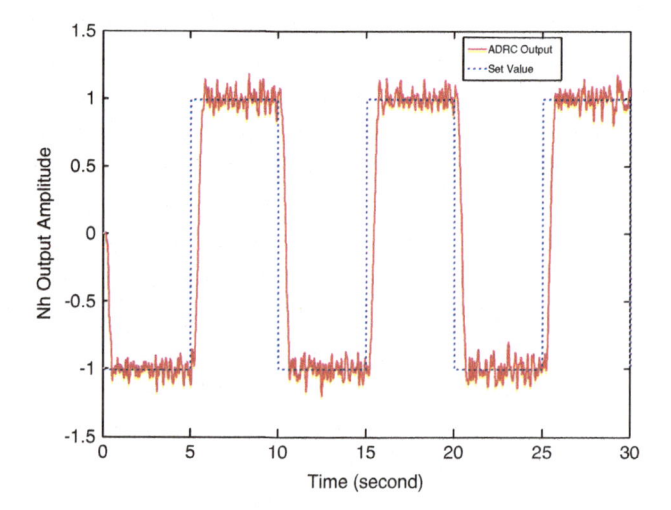

Fig. 7 Applying white noise ($P = 10$), speed tracking response curve

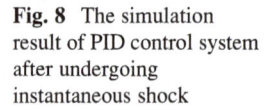

Fig. 8 The simulation result of PID control system after undergoing instantaneous shock

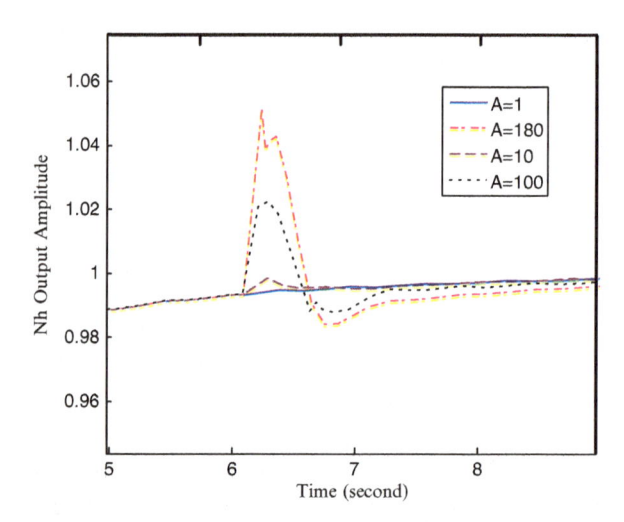

signal was less than 1.06 times. However, for Smith-ADRC control circuit, while undergoing instantaneous shock which was 1,000 times amplitude of input signal, the response of high-pressure rotor output signal was less than 1.08 times (Fig. 9). The results show that ADRC control circuit has stronger shock resistance than PID control circuit, and ADRC control circuit can be faster entering steady-state than PID control circuit after undergoing instantaneous shock.

Fig. 9 The simulation result of Smith-ADRC control system after undergoing instantaneous shock

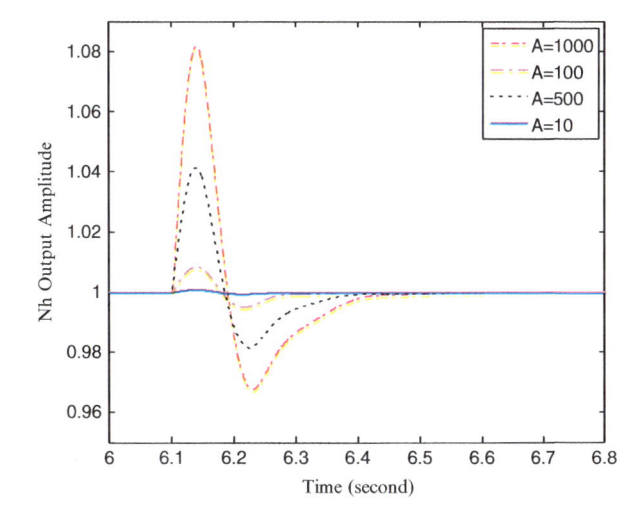

5 Conclusion

In the paper, by taking the mathematical model of three-order time delay speed-fuel control system as the control object, a two-order Smith-ADRC controller was designed, which realized the control of aero-engine steady-state. From the simulation results, although the algorithm of ADRC controller is more complex, parameters tuning are more accurate than PID engineering tuning, and the ADRC controller has a strong anti-interference ability. Two-order Smith-ADRC controller, which can avoid many shortcomings than three-order controller and effectively overcome the time delay of system, is a new controller suitable for aero-engine control.

References

1. Jing-Qing Han (2007) Auto disturbances rejection control technique. Front Sci 1(1):24–31
2. Fan-Dong Meng (2009) Study of design and application for the active disturbance rejection controller. Dissertation, Harbin University of Science and Technology, Harbin
3. Zhao-Jing Zhang (2007) Parameter adjustments of auto-disturbance rejection control systems. Dissertation, Southern Yangtze University, Wuxi
4. Hai-Bo Zhang, Jian-Guo Sun, Li-Guo Sun (2010) Design and application of a disturbance rejection rotor speed control method for turbo-shaft engines. J Aerosp Power 25(4):943–950
5. Hai-Bo Zhang, Li-Guo Sun, Jian-Guo Sun (2010) Robust disturbance rejection control design for integrated helicopter system/turbo-shaft engine. Acta Aeronaut Et Astronaut Sinica 31(5):883–892
6. Hai-Bo Zhang, Jian-Kang Wang, Ri-Xian Wang et al (2012) Design of an active disturbance rejection decoupling multivariable control scheme for aero-engine. J Propuls Technol 33(1):78–83

7. Shu-Qing Li, Sheng-Xiu Zhang, Yi-Nan Liu et al (2012) Simple design and application of auto disturbance rejection controller for aeroengine. Aeroengine 38(3):46–48
8. Xiao-Mei Yao (2002) Two-order ADRC control for general industrial plants. Control Eng China 9(5):59–62
9. Li-Ming Zhang (2009) Research on application of active disturbance rejection technology to AUVheading control. Dissertation, Harbin Engineering University, Harbin
10. Ke Xu (2004) Distributed control systems for aero-engines. Dissertation, Nanjing University of Aeronautics and Astronautics, Nanjing

Asynchronous Motor Vector Control System Based on Space Vector Pulse Width Modulation

YingZhan Hu and SuNa Guo

Abstract In order to improve the control result of asynchronous motor with power supplied by battery, the voltage vector control technology is used to respectively control the asynchronous motor excitation current and the torque current by measuring and controlling the vector of asynchronous motor stator current according to the principle of field oriented. The method to realize the algorithm of the voltage space vector pulse width modulation is introduced in detail, and the simulation model of three-phase asynchronous motor is built. The algorithm is simulated. The results show that the algorithm is reasonable, the control performance is better and application requirements of control for AC motor with power supplied by battery is satisfied.

Keywords Voltage space vector pulse width modulation • Asynchronous motor • Vector control • Simulation

1 Introduction

Space vector pulse width modulation (SVPWM) technique is to control the switching of the inverter in the way of controlling three-phase motor stator to generate the tracking circular rotating field with power supplied by Voltage Source Inverter. This control strategy could improve the utilization of the voltage on the DC side, make the calculation easy [1], reduce the switching loss and thus reduce motor harmonic losses, lower the torque ripple, is especially for the situation with power supplied by battery such as electric vehicles.

Vector control theory develops on the electric machine unification theory, electromechanical energy conversion and coordinate transformation theory, and it

---

Y. Hu (✉) • S. Guo
Department of Electrical Engineering, Henan Polytechnic Institute, Nanyang, China
e-mail: huyz168@163.com

has features such as advanced, novelty and practical etc. It makes induction motor-model become a DC motor-model by the coordinate transformation, decompose the stator current into two DC parts which are orientated towards the rotator magnetic field,control them so as to realize the decouple of magnetic flux and torque and achieve the DC motor effect.

2 Principle of SVPWM

The main circuit structure of the typical three-phase voltage source inverter is shown in Fig. 1 [2, 3].

There are two switching devices on each bridge arm. PWM control is to adjust the average current by adjusting the switch-off time of each bridge arm. The switching rule of the six switching devices must obey the following rules:

1. The numbers of switching devices in the open state and in the off state must be three at any time;
2. The two switching devices of the same bridge arm is controlled by complementary drive signals, and cannot be shoot-through.

The space vector pulse width modulation (SVPWM) is to convert the input voltage of three-phase inverter to the space voltage vector and approximate the voltage circle by using the eight space vectors formed by different switching states of the inverter, and then form the SVPWM trigger wave. The position and size of eight voltage space vectors is shown in Fig. 2 [4].

There are six nonzero vectors U_x (x = 1, 2, 3, 4, 5, 6) and two zero vectors (U_0, U_7) in Fig. 2, The switching states corresponding to the upper bridge of inverter are marked in small brackets after the vector under the action of each vector. "1" indicates the on-state, and "0" indicates the off-state. Each of the six nonzero voltage vectors magnitude is $2U_d/3$, and their phase angle difference is $\pi/3$. The complex plane is divided into six fan-shaped regions, and they are identified as 1, 2, 3, 4, 5, 6 in this paper. In any fan-shaped region, the voltage vector U^* can be made up of the adjacent space voltage vectors (U_x, $U_{x\pm 60}$). The corresponding basic space vector constitute instantaneous command is the purpose of voltage space vector technology. The average input voltage is made equal to the command voltage U^*

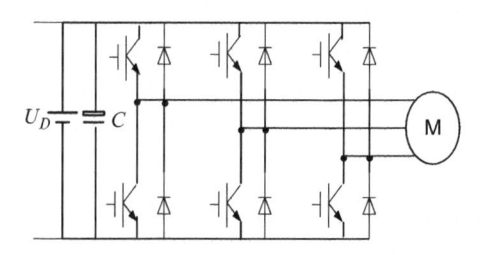

Fig. 1 Three-phase power inverter

Fig. 2 The space vector and switching state

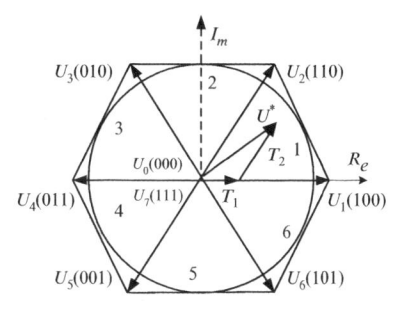

during the pulse period T .so that the space voltage vector trajectory approaches a round.

In other words, in arbitrarily small cycle time T, the output of the inverter is the same as the average instructions voltage, such as Eq. 1.

$$\frac{1}{T}\int_{nT}^{(n+1)T} U^*(t)dt = \frac{1}{T}(T_1 U_x + T_2 U_{X\pm 60}) \tag{1}$$

T_1, T_2 respectively is the action time of U_x, $U_{x\pm 60}$. If the sum of T_1 and T_2 is less than the pulse period T, the zero vector(U_0, U_7) will be used to fill up the remaining time of period T [5] . When the end of the flux vector is stationary under the action of zero vector, and the original rotation frequency of the flux has been changed. As a result, Variable frequency is realized [2, 6].

3 Asynchronous Motor Vector Control System Based on SVPWM

In this system, three-phase windings of Asynchronous Motor adopt Y-connected without zero line, then $i_a + i_b + i_c = 0$ or $i_c = -i_a - i_b$. The diagram of rotor field oriented control system for induction motor based on SVPWM is shown in Fig. 3.

This control system is composed of the outer loop speed control and the inner loop current control. The speed control loop is the speed command value ω_{ref} given by the user compared with the speed feedback signal ω_2 of the optical encoder on the motor shaft, and its deviation is adjusted through the speed PI regulator, and output the torque current component used as the command value of inner torque regulator loop, then compare with the three-phase stator current signal i_q after the Clarke and Park transformations detected by the hall current sensor, then the stator torque voltage component v_q^* in rotating frame is obtained after the torque PI regulator. In the current control loop, the excitation component command value i_{dref} (zero) compared with the signal i_q which is the three-phase stator current signal after the Clarke and Park transformations detected by the hall current sensor, voltage excitation component v^*_d of the stator in rotating frame will been obtained

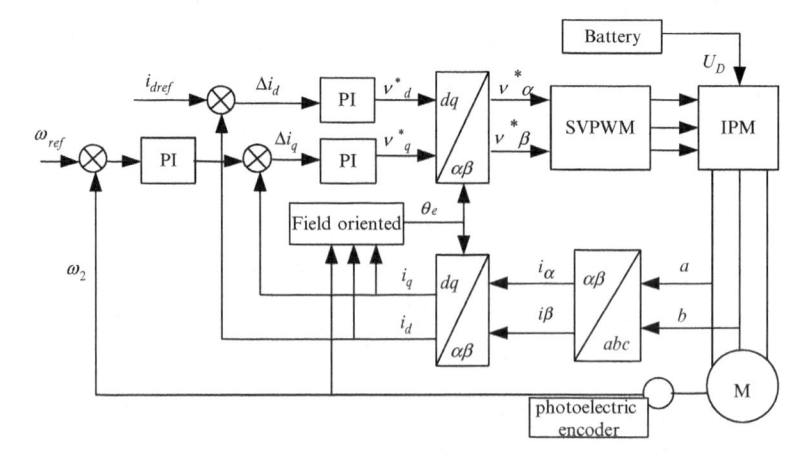

Fig. 3 The diagram of rotor field oriented control system for induction motor based on SVPWM

after the torque PI regulator. Then we can transform v_d^* and v_q^* to the two-phase stationary stator frame by the anti-Park transformation, obtain the two voltage components v^*_α and v^*_β which have the same frame with the inverter voltage space vector. Finally, it uses the space vector pulse width modulation to generate the PWM waveform of the inverter switch on-state. On-state inverter switching PWM waveform is generated by the use of the technique of space vector pulse width modulation [7, 8].

4 The Implementation of SVPWM Algorithm

4.1 Determine the Sector of Space Voltage Vector U*

The command value of space vector voltage U^* is determined by the two voltage components v^*_α and v^*_β acquired by coordinate transformation. But it can be synthesized by adjacent space voltage vectors of a sector only by acquiring the sector. The period of a wave is divided into six intervals according to three-phase voltages with an angle width (60°), and the zero-crossing of voltage is used as the beginning and the end of sector. The sector can be determined by two-phase voltage and another with the opposite sign, as follows [9]:

1. Segment 1 $v_a^* > 0, v_b^* < 0, v_c^* > 0$;
2. Segment 2 $v_a^* > 0, v_b^* < 0, v_c^* < 0$;
3. Segment 3 $v_a^* > 0, v_b^* > 0, v_c^* < 0$;
4. Segment 4 $v_a^* < 0, v_b^* > 0, v_c^* < 0$;
5. Segment 5 $v_a^* < 0, v_b^* > 0, v_c^* > 0$;
6. Segment 6 $v_a^* < 0, v_b^* < 0, v_c^* > 0$.

Table 1 The relationship of sector and P

P value	1	2	3	4	5	6
Sector number	2	6	1	4	3	5

Of course, anti-Clarke transformation should been done before above as follows:

$$\begin{aligned} v_a^* &= v_\beta^* \\ v_b^* &= -\frac{\sqrt{3}}{2}v_\alpha^* - \frac{1}{2}v_\beta^* \\ v_c^* &= \frac{\sqrt{3}}{2}v_\alpha^* - \frac{1}{2}v_\beta^* \end{aligned} \quad (2)$$

Making $P = 4sign(v_b^*) + 2sign(v_c^*) + sign(v_a^*)$, and looking up Table 1 to determine the number of sectors.

4.2 Action Time of the Adjacent Switching Vector

Making

$$\begin{aligned} X &= \sqrt{3}kv_\alpha^* \\ Y &= \frac{\sqrt{3}k}{2}v_\beta^* + \frac{3}{2}kv_\alpha^* \\ Z &= \frac{\sqrt{3}k}{2}v_\beta^* - \frac{3}{2}kv_\alpha^* \end{aligned} \quad (3)$$

There is $k = T/U_d$ in Eq. 3.

The values of T_1, T_2 is defined according to the different values of P as Table 2. If $T_1 + T_2$, then amended T_1, T_2 as follows method:

$$T_1 = \frac{T_1}{T_1 + T_2}T, \quad T_2 = \frac{T_2}{T_1 + T_2}T$$

The values of T_1, T_2 in the right hand side of the equal sign is based on Table 2, the left hand of the equal sign is corrected value. Assume

$$\begin{cases} T_a = (T - T_1 - T_2)/2 \\ T_b = T_a + T_1/2 \\ T_c = T \end{cases} \quad (4)$$

Table 2 Assignment table of T_1, T_2

P value	1	2	3	4	5	6
T1	Z	Y	−Z	−X	X	−Y
T2	Y	−X	X	Z	−Y	−Z

Table 3 T_a, T_b, T_c corresponding to three-phase a,b,c in all sectors

	Sector number					
	1	2	3	4	5	6
Phase a	Ta	Tb	Tc	Tc	Tb	Ta
Phase b	Tb	Ta	Ta	Tb	Tc	Tc
Phase c	Tc	Tc	Tb	Ta	Ta	Tb

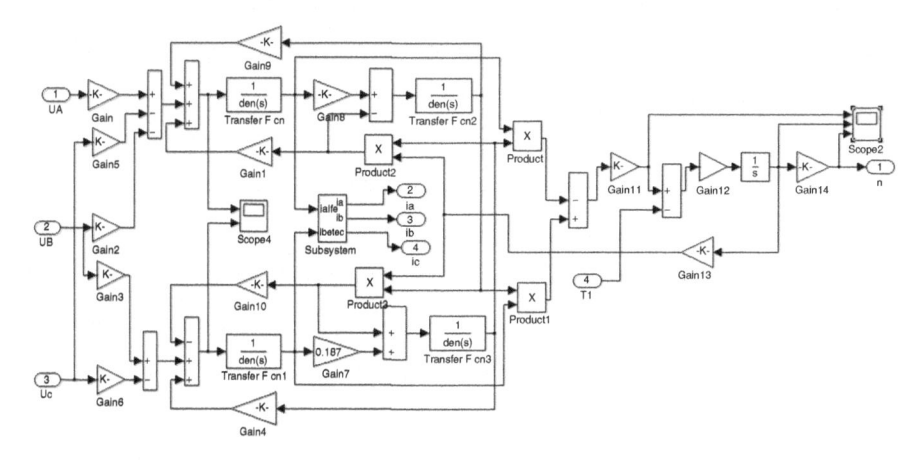

Fig. 4 The simulation model of three-phase induction motor

T_a, T_b, T_c corresponding to three-phase are defined in Table 3. T_a, T_b, T_c are the values to generate PWM waveforms by Comparing with the triangular wave. SVPWM wave is modulated to control the inverter, then to control the motor.

5 Simulation of Motor Vector Control System

The simulation model of three-phase induction motor is shown in Fig. 4, and this model is based on the three-phase induction motor YTSP90L-4.

The Motor parameters: rated power: 1.5 kw, rated voltage: 380 V, rated current:4.0 A, rated frequency: 50 Hz, $R_1 = 0.07\ \Omega$, the self-inductance of stator L1 = 0.066 mH, $R_2 = 0.052\ \Omega$, the self-inductance of rotor L$_2$=0.101 mH, The mutual inductance $L_m = 2.108$ mH, Number of pole pairs $n_p = 2$.

As this algorithm above, a simulink model of the rotor flux oriented induction motor vector control system with simulink is built, shown in Fig. 5.

Figures 6 and 7 are respectively for the motor three-phase speed and the three-phase current waveform. From the figure we can see, the motor is operated steady

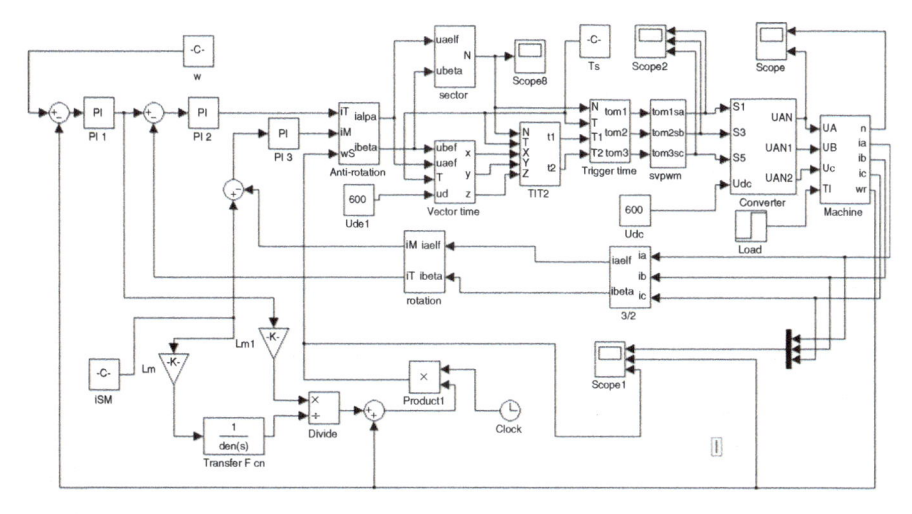

Fig. 5 The simulink model of system

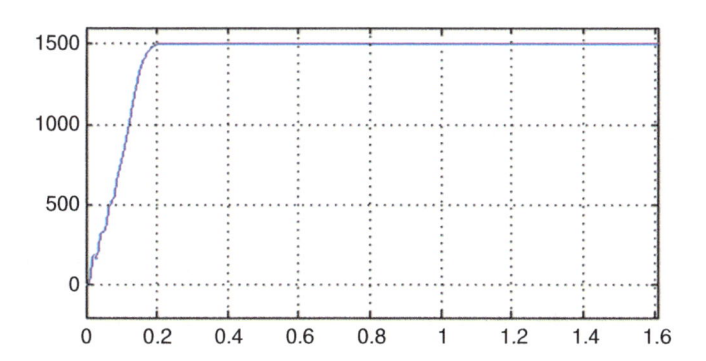

Fig. 6 The waveform of three-phase motor speed (*left*)

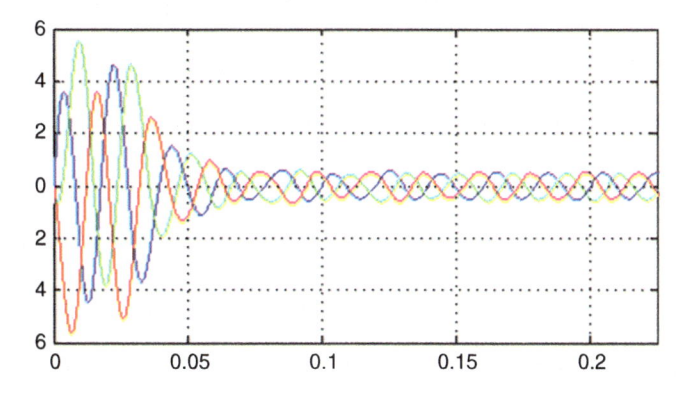

Fig. 7 The waveform of three-phase induction motor stator current (*right*)

and smoothly when the reference speed is 1,500 r/min, and the overshoot is small with fast response. The current harmonic is low. The control results are satisfactory.

6 Conclusion

On the basis of theoretical analysis, the MATLAB simulation model of asynchronous motor vector control system based on SVPWM was realized. The simulation results verified that the algorithm was correct and scientific and the properties of the control systems were fine, and it could satisfy the control requirements of the AC motor with power supplied by battery. The preparations can be made for the realization of control circuit with the core of DSP, SOPC and so on.

References

1. Ag. Wu, Chz. Piao, Rg. Yang (2011) SVPWM simplified algorithm and dead zone compensation. Power Electron 45(7):72–74 (in Chinese)
2. Li J (2012) Implementation of SVPWM control technology for asynchronous motor based on DSP. Min Process Equip 40(3):100–105 (in Chinese)
3. Shr. Huang, Rk. Hao, Zhu J (2010) SVPWM control technology implementation of induction motor based on DSP. Electr Driv Autom Control 32(5):19–22 (in Chinese)
4. Xl. Wen, Xg. Yin, Zhe Z (2009) Unified space vector PWM implementation method for three-phase inverters. Trans China Electrotech Soc 24(10):87–93, in Chinese
5. Yq. Xue, Liu B (2010) Analysis of zero state vector distribution based on space vector PWM. Electr Mach Control 14(8):93–97 (in Chinese)
6. Jc. Fang, Yz. Ling (2006) Simulation of the vector control system with SVPWM for asynchronous motors. Tech Autom Appl 25(9):54–56 (in Chinese)
7. Fl. Shen, Yk. Man, Jh. Wang (2011) The research on space vector control of induction motor system based on rotor field orientation. 2011 Chinese Control and Decision Conference (CCDC2011). IEEE conference Publications, Los Alamitos, CA, USA, pp 3400–3403
8. Qs. Yu. (2010) Study on the SVPWM vector control system for asynchronous motor [D]. Chongqing University, Chongqing, pp 7–34 (in Chinese)
9. Zheng Z, Hj. Tao (2006) Studying of control algorithm for fast space vector PWM. Electr Appl 25(8):38–40 (in Chinese)

Design of Direct Current Subsynchronous Damping Controller (SSDC)

Shiwu Xiao, Xiaojuan Kang, Jianhui Liu, and Xianglong Chen

Abstract First, this article analyzes the basic principles of supplementary subsynchronous damping controller (SSDC) inhibiting SSO caused by the DC control system. Second, based on the principle of phase compensation, a multi-mode SSDC is designed in accordance with the maximum phase compensation method. Third, test signal method is used to analyze SSDC's compensation role in improving the system electrical damping. Finally, the availability of SSDC designed in this paper is simulated and verified in an actual power system model.

Keywords DC transmission • Subsynchronous oscillation • Phase compensation • Test signal method • Electrical damping characteristics

1 Introduction

1.1 Mechanism Analysis of SSO Caused by DC Control System

Interaction of high voltage direct current (HVDC) systems and turbine-generator units may produce subsynchronous oscillation (SSO) to endanger the safe and stable operation of the grid and units. The world's first HVDC causing SSO of turbine-generator unit was found in the debugging of the Square Butte HVDC Transmission Project in 1977 [1]. Rapid control of the HVDC transmission system can cause SSO problem of the system under certain conditions (Fig. 1).

S. Xiao • X. Kang (✉) • J. Liu • X. Chen
School of Electrical & Electronics Engineering, North China Electric Power University,
Beijing, China
e-mail: kangxiaojuan_2006@126.com

Fig. 1 SSO's occurrence schematic

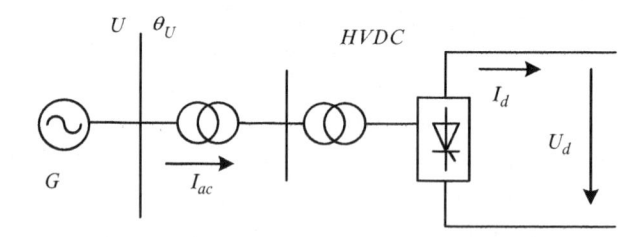

The tiny rotor mechanical perturbation $\Delta\delta = A\sin\mu t$ on the generator strongly coupled with the rectifier station, it will cause the terminal voltage (the rectifier station AC bus voltage $U\angle\theta_U$ in this case) amplitude and phase perturbation ΔU and $\Delta\theta_U$. Among them, $\Delta\theta_U$ makes the same size deviation for the rectifier station trigger angle and expected firing angle, resulting in a perturbation in the DC bus voltage U_d; terminal voltage amplitude perturbation ΔU, also will cause the perturbation of DC bus voltage U_d. The perturbation of U_d would lead to perturbation ΔI_d (and ΔP_d), DC constant current (constant power) control attempting to prevent perturbation of I_d (P_d), which cannot ultimately completely eliminate ΔI_d (ΔP_d), resulting in perturbation of the generator electromagnetic torque ΔT_e. Once the phase is appropriate, this ΔT_e will help to increase the initial mechanical disturbance $\Delta\delta$, namely the electrical damping becomes negative, once it is stronger than the showing mechanical damping in the corresponding shaft frequency, the situation that DC control system causes instability of shaft torsional oscillation will appear, i.e., SSO caused by the DC control system [2].

1.2 The Principle of SSDC Inhibition to DC SSO

The method, using supplementary subsynchronous oscillation damping controller (SSDC) to solve the SSO problems caused by HVDC is mentioned in the EPRI's research report "HVDC System Control for Damping Subsynchronous Oscillation" [3]. This method has been used in many DC projects. The basic principle of the SSDC is to provide positive electrical damping in the subsynchronous frequency range of the turbine-generator shaft [4].

From the mechanism analysis of DC rapid control system causing SSO, the direct cause of the instable shaft torsional vibration is the phase angle difference between generator speed offset $\Delta\omega$ and the electromagnetic torque variation ΔT_e exceeds 90° after the disturbance, which produces an electrical negative damping to help increase the occurrence probability of SSO, as shown in Fig. 2. SSDC usually selects signals can reflect the severity of the generator shaft torsional oscillation as inputs. For example, generator speed deviation is selected as input signal. After the proportion and phase-shift link of SSDC, its output signal through the Constant Current Control loop to provide an additional electromagnetic torque $\Delta T'_e$, and makes the phase angle difference between $\Delta\omega$ and the vector synthesized by ΔT_e

Fig. 2 The relationship of electromagnetic torque and rotation speed after disturbance

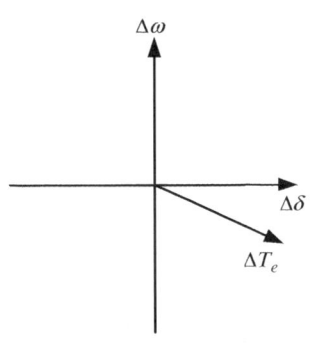

Fig. 3 Synthetic torque of additional electromagnetic torque and the original electromagnetic torque

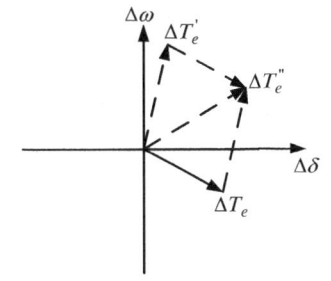

and $\Delta T'_e$ smaller than 90°, Therefore, the system eventually will have a positive damping torque, as shown in Fig. 3. In order to ensure to get the maximum damping effect, additional electromagnetic torque $\Delta T'_e$ should be provided and $\Delta T'_e$ should be in the same phase with $\Delta \omega$.

2 System Model

For islanding operation mode of Suizhong power plant, the SSO problem is most serious. So system model for SSO study is established under this operation mode, shown in Fig. 4. The dynamic response of the speed governor is ignored. DC system using a unipolar 12-pulse structure, the converter valves are triggered at equal intervals, the rectifier side of the DC system uses constant current control mode while the inverter side uses the constant extinction angle control mode [5]. A detailed block diagram of the constant current controller is shown in Fig. 5, the DC current deviation ΔI_d is adjusted by the Proportional Integral (PI) component to output the trigger angle signal α, in which parameters of PI control component are K = 1.0, T = 0.005 s, βmax = 3.054 rad (175°), βmin = 0.52 rad (30°).

Fig. 4 System simulation model

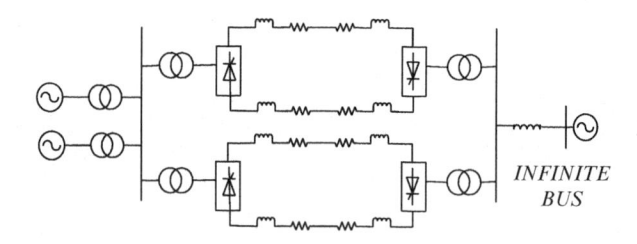

Fig. 5 Structure of constant current controller

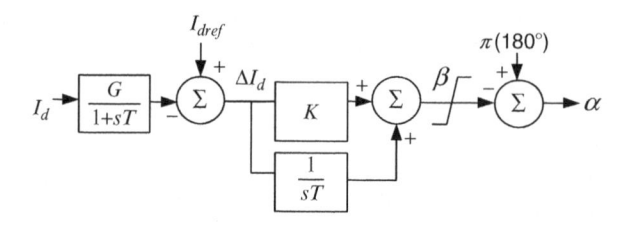

3 The Optimized Design of SSDC

Design methods of SSDC control element can be either the broadband design method or the narrowband design method. In the narrowband design method, narrow-band filters are designed respectively according to several torsional vibration characteristic frequencies of steam turbine with SSO risks, SSO inhibition effect is designed for different torsional vibration modal frequencies. This method requires knowing exact shaft characteristic frequencies of the generator sets, and the designed SSDC link can significantly improve the electrical damping in the specified shaft modal frequency. The disadvantage of this method is additional negative damping may be produced near the specified frequency for units. The broadband design approach is to design a band-pass filter in a frequency range and the band contains characteristic frequencies may which exist SSO risks of the unit, so the electrical damping can be improved in the whole band. Compared with the narrowband design method, additional damping provided by the broadband design method is relatively small, but its advantage is SSDC can play a role in a very wide frequency range [6].

The research system in this article has three characteristic frequencies in the subsynchronous frequency range (see Part IV of this article), and the characteristic frequency of mode 2 is closer to the characteristic frequency of mode 3. If the narrow-band method is adopt to design the SSDC, the controller design will be complex and it is hard to ensure that the designed controller can provide additional positive damping for units in the entire sub-synchronous frequency range, hence, broadband method is applied to design SSDC in this paper.

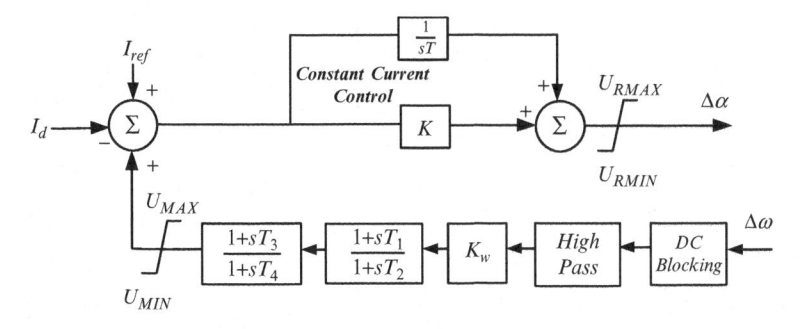

Design of Direct Current Subsynchronous Damping Controller (SSDC)

Fig. 6 SSDC structure diagram

3.1 Determine the Phase Compensation Parameters

SSDC can modulate both the current reference value and the firing angle to provide the positive electrical damping in the subsynchronous frequency range to achieve the purpose of inhibition of synchronous oscillation [7]. And they both can play a good inhibitory effect on the SSO in the case of rational design. In this design, the generator speed deviation is used as input signal of SSDC and SSDC's output signal is the current reference value input signal of the constant current controller in the rectifier side.

Figure 6 shows the structure diagram of SSDC. SSDC's basic principle is that providing positive damping for the turbine-generator units in the subsynchronous frequency range, so it should contains the DC blocking element, the high pass filter element, amplification element, phase compensation elements and limiting element.

The SSDC parameter to be determined most difficultly is the time constants of the phase shift aspects, the compensated phase of shift aspects is determined according to the phase of the current reference value input signal of the constant current controller lagging behind the electromagnetic torque which can be measured in the time domain simulation. This phase can be easily got by the test signal method, the detailed steps are as follows: First, based on the simulation system shown in Fig. 4, put a series of small-signal oscillating current ΔI_0 processed by the DC blocking link of SSDC and low-pass filter in the current reference areas of the DC constant current regulator (containing different frequency components, the frequency range is 7–40Hz), when the system run to steady state, measure the generator electromagnetic torque and get the corresponding output response ΔT_e; then, analyze ΔI_0 and ΔT_e in the public cycle with Fourier decomposition, the phase difference between the electromagnetic torque and the test signal under different frequencies can be calculated in turn, the calculated results are shown in Fig. 7.

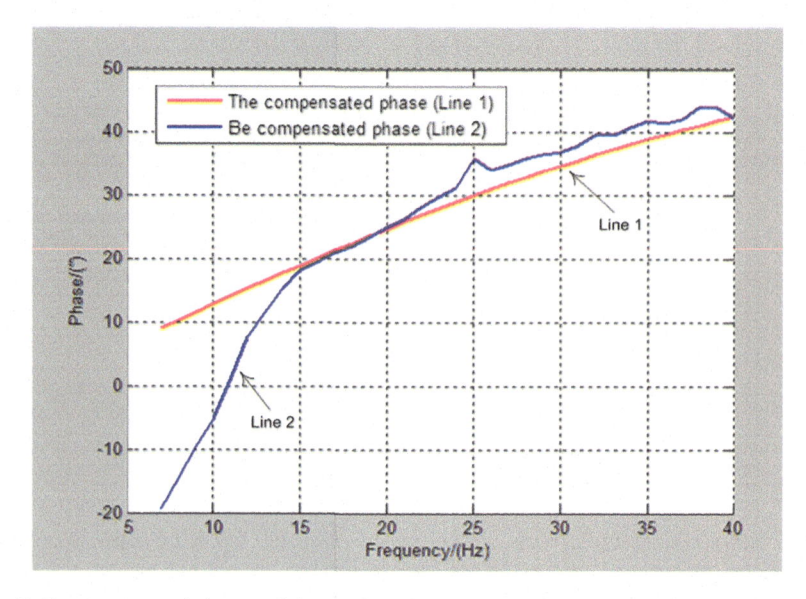

Fig. 7 Be compensated phase and the compensated phase of SSDC's phase shift links

Use two lead links to fit the curve of be compensated phase and then time constants can be calculated out as T1 = T3 = 0.002828, T2 = T4 = 0.001. The lead compensation phases in accordance with fitting function are shown in Fig. 7.

3.2 Determine the Phase Compensation Parameters

Shaft using a single rigid block model, test signal method is applied to simulate the study system, analyze the SSDC effect on SSO electric damping characteristics of the generator set. Curves of generating units' electrical damping coefficient De in the subsynchronous frequency range in different circumstances: without SSDC, SSDC amplification gain Kw = 1 and SSDC amplification Kw = 5 can be obtained respectively, as shown in Fig. 8.

We can see from Fig. 8, electrical damping within the entire synchronization band significantly increases when SSDC is applied, the band width corresponding to the system negative electrical damping is greatly reduced, which helps to avoid the SSO occurrence. The greater controller magnification is, the more beneficial to the SSO inhibition. But taking it into account that applying SSDC should not change the system stability, So the controller magnification should not be selected too large.

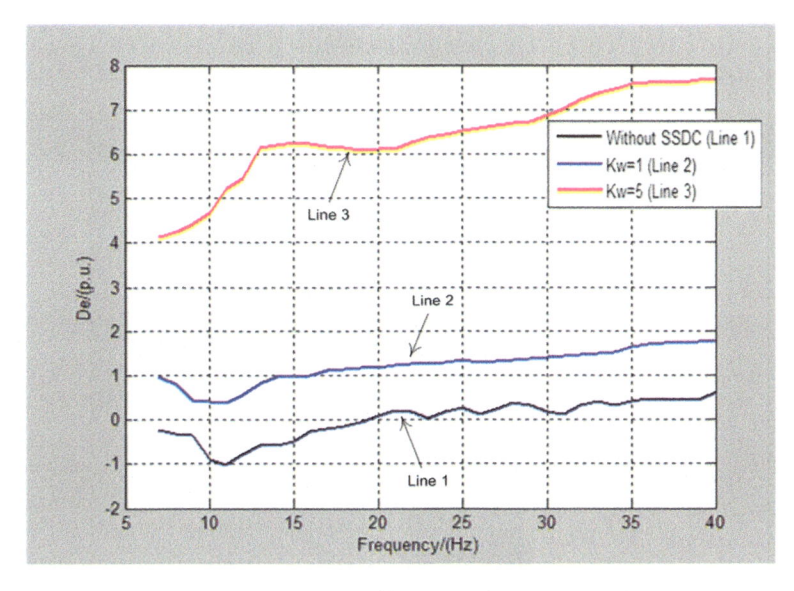

Fig. 8 The influence of magnification Kw on the electrical damping coefficient De

4 Simulation Results

Time-domain simulation of the power system shown in Fig. 4 is conducted with electromagnetic transient simulation software PSCAD/EMTDC. The Suizhong generator shaft using the multi-rigid-body model, the shaft consists of the high-pressure cylinder (HP), the medium-pressure cylinder (IP), the low-pressure cylinder A (LPA), the low-pressure cylinder B (LPB) and the generator (GEN5), totally five concentrated mass blocks. The shaft contains four torsional vibration modes, the torsional vibration frequencies are 13.39, 23.30, 26.74 and 53.89Hz. Observe the electrical damping characteristics curve without SSDC in Fig. 8, we can find that the shaft torsional vibration under the frequency of mode 1 is highly likely to occur when there is no SSDC applied.

The SSDC has a complete control structure and the amplification gain Kw is taken as 5. Observe the changes of modal generator speed differences after the system disturbance when there is no SSDC or not by time-domain simulation. Simulation results are shown in Figs. 9 and 10. A single phase to ground fault is applied in 4.3 s moment in the AC bus of the rectifier side, the fault duration is 0.1 s. Seen from Fig. 9, When the SSDC controller is not applied, the modal 1 generator speed difference is divergent, illustrates the SSO is unstable. Seen from Fig. 10, When the SSDC controller is applied, the modal 1 generator speed difference is decaying rapidly after fault, demonstrates that SSDC can damp the shaft torsional vibration effectively. This conclusion is consistent with the analysis results of the electrical damping characteristics and verifies the validity of SSDC.

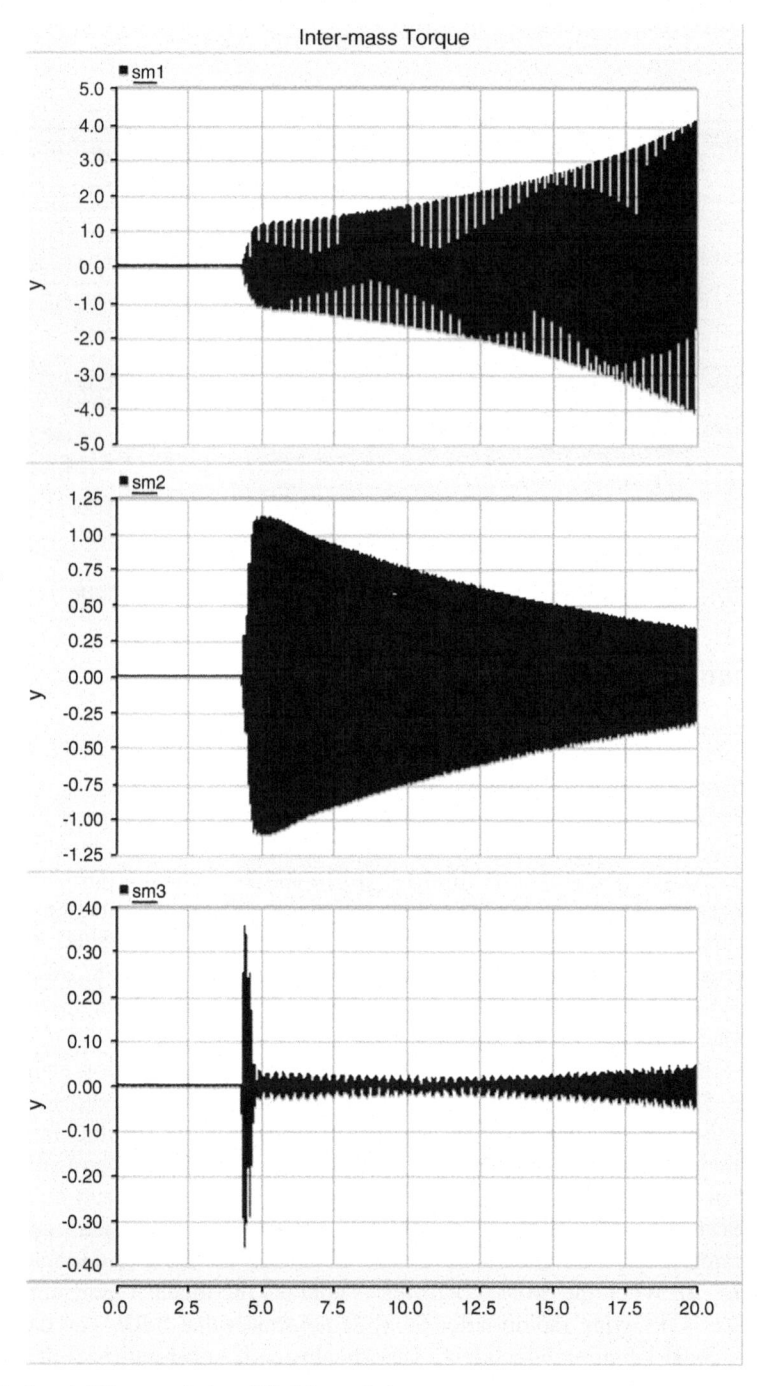

Fig. 9 Speed differences before SSDC is applied

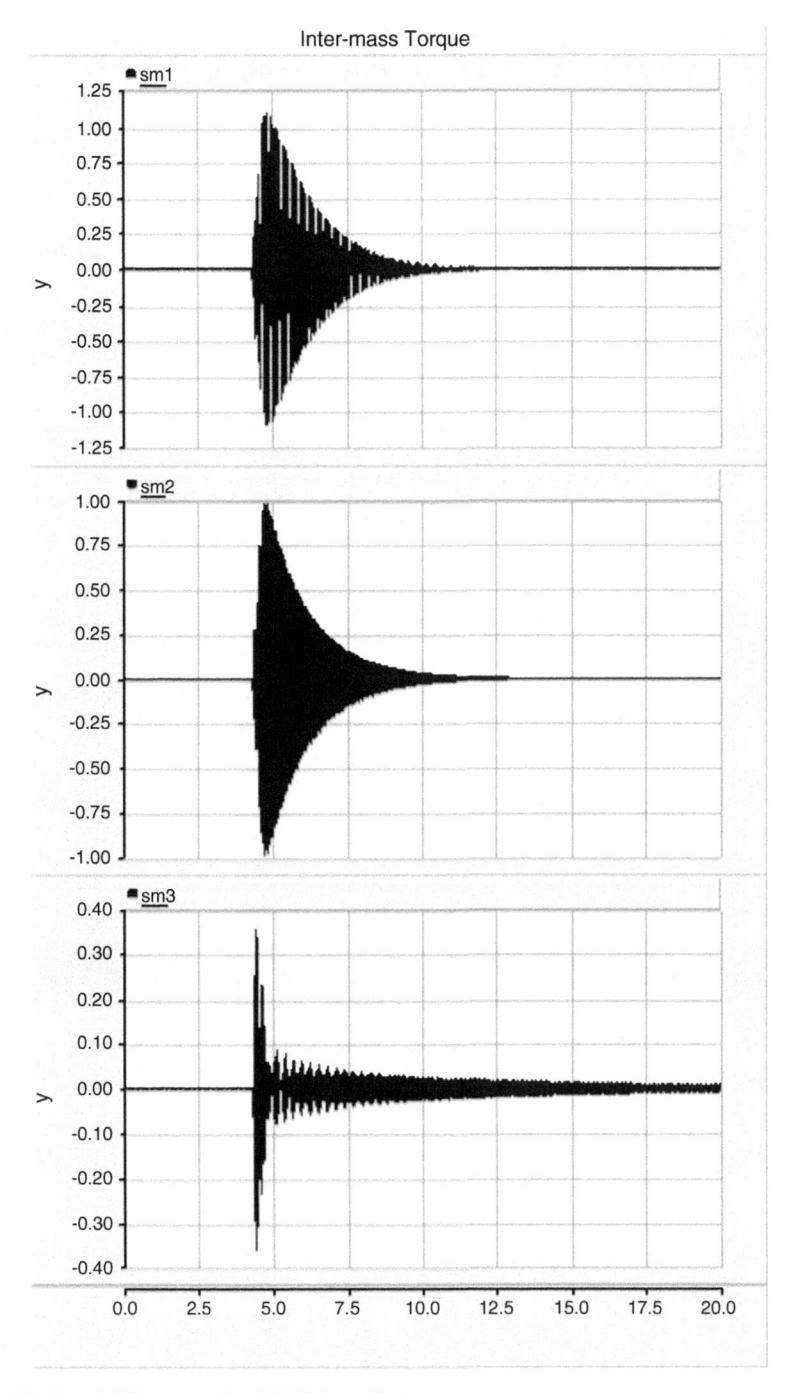

Design of Direct Current Subsynchronous Damping Controller (SSDC)

Fig. 10 Speed differences after SSDC is applied

5 Conclusion

This article designed SSDC is accordance with the broadband design method based on the phase compensation principle. The test signal method was used to analyze the electrical damping characteristics before and after the SSDC is applied. Analysis results demonstrated that the designed SSDC has the ability to significantly increase the electrical damping in the subsynchronous frequency range. Time-domain simulation results also verify the effectiveness of the designed SSDC.

Acknowledgements Thanks to the crew of the subsynchronous resonance (SSR) & subsynchronous oscillation (SSO) research team of North China Electric Power University.

References

1. Bahrnlan M, Larsen EV, Piwko RJ et al (1980) Experience with HVDC-turbine-generator torsional interaction at Square Butte [J]. IEEE Trans Power Appar Syst 99(3):966–975
2. Ni YX, Chen SS, Zhang BL (2002) Dynamic power system theory and analysis [M]. Tsinghua University Press, Beijing, pp 304–309 (In Chinese)
3. EPRI EL-2708 (1982) HVDC system control for damping subsynchronous oscillation. Project 1425-1 final report
4. Gao BF, Zhao CY, Xiao XN et al (2010) Design and Implementation of SSDC for HVDC. High Volt Eng 36(2):501–506 (In Chinese)
5. Xu Z (2004) AC-DC power system dynamic behavior analysis [M]. Mechanical Industry Press, Beijing, pp 34–59 (In Chinese)
6. Thomas R, Pertti J (2010) On feasibility of SSDC to improve the effect of HVDC on subsynchronous damping on several lower range torsional oscillation modes. In: Power and Energy Society General Meeting, Minneapolis, pp 1–8
7. Zhang F, Xu Z (2008) A method to design a subsynchronous damping controller for HVDC transmission system[J]. Power Syst Technol 32(11):13–17 (In Chinese)

Three Current-Mode Wien-Bridge Oscillators Using Single Modified Current Controlled Current Differencing Transconductance Amplifier

Yongan Li

Abstract In order to obtain new current-mode Wien-bridge oscillators using single modified current controlled current differencing transconductance amplifier (MCCCDTA) as the active element, according to the terminal relations of MCCCDTA as well as the basic structure of current-mode Wien-bridge oscillator, three new current-mode Wien-bridge oscillators are presented. The canonical number in proposed oscillators was used for component quantity and the condition and frequency of oscillation can be controlled electronically by means of adjusting bias currents of the MCCCDTA. The oscillators provide current output from high output impedance terminals. Finally, frequency error for one of the proposed oscillators is analyzed. The computer simulation results are given to verify the realizability of the derived circuits.

Keywords Wien-bridge oscillator • Error analysis • Current-mode circuit • MCCCDTA

1 Introduction

Several Wien-bridge oscillator structures made use of different design methods have been reported [1–3]. Although the oscillators based on voltage feedback operation amplifiers (VFAs), current feedback operation amplifiers (CFAs), and second generation current conveyors (CCIIs) possess simple structure, they lack the electronic adjustability. While the oscillators using operation tranconductance amplifiers (OTAs) have good electronic adjustability, they undergo complex structure and a number of active components. The quadrature and multiphase oscillators

Y. Li (✉)
Department of Electronics and Information Engineering,
Shaanxi Institute of International Trade & Commerce, Xianyang, China
e-mail: lya6189@sohu.com

employing current differencing transconductance amplifiers (CDTAs) have also been reported [4, 5]. They not only provide the independent electronic control between the oscillation frequency and the oscillation condition by tuning the bias current of the CDTA, but also possess high out impedances and low input impedances. However, because of the parasitic resistances at input ports, the circuits using CDTA must make use of some external passive resistors. This makes it not suitable for integrated circuits due to take up too much chip area.

Recently, a current controlled current differencing tranconductance amplifier, CCCDTA, has been popularized [6, 7]. As using an input bias current can control its parasitic resistances at two current input ports, the circuits using CCCDTAs are superior to ones using CDTAs. The universal biquad filters and sinusoidal oscillators [6, 7] have supported this viewpoint well.

In this paper, three Wien-bridge oscillators using MCCCDTA [7] are given. They use only one MCCCDTA, two capacitors, and one resistor or resistor less, and are easy to be integrated. The condition for oscillation and oscillation frequency can be varied through controlling bias currents of the MCCCDTA. The outputs of the circuit have high output impedances; the circuit has a low sensitivity. Finally, frequency error for one of the proposed oscillators is analyzed and the results of the circuit simulation are in agreement with theoretical anticipations.

2 Circuit Description

2.1 Three Wien-Bridge Oscillators Using MCCCDTA

According to the terminal relation of MCCCDTA as well as Fig. 2a in [7], three Wien-bridge oscillators using MCCCDTA is given in Fig. 1.

Figure 1a shows the class A oscillator using MCCCDTA. The oscillator consists of one grounded capacitor, one floating capacitors, and one current-controlled current amplifier with the gain $I_{B1}/8I_{B0}$. The parasitic resistance R_p is used as R_1 in the series and parallel RC networks, and the second-stage OTA in MCCCDTA is used as R_2, $R_2 = 1/g_m = 2V_T/I_{B2}$.

Figure 1b shows the class B oscillator using MCCCDTA. The oscillator consists of one grounded capacitor, one floating capacitors, one grounded resistor, and one current-controlled current amplifier. Z-Copy CCCDTA and second-stage OTA in the MCCCDTA are served as the current-controlled current amplifier with the gain I_{B2}/I_{B1}. The parasitic resistance R_p is served as R_1 in the series and parallel RC networks.

Figure 1c shows the class C oscillator using MCCCDTA. The oscillator uses one grounded resistor and two capacitors, one of the ground, the other floating. The parasitic resistance R_p is used as R_1 in the series and parallel RC networks, and second-stage OTA in the MCCCDTA is used as R_2 in ones. Z-Copy CCCDTA is used as current-controlled current amplifier with the gain $R_{Z1}I_{B1}/2V_T$.

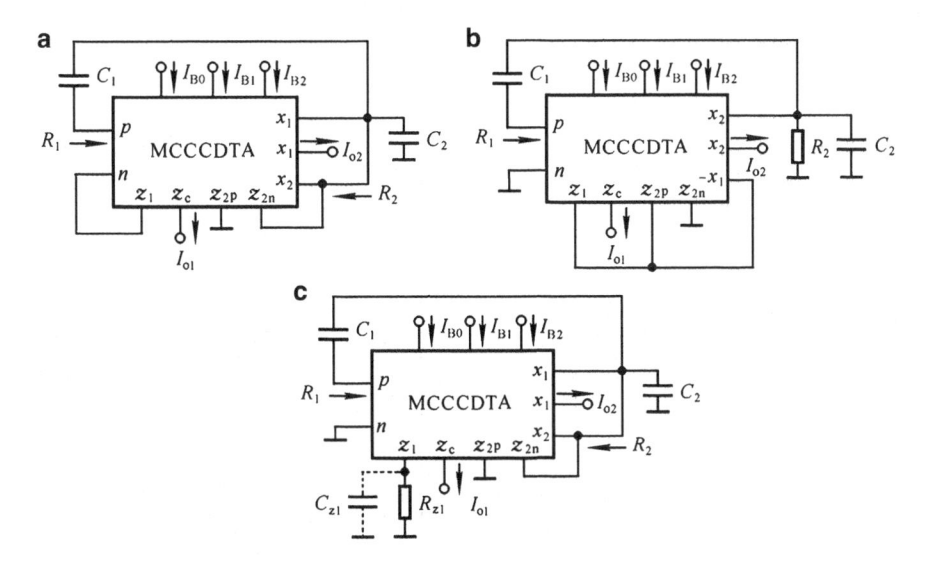

Fig. 1 (a) Class A oscillator, (b) class B oscillator, (c) class C oscillator

Table 1 Oscillation conditions, oscillation frequencies and current ratio of the proposed circuits

Number	Oscillation conditions	Oscillation frequencies	Current ratio	Remarks
A	$I_{B1} \geq 24I_{B0} = 6I_B$	$f_o = I_B/4\pi V_T C$	$I_{o1}/I_{o2} = 1/3$	$I_B = I_{B2} = 4I_{B0}$
B	$I_{B2} \geq 3I_{B1}$	$f_o = 1/2\pi C\sqrt{R_2 V_T/2I_{B0}}$	$I_{o1}/I_{o2} = 1/3$	
C	$I_{B1} \geq 6V_T/R_{z1}$	$f_o = I_B/4\pi V_T C$	$I_{o1}/I_{o2} = 1/3$	$I_B = I_{B2} = 4I_{B0}$

If $C_1 = C_2$, and $R_1 = R_2$, using Eqs. 5, 6, and 7 in [7], the conditions for oscillation, the frequencies of oscillation and the current ratio for three circuits can be given, respectively, as shown in Table 1.

From Table 1, for class A, if $I_B = I_{B2} = 4I_{B0}$, the frequency of oscillation, f_o, varies with bias current, I_B, whereas the oscillation condition separately varies with bias current, I_{B1}. It is clearly shown that both the condition for oscillation and f_o cannot be varied independently by controlling the bias currents.

For class B, f_o can vary with I_{B0} without affecting the condition of oscillation, which can also vary with I_{B1} or I_{B2} without affecting f_o. This means that the oscillation condition and f_o can be varied electronically and independently. However, the adjusting law for f_o is nonlinear.

For class C, if $I_B = I_{B2} = 4I_{B0}$, f_o can be turned electronically and linearly by adjusting I_B without affecting the condition for oscillation, which can also be tuned electronically and linearly by adjusting I_{B1} without affecting f_o. This means that the circuit provides the attractive feature in independent linear current control of f_o and the oscillation condition.

Fig. 2 Multi-transistor proportion current mirror

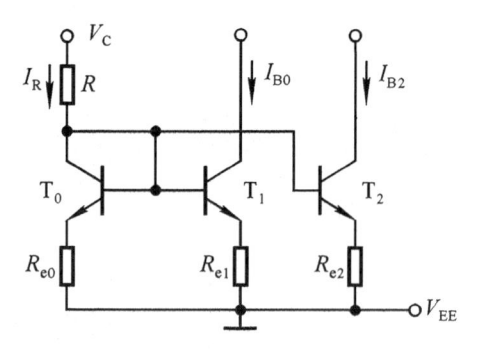

2.2 Adjustment Circuit for the Bias Currents

$4I_{B0}$ and I_{B2} can be varied together by a multi-transistor proportion current mirror, as shown in Fig. 2. Assuming that all transistors are matched and Early voltage is infinite, The relationship between each load current and the reference current can be written as follows:

$$I_R = \frac{V_C - V_{EE} - V_{BE}}{R + R_{e0}}, \tag{1}$$

$$I_{B0} = \frac{R_{e0}}{R_{e1}} I_R, \quad I_{B2} = \frac{R_{e0}}{R_{e2}} I_R. \tag{2}$$

Assuming $R_{e1} = 4R_{e2}$, from (2), we obtain $I_B = I_{B2} = 4I_{B0}$. It is natural that if the circuit parameters R_{e0}, R_{e1}, R_{e2}, V_{EE}, and V_{BE} are available, the load currents, I_{B2} and $4I_{B0}$, can be raised or lowered simultaneously by changing an external control voltage, V_C.

3 Non-ideal Analysis

Since the adjustment rule about the oscillation condition and frequency in Fig. 1c are non-interactive and independent, we only study non-ideal analysis for Fig. 1c. The parasitic capacitances emerging at terminal x_1 can be absorbed into the external capacitor, as they are shunt with the external one. Since the parasitic capacitors at terminal p and n can be neglected, we only take into account parasitic capacitance appearing at terminal z. It is clear that this parasitic capacitance, C_{z1}, is shunt with the external resistor. Again, analyzing the circuit in Fig. 1c produces the following the modified gain of the current-controlled current amplifier

$$A_i = \frac{g_{m1}R_z}{1 + sR_zC_{z1}} = \frac{GB}{s + GB/A_0}. \tag{3}$$

Here, A_0 is DC gain, $GB = g_{m1}/C_{z1}$, it represents the gain-bandwidth of the current-controlled current amplifier. Thus, the finite GB has changed the amplifier from the ideal current amplifier to a first-order low-pass filter. Considering (3), $C_1 = C_2$, and $R_1 = R_2$, we can get the modified characteristic equation:

$$s^3 + s^2 a_2 + s a_1 + a = 0. \tag{4}$$

Here, $a_2 = \dfrac{GB}{A_0} + \dfrac{3}{RC}$, $a_1 = \dfrac{GB}{RC}\left(\dfrac{3}{A_0} - 1\right) + \dfrac{1}{R^2 C^2}$, $a_0 = \dfrac{GB}{A_0} + \dfrac{1}{R^2 C^2}$.

In order to made the circuit to oscillate, the coefficients must satisfy $a_2 a_1 = a_0$. Hence,

$$\left(\dfrac{GB}{A_0} + \dfrac{3}{RC}\right)\left[\dfrac{GB}{RC}\left(\dfrac{3}{A_0} - 1\right) + \dfrac{1}{R^2 C^2}\right] = \dfrac{GB}{A_0 R^2 C^2}. \tag{5}$$

Let $A_0 = 3 + \Delta A$ and simplify:

$$(GBRC)^2 \Delta A + 3(3 + \Delta K)(GBRC)\Delta A - 3(3 + \Delta A)^2 = 0. \tag{6}$$

As long as $GBRC >> 9$, $|\Delta A| << 3$, the above equation reduces to

$$\Delta A = \dfrac{27}{(GBRC)^2} = 27\left(\dfrac{\omega_o}{GB}\right)^2. \tag{7}$$

Hence, the critical value of the gain must be set higher than for the ideal case to sustain oscillations. The higher the frequency of oscillation, the higher is the required gain for oscillation. Consequently, unless additional circuitry is used the oscillator will drop out of oscillation as the frequency is changed to a higher value. Using $\omega_{om} = (a_0/a_2)^{0.5}$, we can obtain the modified frequency of oscillation [8]:

$$\omega_{om} = \dfrac{\omega_o}{\sqrt{1 + 3A_0 \omega_o/GB}} \approx \omega_o(1 - 9\omega_o/2GB). \tag{8}$$

The deviation for oscillation frequency is

$$\dfrac{\Delta f}{f_o} \approx -\dfrac{9\omega_o}{2GB}. \tag{9}$$

Hence, if A_0 is adjusted to the critical value, the higher f_o, the higher becomes $|\Delta f_o/f_o|$.

4 Simulation Results

In order to confirm the performances of the proposed circuit, the circuit model for the MCCCDTA [7] is constructed on transistor QNL and QPL by ELECTRONICS WORKBENCH 5.0 software (EWB5.0), and then Fig. 1c is created. Finally, the circuit of Fig. 1c is simulated with parameters: $V_{CC} = 1.5$ V, $V_{EE} = -1.5$ V, $C_1 = C_2 = 1$ nF, $R_{z1} = 1$kΩ, $I_B = I_{B2} = 4I_{B0} = 250$ µA, and $I_{B1} = 156$ µA. Theoretically, the circuit will oscillate when $I_{B1} = 156$ µA, as mentioned in Table 1. However, in practical behavior, I_{B1} must be 160 µA. It is slightly higher than 156 µA to sustain oscillation, as mentioned in (5). The simulation result is shown in Fig. 3. It goes without saying that the circuit really realizes a sinusoidal oscillation. Using Table 1 can get the design value: $f_o = 765.556$ kHz and using the pointer in EWB5.0 can obtain the actual value: $f_{om} = 571.429$ kHz, so the deviation for f_o is $(571.429-765.556)/765.556 = -25.36$ %. The error results from the parasitic capacitance emerging at terminal z_1. Using the frequency analysis in EWB5.0, we receive the parasitic capacitance $C_{z1} = 0.03006$ nF. Substituting these data into (6) gives that the theoretical expected value for relative error is -21.09 %.

To illustrate the tuning characteristic of f_o by adjusting I_B, letting $C_1 = C_2 = 1$ nF, $R_{z1} = 1$kΩ $I_B = I_{B2} = 4I_{B0} = 125$ µA, and $I_{B1} = 158$ µA, we obtain the simulation result shown in Fig. 4. Similarly, the deviation for f_o is $(308.642-382.778)/382.778 = -19.37$ %, $C_{z1} = 0.04502$ nF, and the theoretical expected value for relative error is -16 %.

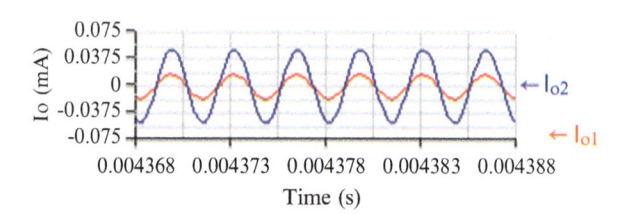

Fig. 3 Waveforms for the circuit of Fig. 1c with IB = IB2 = 4IB0 = 250 µA, C1 = C2 = 1 nF, and IB1 = 160 µA

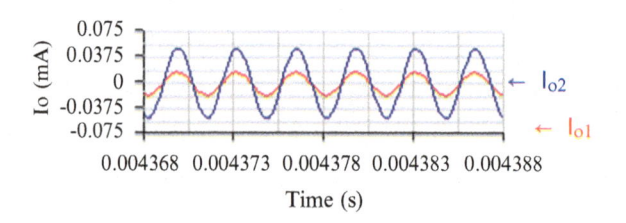

Fig. 4 Waveforms for the circuit of Fig. 1c with IB = IB2 = 4IB0 = 125 µA, C1 = C2 = 1 nF, and IB1 = 158 µA

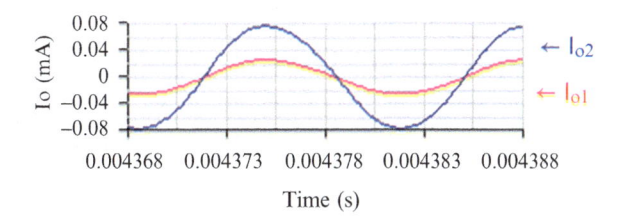

Fig. 5 Waveforms for the circuit of Fig. 1c with IB = IB2 = 4IB0 = 250 μA, C1 = C2 = 10 nF, and IB1 = 156 μA

It is seen that the deviation will be increased when the design value for f_o increases. By setting $C_1 = C_2 = 10$ nF, $R_{z1} = 1$ kΩ $I_B = I_{B2} = 4I_{B0} = 250$ μA, and $I_{B1} = 156$ μA, the simulation result is given in Fig. 5. In like manner, the deviation for f_o is $(76.3359-76.5556)/76.5556 = -0.29\%$, $C_{Z1} = 0.03$ nF, and the theoretical expected value for relative error is -2.1%. It is noted that the circuit simulation results are basically consistent with theory.

When $I_B = I_{B2} = 4I_{B0} = 250$ μA, $C_1 = C_2 = 10$ nF, and $I_{B1} = 156$ μA, the total harmonic distortion of the circuit THD ≈ 1.75 %. The oscillations growing exponentially in amplitude cause the distortion while the OTA and the current mirror in MCCCDTA come near to their saturation regions. The oscillator can maintain output signal with low distortion.

5 Conclusion

Based on the traditional Wien-bridge oscillator, three classes for current-mode Wien-bridge oscillator employing single MCCCDTA are provided, which furnish two current outputs from high output impedance terminals and enjoy low passive and active sensitivities and are canonic in component count. The oscillation condition and frequency of class C oscillator can be tuned linearly and independently by means of controlling bias currents of the MCCCDTA. The main disadvantage of the circuit is that f_o is dependent on temperature. In additional, there is a floating capacitor. Therefore, further research is needed.

References

1. Boutin N (1983) Two new single op-amp RC bridge T oscillator circuits. IEE Proc G Electron Circuits Syst 130(5):222–224
2. Sena R, Kumar BA (1989) Linearly tunable Wien bridge oscillator realised with operational transconductance amplifiers. Electron Lett 25(1):19–21
3. Soliman AM (2010) Generation of CCII and ICCII based Wien oscillators using nodal admittance matrix expansion. AEU-Int J Electron Commun 64(10):971–977
4. Jaikla W et al (2008) A simple current-mode quadrature oscillator using single CDTA. Radioengineering 17(4):33–40

5. Li YA (2010) Electronically tunable current-mode quadrature oscillator using single MCDTA. Radioengineering 19(4):667–671
6. Siripruchyanun M, Jaikla W (2008) CMOS current-controlled current differencing transconductance amplifier and applications to analog signal processing. AEU-Int J Electron Commun 62(4):277–287
7. Li YA (2012) A new single MCCCDTA based Wien-bridge oscillator with AGC. AEU-Int J Electron Commun 66(2):153–156
8. Budak A (1991) Passive and active network analysis and synthesis. Waveland Press, Boston, pp 461–462

Improved Phase-Locked Loop Based on the Load Peak Current Comparison Frequency Tracking Technology

Bingxin Qi, Hui Zhu, Yonglong Peng, and Yabin Li

Abstract Considering that the induction heating power supply controlled by pulse density modulation operated in the free resonance state, it was impossible to precisely detect a very small amount of load current, a conventional PLL (Phase-Locked Loop) circuit does not work properly, so a improved PLL frequency tracking control technology is presented in this paper. When the power stage decreased the discharge power into 1–10 % of the full power, the S/H (Sample and Hold) circuit and the PCD (Peak Current Detector) were used to guarantee the inverter working at the vicinity of the resonant frequency. The output signal from PCD opened and closed the S/H to drive the VCO (voltage-controlled oscillator), and a selected load peak current reference value was compared with the actual load peak current to control the PCD. Finally the control strategy is verified by using the Matlab / Simulink simulation results, the power stage can always work in the state of zero current and zero voltage switching and the phase-locked failure is avoid.

Keywords Induction heating power supply • Pulse density modulation • Frequency tracking control • Improved phase-locked loop

1 Introduction

Induction heating technology is developed in the direction of high-power and high frequency. Different heating process such as smelting, diathermy, annealing and hardening have special requirements for the power and frequency, and then leads to a variety of power conditioning and frequency tracking control method. According to the adjustment link, the power modulation method can be divided into rectifier regulation, DC side control and inverter power regulation [1, 2].

B. Qi (✉) • H. Zhu • Y. Peng • Y. Li
North China Electric Power University, Baoding, China
e-mail: qbingxin@163.com

The inverter side power regulation in line with different control ways can be divided into pulse frequency regulation [3], pulse phase shift modulation [4] and pulse density modulation. Power switching devices using the pulse frequency regulation and pulse phase shift modulation method [5, 6] will work incessantly in the non-zero current and non-zero voltage switching state, it will lead to the increase of switching losses and electromagnetic noise interference [7]. In addition, the pulse frequency regulation and pulse phase shift modulation will change the phase between output voltage and current during regulating the output power, thus affecting the performance of the frequency tracking controller or phase-locked loop controller [8]. The pulse density modulation proposed in literature [9] adopt a reasonable choice of pulse sequence to realize a greater adjustment range in the output power and ensure that the inverter always operating in a quasi- resonance state, thus achieving zero current and zero voltage switching. However, the restoration from the free resonance state to the output power state needs to re-lock the operating frequency of the inverter, which could easily lead to the system out of control.

To avoid phase lock failure after the end of the pulse density modulation period, a improved phase-locked loop control scheme based on the load peak current is proposed in this paper, when the induction heating power supply changes from the powering state to a free resonance state and again returned to the powering state, according to the sampling hold circuit, two different reference current values are analyzed and calculated, a current selection scheme corresponding to the minimum pulse density value is proposed, the simulation results verified the feasibility of the proposed scheme.

2 The Principle of Pulse Density Modulation

According to the main circuit structure shown in Fig. 1, the equation can be expressed as follows:

$$L\frac{di_o}{dt} + \frac{1}{C}\int i_o dt + ri_o = \frac{4U_d}{\pi}\sin \omega t \qquad (1)$$

Only considering the fundamental voltage component added to the load when the load circuit is at the resonance state. Where, $\omega = \omega r = 1/(LC)1/2$, assume that $2\omega rL/r = 2Q \gg 1$, Q represents the quality factor of the resonant circuit, so load current can be given:

$$i_o = \frac{4U_d}{\pi r}(1 - e^{-\frac{r}{2L}t})\sin \omega t \qquad (2)$$

Although the load resonant circuit is a second order system, however, Fig. 2 shows that the envelope line of the load resonant current is fist-order response, and

Fig. 1 The main circuit of series resonant inverter

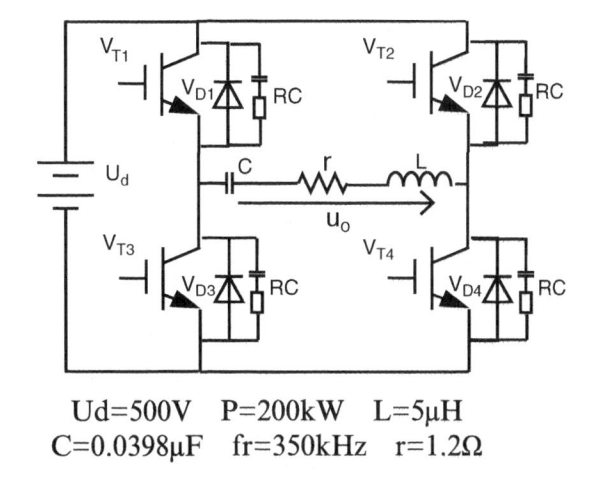

Ud=500V   P=200kW   L=5μH
C=0.0398μF   fr=350kHz   r=1.2Ω

Fig. 2 The load current modulated by pulse density

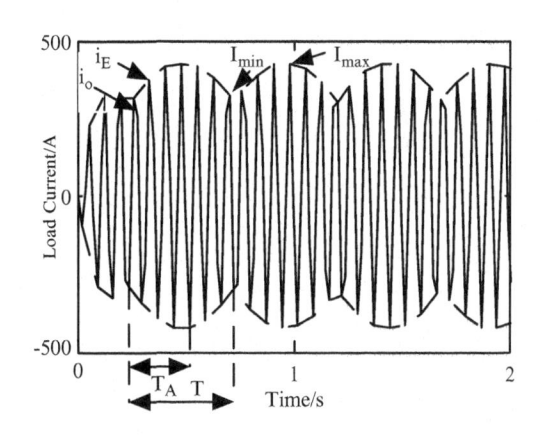

the time constant $\tau = 2L/r$, then the envelope line of the resonant current iE can be expressed as:

$$\begin{cases} i_E(t) = \hat{I}(1 - e^{-\frac{t}{\tau}}) + I_{min}e^{-\frac{t}{\tau}}(0 \leq t \leq T_A) \\ i_E(t) = I_{max}e^{-\frac{t-T_A}{\tau}}(T_A \leq t \leq T) \end{cases} \quad (3)$$

Where:

$$I_{min} = \hat{I}\frac{e^{\frac{T_A}{\tau}} - 1}{e^{\frac{T}{\tau}} - 1} \quad (4)$$

$$I_{max} = \hat{I}\frac{1 - e^{-\frac{T_A}{\tau}}}{1 - e^{-\frac{T}{\tau}}} \quad (5)$$

$\hat{I}$ is the current when the modulation ratio is $D = T_A/T = 1$.

3 Improved Phase-Locked Loop Control Based on the Load Peak Current Comparison

This paper uses the load current comparison control method, introduces the load peak current detection section and compares the load current with the preset reference current value, when the peak current is higher than the set reference value, the sample hold circuit works at the state of sampling, the frequency tracking circuit still operating in the normal phase-locked loop, however, when the peak current detection value is lower than the reference current, sample and hold circuit will no longer receive the output signal of the filter, but keep the signal in the last moment before the state transition to drive the VCO continue working, the control principle is shown in Fig. 3.

However, induction heating surface treatment sometimes needs to reduce the output power to 10 % of the rated power or even lower, namely, when the pulse density modulation ratio decreases, the output load current will be small, according to present resolution and accuracy, the peak current detector have been unable to accurately detect such a small current, thus the frequency tracking circuit will not function properly. So how to select the appropriate current reference value ensure that the frequency tracking control circuit to work properly must be solved.

Further analysis, consider the main circuit parameters shown in Fig. 1, the load quality factor $Q = 10$, the difference of the maximum peak current Imax and the minimum peak current Imin is shown in Fig. 4.

Noted from Fig. 4 that when $D > 0.1$, the difference between the maximum value Imax and the minimum value Imin increases as the modulation ratio significantly increases, when $D = 0.5$ the difference is the maximum, the performance of the phase-locked loop frequency tracking control circuit in the small fluctuations of the sampled signals is much better than the larger signal fluctuations, the smaller the signal fluctuations, the higher the resolution of the phase-locked loop, the more sensitive to changes in the signal phase, while for the large disturbance is not sensitive, which leads to the judgment delay, even in some cases beyond its phase capture range leads to lock failure.

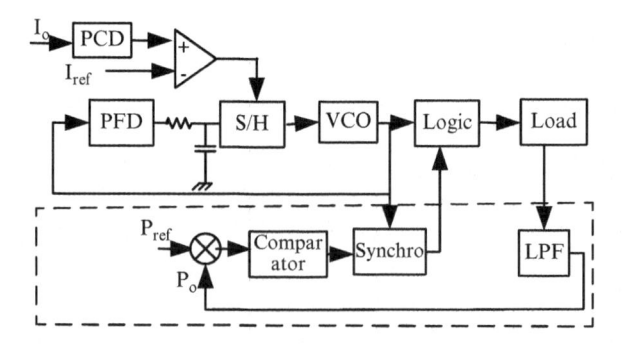

Fig. 3 The frequency tracking circuit based on load peak current detection

Fig. 4 Load current difference Q = 10

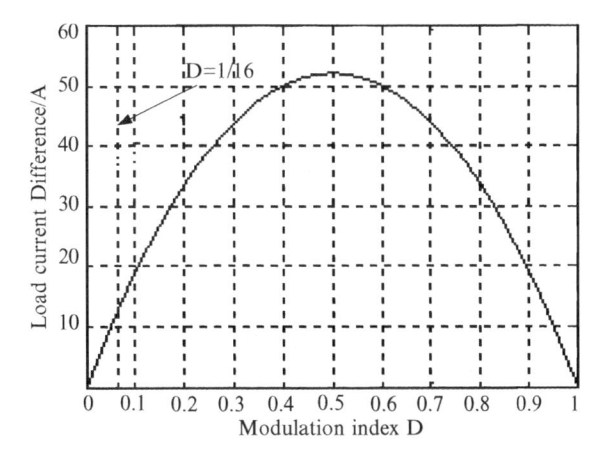

Fig. 5 Phase-locked angle as a function of minimum current

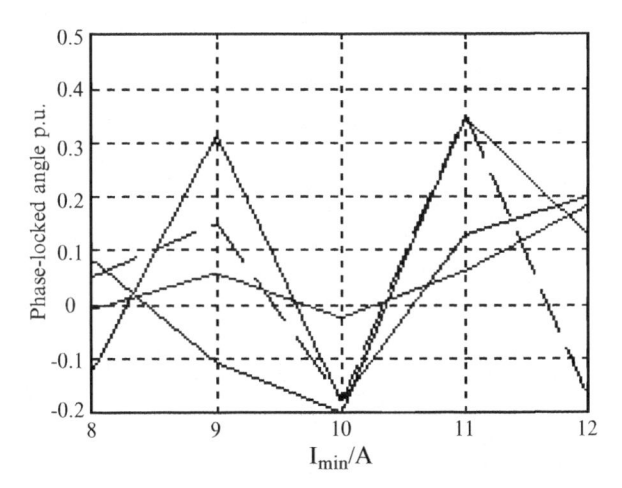

Based on the above analysis, D = 1/16 as the lower bound and 1 ‰ as the gain, infinitely approaching to D = 0.1, the function relationship of the phase-locked angle and the corresponding minimum current of the peak current Imin ranging from D = 1/16 to D = 0.1 is obtained, as shown in Fig. 5.

It can be reduced that when the phase-locked angle is a very small inductive angle ranging from 0 to 0.1, Imin can be equal to 9.5, 10.5, 11.5A. Then the optimal peak current reference value selection scheme will be obtained by simulation analysis.

4 Simulation Analysis

A single phase series resonant inverter model is built by utilizing the Matlab/Simulink tools, main circuit parameter is shown in Fig. 1, DC source voltage is $U_d = 500$ V, the resonant load parameters are $L = 5$ μH, $C = 0.0398$ μF, $r = 1.2\Omega$, the rated power is $P = 200$ kW, the resonant frequency is $f_r = 350$ kHz, RC snubber parameters are $R = 11.5\Omega$, $C = 2200$ pF, the main circuit lead inductance is $L_s = 0.01$ μH, the output power control using pulse density modulation.

Adopting improved phase-locked loop control based on the peak load current comparison, phase-locked switching process is analyzed as follows. Under the condition that the DC voltage $U_d = 500$ V is invariable, regulating the DC side current achieves the purpose of adjusting the output power, Fig. 6 shows that when DC current reduced from 90A to 40A the corresponding output load current and voltage waveform.

1. When the peak load current comparison control is not considered, the pulse density modulation ratio $D = 1:2$ corresponding to the condition that the output power decrease, the output voltage and current waveforms are shown in Fig. 7, the phase-lock tracking control circuit output waveform is shown in Fig. 8.

In Fig. 7, at the free resonance state, the load power flows through the RC snubber and anti-paralleled diode, then the phase angle selection must be accurate, and can predict the phase-locked position angle in the next pulse density modulation cycle, when the modulation ratio is 1:2 or even smaller, as shown in Fig. 8, the inductive angle is so large that a voltage spike is produced

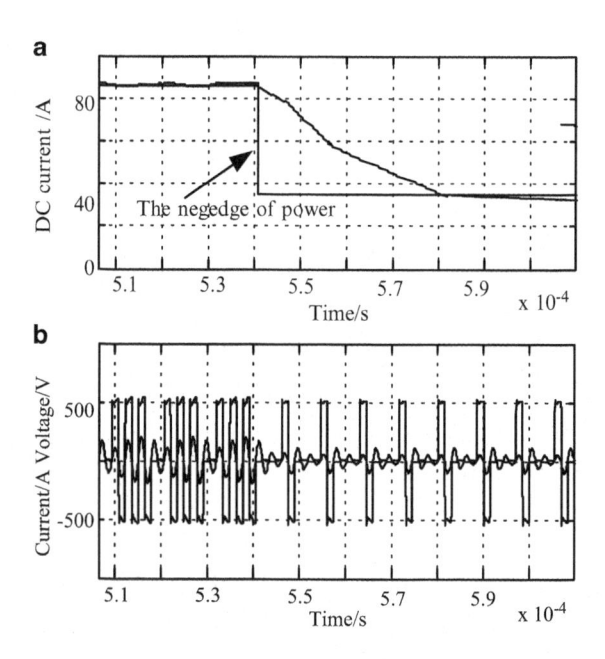

Fig. 6 The output voltage and current when power decrease

Fig. 7 The output voltage and current when D = 1:2

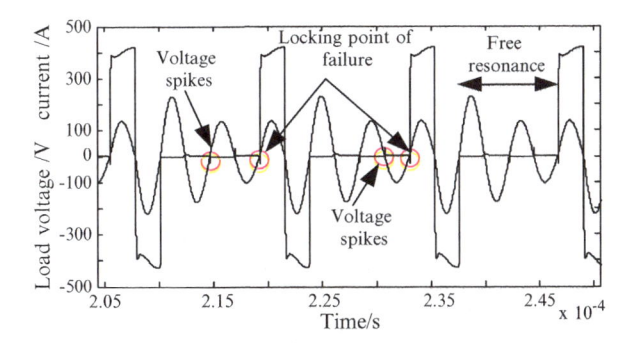

Fig. 8 The change of phase-locked angle when modulation index D decrease

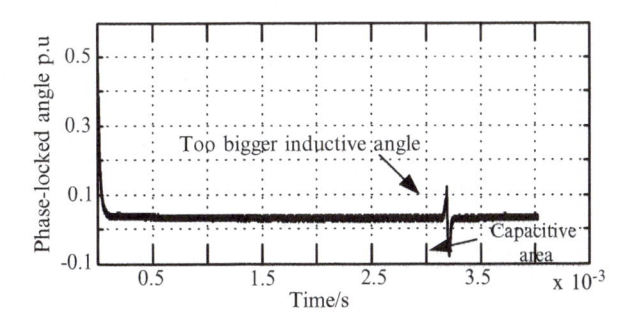

Fig. 9 Added to sample and hold device phased-locked angle

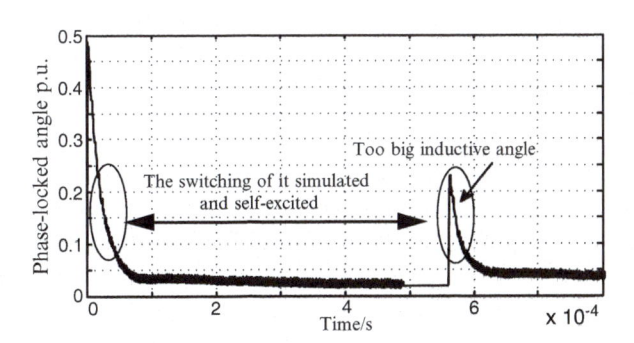

during the free resonance state, the switch loss increases, which leads to the power loss increasing, even at the beginning of the next modulation period, entering the capacitive area, causing the phase lock failure and then burning power electronic devices.

2. Considering the load current comparison control, the pulse density modulation ratio is D = 0.3 or even lower values, we can consider a limit state, when D = 1/16, the corresponding pulse density modulation minimum value is calculated by the formula (6) that Iref = 7.95A, the phase control angle is shown in Fig. 9.

Fig. 10 Phase-locked angle corresponding to selected three current value

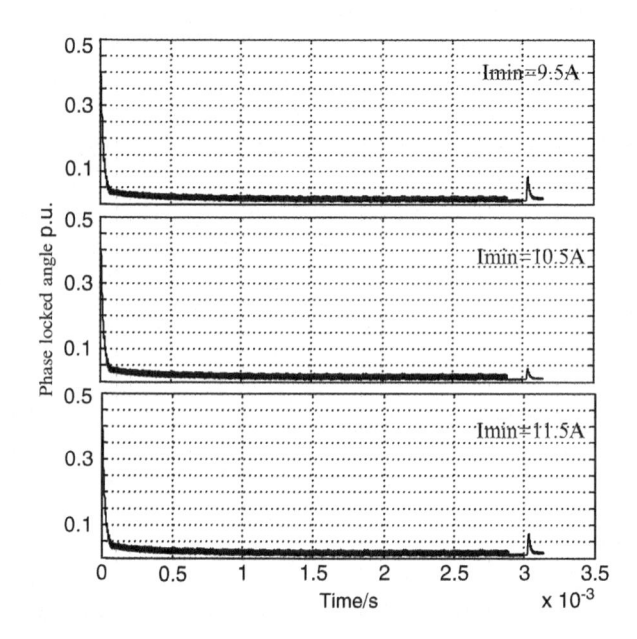

Figure 9 shows, set Iref = 7.95A as the peak current reference value, when the load current works at the free resonance state, the sampling hold circuit works at the hold state, keep the lock phase signal in the last moment before the state transition to drive the VCO to generate the control pulses, as the linear part shown in the graph; but when changes from the free resonance state back to the powering state, if switching phase-locked state at the minimum current reference value Iref = 7.95A, as shown in Fig. 9, a large inductive angle running at an initial state can be observed, repeating the separate-excitation turns to self-excitation process when the inverter startup, at that time the separate-excitation control circuit has stopped the work, which shows the frequency tracking control circuit is out of control, turns into the "free resonance" state, at this time any external electromagnetic disturbance or fault will cause this condition amplifying, switching devices loss increasing even being burned.

According to the above analysis, when the frequency tracking control circuit turns from the powering state to the free resonance state, set a small load current as the reference value of the peak current comparison to ensure reliable work, a load current value is also calculated as the peak current reference value when changes free resonance state back to the powering state to further improve the phase-locked loop load resonant frequency tracking speed. According to the analysis of 3.2 and phase angle and the minimum current Imin distribution relationship in Fig. 5, select Iref = 9.5, 10.5, 11.5A respectively, the corresponding phase angle experimental waveforms are shown in Fig. 10. The Imin = 10.5A is the most appropriate peak current reference value when changing from the free resonance state back into the powering state.

5 Conclusion

In order to meet the requirement of switching frequency constant when the output power drops below 10 % of the rated power in the heating process, an improved phase-locked loop control method based on the load peak current comparison is proposed in this paper. Adding the sampling hold device to the existing phase-locked loop circuit without changing the capture range and accuracy condition, furtherly improving the tracking speed of the phase-locked loop on the load resonant frequency even the load current is quite small, which ensures the inverter switching devices continue working at zero voltage and zero current switching state when switching between the free resonance state and the powering state. Compared with the traditional phase-locked loop control, the proposed control method can further reduce the power loss of power switching devices and improve the operating efficiency of equipment.

References

1. Fujita H, Akagi H (1996) Pulse density modulated power control of a 4 kW, 450 kHz voltage-source inverter for induction melting applications. IEEE Trans Ind Appl 32(2):279–286
2. Hong Lv, Yushui Huang, Zhongchao Zhang (2003) The PWM-PFM control method of induction heating power supply. Power Electron Technol 37(1):8–11
3. Hong Mao, Zhaolin Wu, Zhencheng Hou (1998) Induction heating power supply zero phase difference frequency tracking control circuit. Power Electron Technol 32(2):69–72 (In Chinese)
4. Jun Yang (2004) Research on the IGBT super audio frequency induction heating power supply phase tracking. Tsinghua University, Beijing
5. Liqiao Wang, Yushui Huang, Changyong Wang, Zhongchao Zhang (2001) Research on induction heating power supply based on a novel phase-shifted control. Power Electron Technol 35(1):3–4 (In Chinese)
6. Tianming Pan (1996) Modern induction heating device. Metallurgical Industry Press, Beijing, pp 125–196
7. Tianming Pan (1983) Power frequency and medium frequency induction furnace. Metallurgical Industry Press, Beijing, pp 68–89
8. Yabin Li, Yonglong Peng, Heming Li (2006) The optimal ZVS control of series resonant inverter. Power Electron Technol 40(3):14–16 (In Chinese)
9. Zonggang Qi, Peng Liu, Huiming Chen (2003) Research on power modulation for induction heating[J]. Heat treatment metals (China) 2003(7):54–57 (In Chinese)

Optimization of Low-Thrust Orbit Transfer

Zhaohua Qin, Min Xu, and Xiaomin An

Abstract A Gauss pseudospectral method is used to optimize the interplanetary low-thrust orbit transfer. The Gauss pseudospectral method is utilized to parameterize the orbit transfer. In order to solve the large-scale parameters multi-constraints optimization problem, the sequential quadratic programming (SQP) is used. Then, the optimization method is demonstrated on an Earth-Mars low-thrust orbit transfer application by the solar electric propulsion. The numerical results show that the method is effective and rapid in finding the optimal orbit transfer and does not require particular initial guesses.

Keywords Orbit transfer • Low-thrust • Fast optimization • Gauss pseudospectral

1 Introduction

Historically, the methods to optimize interplanetary low-thrust orbit transfer can be divided into two categories: one is called indirect methods and the other is direct methods [1, 2]. The two-point boundary-value problem can be solved by indirect methods. However, direct methods turn optimal control problem into nonlinear programming problem (NLP) [3–5].

In this paper, we research a method for optimal orbit transfer called the Gauss pseudospectral method. The complicated orbit transfer optimization problem is translated into a large-scale parameters optimization problem with multi-constraints. Thus the NLP can be solved numerically by the SQP.

Z. Qin (✉) • M. Xu • X. An
College of Astronautics Northwestern Polytechnical University, Xian, China
e-mail: xinixn812@163.com

2 The Optimization of Low-Thrust Orbit Transfer

The motion of the low-thrust spacecraft around the sun is described by the Cartesian coordinates:

$$\begin{cases} \dot{r} = v \\ \dot{v} = -\mu r/r^3 + \frac{T}{m}u \\ \dot{m} = -\frac{T}{I_{sp}g_0} \end{cases} \quad (1)$$

Where $m = m_0 - |\dot{m}|t$. m is the spacecraft mass, m_0 is its initial mass, $|\dot{m}|$ is the engine fuel consumption of seconds, T is the engine thrust equal to $2\eta P_0/(I_{sp}g_0r^2)$, t is the flight time. P_0 is its input power in 1 AU, I_{sp} is its specific impulse, and g_0 is the gravitational acceleration. $u = [u_x, u_y, u_z]$ is defined as the unit vector of the thrust.

With the engine continuously working, the optimal-fuel performance index is evaluated at:

$$J_1 = -(m_0 - \int_{t_0}^{t_1} \dot{m}dt) = -m_1 \quad (2)$$

The optimal-time performance index is evaluated at:

$$J_2 = \int_{t_0}^{t_1} dt = t_1 - t_0 \quad (3)$$

In order to make full use of the advantages of the low-thrust propulsion system, we presume the launch of the spacecraft energy be zero, $C_3 = 0$, the trajectory musts satisfy the initial and final boundary conditions:

$$\begin{cases} E_1[x(t_0), t_0] = x(t_0) - x_1(t_0) = \mathbf{0} \\ E_2[x(t_1), t_1] = x(t_1) - x_2(t_1) = \mathbf{0} \end{cases} \quad (4)$$

During the transfer, the thrust vector direction must satisfy the following conditions:

$$C[u(t), t] = u_x^2 + u_y^2 + u_z^2 = 1 \quad (5)$$

The optimal parameter is defined as:

$$Z = [t_0, t_1, x(t), u(t)]^T \quad (6)$$

With the description of the problem, the parameter of the optimal problem concludes the discrete-time variable, the continuous trajectory and control variables. But the continuous variable cannot be recalibrated directly, and it must be processed and then the adjustable discrete variables can be obtained.

It is noted that it can transform the problem from the time interval $\kappa \in [-1, 1]$ to the time interval $t \in [t_0, t_1]$:

$$\kappa = -1 + \frac{2(t - t_0)}{t_1 - t} \tag{7}$$

Orthogonal collocation of the dynamics is performed at the Legendre-Gauss (LG) points in the Gauss pseudospectral method. The state can be approximated by Lagrange interpolating polynomials L,

$$x(\kappa) \approx X(\kappa) = \sum_{i=0}^{N} X(\kappa_i) L_i(\kappa) \tag{8}$$

Where $L_i(\kappa)$ are defined as

$$L_i(\kappa) = \prod_{j=0, j \neq i}^{N} \frac{\kappa - \kappa_j}{\kappa_i - \kappa_j}, i = 0, \ldots, N \tag{9}$$

Additionally, the control can be approximated by Lagrange interpolating polynomials L^* as

$$u(\kappa) \approx U(\kappa) = \sum_{i=1}^{N} U(\kappa_i) L_i^*(\kappa) \tag{10}$$

Where

$$L_i^*(\kappa) = \prod_{j=1, j \neq i}^{N} \frac{\kappa - \kappa_j}{\kappa_i - \kappa_j}, i = 1, \ldots, N \tag{11}$$

Note that the approximation for the control is not the only allowable control approximation, but it has produced very good results when compared with other control approximations.

Base on the differential expression of Eq. 8, we can obtain

$$\dot{x}(\kappa) \approx \dot{X}(\kappa) = \sum_{i=1}^{N} X(\kappa_i) \dot{L}_i(\kappa) \tag{12}$$

Each Lagrange polynomial derivative in the LG points can be expressed as a differential approximation matrix, P, whose elements are determined offline as follows:

$$P_{ki} = \dot{L}_i(\kappa_k) = \sum_{l=0}^{N} \frac{\prod_{j=0, j \neq i, l}^{N} (\kappa - \kappa_j)}{\prod_{j=0, j \neq i}^{N} (\kappa_i - \kappa_j)} \quad (13)$$

Then the dynamic constraint is as follows:

$$F(X_k, U_k, \kappa_k; t_0, t_1) = \sum_{i=0}^{N} D_{ki} X_i - \frac{t_1 - t_0}{2} f(X_k, U_k, \kappa_k) = 0 \quad (14)$$

Where $X_k \equiv X(\kappa_k), U_k \equiv U(\kappa_k)$. Then we can define additional variables in the discretization as follows:

$$\begin{cases} X_0 \equiv X(-1) \\ X_1 \equiv X_0 + \frac{t_1 - t_0}{2} \sum_{k=1}^{N} w_k f(X_k, U_k, \kappa_k; t_0, t_1) \end{cases} \quad (15)$$

Where X_1 is defined in terms of X_k and U_k, f is the variable of the Eq. 1, and w_k are the Gauss weights, $w_k = \int_{-1}^{1} \prod_{i=1, i \neq k}^{N} \frac{\kappa - \kappa_i}{\kappa_k - \kappa_i} d\kappa, k = 1, \ldots, N$.

Finally, it is noted that discontinuities in the state or control can be effectively dealt with by dividing the trajectory into phases, where the dynamics are transcribed at each phase and connected together with the additional phase interface constraints. This procedure has been applied to many fields.

Thus, the interplanetary low-thrust optimization problem can be converted to parameters of the constraint optimization problem. The optimal parameter concludes the launch time, the fair time and the variable of the state and control to each node.

$$Z_1 = [X_0^T, X_1^T, \ldots, X_f^T, U_1^T, U_2^T, \ldots, U_N^T, t_0, t_1]^T \quad (16)$$

The problem is transcribed to the NLP, and it is solved by using the solver of the SQP.

3 Results

An Earth-Mars direct transfer is presented. Its parameters: the solar array power at 1 AU P_0 is 6.5 kW, the constant I_{sp} is 3,100 s, the engine efficiency is 0.65, and the initial spacecraft mass is 1,200 kg. And it uses the astrosphere calendar, DE405, and the Matlab programming environment on a 3.06 GHz Pentium 4 PC.

The launch time window is from January 1, 2014 to December 31, 2014, and the maximum duration is 500 days. The optimal variable and the restriction equation solutions to N = 29 are 291 and 240. The NLP has been used to solve the TOMLABT version of the NLP solver SNOPT [6].

The optimal result is shown in Figs. 1, 2, 3, and 4.

The optimization solution concludes the launch time is April 10, 2014, the flight time is 306.6370 days, the remaining mass is 912.3415 kg, the iterative process is 52 steps and the computing time is 58.467 s.

The optimization result is good, which shows that the method can get the optimal transfer trajectory rapidly and accurately.

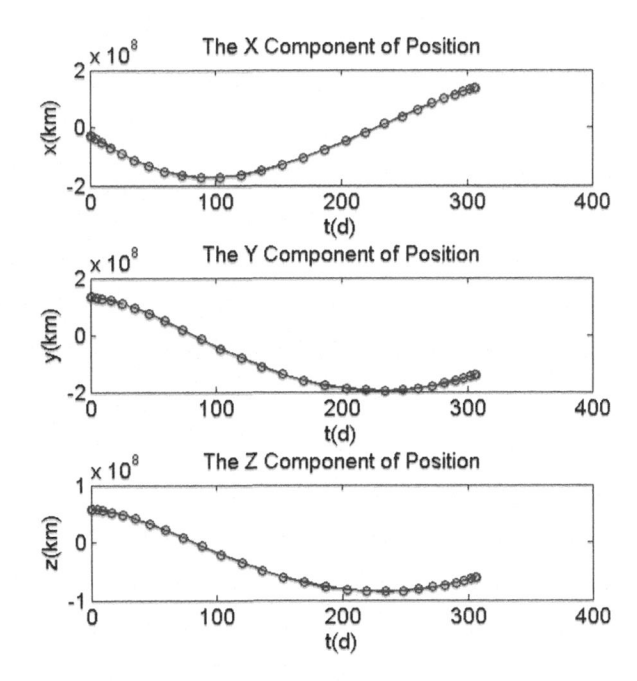

Fig. 1 Optimal position vector

Fig. 2 Optimal velocity vector

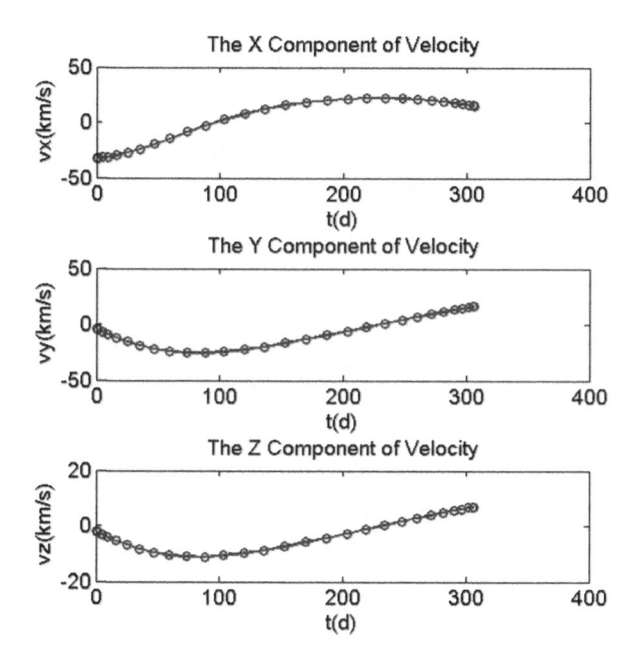

Fig. 3 Optimal control steering

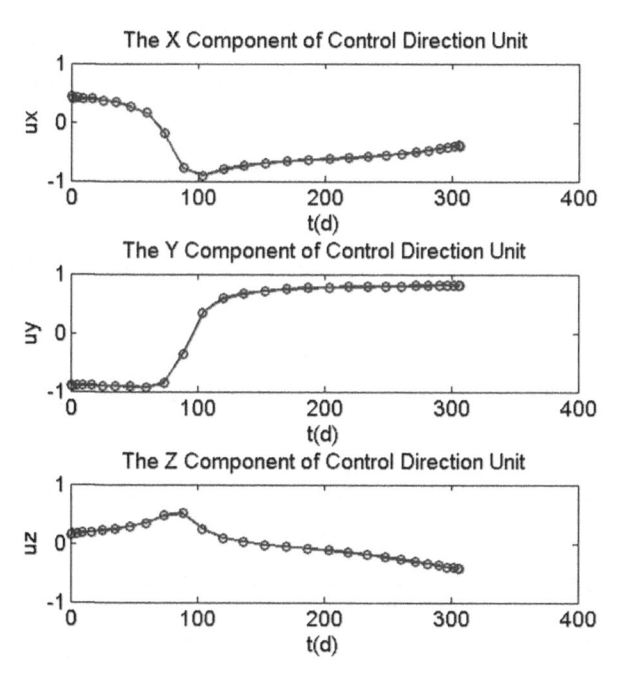

Fig. 4 Optimal Earth-Mars direct transfers

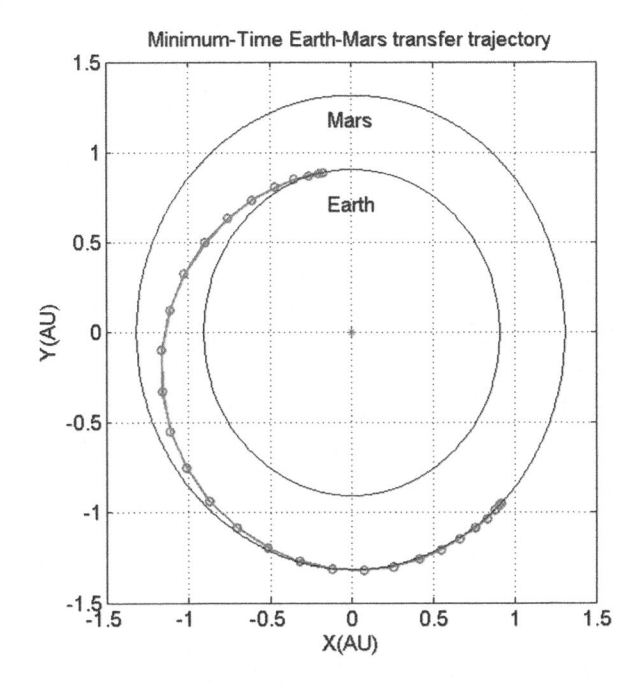

4 Conclusion

This paper studies the rapid optimization of orbit transfer which bases on Gauss pseudospectral method. In the optimal transfer, the method turns the transfer optimal control problem into the NLP, and avoids the huge compute capacity, the real-time identification, the sensitivity to guess value and so on. The results obtained in this paper show that the optimization method adopted for the optimization of the interplanetary low-thrust orbit transfer demands less accurate initial guesses and has good search capability with high accuracy.

References

1. Gao Yang (2003) Advances in low-thrust trajectory optimization and flight mechanics. University of Missouri-Columbia, Columbia
2. Enright PJ, Conway BA (1991) Optimal finite-thrust spacecraft trajectories using collocation and nonlinear programming. J Guid 14(5):981–985
3. Ren Yuan, Cui Pingyuan, Luan Enjie (2007) An earth-mars low-thrust trajectory design based on hybrid method. J Harbin Inst Technol 39(3):359–362
4. Kluever CA, Oleson SR (1998) Direct approach for computing near-optimal low-thrust earth-orbit transfers. J Spacecraft Rockets 35(4):509–515
5. Benson D (2005) A Gauss pseudospectral transcription for optimal control. Massachusetts Institute of Technology, Boston
6. Gill PE, Murray W, Saunders MA (2002) SNOPT: an SQP algorithm for large scale constrained optimization. SIAM J Optim 12(4):979–1006

Self-Healing Control Method Based on Hybrid Control Theory

Qiang Zhao, Meng Zhou, and XinQian Wu

Abstract Due to the overall optimum ability, the hybrid power control technology is particularly useful for self-healing control for the grid. Based on the study of hybrid control theory, this article introduced the WAMS-based (wide area measurement system) real-time information platform to the physical layer of hybrid power control model and applied the modified model into self-healing control. Meanwhile, based on the modified model, the article established an easily-attainable self-healing control model, and produced solutions to some key problems. Since the real-time information alternative to the traditional observer, the self-healing control system based on the modified hybrid power control technology is suitable for the self-healing of large power grid because it is simple, overall optimum and easy to extend.

Keywords Power control • Self-healing control • WAMS • Event-driven

1 Introduction

In recent years, with the constant increase of the demand of grid stability and power supply quality with the mankind, the research on smart grid is intensifying constantly. Although the definition of smart grid remains controversial, however, overall, the smart grid should have the following characteristics [1, 2]: self-healing, interactive, optimization, integration, compatible. Self-healing is not only an important safeguard to ensure automated stable operation of the grid, but also an important symbol of the smart grid. So how to achieve self-healing of the grid has become the focus of the smart grid.

Q. Zhao • M. Zhou (✉) • X. Wu
School of Control and Computer Engineering, North China Electric Power University, Beijing, China
e-mail: zhaoqiang@ncepu.edu.cn; Shenghuo1988@hotmail.com; happyskywxq@qq.com

The existing study of self-healing often emphasis on the reality grid with corresponding characteristics, and then look for a self-healing control method solution which is suitable to this grid, such as reference [3]. These control methods can be implemented on the local grid self-healing controlling, however, they cannot be applied to the overall grid for lacking of compatibility and expandability. Self-healing grid control covers a lot of fields of new technology and new progress, such as automatic control, relay protection, computers and software, Applied Mathematics. It is an integrated software and intelligent device control technology.

As a system composite in parallel or serial of continuous component and discrete component, the hybrid system [4] is well suited for the control of multi-objective large modern power system. At present, hybrid control theory has been successfully used in the construction of the grid voltage control system, and received a good effect.

This paper analyzed the basic conditions of self-healing, modified the theory of hybrid power control, and introduced the concept of hybrid control to grid self-healing control. Based on the modified theory, grid self-healing system model was established.

2 Hybrid Power Control Theory and Its Improvement

The hybrid system [4] contains both discrete behavior (behavior with state and output) and continuous behavior. So how do describe discrete and continuous dynamic characteristics and mutual relations correct and unified has become a research focus on the modeling of hybrid systems. Hybrid power control theory [5], which made use of the concept of the event generator, compare the value of continuous variables in the grid or the value of these variables as independent variables "event function" with the prior defined criteria. Whether the incident consistent or inconsistent, an event will occur.

Based on the WAMS grid real-time information platform, this paper introduces hybrid power system model underlying as an alternative to traditional observer, so that the model of the controlled layer has the capabilities recording network real-time information, and then pass the data have been processed by the real-time data transmission channel to level of decision-making and decision makers only judge event and issue a directive. Improved hybrid power control model has a more cleared structure, while the introduction of WAMS can enhance the real-time of closed-loop control and the entire process of the controlled layer.

In Fig. 1, $u(t)$ represent for the bottom of the power components of the original continuity-based closed-loop control; $x(t)$ represent for the continuous change of the dynamic power system variables; $y(t)$ is the underlying power system state output variables in continuous time; Γ is the formation of discrete events; A is the event used for display; $u_D(t)$ represents for collection operational control inspired by the events.

Fig. 1 Modified hybrid power system model

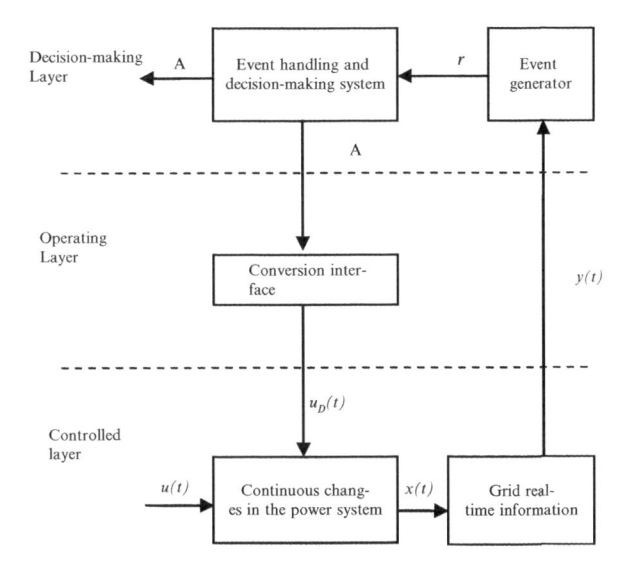

Decision-making Layer — Event handling and decision-making system ← r ← Event generator; A ← (output to Decision-making Layer)

Operating Layer — Conversion interface; $y(t)$

Controlled layer — $u_D(t)$ → ; $u(t)$ → Continuous changes in the power system → $x(t)$ → Grid real-time information

In summary, the hybrid power control system is a collection of continuous process of regulation and discrete scheduling decision-making process, and can achieve the power system multi-goal drive superior purpose, it suited for self-healing control of power systems well, while improved hybrid power control model allows control has better real-time visibility.

3 The System Model for a Self-Healing Grid Based on Hybrid Power Control Theory

This paper adopts this definition for self-healing: to the slight disturbance of the system from the outside and inside, the system should have a good preventive function, timely detection, diagnosis, remove the disturbance; to the fault which cannot be avoided, system should be maintained sustained and stable operation of the power system. Its purpose is on the foundation of less cutting load even not cutting load, to cut off the fault circuit, analysis the fault, alarm the fault, and manually repair the fault.

Based on the above definition of self-healing systems, combined with a hybrid power system control theory, this paper presents a grid self-healing system model, as shown below (Fig. 2).

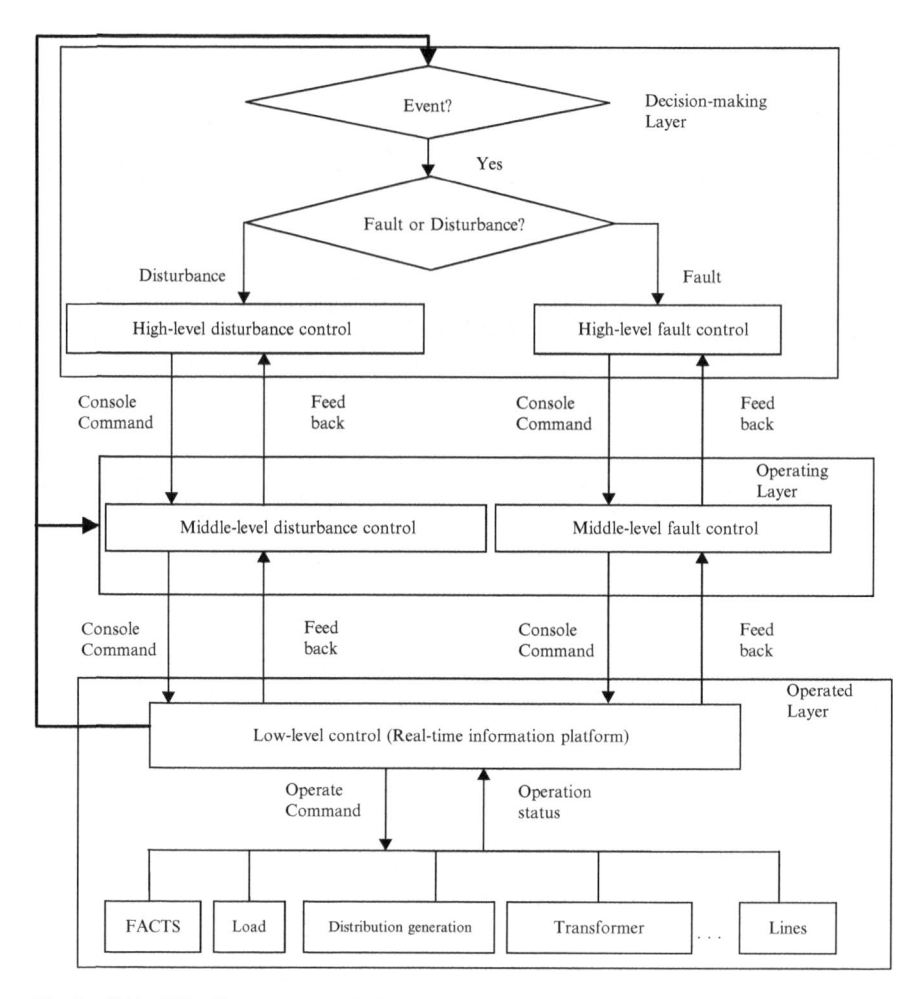

Fig. 2 Grid self-healing system model

3.1 The Task and Design of Decision-Making Layer

First of all, the decision-making layer of the hybrid self-healing control system receive real-time data from the information platform on the bottom of the power system; then accord the definition of the event to judge the formation of the event. The decision-making layer accord node voltage, current, active power, reactive power and other power system real-time parameters, to develop the grid status eligibility standard, and take the failed state of the network parameters as the event; when judge to have the event, the decision-making layer accord the real-time data and the feedback of the operation layer, to determine whether or not this event is disturbance or fault. Finally, the decision-making layer issues a control instruction to the operating layer, to mobilize disturbance middle-layer control or fault middle-layer control. At the same time, the decision-making layer under take the task of the event display, failure notification and other human-computer interaction function.

3.2 The Task and Design of Operating Layer

The operating layer of the hybrid self-healing control system must establish the real-time database, the perturbation model, and the fault control module. First of all, the operation layer of the hybrid self-healing control system receive high-level control commands from the decision-making layer to start the disturbance middle-level control or fault middle-level control, then based on the data transfer from the underlying power system information platform, detail analysis the type of disturbance or fault to give the optimal control program which can correct the disturbance or cut off the fault line and adapt to the current system. Finally, the program formed the instructions to send to the controlled layer, and receive the feedback from the controlled layer. At the same time, the operating layer undertake the task of display optimize control programs, perform and other human-computer interaction function.

3.3 The Task and Design of Controlled Layer

The controlled layer of hybrid self-healing control system is composed of two parts. The one part is the underlying physical grid which contained power plants, substations, FACTS devices and other existing protection devices. The other part is the real-time power system platform based-on the WAMS. The real-time power system platform receive the power system optimization or protection programs from the Operating layer, then send operating instructions to the transformers, power system protection devices, and feedback the implementation to the Operating layer. At the same time, the real-time power system platform also under take the task of real-time record, showing power system operating conditions.

According to the above model, the hybrid self-healing control system can be a brief description of the process as follows: the power system information platform on the bottom record the real-time grid status, upload and synchronization it to the Decision-making layer and Operating layer through the data communication channel. Decision-making layer through the grid state judge whether there is event occurred. If the event occurred, then judge whether it is fault or disturbance, send a control instruction to the operating layer. Operating layer make a specific optimization or fault solutions by the real-time grid data, then send it to controlled layer in the form of a document. Finally the power system information platform control the physical network. In summary, the hybrid self-healing control mechanism is simple, and can achieve the global optimization of the power system, the global sharing of data to fully enhance the automation degree of self-healing power system.

4 Realization of Grid Self-Healing Hybrid Control

The grid of self-healing hybrid control system model according to text, to achieve the grid self-healing, first of all, a clear definition of "event" is needed; second, we need to achieve the control of disturbances and faults, At the same time, the solution should be to ensure the optimal solution; Finally, The real-time and global of overall healing process should be to ensure. The following will detail the key of implementation and necessary technology of the self-healing hybrid control.

4.1 Definition of "Event"

It can be said that the guiding ideology of the grid self-healing system based on hybrid control is defining the grid state which doesn't meet the requirement as "events", then excitation control action by the "event", so how to define the "event" becomes the focus of hybrid self-healing control. Any aspect of the grid state which does not meet the requirements should be regarded as the "event", in other words, for the all real-time network parameters recorded by controlled layer power system, we should define the dissatisfied ones as the "event". The grid has the following important parameters: Power system frequency, harmonics; node voltage of the power system, the amplitude of line current, phase angle; inputs and outputs of node active and reactive power; generators, temperature of power transformers and other hardware state. The above parameters have a certain range of security, that is, the threshold; the state which exceeds or below the threshold is defined as "events". With reactive power of a node for example, the constraint conditions:

$$Q_{Gimin} \leq Q_{Gi} \leq Q_{Gimax} \qquad (1)$$

Q_{Gi}, Q_{Gimin}, Q_{Gimax} are for the reactive power and its upper and lower limits of node i. In this example, "$Q_{Gimin}-Q_{Gi}>0$" and "$Q_{Gimax}-Q_{Gi}<0$" are defined as "events".

"Event set" take the union of the "event" set, in the running grid, once the "incident" occurred, the decision-making control is excited.

4.2 Concrete Realization of Disturbance Control and Fault-Controlled

In recent years, study for disturbance control and fault control of the power system emerges continually, this system use the failure to control and optimal control based on wide area measurement to construct the control layer of hybrid power operation of self-healing.

Wide area measurement technique based on GPS can measure the node voltage magnitude and phase angle directly and has good value in the power system state estimation, fault location and positioning, power system reactive power and voltage optimization, etc. For the failure of the power system control, cutting off the fault line and determining the location of the point of failure is particularly important.

4.3 Real-time and Localized of Self-Healing Control

Due to the complexity of the structure of the grid, it is very easy to cause a chain reaction leading to the large-scale collapse of the power grid, so real-time is required in the grid self-healing control. In this paper, the hybrid self-healing control, take full use of the WAMS systems' advantage of rapid processing of real-time data of the grid ensuring the good liquid of the control process and procedure. The decision-making layer of Self-healing control based on the hybrid control can achieve the global control of the grid. Regarding all the disturbances as "events", when expansion the physical structure of the power grid, we can cover the new physical structure as long as adding a new event definitions without having to re-deploy the software system, this thinking has a good flexibility.

5 Conclusion

This article discussed the possibility of applying hybrid control theory into the grid self-healing control, established a grid self-healing control system based on hybrid control theory and gave the key technology required by concrete realization. Overall, the basic idea of hybrid grid self-healing control is "event driven", regarding the grid state which does not meet the needs as "incident", using discrete event to drive the disturbance control or failure of control. The advantages show in the simplicity in its concept, global optimization, easiness to expand and applicability to global healing of large power grids.

References

1. Wenliang Zhang, Zhuangzhi Liu, Mingjun Wang (2009) Research status and development trend of smart grid. Power Syst Technol 33(13):1–11 (In Chinese)
2. U.S. Department of Energy Office of Electricity Delivery and Energy Reliability, National Energy Technology Laboratory. (2009). Systems view of the Modern Grid, EB/OL. httP://www.netl.doe.gov/moderngrid
3. BR Williams, TD John (2008) System and method for a self-healing grid using demand side management techniques and energy storage. U.S.A., US7389189B2
4. Galan S, Barton PI (1988) Dynamic optimization of hybrid systems. Comput Chem Eng 22:183–190
5. Hu Wei, Lu Qiang (2005) Hybrid power control system and its application. Trans China Electrotech Soc 20(2):11–16

PMSM Sensorless Vector Control System Based on Single Shunt Current Sensing

Hongyan Ma

Abstract To reduce the cost and volume of permanent magnet synchronous motor (PMSM) drive system fed by pulse width modulation (PWM) inverter, this paper presents a single shunt current sensing with rotor-position sensorless control method of PMSM vector control system. The reference voltage of space vector pulse width modulation (SVPWM) inverter is researched to implement the requirements by AC-link phase current reconstruction with single shunt current sensing. By model reference adaptive system (MRAS), speed estimation method is investigated to satisfy rotor-position sensorless control. Simulations are tested on a PMSM vector control system fed by SVPWM inverter. Simulation results demonstrate the feasibleness and the effectiveness of the single shunt current sensing with MRAS sensorless control method.

Keywords Single shunt current sensing · MRAS sensorless · PMSM · PWM inverter

1 Introduction

For permanent magnet synchronous motor (PMSM) having many advantages such as high ratio of torque to weight and high efficiency, PMSM vector control systems, which supplied by pulse width modulation voltage source inverters (PWM-VSI), are widely used in many applications [1]. High performances PMSM vector control systems depended on the precise information of AC-link currents by AC-link current sensors and the rotor position by mechanical sensor. To reduce the cost and volume of inverter, no current sensors control methods based on a single shunt current sensing to reconstruct three phase AC currents have been proposed by

H. Ma (✉)
Department of Electrical Engineering, Beijing University of Civil Engineering and Architecture, Beijing, China
e-mail: m_hy71@163.com

researchers [2–4]. The sensorless rotor-position estimation methods like the Extended Kalman Filter (EKF) algorithm combining with single shunt sensing and the model reference adaptive method combining with no AC-link current sensor have been developed [5, 6].

In sensorless AC drive systems, the practical rotor position/speed estimation method is based on model reference adaptive system (MRAS). In this paper, single shunt current sensing with the MRAS sensorless method is researched in PMSM vector control systems supplied by space vector PWM voltage source inverter (SVPVM-VSI). The validity and feasibility of the researched method are verified by simulation results of PMSM vector control systems supplied by three phase voltage source inverter.

2 Single Shunt Current Reconstruction Based on DC-Link

Single shunt current sensing control scheme is reconstructed the AC-link currents by the measured DC-link current values with single shunt. The voltage vector diagram of SVPWM-VSI shown as Fig. 1, there are six sectors in the voltage vector diagram of SVPWM and six active voltage vectors ($V_1 \sim V_6$) and two zero vectors V_0 (000) and V_7 (111). The reference voltage vector V_r located in sector 1 is only studied in follows.

In sector 1, the reference voltage vector V_r is synthesized by the two adjacent active voltage vectors V_1, V_2. In the linear modulation range, conventional seven segment SVPWM signals distribution strategy which is to synthesize V_r by using two adjacent non-zero vectors and one zero vector in one sampling period T_s is applied, V_r is given as

$$V_r = \frac{T_1}{T_s} V_1 + \frac{T_2}{T_s} V_2 \qquad (1)$$

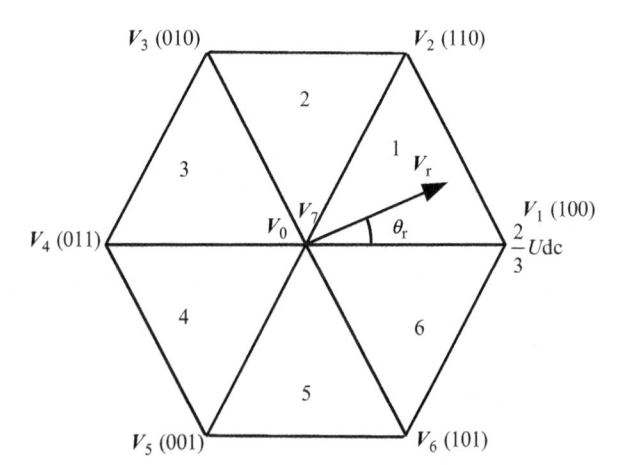

Fig. 1 The voltage vector diagram of SVPWM-VSI

Fig. 2 No AC-link current sensor control in sector 1

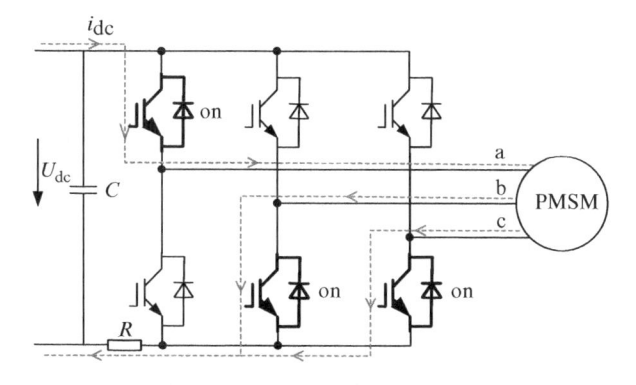

Table 1 Voltage vectors and measured phase currents by i_{dc}

Voltage vector	i_{dc}	Voltage vector	i_{dc}
$V_0(000)$	0	$V_4(011)$	$-i_a$
$V_1(100)$	$+i_a$	$V_5(001)$	$+i_c$
$V_2(110)$	$-i_c$	$V_6(101)$	$-i_b$
$V_3(010)$	$+i_b$	$V_7(111)$	0

T_1 and T_2 are the on-durations of the switching state vectors V_1 and V_2. They can be calculated as

$$\begin{cases} T_1 = \sqrt{3}T_s \dfrac{|V_r|}{U_{dc}} \sin\theta_r \\ T_2 = \sqrt{3}T_s \dfrac{|V_r|}{U_{dc}} \sin(\pi/3 - \theta_r) \end{cases} \quad (2)$$

Where U_{dc} is DC-link voltage, θ_r is the angle of V_r.
The on-duration of zero vector T_0 can be obtained as

$$T_0 = T_s - T_1 - T_2 \quad (3)$$

Used an active voltage vector to PMSM, AC-link phase current is measured by the DC-link current i_{dc}. In Fig. 2, by detecting i_{dc} as active vector V_1 employed, a-phase current i_a of the motor is achieved; as zero vectors employed, i_{dc} equals zero, then the phase current is not measured. In each control period, two phase currents achieved by the DC-link current i_{dc}, the third phase current is determined by the zero sum of three-phase currents.

As shown in Table 1, the applied voltage vector employed, the responding phase current is measured from the DC-link current i_{dc}.

In practice, using single shunt sensing to reconstruct the AC-link phase current, the precision of reconstructed AC-link phase current is determined by the DC-link current. In order to achieve a dependable DC-link current i_{dc}, the minimum sampling time T_{min} has to be less than the operation period of applied active vector.

3 Sensorless PMSM Vector Control

3.1 PMSM Mathematical Model

In the d-q rotor reference frame, PMSM mathematical model of is given by the following equations.

$$\begin{cases} u_d = p\psi_d - \psi_q \omega + Ri_d \\ u_q = p\psi_q - \psi_d \omega + Ri_q \end{cases} \quad (4)$$

$$\begin{cases} \psi_d = L_d i_d + \psi_r \\ \psi_q = L_q i_q \end{cases} \quad (5)$$

$$T_{em} = p_n (i_q \psi_d - i_d \psi_q) \quad (6)$$

Where u_d and u_q stand d-q axis voltages, i_d and i_q express d-q axis currents, ψ_d and ψ_q denote d-q axis flux linkages, R is stator resistance, L_d and L_q are d-q axis inductances, ψ_r is the permanent magnetic flux, T_{em} and T_L are electrical torque and load torque, p_n is numbers of pole pairs of the motor, p is d/dt, ω stands for the rotor speed that is equal to $p\theta$, θ is the actual rotor position.

3.2 Speed Estimation Method Based on MARS

In MRAS method, the current equation of PMSM is chosen as the adjustable model and the actual PMSM as reference model. The error between currents of the adjustable model and the currents of the actual PMSM is used to calculate motor speed.

In the rotating d-q reference frame, the PMSM stator current equations are

$$\frac{d}{dt}\begin{bmatrix} i_d \\ i_q \end{bmatrix} = \begin{bmatrix} -\frac{R}{L_d} & \frac{L_q}{L_d}\omega \\ -\frac{L_d}{L_q}\omega & -\frac{R}{L_q} \end{bmatrix} \begin{bmatrix} i_d \\ i_q \end{bmatrix} + \begin{bmatrix} \frac{u_d}{L_d} & -\frac{\omega \psi_r}{L_q} + \frac{u_q}{L_q} \end{bmatrix} \quad (7)$$

Considering the convenience of stability analysis, the system matrix A is written as

$$A = \begin{bmatrix} -\frac{R}{L_d} & \frac{L_q}{L_d}\omega \\ -\frac{L_d}{L_q}\omega & -\frac{R}{L_q} \end{bmatrix} \quad (8)$$

Let $i_d' = i_d + \frac{\psi_r}{L_d}, i_q' = i_q, u_d' = \frac{u_d}{L_d} + \frac{R\psi_r}{L_d^2}, u_q' = \frac{u_q}{L_q}$. Then the simple reference model form is obtained as

$$\frac{d}{dt}i' = Ai' + u' \tag{9}$$

Speed estimation process described as follows.
The simple parallel connection adjustable model form is

$$\frac{d}{dt}\hat{i}' = \hat{A}\hat{i}' + u' \tag{10}$$

The state variables error is

$$e = i' - \hat{i}' \tag{11}$$

The parallel connection model is

$$\begin{cases} \dfrac{d}{dt}e = Ae \\ v = De \end{cases} \tag{12}$$

If $D = I$, then $v = e$.
By the Popov super stability theory, the estimation equation of $\hat{\omega}$ can be obtained as

$$\hat{\omega} = \int_0^t k_1(i_d'\hat{i}_q' - i_q'\hat{i}_d')d\tau + k_2(i_d'\hat{i}_q' - i_q'\hat{i}_q') + \hat{\omega}(0) \tag{13}$$

Where, $k_1 \geq 0$, $k_2 \geq 0$.
Replacing i_d', i_q' with i_d, i_q, the estimated speed is obtained as

$$\hat{\omega} = \int_0^t k_1[i_d\hat{i}_q - i_q\hat{i}_d - \frac{\psi_r}{L_d}(i_q - \hat{i}_q)]d\tau + k_2[i_d\hat{i}_q - i_q\hat{i}_d - \frac{\psi_r}{L_d}(i_q - \hat{i}_q)] \\ + \hat{\omega}(0) \tag{14}$$

Where, $\hat{i}_d$ and $\hat{i}_q$ are determined by the adjustable model, i_d and i_q are achieved by the transformation of the reconstructed three-phase stator currents with single shunt current sensing control method.
Integrating the estimated speed, the rotor position is

$$\hat{\theta} = \int_0^t \hat{\omega}dt \tag{15}$$

4 Simulation Study

In order to prove the feasibleness and effectiveness of single shunt current sensing with MRAS sensorless method, the diagram of PMSM vector control system is built in Fig. 3. Conventional vector control technique such as $i_d = 0$ is applied to the PMSM drive system.

The simulation parameters are shown in Table 2. The dead time effect is not considered in simulation.

Figure 4 shows the speed curve. The speed steady-state error between the motor speed and the reference speed 1,500 rpm is very small. The motor has good performance under this control strategy. Figure 5 shows PMSM stator current. The phase current can concord with the reconstruction current. Figure 6 shows

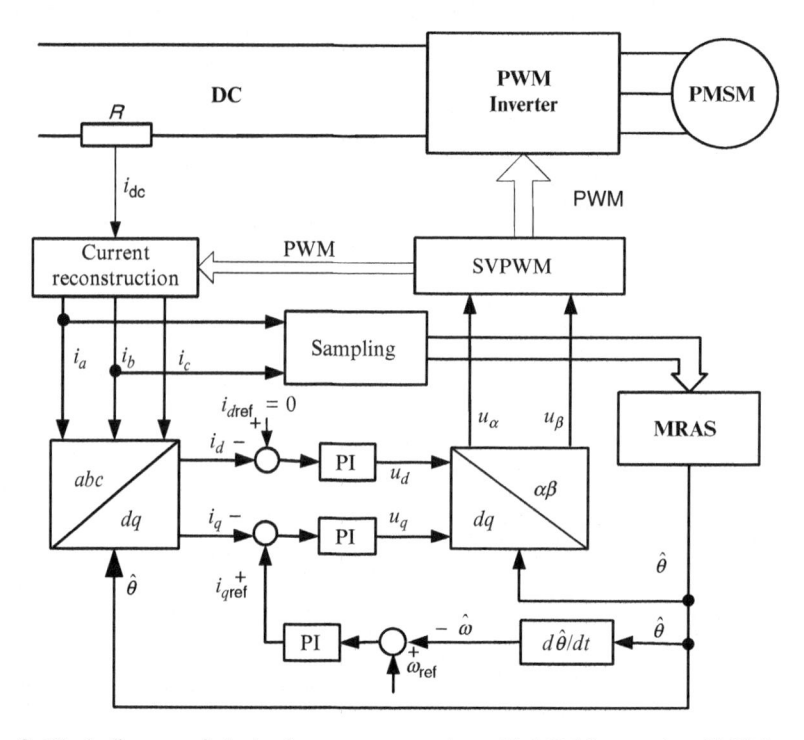

Fig. 3 Block diagram of single shunt current sensing with MRAS sensorless PMSM vector control system

Table 2 The simulation parameters of the motor

Parameter	Value	Parameter	Value
L_d/mH	7.418	L_q/mH	12.285
R/Ω	0.618	ψ_r/V/(rad/s)	0.1128
p_n	2	T_L/Nm	1.5

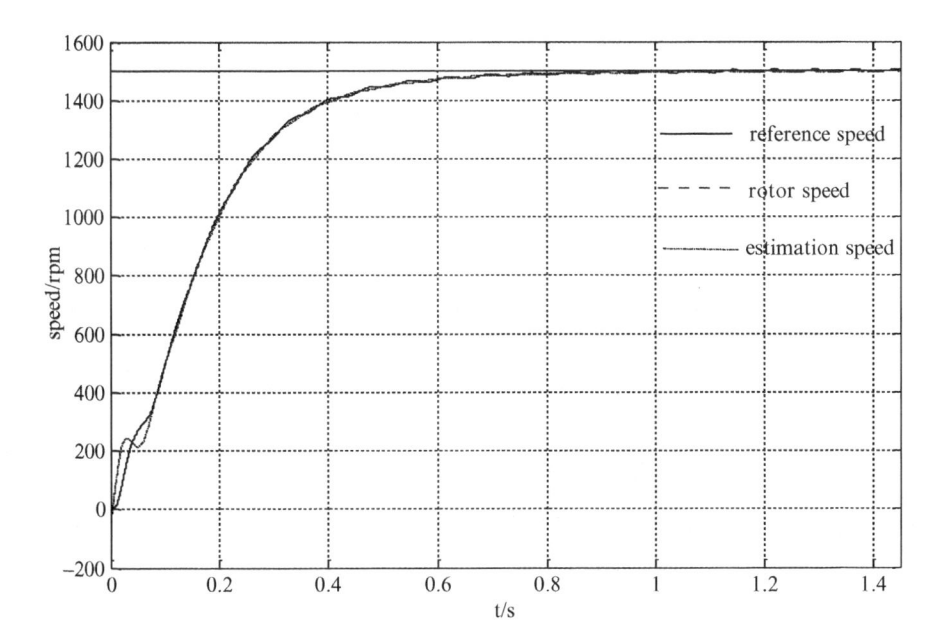

Fig. 4 Speed response curve

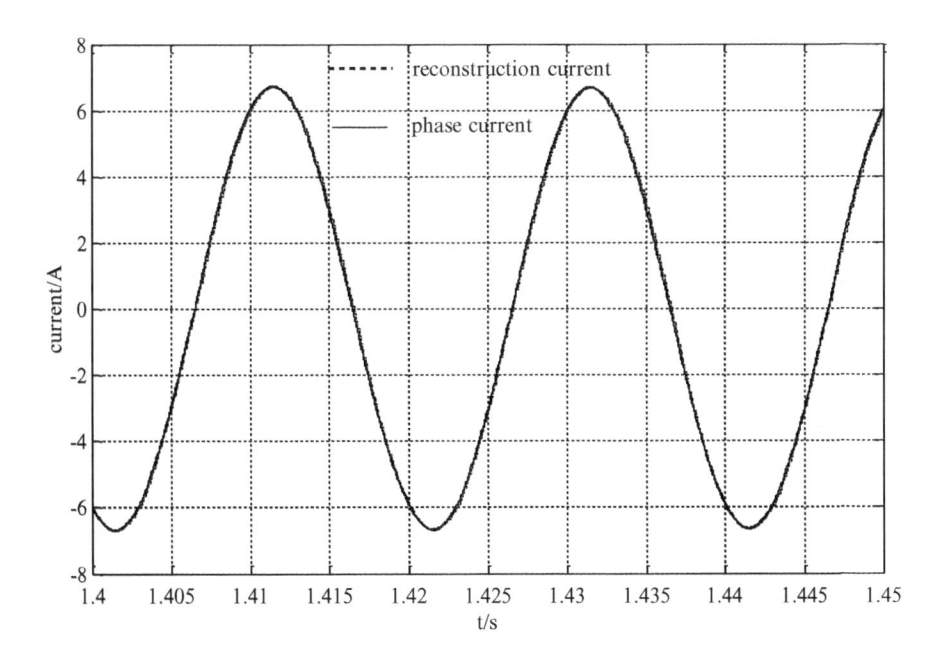

Fig. 5 Waveforms of AC-link phase current

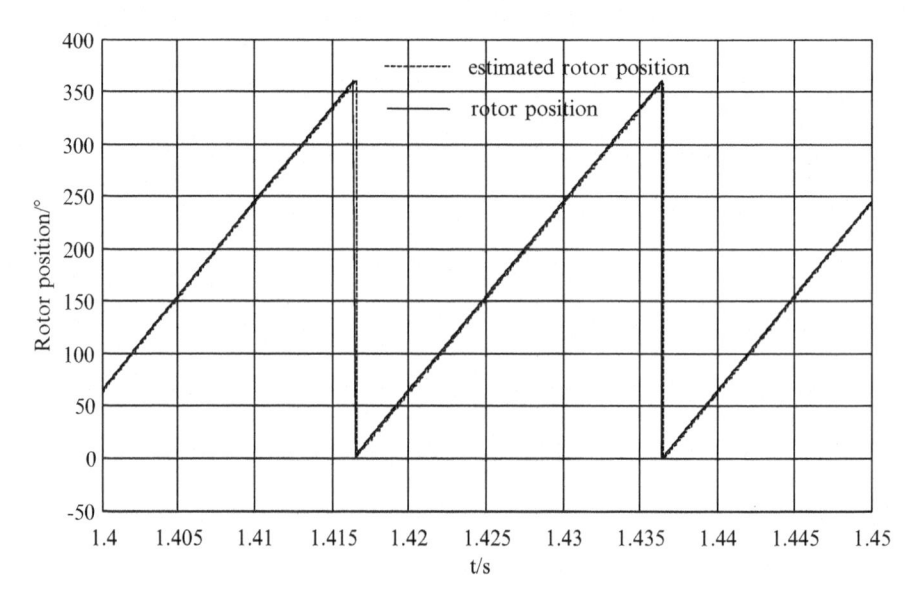

Fig. 6 Waveforms of rotor position

the waveform of rotor position. The real rotor position can concord with the estimated rotor position by MRAS method and reconstructed AC-link phase current.

This verifies that the single shunt current sensing with MRAS sensorless control method is effective in PMSM vector control.

5 Conclusion

In this paper, single shunt current sensing control combined with MRAS sensorless scheme was used for PWM-VSI fed PMSM vector control system. The MRAS sensorless control used the reconstructed AC-link phase currents to estimate rotor position. Simulations demonstrated that, in PWM-VSI fed PMSM vector control, the method that is using single shunt current sensing with rotor position sensorless control based on MRAS method, is valid and feasible.

Acknowledgements The author thanks the financial support by Beijing Municipal Commission of Education of China (PHR201108211) and MOHURD project (2011-k8-3).

References

1. Mohamed B (2005) Implementation and experimental investigation of sensorless speed control with initial rotor position estimation for interior permanent magnet synchronous motor drive. IEEE Trans Power Electron 20(6):1413–1422
2. Liu Yan, Shao Cheng (2007) Reconstruction strategies for phase currents in three phase voltage-source PWM inverters. Info Control 36(4):506–513 (In Chinese)
3. Chu Jianbo, Hu Yuwen et al (2010) Phase current sampling reconstruction for inverter. Trans China Electrotech Soc 25(1):111–117 (In Chinese)
4. Blaabjerg F, Pedersen JK et al (1997) Single current sensor technique in the DC link of three-phase PWM-VS inverters: a review and a novel solution. IEEE Trans Ind Appl 33(5):1241–1253
5. Yuan Xibo, Li Yongdong, Feng Lichao (2009) Low cost sensorless control of PMSM for air-conditioner compressor application. Electr Drive 39(5):15–19 (In Chinese)
6. Sun Kai, Huang Lipei (2010) Sensorless over-modulation control of PMSM-compressor systems. J Tsinghua Univ (Sci Technol) 50(1):18–22 (In Chinese)

Wind Power System Simulation of Switch Control

Yuehua Huang, Guangxu Li, and Huanhuan Li

Abstract In order to find a balance between energy efficiency and reliability of wind power generation system, this paper presents a switch control strategy. This paper establishes a wind power system simulation model by controlling the electromagnetic torque in the Matlab/Simulink environment. The electromagnetic torque consists of three parts: the equivalent control quantity u_{eq}, the changing high-frequency part u_N and the additional low-frequency part u_{Nf}. The u_{eq} can make the system close to the optimal operating point, and u_N、 u_{Nf} make the system operate stably around the optimal operating point. The simulation results show that: the actual speed can track the best speed very well, and the fluctuation range of the electromagnetic torque is very small. This switching control strategy can effectively improve the efficiency of wind power generation system and reduce fatigue load, and it solves the problem of imbalance between energy efficiency and reliability.

Keywords Wind power • Switch control • Matlab/simulink • Torque

1 Introduction

With the rapid development of modern industry, the useful conventional energy sources on Earth become more and more scarce. In order to achieve sustainable development of energy, many countries are trying to develop new energy and renewable energy. Among them, wind energy has many advantages, it has wide distribution, large reserves, and it can be developed and used effectively [1]. As a new type of renewable energy, wind power is the fastest-growing energy in the global.

Y. Huang • G. Li (✉) • H. Li
Institute of Electrical and new energy, Three Gorges University, Yichang, China
e-mail: lgxonly@sina.com

Wind power has following characteristics: good environment, mature technology and strong feasibility, and it is used more and more widely in the world [2]. Wind energy is safe, clean and inexhaustible, and it is different from fossil fuels. This local resource is permanent, and it can give us long-term stable supply of energy. At the same time, it does not produce carbon emissions.

In the case of a certain wind turbine speed, the greater the wind speed, the greater the output power of wind turbine. For a certain wind speed, there is always a existence of biggest power point. Only when the wind turbine works in the optimum tip speed ratio, the system can output the maximum power [3]. The wind speed often changes, so the wind turbine tip speed ratio also changes, which makes the wind turbine deviate from the best working condition, and affects the energy conversion efficiency of wind turbines. In order to ensure the best conversion of energy, we should make the system run in optimal condition. To achieve this goal, this article uses the switch control strategy to make the tip speed ratio close to the optimal tip speed ratio condition, which can realize the biggest energy conversion of wind power generation system.

2 Analysis of Wind Turbine Characteristics

Bates theory shows that the wind turbine power absorbed from the air is [4]:

$$P = \frac{\rho}{2} C_p A V^3 \qquad (1)$$

In this formula, P is useful output power for the wind turbine, the unit is W; ρ is air density, the unit is kg/m^3; A is wind turbine swept area, the unit is m^2; V is wind speed, the unit is m/s; Cp is wind energy utilization factor. We can see from Eq. 1, for a certain wind speed, the greater the wind energy utilization factor Cp, the greater the useful power of wind turbine. Wind energy utilization factor Cp and wind turbine tip speed ratio λare related, λ is operating parameter of the wind turbine:

$$\lambda = \frac{r\Omega}{V} \qquad (2)$$

In this formula: Ω is angular frequency of the wind turbine, the unit is rad/s; r is radius of the wind turbine, the unit is m; For the wind turbine, λ determines the size of the Cp, the relationship between them is curve of the parabolic relationship.

3 The Establishment of the Wind Model

Wind energy is kinetic energy generated by flowing air of the Earth's surface [5]. It is a natural phenomenon on earth, the size of wind energy depends on wind speed and air density, the wind speed and direction are constantly changing with strong randomness in the flowing process. Wind model is shown in Fig. 1, and wind direction is not considered. The purpose of the wind model is describing the randomness of wind, and the wind model is set up by the white noise generator and the shaping filters. The white noise generator generates random signals, and the effect of the shaping filters is filtering. Transfer Fcn 1, Transfer Fcn 2 and Transfer Fcn 3 are all shaping filters, and 7 is the average value of wind speed. Wind velocity waveform is shown in Fig. 2. We can see from the Fig. 2, simulation time is 60 s, and wind speed changes between 6 and 8 m/s.

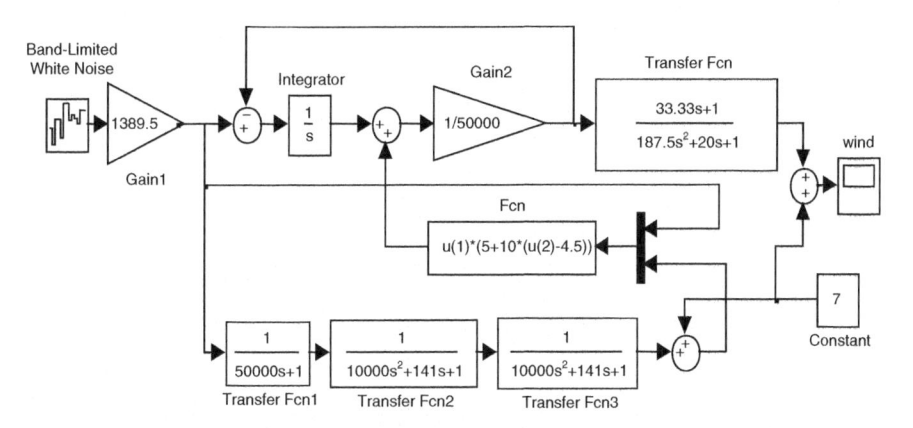

Fig. 1 Wind speed model diagram

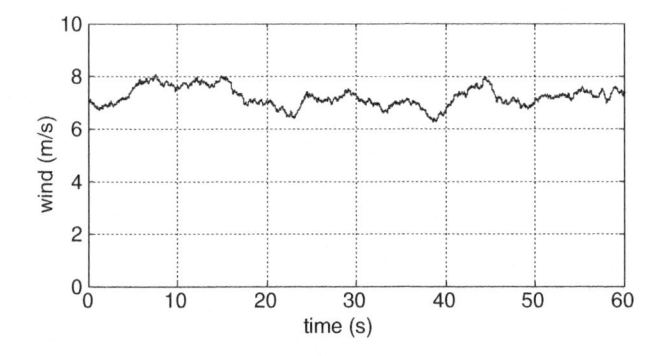

Fig. 2 Wind speed waveform diagram

4 Wind Power System Model

The wind power system model in this article is mainly constituted by the wind, the aerodynamic model, the transmission and the generator model [6]. The role of the aerodynamic model is effectively converting wind energy into useful mechanical energy; the role of the transmission is passing the rotational motion of the wind wheel to the generator, it uses gearbox as the mechanical transmission structure, in this way, the speed of wind wheel can adapt to the needs of generator, which can help generator operate normally. The role of the generator model is generating electricity, the purpose of the model is to convert the mechanical energy transmitted by chain transmission to electricity; the power terminal of the generator is connected to the public grid, in this case, electrical energy is transferred to the power grid. The entire wind power generation system is an organic entity, the various sub-parts must have higher compatibility, which can achieve the efficient conversion of wind energy. The wind power system model is shown in Fig. 3. The whole system is made up by four sections, they are wind model, aerodynamic model, transmission model, and the generator model.

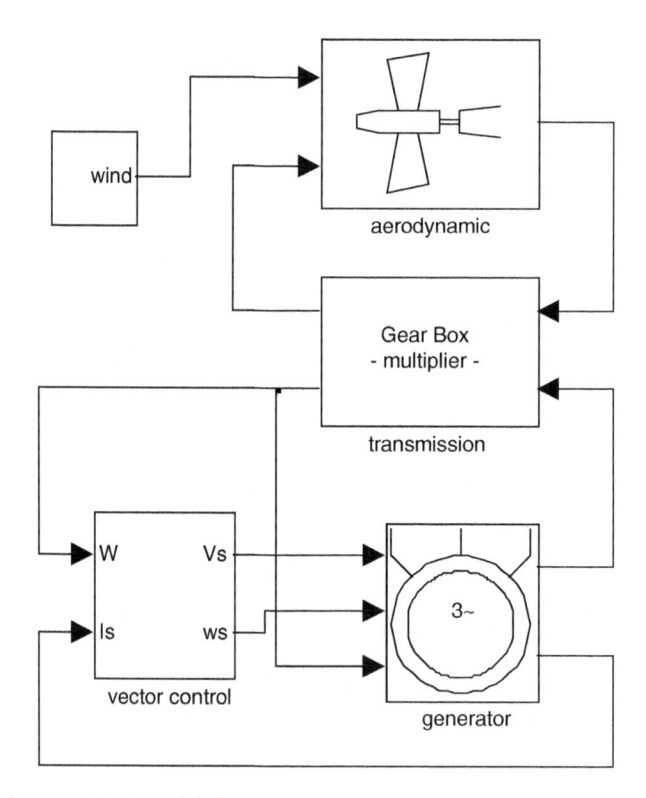

Fig. 3 Wind power system model diagram

5 The Establishment of Switch Control Model

In order to make wind power generation system have the biggest energy conversion efficiency, the paper uses the torque which is relative to the steady-state operating point to control generator, this method can achieve the target. This paper has presented a switch control strategy, the structure of the strategy is shown in Fig. 4.

The control strategy of Fig. 4 shows the steady-state torque reference value U, it consists of three parts:

$$U = Ueq + Un + Unf \tag{3}$$

In this formula, the equivalent control quantity Ueq is a steady-state part which is relative to the optimal operation point, its value is proportional to the square of low-frequency wind speed v_s^2, that is

$$Ueq = Cv_s^2 \tag{4}$$

U_n is the changing high-frequency part, it changes between the two values $+\beta$ and $-\beta$, among them, $\beta > 0$,

$$Un = \beta \bullet \text{sgn}[\sigma(t)] \tag{5}$$

Ueq drives the system to run in the optimal operating point, the role of U_n is to maintain stability of system which is running around the optimal operating point. Unf is received by filtering Un, and it is proportional to the average of Un.

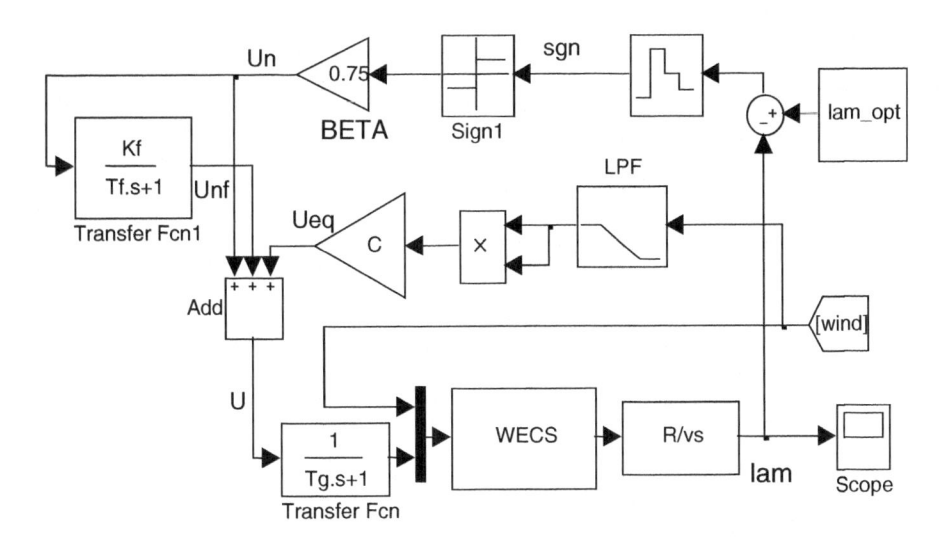

Fig. 4 Switch controller structure diagram

In Fig. 4, lam is tip speed ratio λ, lam_opt is optimal tip speed ratio λ_{opt}, Ueq is obtained by wind speed V, Un is obtained by λ and λ_{opt}, Unf is obtained by filtering, U is obtained by these three values, then we can control the tip speed ratio λ by adjusting U. This control strategy can make tip speed ratio λ close to the optimal value λ_{opt}. The tip speed ratio λ is used to aerodynamic model in Fig. 3, in this case, the whole system can run stably, the output power is the maximum of wind turbine power.

6 Analysis of Simulation Results

The wind power generation system based on switch control strategy is simulated with Matlab/Simulink. The simulation parameters are as follows: $\beta = 3$ N.m, C = 0.25, air density p = 1.25 kg/m^3, efficiency $\eta = 0.95$, optimal tip speed ratio $\lambda_{opt} = 7$, biggest wind energy utilization factor $C_{pmax} = 0.47$, length of wind turbine blade R = 3.1m, drive ratio i = 6.6, low speed shaft inertia J1 = 3.6 kg.m^2, stator resistance Rs = 1.366 Ω, rotor resistance Rr = 1.63 mΩ, stator inductance Ls = 0.172H, rator inductance Ls = 0.162H, mutual inductance Lm = 0.151H, the simulation time is 60 s.

We can see from Fig. 5, the tip speed ratio only has a small amplitude oscillation nearby the optimal value 7, the oscillation amplitude is very small, which explain that the wind turbine can keep a very good tip speed ratio during operation process. Wind energy utilization factor waveform is shown in Fig. 6, wind energy utilization factor is proportional to output power, which can directly reflect the size of the output power. Wind energy utilization coefficient and tip speed ratio have close relationship, when the tip speed ratio is the optimal tip speed ratio, wind energy utilization factor is the biggest wind energy utilization factor, the largest wind energy utilization coefficient of this article is 0.476,

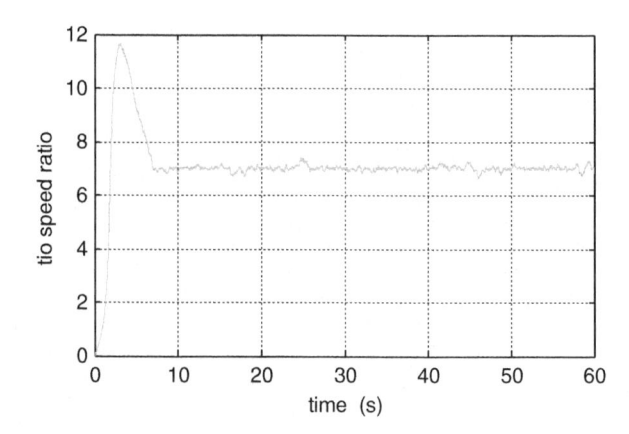

Fig. 5 Tip speed ratio waveforms diagram

Fig. 6 Wind energy utilization factor waveforms diagram

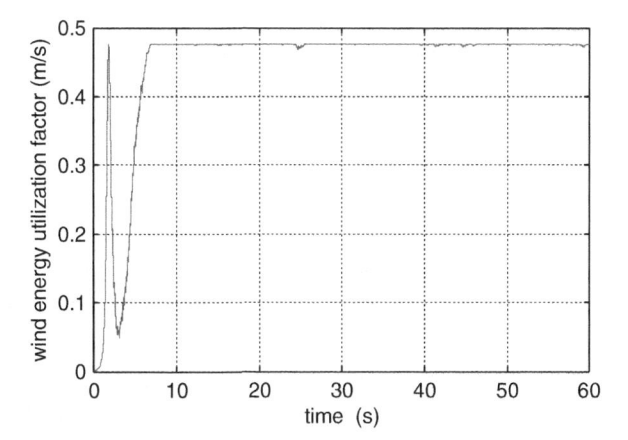

Fig. 7 Wind turbine speed waveforms diagram

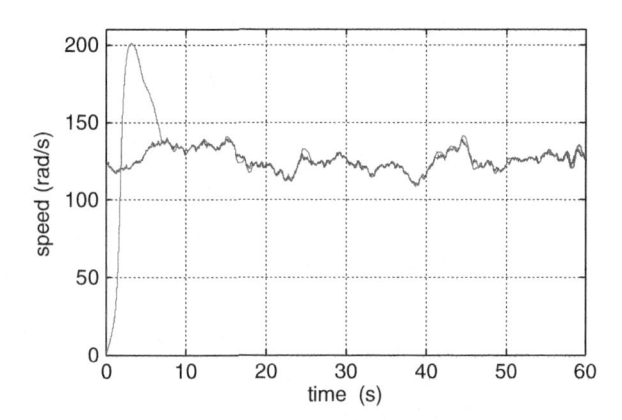

the Cp of Fig. 6 has been near the biggest wind energy utilization coefficient 0.47, the amplitude of fluctuation is very small, therefore, the final output power is the maximum of wind turbine power. The speed simulation waveform is shown in Fig. 7, blue curve represents the optimum rotational speed of the wind wheel, the red curve represents the actual speed of the wind wheel. We can see that the trace of actual speed has been very close to the trace of optimum rotational speed, when the optimum rotational speed increases, the actual speed increases relatively; when the optimum rotational speed reduces, the actual speed also reduces relatively; it can track the optimum speed well. You can explain, in the actual generation process, the wind turbine can achieve maximum power transfer successfully. The waveform of electromagnetic torque is shown in Fig. 8, we can see from Fig. 8 that the amplitude and frequency of the electromagnetic torque pulsation are very small, so the system can reduce the mechanical fatigue.

Fig. 8 Electromagnetic torque waveforms diagram

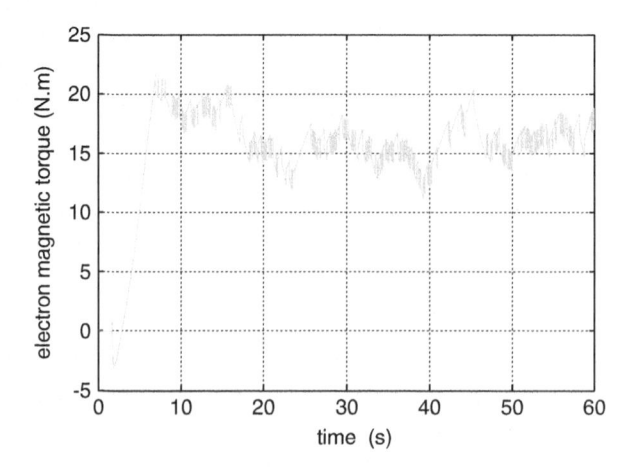

7 Conclusion

This paper establishes a wind power system simulation model by using Matlab/Simulink simulation software, and uses switch control technology to simulate the wind power generation system. Simulation results show that tip speed ratio and wind energy utilization factor curve are in the vicinity of the best value, and the actual speed can track the best speed well. It shows that the wind turbine can get the maximum energy conversion efficiency. In addition, the fluctuation range of the electromagnetic torque is very small, which can reduce mechanical fatigue of the wind power system. It can be concluded that the switch control strategy used in this paper is feasible and effective, and the system can achieve balance between energy efficiency and reliability. The global energy crisis and environmental crisis have become more and more serious, wind power industry has made rapid development, the biggest power point tracking has also become a hot topic of wind power research, and the simulation results in this paper provide a reference for analyzing the efficiency of wind turbines.

References

1. Chen Yidong, Yang Yulin, Wang Liqiao (2010) Maximum power point tracking technology and simulation analysis for wind power generation. High Voltage Engineering 36(5):1322–1326
2. Li Chuantong (2005) New energy and renewable energy technologies. Southeast University Press, Nanjing, China, pp 2–5
3. Li Shaowu (2011) Research on MPPT control algorithm of small- sized wind energy conversion system. Journal of Hubei University for Nationalities 29(2):192–193
4. Huang Jincheng, Yang Ping (2011) An optimize maximum power point tracking algorithm for small scale wind power generator. Electr Mach Control Appl 38(3):44–48
5. Zheng Ping (2010) Maximum power point tracing in 20kw wind power system. Beijing University of Chemical Technology, Beijing
6. Jin Xin (2007) Application of simulation in analysis of performance of wind turbine. J Chongqing Univ 19(24):2823–2836

Simulation of Variable-Depth Motion Control for the High-Speed Underwater Vehicle

Tao Bai and Yuntao Han

Abstract In order to solve the problem of variable-depth motion control of underwater high-speed vehicle, cavitator-fins joint control method and mathematical model simulation analysis method were used in this paper. First, improvement to the model of high-speed underwater vehicle, according to the analysis of relationship between supercavity shape and force for vehicle; Secondly, linearization the motion equations used by state variable feedback exact linearization method and then used pole placement for design controller. Third, use MATLAB/SIMULINK software to design simulation model of high-speed underwater vehicle. The simulations reveal that this controller is proved to be effective. The cavitator-fins joint control method can effectively solve a class of the variable-depth control of high-speed underwater vehicle.

Keywords High-speed underwater vehicle • Supercavity • Variable-depth motion control • Simulation study

1 Introduction

Supercavitation can provide possibility for lessen viscous drag, so that reduction by maintaining a stable single vaporized water bubble around the vehicle, it can making extend the velocity range of the underwater vehicle. However, supercavitation involves complicated hydrodynamic dynamics, that the vehicle experiences strong nonlinear forces and the system dynamics presents challenges for stabilization and maneuver control of the vehicle.

T. Bai (✉) • Y. Han
College of Automation, Harbin Engineering University, Harbin, China
e-mail: baitao812@126.com

Savchenko proposed several stable motion modes of the supercavitation vehicle in the different speed ranges [1] and some related control problem with supercavitation vehicle [2]. Bálint Vanek used low-level longitudinal model to study the control characteristics of the supercavitation vehicle [3, 4].

This paper is devoted to the motion model improvement and design controller for the high-speed underwater vehicle in the longitudinal axis. First, improve motion model of high-speed underwater vehicle on vertical plane; Second, linearization transformation of the nonlinear movement equations based on state variable feedback exact linearization method. Third, design controller for supercavitation vehicle by cavitator-fins joint control method and pole placement. Finally, simulation was performed by using Matlab/Simulink, then discussed about the simulation results.

2 Model Improvement for Supercavitating Vehicle

2.1 Analysis on Supercavity Shape

The cavity entirely enveloping the moving body, which are filled by the water vapor, so it was called "supercavitating". Based on the general theory of similarity of hydrodynamic, the cavitation number σ is the most important independent dimensionless parameters in the similarity criteria of the supercavitating flows, σ may be calculated by formulae [1]:

$$\sigma = \frac{2(p_\infty - p_c)}{\rho V^2} \quad (1)$$

The p_∞ is simply the hydrostatic pressure at the operating depth of the vehicle; p_c is the pressure in cavity; ρ is the fluid density; V is velocity of the vehicle in undisturbed flow.

In the case of the low cavitation numbers ($\sigma < 0.1$), the semi-empirical relations can be derived for shape of supercavity to disk cavitator (Fig. 1).

$$R_c = R_n \sqrt{\frac{c_x}{0.9\sigma}} \quad x_1 = D_n \quad L_c = D_n \frac{2\sqrt{c_{x0}}}{\sigma} \quad (2)$$

The R_c and L_c are the mid-section radius and the length of the supercavity; D_n is the cavitator diameter; c_x is the cavitation drag coefficient; c_{x0} is the cavitatin drag coefficient when $\sigma = 0$. The G.V. Logvinovich's formula [1] is most elementary for compute the shape of axisymmetric and steady cavity, shapes of the part I and part II of the cavity are calculated by different formulae [2].

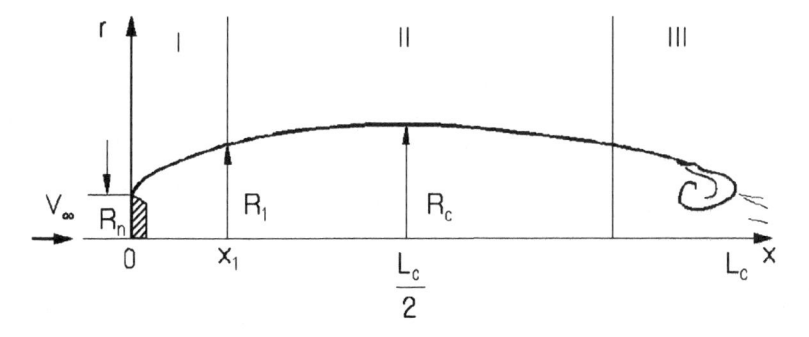

Fig. 1 Shape of the steady axisymmetric cavities

$$\begin{cases} R(x) = R_n \left(1 + \dfrac{3x}{R_n}\right)^{\frac{1}{3}}, & x \leq x_1 \\ R(x) = R_c \sqrt{1 - \left(1 - \dfrac{R_1^2}{R_c^2}\right)\left(1 - 2\dfrac{x - x_1}{L - 2x_1}\right)^2}, & x \geq x_1 \end{cases} \quad (3)$$

2.2 Force Analysis of High-Speed Underwater Vehicle

When vehicles travel at high-speed in underwater, cavitator is contact with water is located front of vehicles. Cavitator can rotate around its axis into a certain angle to produce supercavitation and provide lift for vehicle. The aft plan force have periodic impact on cavity up or down wall in the vertical plane [3] (Fig. 2).

The vehicle can be seen as one rigid body, its reference coordinate is the body coordinate, its reference point O is gravity centre on vehicle, there are five forces acting on the body, the gravity, thrust force, cavitator, fins forces, and planning which is not always present.

Lift force acting on cavitator can be approximated as:

$$F_c = \frac{1}{2}\pi\rho R_n^2 V^2 C_x \alpha_c = C_l \alpha_c \quad (4)$$

Force acting on fins can be expressed as:

$$F_f = nC_l \alpha_f \quad (5)$$

The parameter n is the effectiveness of surfaces in provided lift, it is a function of fins effectiveness relative to the cavitator effectiveness. In both force equations, the angle-of-attack terms are assumed to be small so that small angle approximations

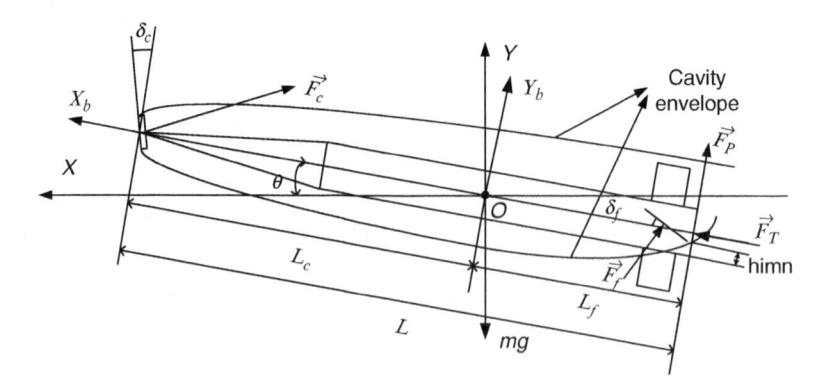

Fig. 2 Schematic of geometry and force of supercavitating vehicle

apply to trigonometric functions. The fins effectiveness n = 0.5 in the whole proposed for Dzielski's model.

The planning force is provided by interaction of vehicle aft and cavity wall. If the diameter of cavity at the planning location is R_c, then the planning force can be approximated as Eq. 6 in small immersion angle:

$$F_{plane} = -\rho R^2 \pi V^2 \left(1 - (R'/(h' + R'))^2\right)\left(\frac{1+h'}{1+2h'}\right)\alpha_p \quad (6)$$

Where, R' denotes the normalized distance from cavity radius to vehicle radius, h' is the immersion depth and α_p is the immersion angle.

Due to the existence of vehicle aft planning force, the system has nonlinear terms, hence the mathematic model of vehicle longitudinal motion is expressed as the affine nonlinear systems [4]:

$$\dot{x} = f(x) + G(x)u$$
$$y = H(x) \quad (7)$$

Where $x \in R^4$, $u \in R^2$, $y \in R^2$, $f(x)$ is four dimensions full smooth vector field; and $G(x)$ is 4×2 dimension full smooth matrix.

$$H(X) = [h_1(x), h_2(x)]^T = [x_1, x_3]^T \quad x = [x_1 \ x_2 \ x_3 \ x_4]^T = [y, V_y, \omega_z, \theta]^T \ y$$
$$= [y_1 \ y_2]^T \quad (8)$$

$$f(X) = [f_1 \ f_2 \ f_3 \ f_4]^T$$

$$= \begin{bmatrix} x_2 - Vx_4 \\ \frac{(C_1 - C_2)}{(MV)}x_2 + \frac{(C_1 L_c - C_2 L_f)}{(MV)}x_4 + g \\ \frac{(-C_1 L_c + C_2 L_f)}{(I_{yy}V)}x_2 + \frac{(C_1 L_c^2 - C_2 L_f^2)}{(I_{yy}V)}x_4 \\ x_3 \end{bmatrix} + \begin{bmatrix} 0 \\ \frac{C_p}{M} \\ -\frac{C_p}{I_{yy}}L_f \\ 0 \end{bmatrix} F_{plane} \quad (9)$$

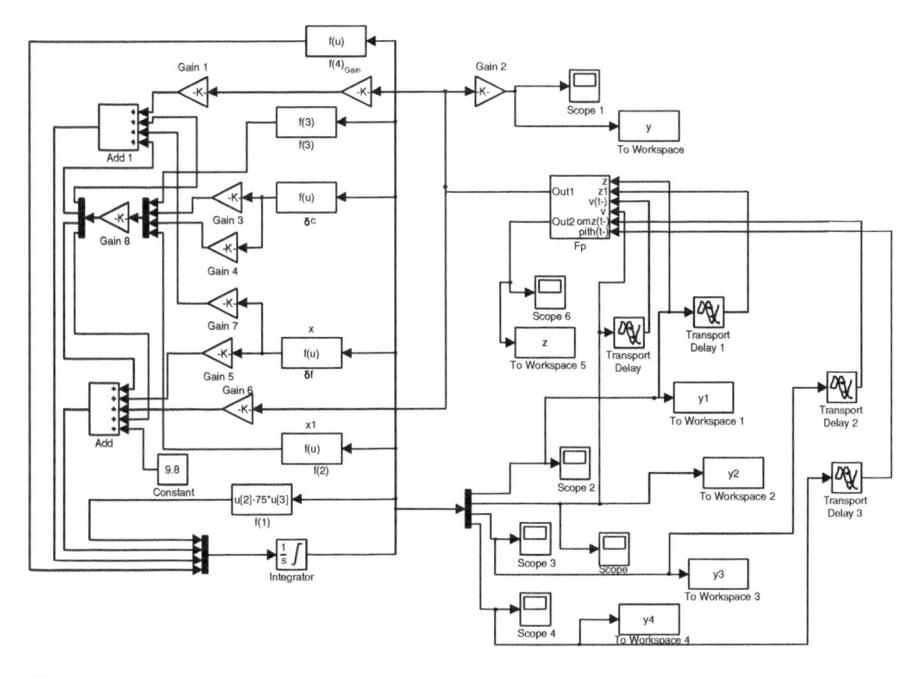

Fig. 3 Simulation model of high-speed underwater vehicle

$$G(x) = [g_1 \quad g_2]$$
$$u = [u_1 \quad u_2]^T = [\delta_c, \delta_f]^T \qquad (10)$$

Where

$$g_1(x) = \begin{bmatrix} g_{11} \\ g_{12} \\ g_{13} \\ g_{14} \end{bmatrix} = \begin{bmatrix} 0 \\ C_1/M \\ -C_1 L_c / I_{yy} \\ 0 \end{bmatrix}, g_2(x) = \begin{bmatrix} g_{21} \\ g_{22} \\ g_{23} \\ g_{24} \end{bmatrix} = \begin{bmatrix} 0 \\ C_2/M \\ -C_2 L_c / I_{yy} \\ 0 \end{bmatrix} \qquad (11)$$

The simulation model of high-speed underwater vehicle is established by MATLAB/SIMULINK (Fig. 3).

3 Design Controller of High-Speed Underwater Vehicle

3.1 Model Linearization

Based on differential geometry theory, first verify whether or meet the conditions of state variable feedback exact linearization for the affine nonlinear system (9). If a matrix $g(X_0)$ rank is m, then state-space exact linearization problem is solvable, if and only if [5]:

- For each $0 \leq i \leq n - 1$, distribution G_i dimension is invariant near the X_0.
- Distribution G_{n-1} have dimension n.
- For each $0 \leq i \leq n - 2$, the distribution G_i is involution.

In the system (9), $G_0 = \text{span}\{g_1, g_2\}$ in neighborhood X_0 has dimension $m = 2$, because $[g_1, g_2](x) = 0$ and $rank(g_1, g_2) = rank(g_1, g2, [g_1, g_2])$, so distribution G_0 is involution.

Calculation of the dimensions of distribution G_1, G_2, G_3, so G_1, G_2 is involution. G_3 is 4, that equal to n.

Based on the above calculation, the multi-input affine nonlinear Eq. 9 satisfied the conditions for exact linearization, could transformed into linear and control system based on feedback and appropriate coordinate transformation.

3.2 Control System Design

Based on above calculation and analysis, that in the neighborhood U of X0 have $m = 2$, so $\lambda m\ (x)$ is smooth real output functions. The Eq. 9 is considered as multiple-input multiple-output (MIMO) affine nonlinear system [6]:

Assumption that output functions are: $\lambda_1(x) = x_1$, $\lambda_2(x) = x_4$.

$$\begin{cases} \dot{z}_1 = \dot{x}_1 = f_1(x) = z_2 \\ \dot{z}_2 = f_2(x) - Vf_4(x) = v_1 \\ \dot{z}_3 = \dot{x}_4 = f_4(x) = z_4 \\ \dot{z}_4 = f_3(x) = v_2 \end{cases} \quad (12)$$

So chosen output functions $\lambda m\ (x)$ is meet conditions, because $A(x)$ is nonsingular matrix in X0. Coordinate transform: $z_1 = x_1$; $z_2 = x_2 - V\ x_4$; $z_3 = x_4$; $z_4 = x_3$. Let input of system: $\dot{z}_2 = v_1$; $\dot{z}_4 = v_2$.

Design a control law be used by MIMO pole placement method and coordinate transforming, and obtained controller Eq. 13.

$$u = \begin{bmatrix} \delta_c \\ \delta_f \end{bmatrix} = \begin{bmatrix} k_{11} & k_{12} & k_{13} & k_{14} \\ k_{21} & k_{22} & k_{23} & k_{24} \end{bmatrix} \begin{bmatrix} x_1 \\ x_2 \\ x_3 \\ x_4 \end{bmatrix} \quad (13)$$

4 Numerical Simulation

The following conditions to ensure the cavity integrated and the vehicle moved stably:

- Immersion depth $h < R/6$, R is vehicle radius.
- $0.02 < \sigma < 0.036$.
- θ, δ_c, δ_f is not more than 15 ° (0.26 rad).

Fig. 4 Simulation to variable-depth motion of vehicle

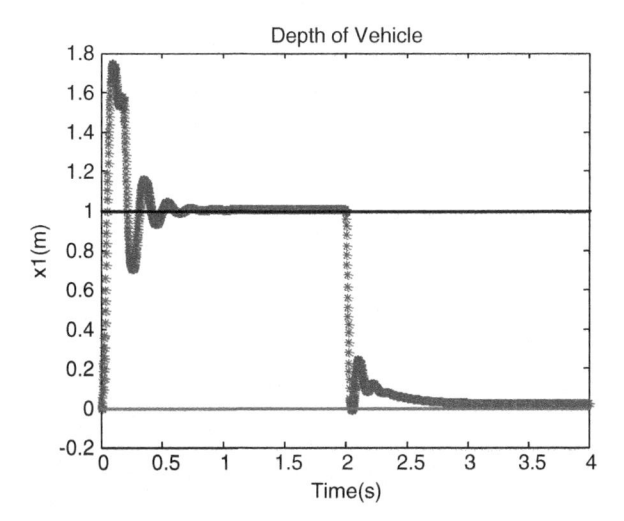

Parameters of vehicle in simulation model follow:

Depth objective: $y_1 = 1$ m, $y_2 = 0$ m; Gravitational acceleration g = 9.8; Density ratio (water) m = 2; Vehicle mass $M = 22.7$ kg; Vehicle length L = 1.8 m; efficiency coefficient n = 0.5; Cavitator radius $R_n = 0.0191$(m); Vehicle radius R = 0.0508(m); Velocity V = 75(m/s); Drag coefficient $C_x = 0.82$.

Motion simulations of vehicle are performed in Matlab/Simulink, all parameters is respect to the above setup. The move is a variable-depth motion: the horizontal speed is 75 m/s, while the vehicle moves down 1 m under only cavitator control, then cavitator-fins joint control to returns to continue its straight path as seen in Fig. 4. The cavitator-fins joint control method can ensure vehicle moving stably in different depth.

In Fig. 5, θ, δ_c, δ_f is not more than 0.26 rad, the immersion depth not more than 0.008 m in most of the time, which satisfy control demand, meet the requirement for the vehicle stable motion control.

Based on analysis of the simulations result, it can be seen that both of the different depth control via the cavitator-fins joint control are efficacious and the motion stability can be ensured for supercavitation vehicle. Consequently, the control method and algorithm in paper is simple and well implemented.

5 Conclusion

Through the motion analysis to high-speed underwater vehicle, conclusions are achieved as the following: the improved system longitudinal motion model, it accords with the real motion law of the high-speed underwater vehicle. It can be seen from the simulations result, that the vehicle depth can quickly track settings point, the controller designed has good tracking performance. At the same time,

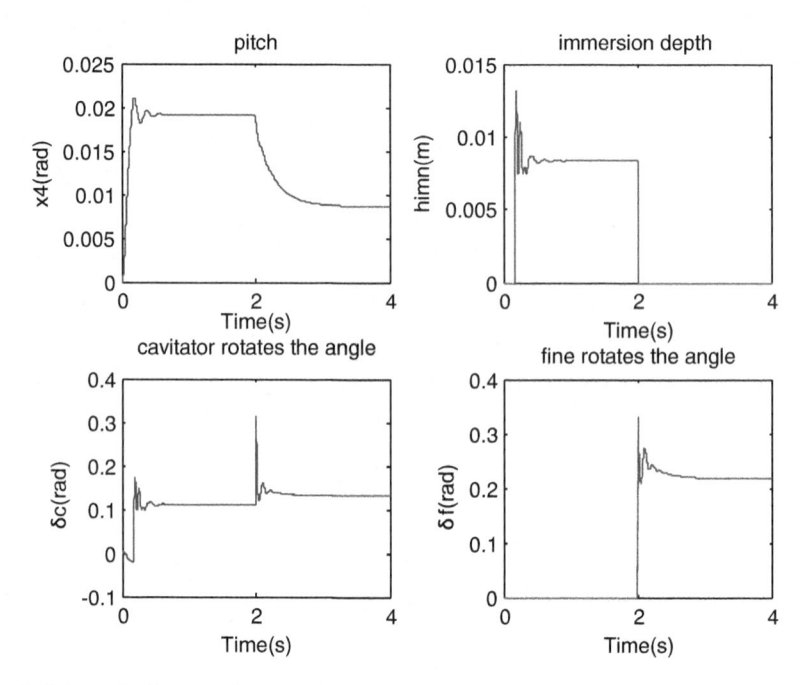

Fig. 5 Schematic diagram of parameter variation

when vehicle is controlled by the designed controller, all state variables return to steady-state value after changing in the allowed range, that can meet the control requirements of system. It is verified that suitable controller can successfully to eliminate motion oscillations and strengthen the stability of the system of high-speed underwater vehicle.

Acknowledgements The work was supported by "the Fundamental Research Funds for the Central Universities" No. HEUCF041212, HEUCFR1211, and "Province in Heilongjiang youth science fund" No. QC2011C031.

References

1. Savchenko Yu N (2001) Control of supercavitation flow and stability of supercavitating motion of bodies, vol 13, VKI special course on supercavitating flows, Brussels. RTO-AVT and VKI. RTO-EN-010, Brussels, pp 212–221
2. Savchenko Yu N (2001) Control of supercavitating motion of bodies, vol 11, VKI special course on supercavitating flows. RTO-AVT and VKI. RTO-EN-010, Brussels, pp 313–329
3. Vanek B, Bokor J, Balas G (2006) Theoretical aspects of high-speed supercavitation vehicle control. In: Proceedings of the 2006 American control conference, IEEE, Minneapolis, pp 5263–5268

4. Vanek B, Bokor J, Balas GJ (2007) Longitudinal motion control of a high-speed supercavitation vehicle. J Vib Control 13(2):159–184
5. Isidori A (1995) Nonlinear control system, 3rd edn. Springer, London, pp 101–112
6. Yuan Lei, Wu Han-song (2010) Multiple sliding mode adaptive fuzzy controller for nonlinear marine autopilot systems. Trans Intell Syst 5(4):308–312

A Low Power Received Signal Strength Indicator for Short Distance Receiver

Qianqian Lei, Erhu Zhao, Min Lin, and Yin Shi

Abstract A low power received signal strength indicator (RSSI) with limiter amplifier (LA) and an integrated automatic gain control (AGC) loop are fabricated in SMIC 0.13 μm technology for short distance receiver. The proposed limiter achieves minimum power dissipation which uses six-stage amplifier architecture. The RSSI has larger than 58 dB dynamic linear range, and the linearity error of the RSSI is less than 1 dB with the input power of limiter chain from −65 to −8 dBm. The voltage of RSSI output is from 0.13 to 0.98 V, the slope of output curve is 14.65 mV/dB, and the RSSI consumes 1.5 mA under 1.2 V supply voltage. By the proposed AGC loop, the receiver senses the input signal strength automatically and modifies the low noise amplifier (LNA) gain in order to make the strength of the output signal stable.

Keywords Limiter • RSSI • AGC

1 Introduction

System of wireless communication, propagation loss or multi-path fading causes the strength of the received signal to change several decades in the receiver. The variation of magnitude increases the loading of dynamic range, which makes the receiver synchronization mechanism complicated. In order to ease the proposed problem, at the last stage of intermediate frequency (IF) processor, the control mechanism of the signal magnitude needs to be provided, which is used to make the constant output signal for being demodulated further.

Q. Lei (✉) • E. Zhao
Department of physics, Xi'an Polytechnic University, Xi'an, China
e-mail: leiqianqian@163.com

M. Lin • Y. Shi
Suzhou-CAS Semiconductors Integrated Technology Research Center, Suzhou, China

Generally, magnitude control circuit has two types, namely LA and AGC. An AGC is configured as a closed loop structure which consists of a variable gain amplifier and a detector of magnitude to adjust the receiver gain [1, 2] However, a limiter amplifier is made of a gain stages chain saturates the input scope to a constant value, which is an open-loop structure. With less current dissipation, LA can achieve a very larger dynamic range, so the LA is broadly used than an AGC in wireless transceiver. Besides, the RSSI is generally realized in logarithmic-linear form which is used to show the strength of the received signal.

This paper is organized as follows: System requirements and circuit architecture designs are depicted in Sect. 2. Detailed circuit design of the RSSI, LA and comparator are presented in Sect. 3. Measurement results are supplied in Sects. 4 and 5 gives conclusion, finally.

2 System Requirements and Circuit Architecture

Figure 1 shows the AGC loop architecture which is mainly composed of RSSI, LA, comparator, FSM (Finite State Machine) circuit and voltage generator. From Fig. 1, the LA amplifies the input signal to provide a signal of constant amplitude. RSSI circuit indicates the strength of the received signal is employed, be compared with the given voltages V_{IH} and V_{IL}, the RSSI output voltage can generate corresponding values, then, the FSM circuit produces a DC control voltage to change the low noise amplifier (LNA) gain according the compared results. Using this feedback loop, the system can automatically sense the receiver input signal strength and change the LNA gain to maintain constant strength of limiter output.

The RSSI characteristic can be seen in Fig. 2a and the state diagram of LNA gain can be seen in Fig. 2b. The principle of LNA gain tuning can be described as follows: When the input signal of LNA is larger enough, the LNA circuit should work under the set of low gain mode finally. However, if LNA works on the state of high gain initially, the control word of LNA is 11. On the basis of the larger input signal, the RSSI can output a smaller voltage which is compared with the first low reference voltage V_{AL} and exports a new product to vary the state of FSM, and the new state 10 through the loop feedback to the control port to vary the LNA gain, the LNA works under the gain state of moderate at present. On this state, if the output of RSSI is also a smaller voltage compared with second reference voltage V_{BL}, then, the present state will be changed by FSM, and new control word 01 should be supplied to LNA circuit to make the circuit works on the state of low gain, which forces the AGC loop to stability. On the other hand, if the input signal is smaller enough, using the same feedback and comparison process, the AGC loop can force LNA working on the state of high gain at last.

The block diagram of logarithmic amplifiers and limiters is proposed in Fig. 3. While the limiter input signal is very small, all limiter stages provide full gains. If A is each stage gain and n is stage numbers, thus, the overall gain of the chain becomes A^n. When input signal of chain raises and achieves a special level, the

A Low Power RSSI with Integrated AGC Loop

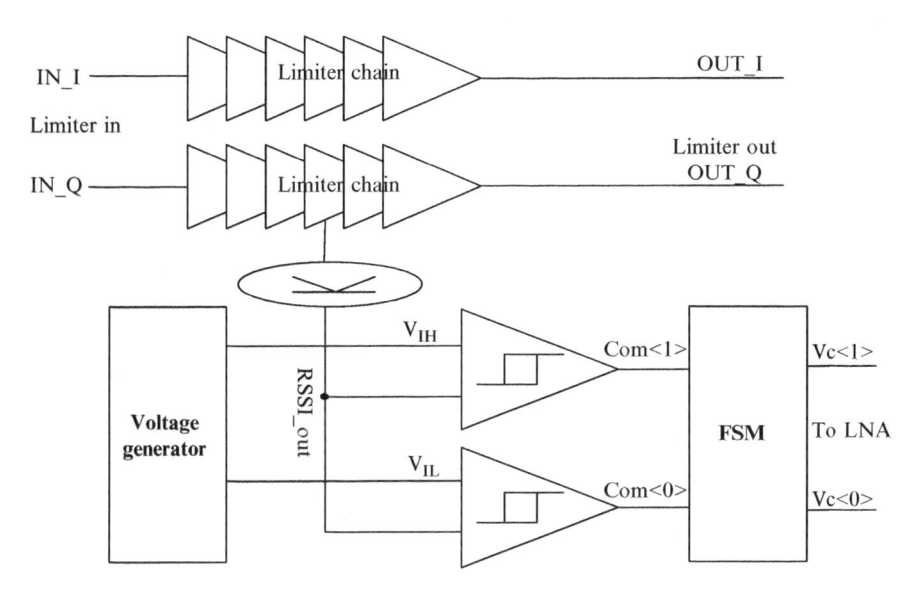

Fig. 1 The system architecture of AGC loop

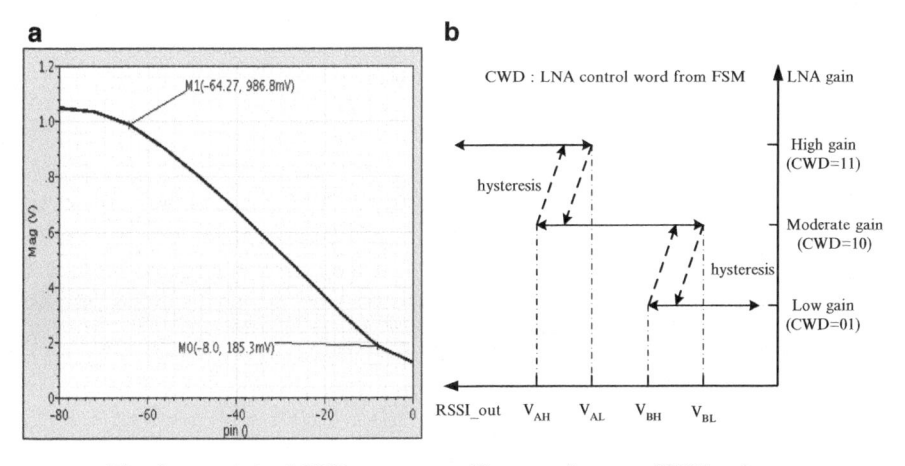

Fig. 2 (a) The characteristic of RSSI (b) The state diagram of LNA gain

end stage of the limiter chain starts to clip, so the full gain of the chain converts A^{n-1}. As increasing of the LNA input signal, more limiter gain stages clip. While the LNA input rises farther, every limiter stages clip, finally, so the gain of limiter chain turns one. Thus, the strength of RSSI input determines all gain piece wisely.

Also, the proposed amplifiers chain in Fig. 3 can extract the RSSI information. The input stage of rectifiers are carried out to transform voltage into current signal at every stage node, thus, each current of rectifier is summed and transformed into

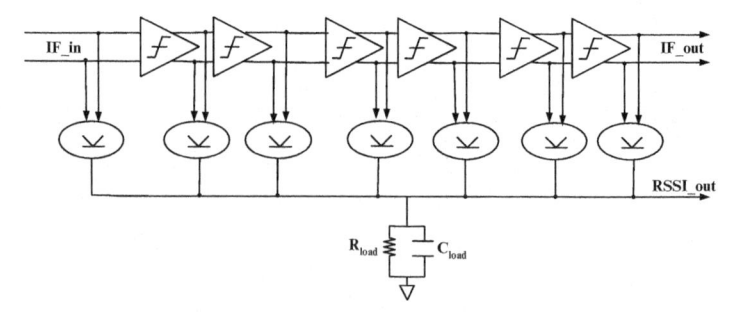

Fig. 3 Block diagram of RSSI

voltage with a resistor R_{load}. And AC component can be filtered by paralleled capacitor C_{load} and resistor R_{load}.

In order to achieve gain, allowable errors, bandwidth as well as low power consumption, it is significant to select the number of stages in the limiter and RSSI design. Increasing the limiter chain stage numbers, there are more poles are generated from every LA, and the bandwidth of the chain will reduce [3, 4]. Furthermore, if the stage number is six or seven, the chain obtains minimum power consumption with the overall bandwidth is 10 MHz and overall gain is 60 dB.

3 Circuit Designs

Figure 4 shows the individual LA circuit, the limiter uses current mirrors to eliminate the load transistors body effect [4]. The circuit configuration can use in low supply voltage. As can be seen in Fig. 4, if the size of transistors M_{3-4} and M_{7-8} has the same values, and the input transistors have the identical bias current. So, the amplifier gain can be given as

$$A_V = \frac{g_{M1}}{g_{M5}} = \frac{\sqrt{(W_1/L_1)}}{\sqrt{(W_5/L_5)}} \quad (1)$$

From Eq. 1, we can see the gain of the limiter amplifier circuit is determined by the transistor ratio, so, the amplifier gain is insensitive to thermal variations and process during prospered design.

Figure 5 shows the rectifier circuit [5]. The rectified signals are generated from differential pairs with unbalanced source-coupled. And one differential pair size is k times as big as the other differential pair size.

The rectifier current I_{out} can be shown as

$$I_{out} = (I_2 + I_3) - (I_1 + I_4) \quad (2)$$

Fig. 4 The limier gain cell

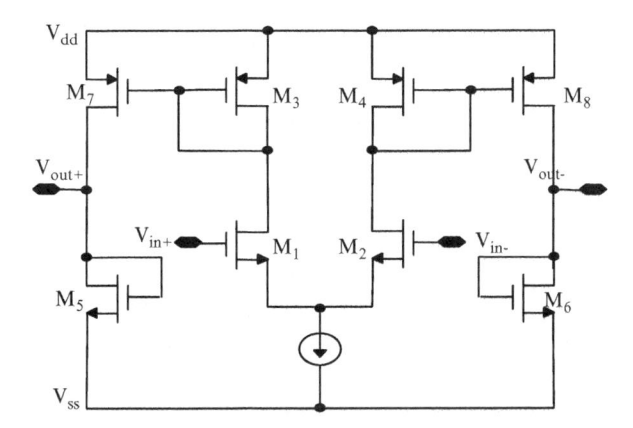

Fig. 5 The rectifier cell

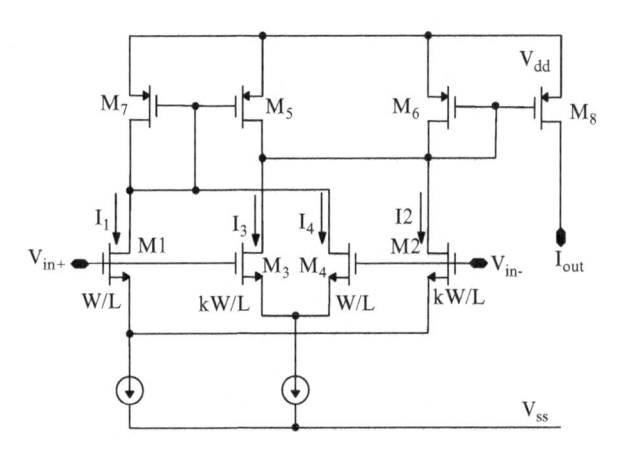

As the input voltage of rectifier has no differential, the larger size transistor M_2, M_3 current will be K-times bigger than the current of smaller size transistor M_1, M_4. As input voltage is very small, the M_2 and M_3 is flowed by most of current. As a result, the current I_{out} will achieve maximum value based on Eq. 2, because the current flow into M_6 and M_8 which is the right side current mirror is larger than the current mirror (M_5 and M_7) of the left side. At the same time, the current I_{out} will decrease, when the input voltage differential increases.

As input voltage of the schematic in Fig. 5 increases, transistors M_1 and M_4 begin to get the current flow into the left side current mirror. Thus, the current of the right side decreases, thus, the input voltage determines the output current. When the input of the differential is large enough, the current flowing into M_{2-4} or M_{1-3} equivalent to the maximum tail currents, and the output current I_{out} will be nearly zero. The proposed rectifier can be uses in lower application because it has only three transistors stacked.

From Fig. 3, we can see each FWR current at output is summed and filtered to a filter which is a first-order passive LPF. We use internal capacitor and resistor as the passive LPF, and use variable resistors to adjust the filter bandwidth and RSSI linear slope.

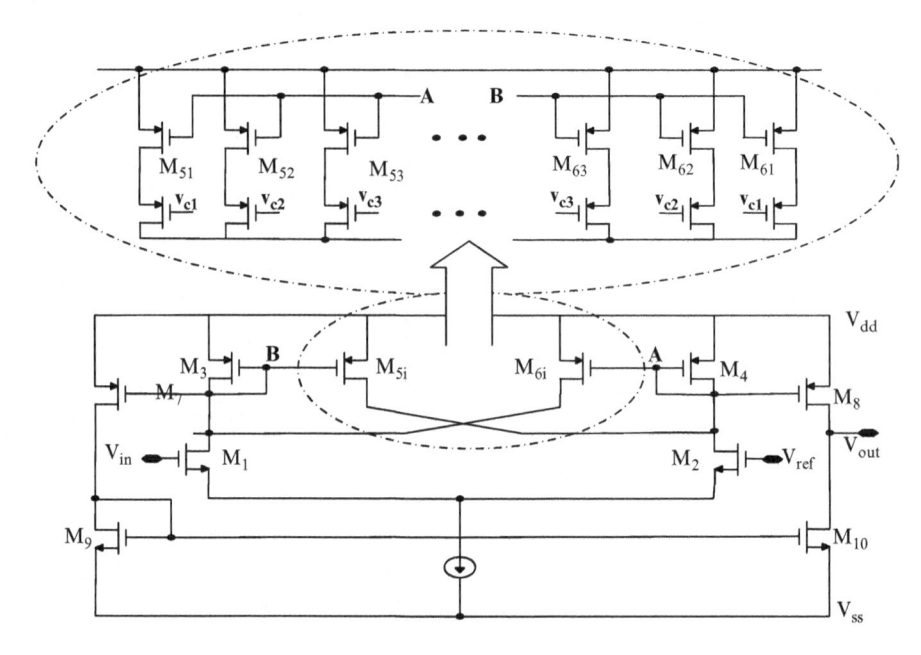

Fig. 6 Schematic of comparator with hysteresis

Figure 6 shows the schematic of comparator circuit which is used to transform the output voltage of RSSI into binary data. Considering the output of RSSI is not a very clean voltage, so we need a comparator with hysteresis. From Fig. 6, we can see the comparator contains a differential source-coupled pair and a differential to single ended converter. The hysteresis of this circuit is

$$V_H = 2 \times \sqrt{\frac{I_0}{k(W/L)_1}} \frac{\sqrt{\alpha}-1}{\sqrt{1+\alpha}} \qquad (3)$$

From the Eq. 3, we can see the comparator hysteresis is dependent on transistors size W/L, voltages of V_{GS} and V_{TH}, so, if the circuit process or supply voltage or temperature (PVT) has changed, the hysteresis of proposed comparator is also varied. In order to compensate the variation, we use control word of three bits to realize changeable hysteresis. Figure 6 gives the realization circuit of variable hysteresis in the dashed line.

4 Measurement Results

The proposed LA and RSSI with internal AGC loop have been implemented in SMIC 0.13 μm technology under $V_{DD} = 1.2$ V, the total power is 1.8 mW (without LNA). The die micrograph of the proposed circuit is shown in Fig. 7, and the die area with pads is 0.4 mm².

Fig. 7 Die photograph of the proposed RSSI with AGC loop

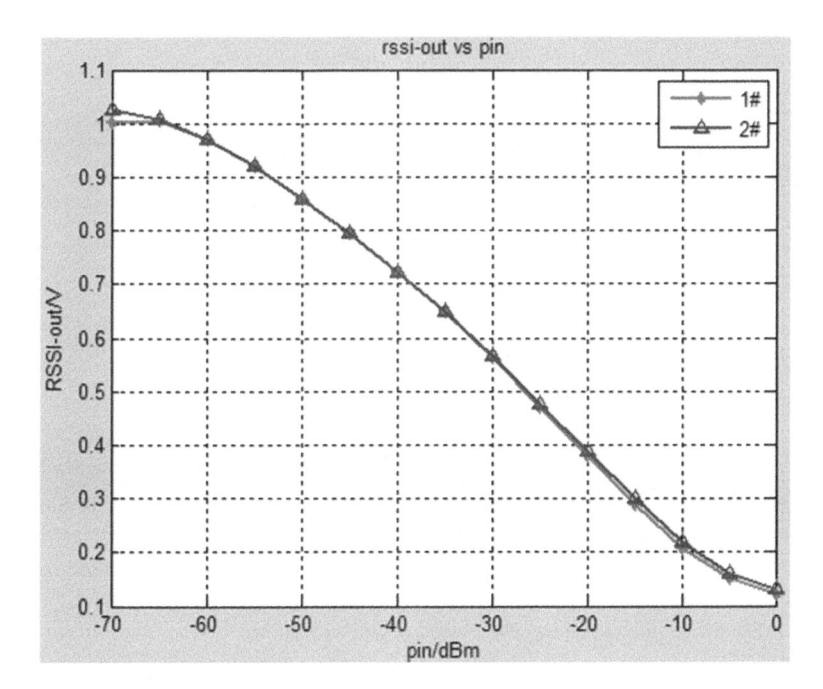

Fig. 8 The characteristic of RSSI

The characteristic of RSSI is tested with 22 Kohm load resistance and 10 pF internal capacitance. We use variable load resistance to achieve the desired value. Figure 8 gives the tested RSSI transfer functions which is graphed with two randomly chips. From Fig. 8, we can seen the linear scope of RSSI is about 58 dB, the linearity error is less than 1 dB, the linear voltage of RSSI is from 0.13 to 0.98 V, and the slope of linear range is achieved as 14.65 mV/dB.

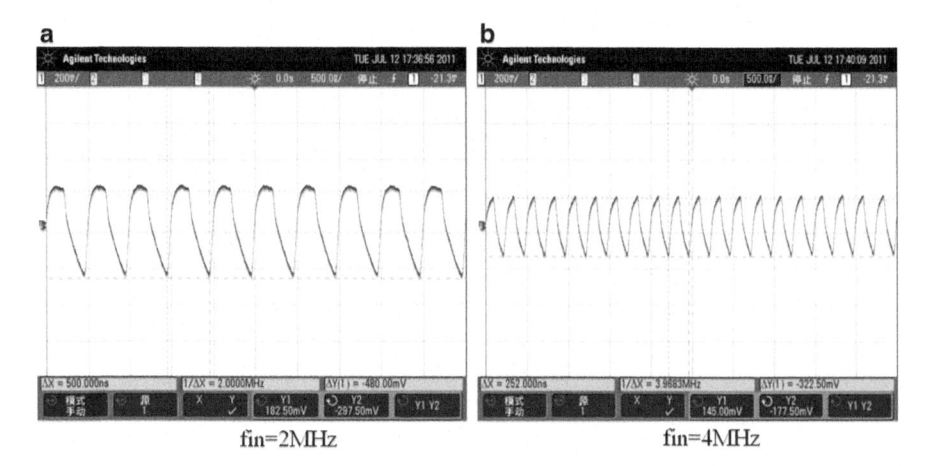

Fig. 9 Measured limiter outputs (**a**) fin = 2 MHz, (**b**) fin = 4 MHz

Table 1 The overall performance of the proposed RSSI

Parameters	Values
Technology	SMIC 0.13 μm technology
Die areas	0.4 mm^2
Output of limiter	480 mV
output voltage of RSSI	0.13~0.98 V
Linearity range	58 dB
Error of RSSI	<1 dB
Slop of RSSI	14.65 mV/dB
Variation of RSSI with temperature	30~40 mV
Supply voltage	1.2 V
Two path currents	1.5 mA

Figure 9 shows the limiter tested outputs under the 4 and 2 MHz frequencies of. As can be seen from Fig. 9, as increasing of frequency, the severely distortion of output waves. In our simulation, the limiter output wave does not have distortion at 4 MHz. The reason is that we use 2 pF capacitance as the simulated load, but use 10 p ~ 20 pF capacitance as the tested load, which distortion the output. The measured limiter output is 480 mVpp. And the measurement results of the proposed RSSI is listed in Table 1.

5 Conclusion

A larger dynamic range lower power RSSI with integrated AGC loop is presented in this paper. The proposed RSSI with AGC loop can automatically control the LNA gain to make stable output. Tested results prove that the proposed RSSI linear range is larger than 58 dB, the output of the limiter peak to peak is 480 mV. Without the front-end LNA, the total loop with two paths consumes 1.5 mA under the 1.2 V supply voltage.

Acknowledgements This work was supported by the Doctoral Scientific Starting Research from the Xi'an Polytechnic University (Grant No. BS1209), and also was supported by Scientific Research Program Funded by Shaanxi Provincial Education Department (Grant No. 12JK0546) and (Grant No. 12JK0981).

References

1. Alegre JP, Calvo B, Celma S (2008) A high performance CMOS feedforward AGC circuit for wideband wireless receivers. In: IEEE international symposium on industrial electronics, Cambridge, UK, pp 1657–1661
2. Jeon O, Fox RM, Myers BA (2006) Analog AGC circuitry for a CMOS WLAN receiver. IEEE J Solid State Circ 41(10):2291–2300
3. Huang P-C, Chen Y-H, Wang C-K (2000) A 2-V 10.7MHz CMOS limiting amplifier/RSSI. IEEE J Solid State Circ 35(10):1474–1480
4. Wu C-P, Tsao H-W (2005) A 110MHz 84-dB CMOS programmable gain amplifier with integrated RSSI function. IEEE J Solid State Circ 40(6):1249–1258
5. Kimura K (1993) A CMOS logarithmic IF amplifier with unbalanced source-coupled pairs. IEEE J Solid State Circ 28(1):78–83

Ship Shaft Generator Control Based on Dynamic Recurrent Neural Network Self-Tuning PID

Ming Sun

Abstract Dynamic recurrent neural network (DRNN) self-tuning PID is proposed to control the unknown dynamic nonlinear function that includes nonlinearity, parameter uncertainty, load disturbance, etc. This control scheme is applied to the hardware-in-the-loop ship shaft generator (SSG) system; it can be mainly utilized to test and verify a real SSG control system on a simulating plant. The performance of DRNN self-tuning PID controller is shown assuming that the measurements available are the AC circuit voltage and current. This controller was also compared with a conventional PID controller for SSG. Simulation and experiment results show that the DRNN self-tuning PID controller is modified to track a desired reference signal.

Keywords Marine power station • Ship shaft generator • Self-tuning PID • DRNN

1 Introduction

Marine main engines (ME) are widely used in ocean-going ships, due to their high efficiency [1]. Optimization methodologies have been introduced into the ocean-going ships. An attractive option for ocean-going ships consists of electricity producing units, such as the ship shaft – generator (SSG) and the exhaust-gas turbo-generator (TG). The SSG is coaxially connected with ME and makes use of residual power of ME and raises its efficiency and economy. The application of SSG significantly reduces the fuel consumption and forms an effective energy-saving strategy [2]. A doubly fed induction generator was developed for a stand alone shaft alternator supplying constant frequency power to more effectively utilize the SSG to save energy [3].

M. Sun (✉)
Qingdao Ocean Shipping Mariners College, Research Center of Marine Energy Efficiency and Emission Reduction, Qingdao, China
e-mail: sunming@coscoqmc.com.cn

PID controllers have dominated control applications for marine power station, although there has been considerable research achievements in the application of advanced controllers [4]. SSG processes have nonlinearity in the system dynamics in a wide operating range and DRNN self-tuning PID control strategies have been investigated in [5]. DRNN can evaluate online the unknown dynamic nonlinear functions that include nonlinearity, parameter uncertainty, load disturbance et al. [6]. Adaptive laws of the adjustable parameters and the evaluation error bounds of DRNN are formulated based on Lyapunov stability theory and a stable adaptive controller is synthesized [4, 6].

In this paper, the SSG of static frequency converter type is introduced, which includes a Thyristor inverter and converter, first used to direct current and then to alternate current at a constant frequency [7]. The mathematical model of the variable speed constant frequency SSG is used to develop a stator flux oriented vector control for a stand-alone system that maintains the output voltage constant [8]. Double self-tuning PID controller gives good stability and transient behavior and the SSG can supply the entire range of power demands while keeping a constant voltage and frequency for variations in both load and speed.

2 Description of the Ship Shaft Generator

The ship shaft generator system is shown as follows [9], which consists of SG (shaft generator), SC (synchronous condenser), PM (run-up motor), FM (fan motor), DCL (DC reactor), ACL (AC reactor), TH1 (thyristor converter), TH2 (thyristor inverter), ACB (air circuit breaker), MCCB (model case circuit breaker), BT (battery) and AVR (automatic voltage regulator).

In Fig. 1, SG is driven by the ME, which takes ME as its prime mover and makes full use of the remaining power of the ME. There is typical thyristor SSG system in Fig. 1, in which voltage, frequency and power can be automatically adjusted to ensure the power system voltage and frequency with high accuracy, while its energy saving effect is more prominent, it is widely used in ocean-going vessels.

3 Control Function

PID1 controller is the current controller, while PID2 controller is the frequency controller. The control circuit droops the frequency instruction signal with the increase of the output power, providing drooping characteristics of the power and eliminating any trouble for parallel running. Function generator is margin angle constant controller. The fire angle of the inverter side thyristor is normally set to approx. 140 (constant). However, while the inverter current increases, the voltage drops due to the revolution of the converter, and the overlap angle enlarges due to

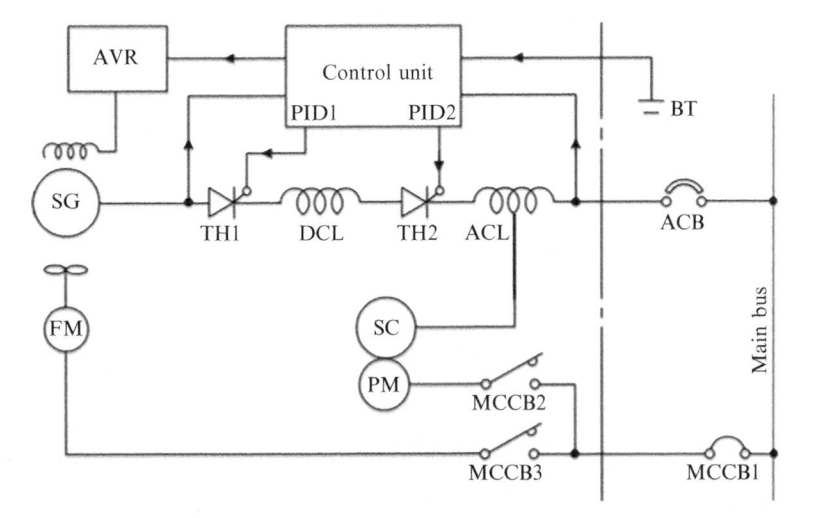

Fig. 1 The skeleton diagram for ship shaft generator system

the increase in S/G revolution, causing the reduction of the margin angle. This circuit, therefore, is to provide the signal to keep this margin angle constant.

The discrete-time PID of the following form:

$$u(k) = K_p(k)x_1(k) + K_i(k)x_2(k) + K_d x_3(k) \tag{1}$$

Where $x_1(k) = e(k), x_2(k) = \sum_{i=1}^{k}(e(i) \times T), x_3(k) = \frac{e(k)-e(k-1)}{T}$. $e(k)$ is the process tracking error defined as $e(k) = y_d(k) - y(k)$. $y_d(k)$ is the desired trajectory, T is the sampling time, (K_{p1}, K_{i1}, K_{d1}) and (K_{p2}, K_{i2}, K_{d2}) are PID1 and PID2 controller parameter matrices, respectively.

A parameter vector $\theta_{\text{pid}}(k)$ is formulated as

$$\theta_{\text{pid}}(k) = \begin{bmatrix} [k_{p1}(k) & k_{i1}(k) & k_{d1}(k)]^T \\ [k_{p2}(k) & k_{i2}(k) & k_{d2}(k)]^T \end{bmatrix} \tag{2}$$

The PID self-tuning algorithm is defined as

$$\theta_{\text{pid}}(i+1) = \theta_{\text{pid}}(i) + \Delta\theta_{\text{pid}}(i) = \theta_{\text{pid}}(i) + K_{\text{pid}}(i)e(i) \tag{3}$$

Where i is the iteration steps, $K_{\text{pid}}i)$ is the gain matrix and is designed as

$$K_{\text{pid}}(i) = \frac{1-\varepsilon}{\varepsilon} \left[\frac{\partial y(i)}{\partial \theta_{\text{pid}}(i)}\right]^T \left\{\frac{\partial y(i)}{\partial \theta_{\text{pid}}(i)} \left[\frac{\partial y(i)}{\partial \theta_{\text{pid}}(i)}\right]^T\right\}^{-1} \tag{4}$$

ε is a positive constant and defined as a discrete-time Lyapunov function $V(i) = \varepsilon e$ $(i)^T e(i) > 0$.

$K_p(k)$, $K_i(k)$ and $K_d(k)$ tuned by DRNN is designed as

$$E(k) = \frac{1}{2}(e(k))^2 \tag{5}$$

$$K_p(k) = K_p(k-1) - \eta_p \frac{\partial E}{\partial K_p} = K_p(k-1) + \eta_p e \frac{\partial y}{\partial u} x_1(k) \tag{6a}$$

$$K_i(k) = K_i(k-1) - \eta_i \frac{\partial E}{\partial K_i} = K_i(k-1) + \eta_i e \frac{\partial y}{\partial u} x_2(k) \tag{6b}$$

$$K_d(k) = K_d(k-1) - \eta_d \frac{\partial E}{\partial K_d} = K_d(k-1) + \eta_d e \frac{\partial y}{\partial u} x_3(k) \tag{6c}$$

η_p, η_i and η_d are the learning rates of proportional, integral and derivative, respectively.

$$\frac{\partial y}{\partial u} \approx \frac{\partial y_{DRNN}}{\partial u} = \sum_j \omega_j^o f'(S_j) \omega_{ij}^I \tag{7}$$

$$\omega_j^o(k) = \omega_j^o(k-1) + \eta_{out} \Delta \omega_j^o(k) + \alpha \left(\omega_j^o(k-1) - \omega_j^o(k-2) \right) \tag{8a}$$

$$\omega_{ij}^I(k) = \omega_{ij}^I(k-1) + \eta_{in} \Delta \omega_{ij}^I(k) + \alpha \left(\omega_{ij}^I(k-1) - \omega_{ij}^I(k-2) \right) \tag{8b}$$

$$f(x) = (1 - e^{-x})/(1 + e^x) \tag{8c}$$

Where η_{in} and η_{out} are the learning rates of DRNN input layer and output layer, respectively, α is the inertia coefficient. S is the hyperbolic functions of recurrent layer neural networks. w^I is the NN weight vector of input layer, w^o is the NN weight vector between recurrent layer and output layer.

4 Modeling of SSG

Suppose the SSG is connected to an infinite bus, then SSG rotating at a synchronous speed ωs and capable of absorbing or delivering any amount of energy, the SSG can be modeled by the following equations

$$\zeta \ddot{\delta}_m + \xi \dot{\delta}_m + P_g = P_m \tag{9}$$

$$T'_{D0}\left|\dot{E}'_a\right| = -\frac{X_D}{X'_D}|E'_a| - \left(\frac{X'_D - X_D}{X'_D}\right)|V_\infty|\cos(\delta_m) + E_{fD} \quad (10)$$

Where $\delta_m = \angle E'_a - \angle V_\infty$ is the SSG power angle with reference to the infinite bus, $\omega = \dot{\delta}_m$ is the rotor angular speed, E'_a – transient voltage, ζ – inertia constant, ξ – damping constant, P_m is the constant mechanical power supplied by the ME, and T'_{D0} – transient time constant. $X_D (= x_D + x_L)$ is the augmented reactance, x_D – the direct axis reactance and x_L – the line reactance, X'_D – the transient augmented reactance, V_∞ – the infinite bus voltage. P_g is the generated power while E_{fD} is the field excitation voltage given by

$$P_g = \frac{1}{X'_D}|E'_a||V_\infty|\sin(\delta_m) + \frac{1}{2}\left(\frac{1}{X_Q} - \frac{1}{X'_D}\right)|V_\infty|^2 \sin(2\delta_m) \quad (11)$$

$$E_{fD} = \frac{\omega_s M_f}{\sqrt{2}r_f} v_f \quad (12)$$

Where v_f is the field excitation voltage, X_Q is the quadrature axis augmented reactance, M_f is the mutual inductance between stator and rotor windings, r_f is the field resistance.

SSG model can be rewritten as

$$\dot{\delta}_m = \omega - \omega_s \quad (13)$$

$$\dot{\omega} = \frac{P_m}{\zeta} - \frac{|V_\infty|}{\zeta X'_D}|E'_a|\sin(\delta_m) - \frac{1}{2}\frac{|V_\infty|^2}{\zeta}\left(\frac{1}{X_Q} - \frac{1}{X'_D}\right)$$
$$\cos(\delta_m)\sin(\delta_m) - \frac{\xi}{\zeta}(\omega - \omega_s) \quad (14)$$

$$\left|\dot{E}'_a\right| = -\frac{X_D}{T'_{D0}X'_D}|E'_a| - \left(\frac{X'_D - X_D}{T'_{D0}X'_D}\right)|V_\infty| + \frac{E_{fd}}{T'_{D0}} \quad (15)$$

$$\dot{\psi}_D = -\frac{\omega_s R_s}{X'_D}\psi_D + \omega\psi_Q + \frac{\omega_s R_s}{X'_D}E'_Q + \omega_s V \sin(\delta_m) \quad (16)$$

$$\dot{\psi}_Q = -\omega\psi_D - \frac{\omega_s R_s}{X'_D}\psi_Q + \omega_s V \cos(\delta_m) \quad (17)$$

Where ψ_D and ψ_Q are the direct and quadrature-axes stator flux linkages, respectively, R_s is an equivalent transmission line resistance.

Suppose the above parameters are constant limited in only equilibrium point and are bounded, in other cases, they are time-variant [10, 11].

Fig. 2 Step response with PID1 controller parameters

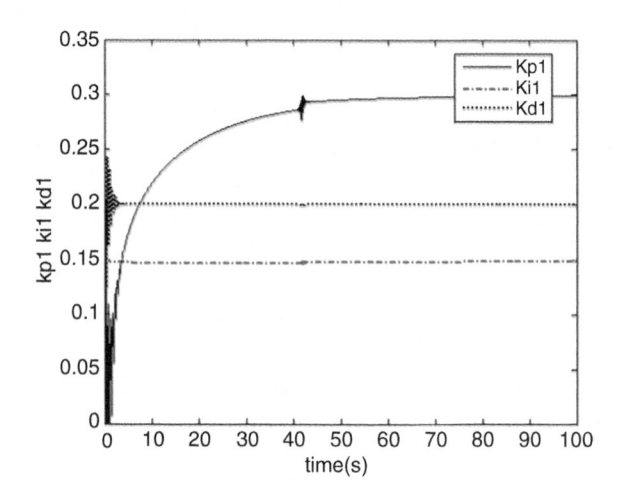

5 Experiments and Results

PID controller is based on DRNN self-tuning algorithm, its parameters are originally obtained as follows: NN input $I = \{u(k-1), y(k), 1.0\}, \eta_{\text{in}} = 0.4, \eta_{\text{out}} = 0.4$, $\alpha = 0.04$, w^o and w^I are random numbers and its span is $[-1,1]$. Initial $K_p = 20$, $T_I = 4$ s, $T_D = 1.75$ s, $\varepsilon = 0.6$; $\eta_{p1} = 0.5$, $\eta_{d1} = 0.3$, $\eta_{i1} = 0.001$; $\eta_{p2} = 0.5$, $\eta_{d2} = 0.3$, $\eta_{i2} = 0.0001$. Sampling time $T = 0.1$ s. Step response results of K_p, K_i and K_d shown in Figs. 2 and 3.

The SSG process controlled by the developed DRNN self-tuning PID control and the fixed parameter PID control is made as follows, the tracking performance for a process is shown in Figs. 4 and 5.

In Fig. 4, it can be seen that the performance of the DRNN self-tuning PID is superior to the fixed parameter PID. In Fig. 5, the SSG hardware-in-the-loop system is detailed in [10, 11], it is able to track the desired reference signals with minimum error. The effects of parameters, such as ε, α and η on the SSG control model, will compensate for the PID parameters variation through DRNN self-tuning. Hence, DRNN self-tuning PID multivariable control has a fast response, adaptive ability, strong anti-disturbance capability.

6 Conclusion

In this paper, a DRNN self-tuning PID control was proposed for SSG system with unknown dynamic state. As the DRNN self-tuning algorithm of the PID was derived using the Lyapunov method, convergence of the model tracking error was guaranteed stable and maintained minimum. Therefore, the DRNN self-tuning PID controller can exert adaptive compensation effect for the system parameter

Fig. 3 Step response with PID2 controller parameters

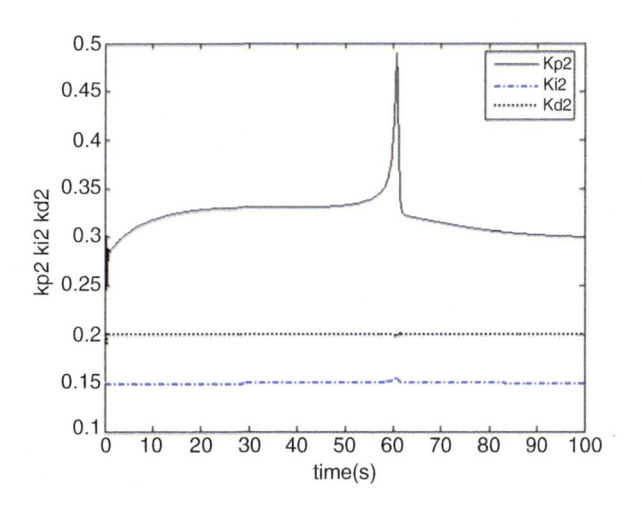

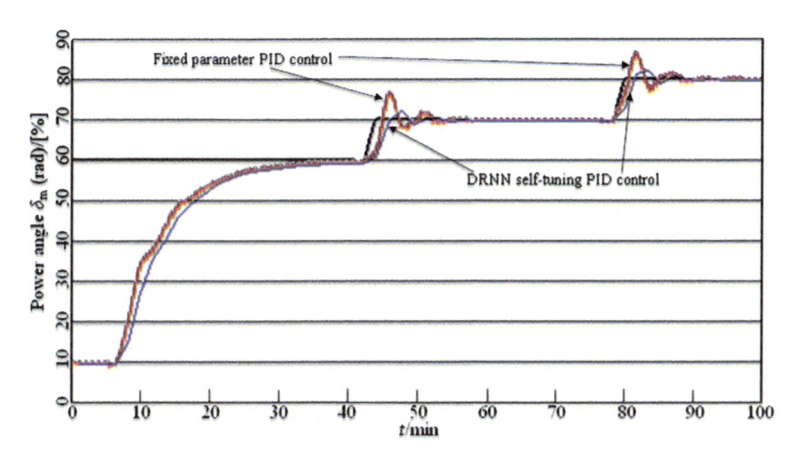

Fig. 4 Process response with PID controller

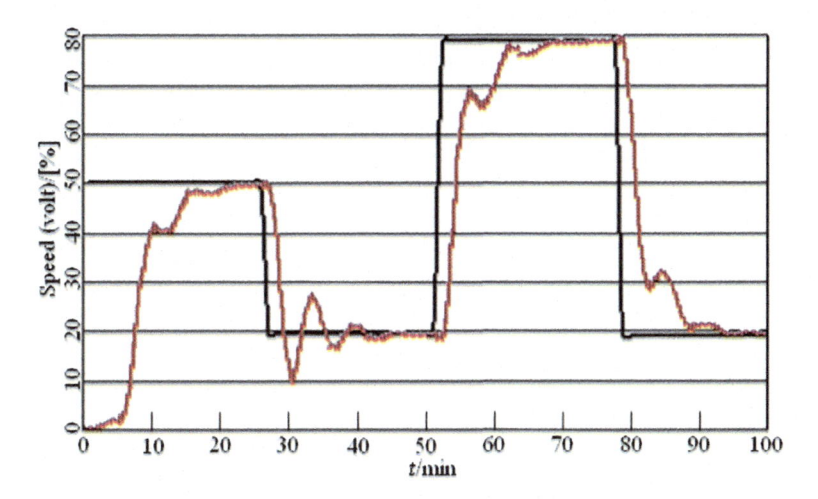

Fig. 5 Hardware-in-the-loop system speed response

uncertainty and disturbance, and SSG system is capable of robustness and good tracking performance. The developed controller is applied to the improvement of hardware-in-the-loop SSG system. The features described above were verified by the simulation results.

Acknowledgements This work was supported by the National Natural Science Foundation of China (51179102).

References

1. Lamaris VT, Hountalas DT (2010) A general purpose diagnostic technique for marine diesel engines-application on the main propulsion and auxiliary diesel units of a marine vessel. Energy Convers Manage 51(4):740–753
2. Gui-chen Zhang, Jie Ma (2011) A general purpose diagnostic technique for marine diesel engines-application on the main propulsion and auxiliary diesel units of a marine vessel. J Harbin Inst Technol 43(11):682–687 (In Chinese)
3. Peng Ling, Li Yong Dong et al (2009) Vector control of a doubly fed induction generator for stand-alone shaft generator systems. J Tsinghua Univ (Sci Tech) 49(7):938–942 (In Chinese)
4. Yu DL, Chang TK, Yu DW (2007) A stable self-learning PID control for multivariable time varying systems. Control Eng Pract 15(12):1577–1587
5. Wei-der Chang, Jun-juh Yan (2005) Adaptive robust PID controller design based on a sliding mode for uncertain chaotic systems. Chaos Soliton Fract 26(1):167–175
6. You-wang Zhang, Wei-hua Gui, Quan-ming Zhao (2005) Adaptive electro-hydraulic position tracking system based on dynamic recurrent fuzzy neural network. Control Theory Appl 22(4):551–556 (In Chinese)
7. Gui-chen Zhang (2010) Application of energy-feedback and energy-saving technology based on SINAMICS in the marine shaft generating. Ship Eng 32(2):43–46 (In Chinese)
8. De Leon-Morales J, Busawon K, Acha-Daza S (2001) A robust observer-based controller for synchronous generators. Elec Power Energy Syst 23(3):195–211
9. Gui-chen Zhang (2012) Modern marine power station. Dalian Maritime University Press, Dalian, p 144 (In Chinese)
10. Gui-chen Zhang (2012) Hardware-in-the-loop for on-line identification of SSP driving motor. Lect Notes Elec Eng 135(31):251–256
11. Saleem A, Issa R, Tutunji T (2010) Hardware-in-the-loop for on-line identification and control of three-phase squirrel cage induction motors. Simul Model Pract Theory 18(3):277–290

Control System Design and Simulation of Microelectromechanical Hybrid Gyroscope

Haiyan Xue, Bo Yang, and Shourong Wang

Abstract In order to achieve the closed-loop control of those microelectromechanical hybrid gyroscopes with large open-loop zero outputs, a self-tuning control system is designed. Firstly, the gyroscope tuning principle is illustrated and the dynamic system is presented. Next, the whole open-loop system is analyzed and the tuning simulation is performed. The tuning simulation results confirm the feasibility of the tuning principle. Finally, the self-tuning control system is designed. The scheme of the closed-loop system is constructed and the corresponding simulation is carried out. The self-tuning simulation results indicate that the closed-loop system can lock the positions of the rotors, showing that the self-tuning control system is feasible.

Keywords Microelectromechanical hybrid gyroscope • Tuning • Self-tuning control • Closed-loop system

1 Introduction

Microelectromechanical hybrid gyroscope evolves on the basis of the dynamically tuning technology and the silicon micro-machined technology [1, 2]. It is featured in the theoretical accuracy of 0.005 °/h, as accurate as Dynamically Tuned Gyroscope (DTG), and the smaller volume compared to the silicon micro-gyroscope [3, 4]. By means of the dynamically tuning technology, the classical decoupling control system can theoretically achieve the closed-loop control on the hybrid gyroscope [5, 6]. However, since the structure design and the assembly technology of the gyroscope is still not imperfect, zero outputs of some gyroscopes are large in

H. Xue • B. Yang (✉) • S. Wang
Key Laboratory of Micro-inertial Instrument and Advanced Navigation Technology of Ministry of Education, Southeast University, Nanjing, China
e-mail: yangbo20022002@163.com

open-loop mode, which makes the integrators saturation in the decoupling network. As a result, the decoupling control system is difficult to achieve the closed-loop control of gyroscopes [7, 8]. It's necessary to seek an innovative method to control these gyroscopes steadily.

2 The Dynamic Analysis of Microelectromechanical Hybrid Gyroscope

The schematic of microelectromechanical hybrid gyroscope is shown in Fig. 1. Traditionally, the tuning in the DTG is achieved by adjusting its stiffness coefficient of torsion bars, the gimbal inertia moment and the motor's speed. However, the rotor structure is not easy to be adjusted repeatedly by means of the bulk silicon process. Besides, the gimbal negative stiffness coefficient is only 10^{-3} times of the torsion bar. In this way, a high speed motor which can be accurately adjusted is required in the hybrid gyroscope. But this is difficult to realize. At the same time, in order to make full use of the static electricity feedback of the silicon micro mechanical structure, a preload voltage is exerted on torque feedback electrodes in the hybrid gyroscope, generating the electrostatic torque compensate stiffness moment of torsion bars. Referring to the dynamics equations of DTG, the hybrid gyroscope dynamics equations can be deducted [9, 10]:

$$J\ddot{\beta} + \delta\dot{\beta} + (\Delta K - K_C)\beta + H\dot{\alpha} + \lambda\alpha = M_{fx} - J\ddot{\varphi}_x - H\dot{\varphi}_y \\ J\ddot{\alpha} + \delta\dot{\alpha} + (\Delta K - K_C)\alpha - H\dot{\beta} - \lambda\beta = M_{fy} - J\ddot{\varphi}_y + H\dot{\varphi}_x \quad (1)$$

Where, $\Delta K = K_p - K_N$. K_p is the stiffness coefficient of torsion bars. K_N is the negative stiffness coefficient of gimbals. K_C is the electric stiffness coefficient. J is the inertia moment of gyroscope spindle, H is the rotation angular momentum of gyroscope. $\alpha, \dot{\alpha}, \ddot{\alpha}, \beta, \dot{\beta}, \ddot{\beta}$ are angles, angular velocities, angular accelerations of gyroscope spindle in the oy-axis and the ox-axis of the shell coordinate. $\varphi_x, \dot{\varphi}_x, \ddot{\varphi}_x, \varphi_y, \dot{\varphi}_y, \ddot{\varphi}_y$ are angles, angular velocities, angular accelerations of the gyroscope shell in ox-axis and oy-axis of inertial space. M_{fx}, M_{fy} are feedback torques exerted on the

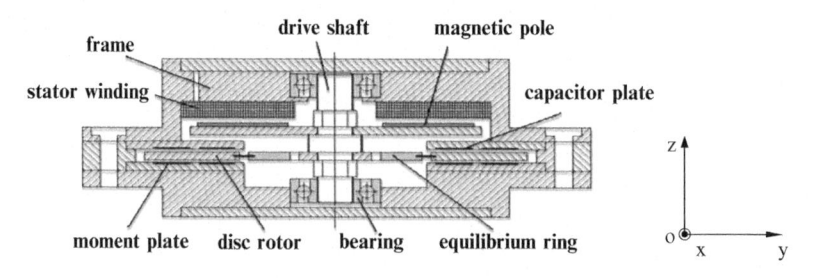

Fig. 1 Schematic diagram of microelectromechanical hybrid gyroscope

ox-axis and the oy-axis of the shell coordinate. λ is the orthogonal damping elasticity coefficient, $\lambda = (\delta + D)\dot{\theta}$. δ is the viscous damping coefficient of flexible torsion bar. D is the damping coefficient of rotor rotation axis. The Laplace transformation of Eq. 1 is as follows:

$$\begin{bmatrix} \beta \\ \alpha \end{bmatrix} = \begin{bmatrix} G_{11} & G_{12} \\ G_{21} & G_{22} \end{bmatrix} \begin{bmatrix} M_{fx} \\ M_{fy} \end{bmatrix} - \begin{bmatrix} C_{11} & C_{12} \\ C_{21} & C_{22} \end{bmatrix} \begin{bmatrix} \varphi_x \\ \varphi_y \end{bmatrix} \quad (2)$$

Where,

$$\begin{bmatrix} G_{11} & G_{12} \\ G_{21} & G_{22} \end{bmatrix} = \begin{bmatrix} [Js^2 + \delta s + (\Delta K - K_C)]/\Delta(s) & -(Hs + \lambda)/\Delta(s) \\ (Hs + \lambda)/\Delta(s) & [Js^2 + \delta s + (\Delta K - K_C)]/\Delta(s) \end{bmatrix} \quad (3)$$

$$\begin{bmatrix} C_{11} & C_{12} \\ C_{21} & C_{22} \end{bmatrix} = \begin{bmatrix} z_1(s)/\Delta(s) & z_2(s)/\Delta(s) \\ -z_2(s)/\Delta(s) & z_1(s)/\Delta(s) \end{bmatrix} \quad (4)$$

$$z_1(s) = J^2 s^4 + J\delta s^3 + (J(\Delta K - K_C) + H^2)s^2 + H\lambda s$$

$$z_2(s) = (H\delta - J\lambda)s^2 + H(\Delta K - K_C)s$$

$$\Delta(s) = J^2 s^4 + 2J\delta s^3 + (\delta^2 + 2J(\Delta K - K_C) + H^2)s^2 \\ + (2\delta(\Delta K - K_C) + 2H\lambda)s + (\Delta K - K_C)^2 + \lambda^2$$

Where, G_{11} and G_{22} indicate the damping effects on the inertia axis. G_{12} and G_{21} show the damping effects on the orthogonal axis. In the closed-loop state, C_{11}, C_{12}, C_{21} and C_{22} also have damping effects on outputs. Neglecting damping effects, C_{12} and C_{21} characterize the precession of the microelectromechanical hybrid gyroscope, C_{11} and C_{22} show the rigid body characteristic.

3 Microelectromechanical Hybrid Gyroscope Tuning Principle

The hybrid gyroscope structure diagram is shown in Fig. 2 [11]. The upper and lower discs are fixed electrode plates. These electrodes are divided into the inner and outer rings. The inner ring is a detecting electrode and the outer ring is a torque feedback electrode. The intermediate disc is a silicon structure rotor module. The rotor module consists of the inner ring, gimbals, the rotor and torsion bars. In Fig. 2, V_o is the preload voltage, V_f is the feedback voltage in closed-loop condition, V_1 is the carrier voltage, I_{out1} and I_{out2} are the detecting signal outputs of the inner rings.

On one hand, by applying a proper preload voltage on the torque feedback electrodes, the electric stiffness moment is generated to balance the stiffness

Fig. 2 The structure diagram of microelectromechanical hybrid gyroscope

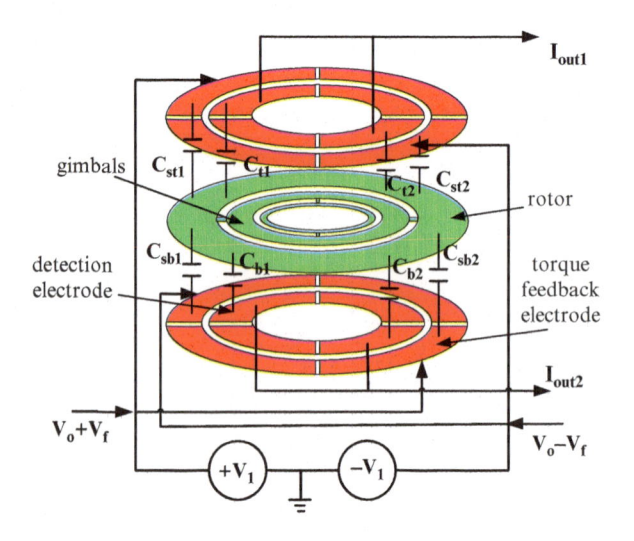

moment of torsion bars. On the other hand, the feedback voltage balances the Coriolis torque caused by the angular velocity.

When applying $V_o + V_f$, $V_o - V_f$ on two normal torque feedback electrodes of y axis, The total moment, made up of the generated electrostatic torque and the torque component in y-axis of the shell coordinates from inside to rotor, is as follows:

$$T'_y + M_y = -(\Delta K - K_C)\alpha + \delta\beta\dot{\theta} + K_{VF}V_f \tag{5}$$

Where the electric stiffness coefficient is $K_C = \varepsilon_s \gamma (R^4_M - r^4_M) V^2_o / d^3$. ε_s is the dielectric constant in air. γ is the single feedback electrode angle. R_M and r_M are outer and inner radiuses of the force feedback electrodes. d is mounting distance between rotor and electrodes.

Thus, a proper preload voltage is exerted to make $\Delta K - K_C = 0$. When the gyroscope is tuned, the preload voltage is as follows:

$$V_o = \sqrt{\Delta K d^3 / \varepsilon_s \gamma (R^4_1 - r^4_M)} \tag{6}$$

Because $\Delta K = 6.858 \times 10^{-4}$ N · m/rad, $d = 40 \times 10^{-6}$ m, $\varepsilon_s = 8.86 \times 10^{-12}$, $\gamma = 1.309$ rad, $R_M = 10 \times 10^{-3}$ m, $r_M = 7.05 \times 10^{-3}$ m, the tuning voltage is $V_o = 25.77$ V.

4 Microelectromechanical Hybrid Gyroscope Open-Loop System Analysis and Simulation

In open-loop condition, ignoring the external disturbances and the gyro nutation, the steady-state solutions of hybrid gyroscope dynamic equations are as follows:

Fig. 3 Hybrid gyroscope open-loop system simulation diagram with variable preload voltage

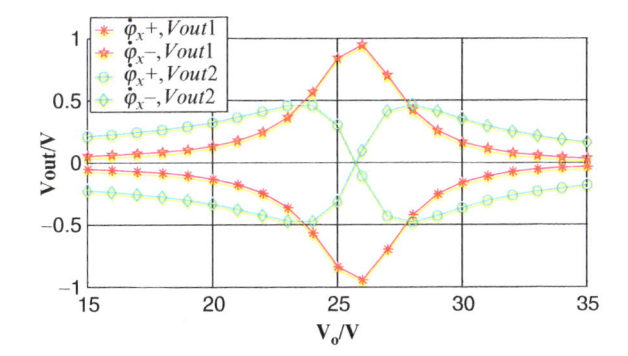

$$\alpha^* = \frac{H\dot{\varphi}_x(\Delta K - K_C) - \lambda H\dot{\varphi}_y}{\lambda^2 + (\Delta K - K_C)^2}$$

$$\beta^* = \frac{-H\dot{\varphi}_y(\Delta K - K_C) - \lambda H\dot{\varphi}_x}{\lambda^2 + (\Delta K - K_C)^2} \quad (7)$$

When $\delta = 1 \times 10^{-7}$, $D = 5 \times 10^{-9}$, angular rate input of y axis $\dot{\varphi}_y = 0$, angular rate inputs of x axis $\dot{\varphi}_x = 1$ °/s and $\dot{\varphi}_x = -1$ °/s, considering the angle-capacitor transforming coefficients $K_{\beta C}$ and the capacitor-voltage transforming coefficients K_{CV}, the inertia axis output voltage *Vout1* and orthogonal axis output voltage *Vout2* are shown as Fig. 3.

From the simulation of variable preload voltages, it is evident that the tuning voltage is 25.77 V which is accordant with the calculation in the upper segment. At the tuning point, the output of the inertia axis reaches a maximum value and the output of the orthogonal axis is zero. Two axis outputs can be obviously distinguished. When off the tuning point, effected by the torsion bars rigidity, the outputs of two axes get closer and closer, and even crossed.

5 Closed-Loop Control System Design and Simulation

The cross-coupling component is expected not to exist in the gyro rotor's output. In other words, the input angular velocities along their own measuring axis have a one-to-one relationship with their corresponding outputs, and won't affect each other. However, confined by the mechanical structure, the cross-coupling between the two measuring axes is inevitable. Since the zero outputs of some gyros are large, the integrator in the decoupling network tends to saturate, which restricts the application of whole decoupling control system. Thereby, the system is designed to be self-turned only.

When the gyroscope works in the tuning state, ignoring the damping, the open-loop transfer functions of the inertia axis and the quadrature axis are respectively as follows:

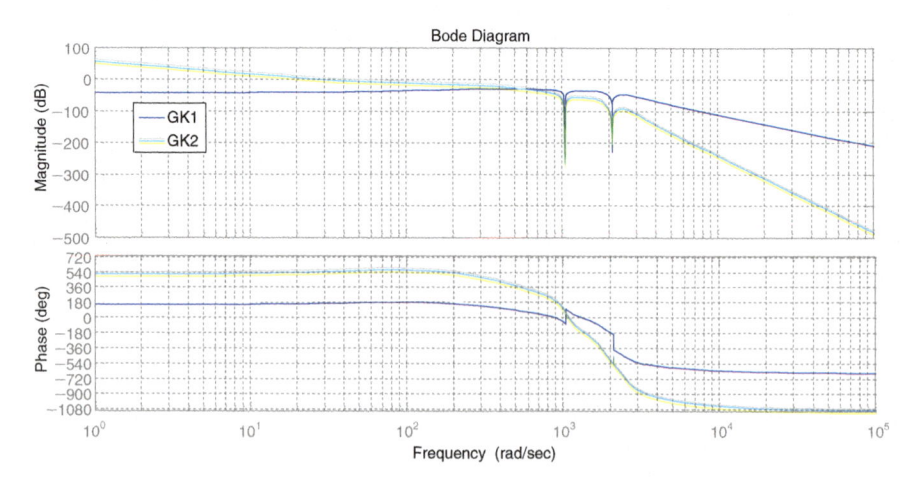

Fig. 4 Open-loop Bode diagram of hybrid gyroscope with self-tuning element

$$G'_{K1} = K_{\beta C}K_{CV}K_{VF} \Big/ \left\{ J[s^2 + (H/J)^2] \right\} \quad (8)$$

$$G'_{K2} = K_{\beta C}^2 K_{CV}^2 K_{VF}^2 H^2 \Big/ \left\{ J^2 s[s^2 + (H/J)^2] \right\}^2 \quad (9)$$

Where, K_{VF} is the torque feedback coefficient. Considering adopting the band-stop filter to eliminate the motor disturbance and nutation, the system with the proper configuration of zeros and poles should be fast, accurate and steady in response. The open-loop transfer functions are as follows:

$$G_{K1} = \frac{2.4456 K_{\beta C}K_{CV}K_{VF}(1/60\ s + 1)}{J[s^2 + (H/J)^2](1/1,500\ s + 1)(1/1,500\ s + 1)} \quad (10)$$

$$G_{K2} = \frac{5.9811 K_{\beta C}^2 K_{CV}^2 K_{VF}^2 H^2 (1/60\ s + 1)^2}{\left\{ J^2 s[s^2 + (H/J)^2](1/1,500\ s + 1)(1/1,500\ s + 1) \right\}^2} \quad (11)$$

The Bode diagram of the open-loop system is shown in Fig. 4. It can be seen from the diagram that when the inertia axis amplitude is under 0 dB, the intermediate region of the precession axis is gentle. The crossover frequency is 48 rad/s. The amplitude margin is 10.4 dB and the phase margin is 45°. The system can lock the rotor position quickly, accurately and stably.

Based on the design of the self-tuning system, the closed-loop system block diagram is built, as Fig. 5. Where, k is the self-tuning element $(1/60\ s + 1)/((1/1,500\ s + 1)(1/1,500\ s + 1))$, $k_x = k_{\beta C} \times k_{cv}$, k_l is the low pass filter and band-stop filter.

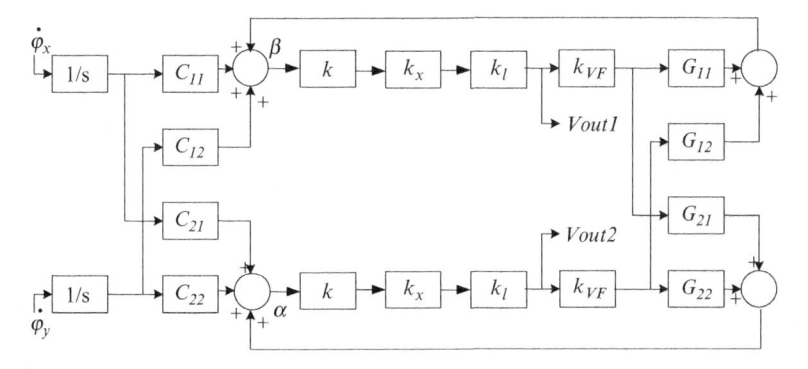

Fig. 5 Hybrid gyroscope closed-loop system block diagram

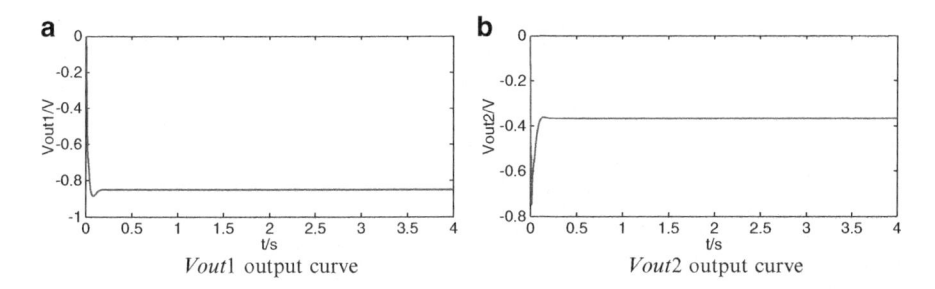

Fig. 6 When $\dot{\phi}_x = 0°/s$ and $\dot{\phi}_y = 1°/s$, the system outputs (**a**) *Vout*1 output curve, (**b**) *Vout*2 output curve

Considering the damping, the tuning voltage is utilized in the system. When $\dot{\phi}_x = 0°/s$, $\dot{\phi}_y = 1°/s$, Vout1 and Vout2 are shown in Fig. 6. From the graph, the system realizes the closed-loop control for the gyroscope. Due to the effect of damping, the response time is 0.15 s. The output of the orthogonal axis not only tracks the input angular velocity but also is affected by the damping. The response amplitude is 0.85 V. The output of the inertia axis is 0.37 V under the effect of damping. It should be explained that the damping coefficients of microelectromechanical hybrid gyroscope are the valuations in reference to DTG.

6 Conclusion

This paper introduces a microelectromechanical hybrid gyroscope dynamic model. It briefly illustrates the tuning principle, deduces the corresponding tuning voltage, analyzes the open-loop system and performs the tuning simulation. In view of the large open-loop zero outputs of some gyroscopes, in order to avoid saturation failure of the integrators in the traditional decoupling control system, the self-tuning control system is designed. The simulation results indicate that the self-tuning control system is feasible. When considering damping, the closed-loop system realizes the locking function in controlling the rotor position, but the outputs are coupled under the effect of damping. It is necessary to further probe how to eliminate the damping effect.

References

1. Jenkins LJ, Hopkins RE, Kumar K et al (2003) Hybrid wafer gyroscope. US Patent 6615681
2. Arif SM, Raghunandan PS, Andhra TA et al (2003) A novel method to measure the decay frequency of a dynamically tuned gyroscope flexure. Meas Sci Technol 14:2081–2088
3. Ren Wang, Peng Cheng, Fei Xie et al (2011) A multiple-beam tuning-fork gyroscope with high quality factors. Sensor Actuator A 166:22–33
4. Dong Y, Kraft M, Hedenstierna N et al (2008) Microgyroscope control system using a high-order band-pass continuous-time sigma-delta modulator. Sensor Actuator A 145–146:299–305
5. Jin Woo Songa, Jang Gyu Leea, Taesam Kang (2002) Digital rebalance loop design for a dynamically tuned gyroscope using H2 methodology. Control Eng Pract 10:1127–1140
6. Yen-Cheng Chen, M'Closkey RT, Tuan A et al (2005) A control and signal processing integrated circuit for the JPL-Boeing micromachined gyroscopes. IEEE Trans Control Syst Technol 13:286–300
7. Acar C, Shkel AM (2005) An approach for increasing drive-mode bandwidth of MEMS vibratory gyroscopes. J Microelectromech Syst 14:520–528
8. Xu Guoping, Weifeng Tian, Zhihua Jin et al (2007) Temperature drift modelling and compensation for a dynamically tuned gyroscope by combining WT and SVM method. Meas Sci Technol 18:1425–1432
9. Bailin Zhou (2002) Design and manufacturing of dynamically tuned gyroscope. Southeast University press, Nanjing, pp 30–35 (In Chinese)
10. Shie Lin et al (1983) Dynamically tuned gyroscope. National Defense Industry Press, Beijing, pp 40–47 (In Chinese)
11. Bo Yang, Shourong Wang, Kunyu Li et al (2009) System design and simulation of microelectromechanical hybrid gyroscope. J Southeast Univ 39(5):951–955 (In Chinese)

Part V
Data Mining and Application

New Detection Technology Based on the Theory of Eddy Current Loss

Yafei Si and Jianxin Chen

Abstract The equivalent resistance of eddy current power losses was proposed in the paper as characteristic parameter of metal target information, which was in direct proportion with sixth power of detection distance. Using the method of pulse width modulation for detection coil and by detecting the change of coil's freewheeling process, the equivalent resistance of eddy current losses could be measured. It could not only determine the distance but also eliminate environmental interference and improve anti-jamming capability by measuring area difference of coil's freewheeling process between whether there is a target. Target signal detection circuit was designed and target properties measurement results were given.

Keywords Magnetic induction • Eddy current • Detection equivalent resistance

1 Introduction

The electromagnetism detection technology is applied to many fields such as industry, agriculture and national defence [1]. Since the magnetism detection technology is not interfered by cloud, fog, dust and smog in wars and has high sensibility, it is gradually applied to many related products. The theory of the magnetism detection technology is based on Fluxgate Magnetic Technology, Reluctance Technology and Magnetic Film Technology [2]. There is already a paper which investigates a method using vector impedance characteristics of the metal target to detect the changes of probe coil impedance vector when the target approaches [3]. Since we are supposed to detect the changes of voltage with the same direction and different direction while detecting the impedance vector, it is difficult to apply this method to the products. This paper proposes a method based

---

Y. Si (✉) • J. Chen
Beijing University of Technology, Beijing, China
e-mail: debby.1992@yahoo.com.cn; chjx@bjut.edu.cn

on the eddy current power target: The equivalent resistance of the target eddy current power is used as the characteristics of the target amount of information detection. By applying this method, we are able to detect a circuit in a more simple way. This method has a high anti-interference ability also. The result of the intersection of the detector and the target is given in the paper.

2 Researches on the Theory of the Magnetic Sensor Section Heading

A metal conductor is placed next to a detection coil, and when the alternating current passes through the coil, space around the coil will produce an alternating magnetic field H_1. The metal conductor in magnetic field will produce an eddy and an alternating magnetic field H_2. H_1 and H_2 are in different direction. Since the reaction of magnetic field H_2 will reduce the magnetic flux of the detection coil, which means we can detect the induced EMF produced by eddy current magnetic field in the detection coil or the effective impedance of the detection coil will change. The essence of studying the impact of metal target against the detector is to investigate the distribution of the eddy on the surface of metal. Since the working fluency is between 102 and 105 Hz, circuit size is much smaller than operating wavelength, electromagnetic field distribution and the current changes in the distribution at the same time is quasi-static electromagnetic field. The eddy current magnetic field induces by the metal surface follows the equations below [4]:

$$\nabla^2 H(x,y,t) = \mu\sigma \frac{\partial H(x,y,t)}{\partial t} \qquad (1)$$

μ stands for the permeability of the metal target. σ stands for the electrical conductivity of the metal target. Equation 1 is called the electromagnetic penetration equation. Eddy current loss power P_w is [5]:

$$P_w \propto \sum_{m=1}^{\infty}\sum_{n=1}^{\infty} J^2{}_{mn} \propto H_0^2 \propto \frac{1}{l^6} \qquad (2)$$

H_0 is the alternating magnetic field produced by detection coil. L is the distance between the detection coil and the target.

Magnetic sensor is also called magnetic probe, which is made by the core of the probe and some coils. The core of the probe is rolled by Fe-Co-based amorphous ribbons. The profile is shown in Fig. 1. Figure 2 is the circuit diagram of the metal target.

The detection coil in Fig. 1 is regarded as the collector load of transistor T in the Fig. 2. Resistance R_1 and diode D are used as the freewheeling of detection coil B. U_i is a square wave excitation signal with frequency f and duty cycle δ. As the

Fig. 1 Section of detection coin

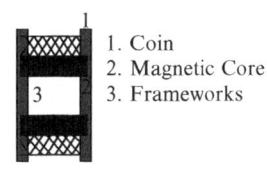

1. Coin
2. Magnetic Core
3. Frameworks

Fig. 2 Detection current of iron

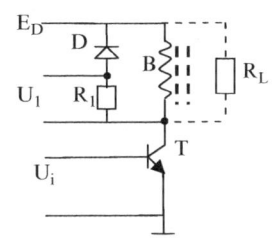

Fig. 3 Detection signal wave

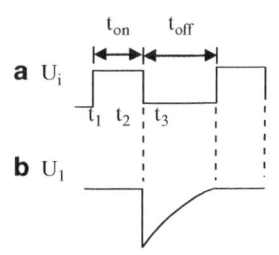

waveform 3a shows, it is used to drive the base of transistor T. U_1 is the voltage waveform of freewheeling resistor R_1 produced by the freewheeling current of the detection coil B during the transistor turns off (Fig. 3b). R_1 is an introduction physical element. It equals the resistance lead by the eddy current loss when the target appears.

Take the oscillation cycle as an example to investigate its working principle. U_i is the PWM(Pulse-Width Modulation) signal produced by oscillator. Its period is $T = T_{on} + T_{off}$ and the duty cycle is $\delta = t_{on}/(t_{on} + t_{off})$. From Fig. 3 we can know that $t_{on} = t_2 - t_1$, $t_{off} = t_3 - t_2$. During $t_1 \sim t_2$, the base of transistor is driven, the transistor turns on, the DC E_D initiates detection coil through transistor T and the current in the coil increases. The transistor turns off in t_2. From Lenz's Law we know that reverse voltage will be produced in both sides of the detection coil. The reverse voltage releases the magnetic energy stored during detection coil turns on via the freewheeling R_1 and D. And after freewheeling process, the core flux will return to zero to ensure that the core will work normally during next period. When the transistor turns on, it actually means to put a battery at both sides of the detection coil. According to the circuit theory, the maximum current in the detection coil is determined by the equation below.

$$i_{L\max} = \frac{1}{L} \int_{t_1}^{t_2} E_D dt = \frac{E_D}{L}(t_2 - t_1) = \frac{E_D}{L} t_{on}$$

L stands for the inductance of the detection coil. E_D stands for the battery's voltage. t_{on} stands for the conduction time.

During $t_2 \sim t_3$, the transistor turns off and the freewheeling is conducted. The current in freewheeling is determined by the equation below

$$i = i_{L\max} e^{-\frac{t}{\tau_1}} = \frac{E_D}{L} t_{on} e^{-\frac{t}{\tau_1}} \tag{3}$$

$\tau_1 = \frac{L}{R_1}$ is the time constant (we ignore the resistance of the diode). R_1 is the resistance of the freewheeling. $U_1(t)$ (the voltage of R_1) is

$$U_1(t) = i \cdot R_1 = i_{L\max} \cdot R_1 \cdot e^{-\frac{t}{\tau_1}} = \frac{E_D R_1}{L} \cdot t_{on} \cdot e^{-\frac{t}{\tau_1}} \tag{4}$$

It equals the equation $U_1(t) = U_{1\max} \cdot e^{-\frac{t}{\tau_1}}$.

In the equation above, $U_{1\max} = \frac{E_D R_1}{L} \cdot t_{0n}$, which is the highest reverse voltage of the freewheeling. But actually, the reverse peak of the detection coil is much higher than when the transistor turns off due to the impact of the distribution of the capacitance and the disclosure of the magnetism. To guarantee the safety of the transistor, we use some method such as connecting two voltage regulator tubes and two varistors in different direction in order to limit the maximum voltage to an extent which is a little bit higher than $U_{1\max}$. If the external factors have no influence, the initiation and the freewheeling process of the detection coil and the changes of the current and the voltage due to the time follow Figs. 3 and 4.

During t_{on}, the battery will provide energy P_{on} to the detection coil when the metal target accesses to the detector. P_{on} is made up of two parts: one is the excitation power P_{on1} produced by the detection coil, the other is the power of eddy current loss produced by the target. Since P_{on} is provided by the battery, and $P_{on1} > P_{on2}$, it is difficult to distinguish the energy of the eddy current loss from the whole energy during the positive half cycle of the detector generation.

However, in the negative cycle of the detector (during t_{off}), we use the energy stored in the detection coil instead of the energy provided by the DC battery to guarantee the continuation of the currency. Then we can equal the value of the eddy current loss provided by the freewheeling current on the surface of the target to the power provided by resistance R_L hang in parallel at both sides of the detection coil. If U stands for the voltage produced by the detection coil in the freewheeling process, the eddy current loss equals to:

$$P_w = \frac{U^2}{R_L} \tag{5}$$

Compared Eq. 2 with Eq. 5, the eddy current loss of the target and the distance follow the equation below:

Fig. 4 Voltage wave of loop current

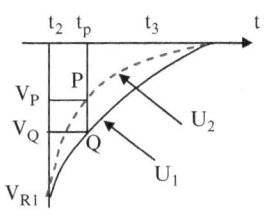

$$R_L \propto l^6 \tag{6}$$

If we ignore the resistance of the diode when it is conducted, the resistance of the freewheeling equals to the resistance of R_1 and R_L in parallel. The time constant of the freewheeling changes and the freewheeling voltage is:

$$U_2 = U_{1\max} \cdot e^{-\frac{t}{\tau_2}} \tag{7}$$

In the equation above, $\tau_2 = \frac{R_1 + R_L}{R_1 R_L} L$.

Figure 4 is used to describe Eqs. 4 and 7. Solid curve represents freewheeling decay U_1, which is without the metal target. Virtual curve represents the freewheeling decay U_2 that is with the metal target. We select a moment t_p from t_2 to t_3, the points of intersection of the two curves and t_p are P and Q. In t_p, the difference of the freewheeling voltage of whether there is the target near the detection is:

$$\Delta U = U_Q - U_P = U_1 - U_2 = U_{1\max}\left(e^{-\frac{t_p}{\tau_1}} - e^{-\frac{t_p}{\tau_2}}\right)$$

After using Taylor series expansion, the equation is:

$$\Delta U = U_{1\max} t_p \frac{\tau_2 - \tau_1}{\tau_2 \tau_1} \approx U_{1\max} t_p \frac{L}{\tau_1^2}\left(\frac{R_1 + R_2}{R_1 R_2} - \frac{1}{R_1}\right) = U_{1\max} \frac{t_p}{\tau_1^2} \cdot \frac{L}{R_L}$$

Taking Eq. 6 into account,

$$\Delta U \propto \frac{1}{l^6} \tag{8}$$

The theory indicates the target position by measuring the scale factor of the equation above or adopting the multi-probe measurement system. However, due to the constant changes of the signal, it is difficult to measure all the voltage and do the calculation. Therefore, in the real situation, we are supposed to find a new way.

After analyzing the information in Fig. 4, we can find that the area surrounded by curve U_1 and U_2 is the time integral of ΔU. It means the change of freewheeling area is:

$$\Delta S_{\Delta U \Delta t} = \int_{t_2}^{t_3} \Delta U dt_p = \int_{t_2}^{t_3} \left(U_{1\max} \frac{L}{R_L} \cdot \frac{t_p}{\tau_1^2} \right) dt_p = \frac{U_{1\max} L}{2\tau_1^2} \cdot \frac{t_3^2 - t_2^2}{R_L} \tag{9}$$

Taking Eq. 6 in to account, Eq. 9 changes into:

$$\Delta S_{\Delta U \Delta t} \propto \frac{U_{1\max} L}{2\tau_1^2} \cdot \frac{t_3^2 - t_2^3}{l^6} \propto \frac{U_{1\max} L}{l^6} \tag{10}$$

According to Eq. 4, $U_{1\max} = \frac{E_D R_1}{L} \cdot t_{0n}$. $U_{1\max}$ is the highest reverse voltage of the freewheeling branch. Put $U_{1\max}$ into above equation, then

$$\Delta S_{\Delta U \Delta t} = K \cdot \frac{E_D R_1 t_{on}}{l^6} \tag{11}$$

K is a coefficient, which can be got by the experiment.

Equation 11 is called Target Detection Equation, which shows the difference of the freewheeling curve area between whether there is the target near the detection or not is proportional to the voltage E_D supplied by the power, the resistor R_1 of the freewheeling circuit and the oscillation pulse width ton. But R_1 and ton are limited by other factors. Since R_1 has a direct effect on the radiation field of the detection coil during the transistor turns off, it is not a good way to improve the detection sensitivity by increasing R_1. Besides, it is also a bad way to improve the detection ability by improving the pulse width ton since the pulse width is inversely proportional to the working frequency.

According to the theory and the experiment, raising battery voltage E_D is the most efficient way to increase detection range. This conclusion can be drawn from Eq. 11. Besides, when the voltage of working battery rises, the electromagnetic energy that detection coil radiates increase and the eddy current loss increases. Therefore, the detection range increases as well.

3 The Target Signal Detection Principle

Figure 5 is a circuit principle diagram based on the theory of eddy current loss. The circuit consists of three parts: PWM pulse width modulation oscillator; detection coil, reference coil and comparator. The working principle is analyzed below.

If there is no metal target around the detection coil, the detection coil and the reference coil output the same waves, and the comparator outputs zero. When a metal target exists around the detection coil, according to Eq. 11, the detection coil and the reference coil output different waves. This signal goes through the comparator and is output as the target signal after being filtered by the filter circuit. When the ferromagnetic metal targets go through the detection area at a certain speed along the X direction as Fig. 6 shows, we suppose the maximum detection range of

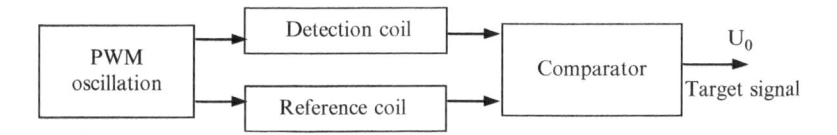

Fig. 5 Target signal detecting electrical principle of detector

the detector is r_{max}. Detection zone is a circle of radius r_{max}, and the detector is at the center point O. In t_1, the target moves to point A and starts to enter the target zone. In t_m, the target moves to point C, the nearest position from the detection. In t_2, the target goes through point B and then, leaves the target zone. During the intersection of the bomb project, the target waveform output by detector circuit is shown as Fig. 7. According to Fig. 7, detection range is the closest in t_m, and the maximum signal is U_{0max}.

4 Measurements of Target Characteristics

4.1 Target Characteristics in Different Frequency

After analyzing the experimental results, we can know that the sensitivity of detector is related to working frequency. Take ferromagnetic metal targets as the example, it has the highest sensitivity under the frequency ranging from 5.0 to 6.5 kHz, and 5.565 kHz is the best.

4.2 Target Characteristics in Different Detection Range

Figure 8 shows the response waveform of detector and ferromagnetic metal targets (the size is 4 × 100 × 150 mm/A_3) in different intersection distance. It shows that as the detection range raises, the margin of response waveform decreases. And when the intersection distance is close to a certain degree, there will be no response of the detector.

4.3 Target Characteristics in Different Transmit Power

The theory above shows that it is practical to increase detection range by raising the working voltage of detector. To confirm this conclusion, we make the response waveform of one detection target in the same detection range and in two different voltages. Figure 9a shows the detection waveform in the excitation parameters 6 V/23 mA.

Fig. 6 Detector output wave to target

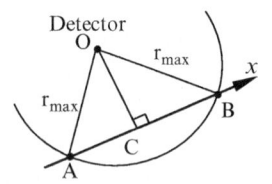

Fig. 7 Test wave of target

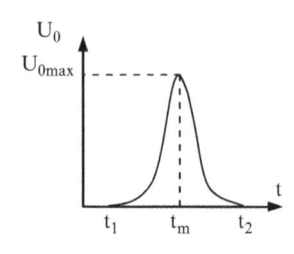

Fig. 8 Detector output wave to target

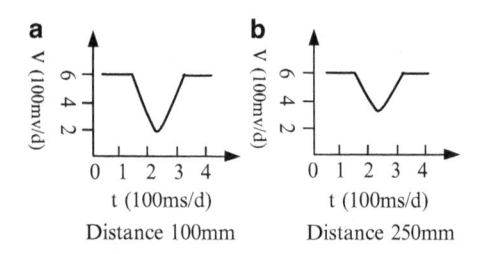

The peak of waveform is 500 mV × 1.5 = 750 mV. Figure 9b shows the detection waveform in the excitation parameters 9 V/61 mA. The peak of waveform is 500 mv × 3.4 = 1,700 mv. The consequence conforms that rising transmit power of the detector can increase detection range.

5 Conclusion

The paper proposes the characteristic quantity based on the theory that regarding target eddy current loss as the detection of target information from the perspective of conservation of energy, and introduces the concept of target loss equivalent resistance and derive the result that equivalent resistance is proportion with six power of detection distance. By using pulse width modulation excitation coil detection method and detecting the changes of freewheeling process in the coil, the equivalent resistance can be measured by the eddy current loss. It can not only determine the distance, but also exclude environmental interference. As the experimental data shows, the best detection frequency is 5.565 kHz and raising excitation voltage can increase detection range.

Fig. 9 Detector output wave to target

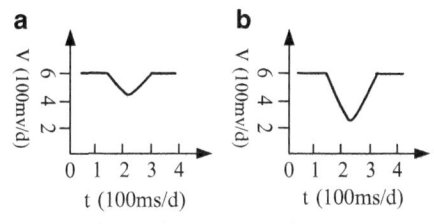

a — Excitation voltage 6v
b — Excitation voltage 9v

References

1. Aderin ME, Burch IA (1998) Countermine: hand held and vehicle mounted mine detection. Deflection of abandoned land mines, 1998 second international conference. IEE Conference Publication, Dera Chertsey, pp 198–202
2. Kildishev AV, Nyenhuis JA (1999) Application of magnetic signature processing to magnetic center primpointing in marine vehicles. OCEANS'99MTS/IEEE 3:1532–1536
3. Si Huaiji, Cui Zhanzhong, Zhang Yanmei (2005) Study of electronmagnetic fuze detection principle. J Beijing Inst Technol 25(1):79–82 (in Chinese)
4. Feng Cizhang (1985) Static electromagnetic field. Xi'an Jiaotong University Press, Xi'an, pp 73–75 (in Chinese)
5. Si Huaiji (2005) Research on technology of magnetic detection based on the theory of the eddy current loss. School of Mechatronics Engineering, Beijing Institute of Technology, Beijing (in Chinese)

Pedestrian Detection Based on Road Surface Extraction in Pedestrian Protection System

Hao Heng and Huilin Xiong

Abstract Pedestrian protection system (PPS) in the advanced driver assistance system (ADAS) to improve traffic safety, has become an important research area. The major challenge in PPS is to develop a reliable on-board pedestrian detection system. Compared to the pedestrian detection on static images, on-board pedestrian detection is facing some new difficulties, such as high real-time demand, wide range of illumination variation and so on. In order to deal with these challenges, we presented a method in this paper by combining the road surface extraction technique and the histogram of oriented gradient (HOG) feature based classification, so that the search regions are only limited within the extracted road surface. Experiment results show that this method can remarkably reduce the false alarm rate, improve the detection speed, and significantly improve the small pedestrian detection rate.

Keywords Pedestrian detection • Road surface extraction • Histogram of oriented gradients • Support vector machine

1 Introduction

Road accidents have become the main cause of road traffic casualties over the last century. In the recent decade, researchers have turned their attention to the more intelligent on-board systems, aiming to avoid accidents or reduce the severity of accidents. For example, the intelligent on-board systems can remind the driver of a potential dangerous situation, assist him/her in taking action immediately to avoid the accident. These systems are usually referred to as advanced driver assistance systems (ADAS).

H. Heng (✉) • H. Xiong
Department of Automation, Shanghai Jiao Tong University, and Key Laboratory of System Control and Information Processing, Ministry of Education of China, Shanghai, China
e-mail: haoheng@outlook.com

In this paper, we focus on the pedestrian detection, which is the most important part of the pedestrian protection system (PPS). Accident statistics indicate that 70 % vehicle-to-pedestrian accidents occur in front of vehicle; therefore, using a forward-facing on-board camera to capture the driving scenarios is necessary.

In recent years, a number of methods have been proposed to tackle these problems. Papageorgiou and Poggio [1] employed Haar features in combination with a polynomial support vector machine (SVM) classifier to detect pedestrians. Tran and Forsyth [2] used local histograms of gradients and local PCA of gradients as features to establish a model of human body. Dalal and Triggs [3] employed histograms of oriented gradients (HOG) features in combination with SVM classifier to detect pedestrian and achieved good performances. Felzenszwalb [4] presented a flexible part model by adding the position of parts of the body as a latent variable into the SVM learning algorithm. Wu and Nevatia [5] proposed a system to automatically construct tree hierarchies for the problem of multi-view pedestrian detection. They use a boosting framework in combination with edge-let features. Dollár [6] presented an approach that automatically learns flexible parts from training data and uses a boosting framework with wavelet features.

We employ a road extraction technique as foreground segmentation, together with HOG feature-based classification to detect pedestrian. The rest of this paper is organized as follows. Section 2 introduces the algorithm that used to extract road surface in the illumination invariant color space. Section 3 describes pedestrian detection in the extracted ROIs. The experiment results are presented in Sects. 4 and 5 is the conclusion.

2 Road Surface Extraction

2.1 Theory Introduction

Finlayson showed that if an object has Lambertian surfaces and is imaged by three narrow band sensors under Planckian illumination (*PLN*), it is possible to convert the RGB color image into a shadow free gray scale image, on which the influence of lighting variations could be greatly reduced [7]. This theory is derived from the physics behind color formation in the presence of a Planckian light source, Lambertian surface, and narrow band camera sensors. Experiment results show that this theory holds even for real world situation.

Experiment results show that this theory holds even for real world situation. Figure 1 shows the process of computing a shadow free image I from a given color RGB image I_{RGB}. Let R, G and B be the standard color channels, $r = \log(R/G)$ and $b = \log(B/G)$ are the corresponding log-chromaticity values by using the G as the normalizing channel. Then, different colors under different illuminations form the parallel lines in the log-chromaticity space. There is a line, donated by l_θ in Fig. 1, being orthogonal to these parallel lines. On l_θ the same color with different

Fig. 1 Obtain the shadow free *gray-scale* image

illuminations is represented by the same point, and moving along l_θ means changing the chromaticity. In other words, l_θ is a gray level axis, where each gray level corresponds to a surface chromaticity. Therefore, projection χ' onto l_θ can remove the influence of illumination variations.

2.2 Camera Calibration

According to the theory above, obtaining the direction (θ) of l_θ is crucial. In fact, θ is a parameter of the camera sensor, independents to the lighting condition or the surface material. So, a calibration process for a given camera can be computed offline and just needs to be done once.

Different to the ways [8, 9] for calibrating the camera parameter θ, in this paper, we present a new standard-deviation based calibration algorithm. The idea behind our calibration is pretty simple, with a wrong θ, the similar chromaticity values of I_{RGB} will scatter with a large value of the standard-deviation, denoted by *Var*; on the other hand, with the right θ, the similar chromaticity values of I_{RGB} will reach the minimal value of *Var*. Hence, the minimal value of *Var* is expected. The algorithm is summarized below:

Fig. 2 The value of $mVar(\alpha)$ changes according to θ

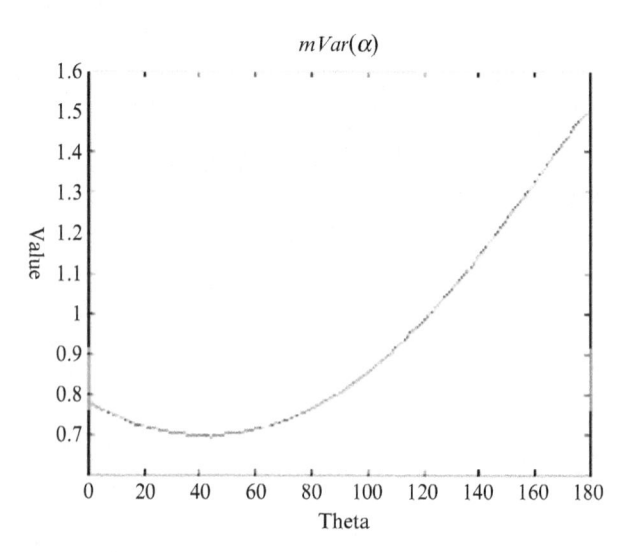

1. Select K images $\{I_{RGB}^1, \cdots, I_{RGB}^K\}$ from the database together with their pre-annotated road region images $\{I_{road}^1, \cdots, I_{road}^K\}$.
2. Select an image, I_{RGB}^k, and its corresponding I_{road}^k. Initialize the projection angle $\alpha = 0°$.
3. Projecting the log-chromaticity pixel values of I_{RGB}^k onto l_α to obtain a grayscale image I_α, and extract the road surface image I_{road} from I_α using I_{road}^k.
4. Compute the standard deviation $Var^k(\alpha)$, using the pixels whose gray values are in the $[0.05, 0.95]$ range of I_{road}.
5. Calculate the standard deviation values for I_{RGB}^k as steps 3–5 with $\alpha \in [0°, 1°, \cdots, 179°]$.
6. Repeat steps 2–6 for each image. $k \in [1, 2, \cdots K]$.
7. For each discrete value of α, calculate the average value of $Var^k(\alpha)$ with $mVar$ $(\alpha) = \frac{1}{K} \sum_{k=1}^{K} Var^k(\alpha)$.
8. Then the value of the parameter θ is chosen as $\theta = \arg\min_\alpha mVar(\alpha)$.

Figure 2 shows the curve of $mVar(\alpha)$ using the same database as Finlayson [8] used, from which we calibrate the camera parameter θ to be $44°$. The value of θ obtained by our calibration algorithm is the same as Finlayson's. However, our algorithm is more stable and robust.

2.3 Road Detection Algorithm

The shadow free gray-scale image I can be calculated by θ. According to the illuminant invariant property of I, pixels are of the similar gray values, even in

Input Image

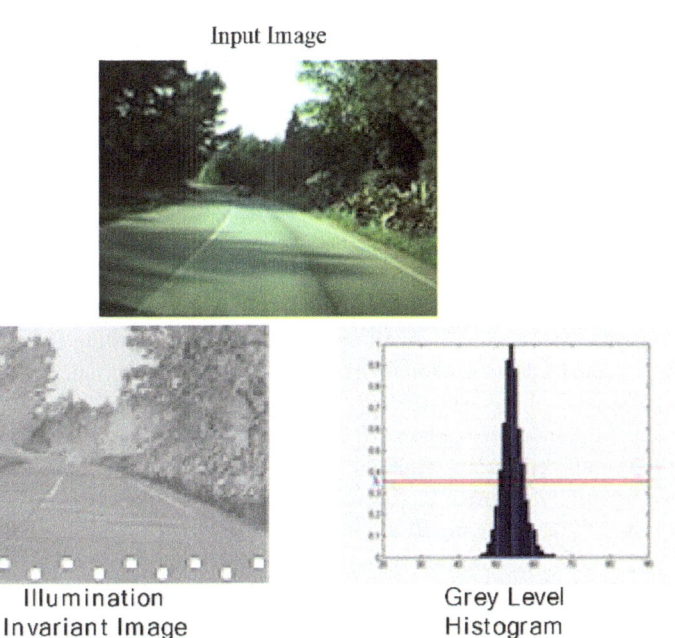

Fig. 3 Obtain the *gray-level* distribution of road surface

shadow regions. Based on this property of this illuminant invariant image I, we extract the road surface by employing a statistic technique and the flood-fill based morphological processing.

We employ a conditional function $p(I(p)|road)$ to decide whether a gray value, $I(p)$, belongs to road or not. A pre-defined threshold, λ, is introduced to eliminate the influence of the tiny variants of the road pixels and other interference factors. Our experiments find that $\lambda = 0.35$ can lead to a better performance.

$$\begin{cases} I(p) \text{ belongs to road,} & \text{if } p(I(p)|road) \geq \lambda \\ I(p) \text{ doesn't belong to road,} & \text{otherwise} \end{cases} \quad (1)$$

The conditional probability, $p(I(p)|road)$, can be estimated using a set of "seeds", which are a number of small square blocks on I, usually located at the bottom of the image (see Fig. 3). The regions that the seeds located are assumed to be on the road, since they correspond to the nearest front of the vehicle. In this paper, we select 11 small square blocks in size of 9×9 at the bottom of I (see Fig. 3), to estimate $p(I(p)|road)$.

Our road detection algorithm can be summarized as follows:

1. Compute the illuminant invariant gray-scale image, I, from I_{RGB} using the calibrated parameter, θ.
2. Estimate $p(I(p)|road)$ from the normalized histogram of the selected "seed" blocks using threshold λ, and then get the initial road pixels according (10).

3. From each seed, using the Flood-fill method to generate a coarse road surface.
4. Use a morphological operation to fill some small holes on the road surface.

3 Pedestrian Detection

After the road detection, we treat the extracted road surface as foreground, on which a series of parallel rectangular boxes are placed. These rectangular boxes are viewed as our ROI regions, on which the pedestrian searching is performed. The heights of these ROIs are established using the distance information and the pinhole imaging model.

To detect the small pedestrian effectively, we resize these ROIs to a unified scale so that the small pedestrian, which is far from the cameras, will be enlarge, and this could substantially improve the detection rate of small pedestrian. Figure 4 shows the procedure of resizing the ROIs. In our experiments, we resize all the heights of ROIs to 128 pixels. Another advantage of resizing ROIs is that, in pedestrian detection on each ROI, we do not need to perform a time-consuming search of targets in different sizes, which is the conventional way to detect the targets.

Then, on each ROI, we use the HOG features [4] and a linear support vector machine (SVM) classifier to detect pedestrian.

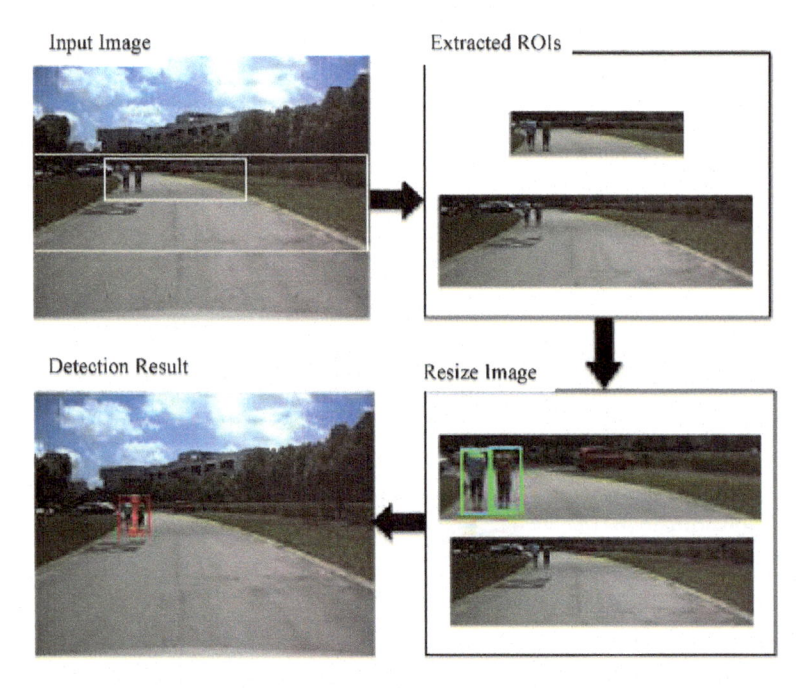

Fig. 4 Extract ROIs based on road surface, and then resize them

4 Experiment Results

We tested our algorithm on a public available dataset, CVC-02 Pedestrian Dataset,[1] and a video dataset collected by ourselves. We select four sequences (the third, fifth, sixth and tenth) of 1,173 frames with 2,347 pedestrian from CVC-02 and 1,026 frames with 2,148 pedestrian from our dataset.

In our experiment, the HOG features of the ROIs are sent to the 64×128 size based SVM classifier to detect pedestrians. We use detection rate, false alarm rate and detection time to evaluate and compare the performances of our method and the conventional sliding window based searching approach.

Figure 5 shows part of detection results on three video frames. Table 1 gives the statistic results concerning the detection rate, false alarm rate and detection time per frame (in seconds) on the testing video data. We can see that:

- The 64×128 HOG detector works well for the large pedestrians; however, it often misses the small pedestrians.
- The 32×64 HOG detector performs well in detecting small and large targets; however, it tends to produce many false alarms.

HOG (64*128)   HOG (32*64)   HOG + road detection

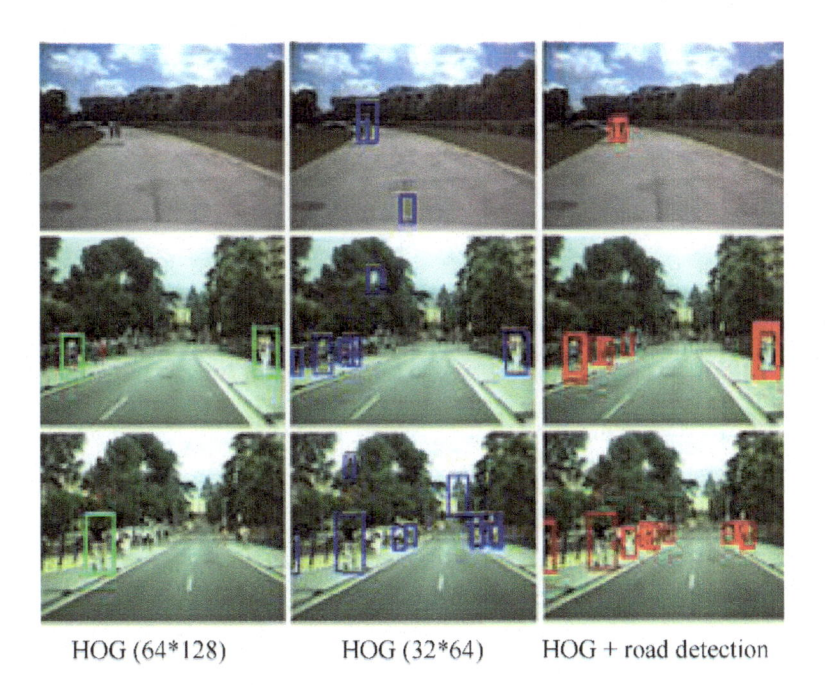

Fig. 5 Part of experiment result images

---

[1] CVC-02 dataset: http://www.cvc.uab.es/adas/site/

Table 1 The statistical results of experiment

Experiment result	HOG (64 × 128)	HOG (32 × 64)	HOG + road detection
Detection rate	36.14 %	79.08 %	77.82 %
False alarm rate	15.24 %	78.12 %	7.76 %
Detection time (s)	3.23	18.89	1.52

- Our method presents the best performance, either in detecting small targets or large targets with a low false alarm rate. Meanwhile, the proposed method works faster than the other two detectors.

5 Conclusion

In this paper, we have presented a method by combining the road surface extraction technique with the HOG feature based classification. Experiment results have shown that this method can remarkably reduce the false rate, improve the detection speed, and significantly improve the small pedestrian detection rate.

References

1. Papageorgiou C, Poggio T (2000) A trainable system for object detection. Int J Comput Vision 38:15–33
2. Tran D, Forsyth D (2007) Configuration estimates improve pedestrian finding. Adv Neural Inform Process 644:1529–1536
3. Dalal N, Triggs B (2005) Histogram of oriented gradient for human detection. Comput Vision Pattern Recogn 1:886–893
4. Felzenszwalb P, McAllester D, Ramanan D (2008) A discriminatively trained, multi-scale, deformable part model. Computer vision and pattern recognition, Chicago, pp 1–8
5. Wu B, Nevatia R (2007) Cluster boosted tree classifier for multi-view, multi-pose object detection. In: International conference on computer vision, Los Angeles, pp 1–8
6. Dollár P, Babenko B, Belongie S, Perona P (2008) Multiple component learning for object detection. In: European conference on computer vision, vol 5303, Marseille, France, pp 211–224
7. Finlayson G, Hordley S, Lu C, Drew M (2006) On the removal of shadows from images. IEEE Trans Pattern Anal Mach Intell 28:59–68
8. Finlayson G, Hordley S, Drew M (2002) Removing shadows from images. In: Proceedings European conference on computer vision, Comenhagen, Denmark, pp 129–132
9. Alvarez JMA, López AM (2011) Road detection based on illuminant invariance. IEEE Trans Intell Transport Syst 12(1):184–193

Methods on Reliability Analysis of Friction Coefficient Test Instrument

Tianrong Zhu, Xizhu Tao, Xinsheng Xu, and Tianhong Yan

Abstract In order to analyze the reliability of instrument with the consideration of experiment data shortage or even no, reliability element method is proposed in this paper. The reliability indexes corresponding to serial system and parallel system are investigated respectively. The steady-state availability and the reliability degree are introduced as the key reliability index for friction test instrument based on the analysis of the function and structure characteristics of its. In addition, the reliability diagram of sliding friction test instrument is constructed. A program module based on MATLAB is developed to calculate the reliability index effectively. Finally, an example was given out to verify the feasibility of methods mentions above.

Keywords Reliability index • Reliability element method • Steady-state availability • Degree of reliability • Sliding friction test instrument

1 Introduction

With the rapid development of manufacturing technologies for test instrument, the accuracy and veracity of test instrument has been improved significantly. As we all know a product which is going to be produced in batch should ensure its quality [1]. Nowadays, the quality characteristics of product, besides accuracy and reliability, also include reliability, safety and economy and so on [2–4].

Friction tester is a kind of instrument used to measure the friction coefficient between the conveyor and the armrest of elevator [5]. The reliability analysis of

---

T. Zhu • X. Tao • X. Xu (✉)
Institute of Industrial Engineering, China Jiliang University, Hangzhou, China
e-mail: zhu_tianrong@163.com; taoxizhu008@126.com; lionkingxxs@cjlu.edu.cn

T. Yan
Institute of Mechatronics Engineering, China Jiliang University, Hangzhou, China
e-mail: thyan@cjlu.edu.cn

friction test instrument is to obtain the primary information of the reliability index of it. Furthermore, it is to find out that the defect or weak link of it as well as the parts that affect its reliability. Thereupon, these can provide evidences to improve the friction test instrument itself. In reliability engineering, the reliability analysis of system equipment, based on large amount of experiment data, is to use the mathematical theories such as probability and statistics to investigate the exact reliability index [6]. And on the basis of this, the reliability of system equipment can be evaluated in the end. However, it always leads to low assessment accuracy when experiment data of equipment itself is little or even no [7]. Obviously, it cannot meet the analysis requirement of some instrument products. In engineering practice, an instrument always consists of a lot of components. And the reliability level of instrument depends on the reliability of components constituting instrument [8]. The reliability model of instrument can be constructed by analyzing the functional diagram of the instrument.

2 Reliability Index and Its Calculation

Usually, the reliability level of instrument is the comprehensive reflection by several parameters (namely reliability index). The reliability is a kind of statistical index, and it can be used to analyze the reliability of instrument quantitatively.

In general, the reliability index of instrument consists of many types of elements including degree of reliability, failure rate, cumulative failure probability, mean repair time, reliable life, and steady-state availability and so on. In this work, the degree of reliability and steady-state availability are used as the key reliability index of friction test instrument with the consideration of the function characteristics and the structure features of friction test instrument.

2.1 Reliability Calculation of Friction Test Instrument

Steady-state availability A is an important index to measure the degree of reliability $R(t)$ of friction test instrument. It refers to the probability of friction test instrument keeping its stability normal function in long-term work. And the degree of reliability refers to the probability of friction test instrument fulfilling the specified function under the stipulated conditions within prescribed period. The state transferring process of instrument is a kind of Markov random process when the life and the repair time of all components constituting instrument are all exponential distribution [9]. Otherwise it is non-Markovian stochastic process. To the large complex instrument consisting of many components, the steady-state availability can be calculated through two steps. The first step is to calculate the component failure rate and component repair rate respectively, and the second step is to

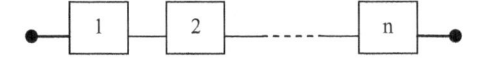

Fig. 1 The component structure of serial system

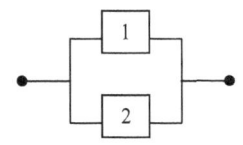

Fig. 2 The component structure of parallel system

calculate the degree of availability of the whole instrument based on the logic constraint relationships among components constituting instrument.

Based on stochastic process theory, the steady-state availability A and the degree of reliability $R(t)$ of each component is as the following respectively.

$$A = \frac{\mu}{\mu + \lambda}, \quad R(t) = P(T) \tag{1}$$

Where λ is the failure rate of components, μ is the repair rate of components, T is the life of instrument, and t is specified time.

- For the serial system consisting of n components and repair equipment illustrated in Fig. 1, the parameters of reliability index are as follows.

$$\lambda_s = \sum_{i=1}^{n} \lambda_i \tag{2}$$

$$\mu_s = \sum_{i=1}^{n} \lambda_i \bigg/ \sum_{i=1}^{n} \frac{\lambda_i}{\mu_i} \tag{3}$$

$$R_S(t) = \prod_{i=1}^{n} e^{-\lambda_i t} \tag{4}$$

Where λ_i is the failure rate of the ith component, μ_i is the repair rate of the ith component, and $R_i(t)$ is the degree of reliability of the ith component. The exponential form of $R_i(t)$ indicates that it obey Markov distribution.

- For the parallel system consisting of two different components and a repair equipment illustrated in Fig. 2, the parameters of reliability index are as the following respectively.

$$\lambda_s = \frac{\mu_1 \pi_3 + \mu_2 \pi_2}{\pi_0 + \pi_1 + \pi_2} \tag{5}$$

$$\mu_s = \frac{\mu_1 \pi_3 + \mu_2 \pi_4}{\pi_3 + \pi_4} \tag{6}$$

$$R_s(t) = 1 - \prod_{i=1}^{n}[1 - e^{-\lambda_i t}] \tag{7}$$

Where π_0, π_1, π_2, π_3, and π_4 are the steady-state distribution of components respectively.

$$\pi_1 = \frac{\lambda_2(\lambda_1 + \lambda_2 + \mu_1)}{\lambda_1 \mu_1 + \lambda_2 \mu_2 + \mu_1 \mu_2} \pi_0 \tag{8}$$

$$\pi_2 = \frac{\lambda_1(\lambda_1 + \lambda_2 + \mu_2)}{\lambda_1 \mu_1 + \lambda_2 \mu_2 + \mu_1 \mu_2} \pi_0 \tag{9}$$

$$\pi_3 = \frac{\lambda_1 \lambda_2(\lambda_1 + \lambda_2 + \mu_2)}{\mu_1(\lambda_1 \mu_1 + \lambda_2 \mu_2 + \mu_1 \mu_2)} \pi_0 \tag{10}$$

$$\pi_4 = \frac{\lambda_1 \lambda_2(\lambda_1 + \lambda_2 + \mu_1)}{\mu_2(\lambda_1 \mu_1 + \lambda_2 \mu_2 + \mu_1 \mu_2)} \pi_0 \tag{11}$$

$$\pi_0 = \frac{\psi}{(\lambda_1 \mu_2(\lambda_2 + \mu_1) + \lambda_2 \mu_1(\lambda_1 + \mu_2)) \times (\lambda_1 + \lambda_2 + \mu_1) + \psi} \tag{12}$$

$$\psi = u_1 u_2 (\lambda_1 u_1 + \lambda_2 u_2 + u_1 u_2)$$

It can be seen that the calculation models of the degree of reliability $R(t)$ are different for different instrument systems. However, the steady-state availability of the whole instrument and the steady-state availability of single component are the same at calculation models. They are all expressed as follows.

$$A_s = \frac{\mu_s}{\mu_s + \lambda_s} \tag{13}$$

Where λ_s is the failure rate of the whole instrument, and μ_s is the repair rate of the whole instrument.

2.2 Reliability Analysis Method of Friction Test Instrument

The experiment data concerning friction test instrument is little or even no in practice. It would lead to very low assessment accuracy if only use the data of instrument itself. As a result, reliability unit method is adopted to calculate the reliability index of instrument in this work. The analysis processes are as follows.

- Analyzing the functions structure of instrument and drawing its diagram of reliability. The reliability association among components is described based on the diagram of reliability.
- Calculating the failure rate and repair rate of a single component. On the basis of these, the steady-state availability of its would be calculated.
- According to the logic structure among components such as parallel or serial, and selecting reliability unit method, the reliability index of the whole instrument can be calculated by using unit reliability.

3 Reliability Model of Sliding Friction Test Instrument

The sliding friction test instrument illustrated in Fig. 3 is a kind of pin disc structure which was developed by our research team. It is used to measure the sliding friction coefficient between conveyor belt and armrest of elevator. This sliding friction test instrument is designed with the consideration of different friction conditions. It can show friction coefficient in time and implement measure online.

3.1 Functional Structure of Friction Test Instrument

3.1.1 Sliding Friction Device

This device is mainly composed of test platform, main shaft, loading mechanism, motor, and reducer and so on. All these are used to simulate the mechanical equilibrium when friction is generated between specimens. Based on these, the

Fig. 3 The pin disc sliding friction test instrument

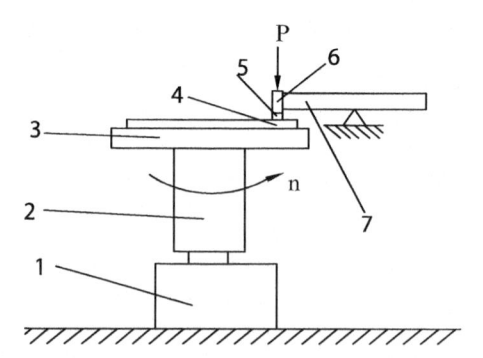

Fig. 4 Working principle diagram of sliding friction device (*1* – motor; *2* – axis; *3* – Test bench; *4* – down-specimen; *5* – up-specimen; *6* – round pin; *7* – support)

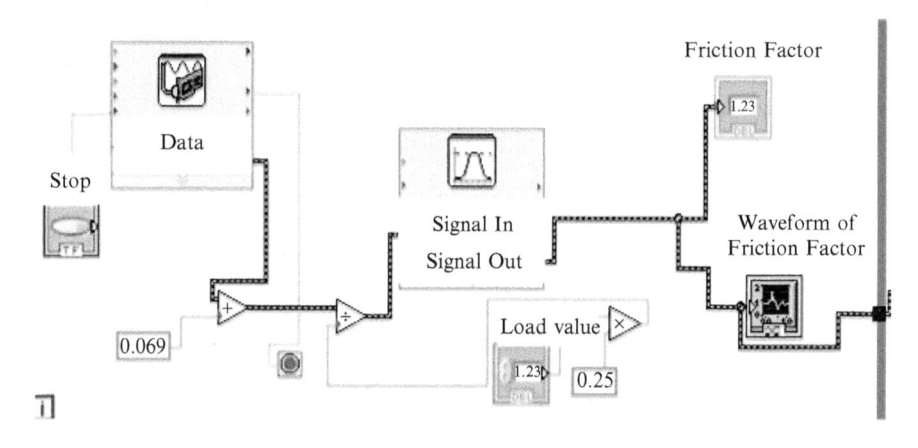

Fig. 5 The test program diagram of friction factor

acquisition environment of friction value is constructed. The working principle is presented in Fig. 4.

3.1.2 Signal Acquisition System

Signal acquisition system makes analog signal into signal forms that computer can read. It consists of sensor, amplifier and data acquisition card.

Fig. 6 Reliability diagram of friction test instrument

```
speed controller — electric motor — reducer
force transducer — test desk — the loading axis
balancing weight — coupling — principal axis
key — amplifier — speed sensor
data acquisition card — upper monitor
```

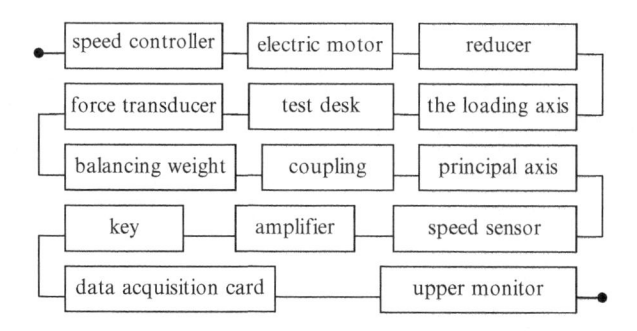

Table 1 The list of parameter distribution of failure time parameters and the working time parameters of each component

Number	Name	Failure rate	Repair rate
1	Speed controller	0.00903	0.8903
2	Electric motor	0.01026	0.9029
3	Reducer	0.00891	0.8894
4	Balancing weight	0.00073	0.6491
5	The loading axis	0.00513	0.8675
6	Force transducer	0.00729	0.8853
7	Coupling	0.00627	0.8768
8	Principal axis	0.01104	0.9113
9	Key	0.00187	0.7186
10	Speed sensor	0.00774	0.8879
11	Amplifier	0.00682	0.8779
12	Data acquisition card	0.00701	0.8842
13	Upper monitor	0.00145	0.8391
14	Test desk	0.00132	0.8087

3.1.3 Computer Control System

Computer control system use software such as LabView to compile the PC software. This environment provides control method to process data. And then a automatic measure process would be implemented. This computer control system can display the curve of friction coefficient after experiment finish as well as save and read experiment data (Fig. 5).

3.2 Reliability Diagram of Friction Test instrument

After analyzing the function and structure of friction test instrument, the reliability diagram of friction test instrument are obtained shown in Fig. 6.

4 Experiments and Applications

4.1 Component Parameter Description of Friction Test instrument

It can be seen that this friction test instrument is a kind of serial system composed of many components. Table 1 gives out the failure time parameter and working time parameter of each component according to historical data. Its reliability index can be calculated based on formulas (2), (3), (4), and (5) mentioned above.

4.2 Reliability Index Calculation Based on MATLAB

In order to calculate the reliability index of friction test instrument introduced in Sect. 2.2, a software module was developed based on MATLAB. Based on this MATLAB program module, the steady-state availability of friction test instrument, after plugging the experiment data of friction test instrument into formulas above, can be obtained which is 0.912.

5 Conclusion

Aiming at the facts of experiment data shortage or even no, reliability element method was proposed to analyze the reliability of friction test instrument namely steady-state availability. And a program module was developed based on MATLAB to calculate these reliability indexes efficiently. The methods presented in this paper not only reduce the amount of calculation, but also increase the accuracy of reliability index. It is an effective way to analyze the reliability issues of customized instrument.

References

1. Liu p (2008) Base of reliability engineering. China Metrology Publishing House, Beijing, pp 35–46, In Chinese
2. Cao JH, Cheng K (2012) A mathematical introduction to reliability. Higher Education Press, Beijing, pp 117–132, In Chinese
3. Zhang S, He MG (2012) Reliability analysis method of handling system of specialized bulk cargo terminal. Logist Eng Manag 4(32):60–62
4. Li Y (2002) Reliability analysis of several Markov repairable system. Nanjing University of Aeronautics and Astronautics, Nanjing, pp 31–69, In Chinese

5. Shi HL (2012) Development of pin-on-disc sliding friction coefficient tester. China Jiliang University, Hangzhou, pp 13–56, In Chinese
6. Kang R, Wang ZL (2005) Framework of theory and technique of reliability of system engineering. Chin J Aeronaut 26(5):633–636, In Chinese
7. Cheng ZJ, Guo B (2006) System availability analysis under repair strategy opportunity. Pract Underst Math 36(10):137–140
8. Mi J (2006) Limiting availability of system with non-identical lifetime distributions and noneidentical repair time distributions. Stat Probab Lett 76(7):729–736
9. Cao JH, Chen K (2012) Introduction of reliability mathematics. Higher Education Press, Beijing, pp 126–141, In Chinese

Design and Implementation of a New Power Transducer of Switch Machine

Yanli Wang and Bin Li

Abstract The design and implementation of an on-line monitoring system which can improve anti-interferences and stability of the railway switch machine in the railway industry is illustrated in this paper. The on-line monitoring system can detect three-phase electrical power by collecting the current and power of railway switch throughout its startup and working periods. This scheme adopts the multi-parameter power collecting chip ATT7022A to collect and process data, and uses the 485 bus line to implement communication between C8051F410 and PC. Field practice shows that the system possesses advantages of high-precision, low-energy consumption, and high reliability.

Keywords Switch machine • Power transducer • Online monitoring • Fieldbus

1 Introduction

Railway switch is one of important railway electrical signal devices. Generally, it is used to change the switch of railway tracks and indicate the switch's position at cross ways. It plays a crucial role in improving efficiency and ensuring the transportation safety of railway [1].

When a train arrives, it is crucial that whether the railway switch can switch on time and reliably or not for the safety of train running. However, there are some disadvantages in the present railway switch products, such as easy to reset, weak anti-interferences, instability. Therefore, designing a product with high anti-interferences and stabilities is very important. This paper demonstrates a new power transducer, which can improve the performance of the railway switch machine. The system includes fieldbus, sensor, computer network, and data communicational technologies.

Y. Wang (✉) • B. Li
Henan Technical College of Construction, Zhengzhou, China
e-mail: yanli1015@sina.com

Fig. 1 Photo of prototype

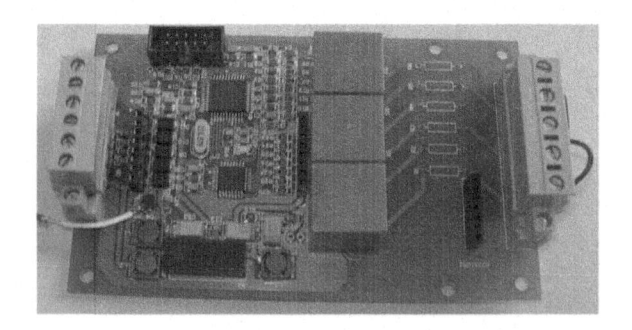

The prototype is shown in Fig. 1. As adoption of high reliable isolation technology in the whole system, the system's security, stability, anti-interferences capability, and reliability have been improved greatly. From the test results of the prototype, one can see that the switch power sensor has stable performance, superior anti-lightning strike and anti-electromagnetic interference capabilities. Consequently, the maintenance and management works would achieve a new level if the system is applied in the railway widely.

2 The Whole System Design Scheme

The system is used to complete the on-line monitoring for three-phase motor power of industrial occasions, which mainly includes lower single chip micyoco (SCM) system, upper measurement, control software and the corresponding buffer circuit. The design includes systematic function, the overall structure design, and selection of upper computer communication mode, peripheral circuit design, and systematic anti-interference and so on. In terms of hardware design, we choose the C8051F410 SCM produced by the American Silabs company as the main controller micro programmed control unit (MCU) [2]. The signal acquisition and processing are completed by precise multi-function dedicated energy metering chip ATT7022A. The signal processing results are delivered to MCU by serial peripheral interface (SPI) of ATT7022A, and the MCU communicates with upper computer are completed through RS485. Hardware design program is shown in Fig. 2.

3 System Hardware Design

Data acquisition is completed by the corresponding transducer. Therefore, system designs must satisfy and match test requirement. We adopt special isolate modules (voltage and current transducer) to isolate and convert the measured AC voltage (current) into the Ac current (voltage) maintaining the same frequency and phase. Because that the current transducer connects two sides of the large current or

Design and Implementation of a New Power Transducer of Switch Machine   813

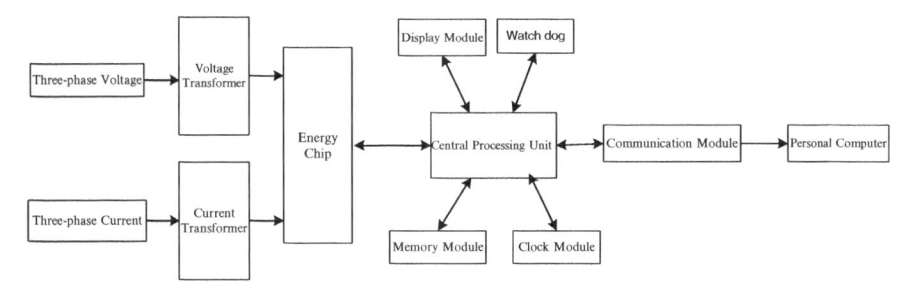

Fig. 2 Design program of hardware

current rectifier (the maximum input current required by the design is 10 A and the effective value of the maximum input sinusoidal signal offered to ATT7022 is 1 V), the linear current mutual inductor, with work function range 0–10 A and specification 5 A/2.5 mA is chosen.

The required input voltage (0–500 V) and current (0–10 A) are simulation signals, but the computer can only process the digital signal. So the collected simulation signal needs to be changed into digital signal (A/D change). For the same reason, when the control signal which comes from the computer drives the external device, the digital signal must be changed into simulation signal (D/A change), and then is magnified to control the electrical machine. By comparison, we select ATT7022A produced by Zhuhai Juli integrated circuit Design Company as sigma-delta ADC. ATT7022A is a digital signal processor (DSP) inserted with high velocity compute component, and it can be used to high accurate electrical gauging of three phase multi-parameter system. It includes six sigma-deltas ADC, reference voltage circuit and all of power, energy, effective value, power factor, frequency and digital signal dispose circuits. Consequently, it can gauge effective power, ineffective power, apparent power, and apparent current, effective voltage both each phase and mixed phase, which fully satisfies the development standard of the switch machine power transducer.

The main control module of C8051F410 is adopted as CPU to finish compute, storage, manifestation and communication of data. Combined with anti-interference circuit, the main control module possesses high ability in anti-interference, integration and system expansion. C8051F410 is a kind of fully integrated system-on-chip mixture signal MCP systems with low power consumption [3]. The functions of power on reset, VDD monitor, watchdog and clock oscillator in the system-on-chip system C8051F410 can work independently at the factory temperature (-40–$80\,^\circ$C).

Three-phase voltage outputted from the main control circuit converts into the output current of 4–20 mA through filter rectification of circuit. The circuit diagram is shown in Fig. 3.

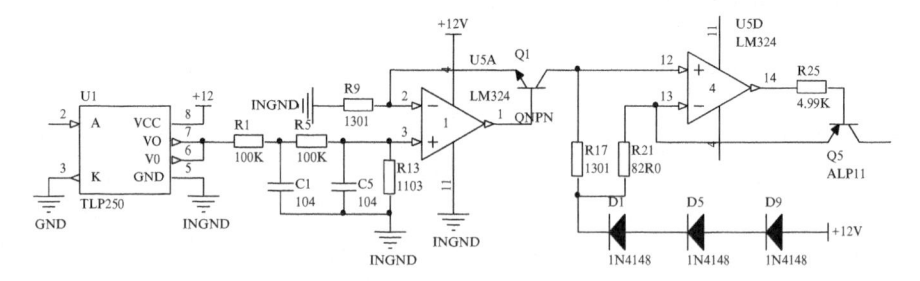

Fig. 3 Circuit diagram of current source

4 The Anti-interference Design of the Data Adoption System

Data acquisition system is usually applied at the complicated work site, with various kinds of interferences. Due to internal or external interferences, there will be interference signals in the measured voltage and current. As the main interferences of the electrical net are high-order harmonic, low pass filter was adopted at the output of mutual inductor to dispose high order wave with frequency higher than 50 Hz. Consequently, the wave shape has been ameliorated. To guard the stability of AC power supply, voltage-stability source between the electrical net and the mutual inductor was adopted to increase the anti-interferences of the system, which mainly contains isolation, wave filter technologies. For the isolation part, electromagnetic coupling of isolated magnify are used to isolate and transmit the captured message (after A/D change) before entering data gathering system [4]. Then the signal (after A/D change) is isolated by photo-coupler, and then inputted to the total line of the computer using photo coupling component. RC filter is used to dispose interference of the power frequency signal at 50 Hz.

5 System Software Design

The lower computer is mainly used to collect, manage and send the gathered data to the control room by RS-485 total line transmission [5]. For the overall design, the system is divided into main program and many sub modules. In the programming process, every module should be as independent as possible to compile, modify and debug software to reduce error easily. With the recycle property, the development time is reduced greatly. Besides, because the modules read in internal memory only if they are used, the occupation of the system resources and the running of the program space are reduced, and thus the running speed is enhanced greatly.

The upper machine mainly completes the following functions: communicate with the RS-485 serial module of master; receive, dispose, analyze and compute the data; manifest multi-switch machine real-time waveform; inquire history data and so on. Firstly, programs receive the data package transferred from lower machine and unpack the data package according to the custom format. Secondly, after analyzing and computing the data, the programs compress and store the data into data bank for inquiry. If any debugging appears, it will give alarms and present debugging information. The program can display the power variation of multiple switch machines in the form of wave shape on the displayer.

6 Communication Method and Debugging Results

RS-485 communication interface is adopted to achieve remote data transmission in the system. Doublet RS-485 communication method is chosen to control the center machine to achieve the connectivity of multi-smart nodes and delivery of the remote data in the design. The system of main machine serial port uses RS-232 treaty. And the isolated RS-485/232 converter is chosen to complete the level conversion which needed in the program. The subject of the whole system includes three parts: front smart nodes, central monitor, and analyzing and diagnosing work station of the central controller [6].

The switch machine power transducer have been debugged at spot, and the debugging precision achieves the design requirement of 0.5 level (the basic cite error at any point of the linear measurement scope is less than 0.5 % of the output standard). On the other hand, the transducer functions required in the design are checked. The on-spot debugging shows that the system has high reliability, high anti-interference ability and can be applied to the railway line. The switch machine power transducer can provide technique guarantees for the steady and safety of train running. The linearity of test results is shown in Fig. 4. The linearity of the power transducer can meet the design requirement.

7 Conclusion

The design and implementation of an on-line monitoring system in the railway industry was illustrated in this paper. The multi-parameter power collection chip ATT7022A, with advantages of high precision (achieved 0.5 level), convenient function expansion and easy to achieve of software, is adopted in system, which simplifies the hardware and software structures. This system has many advantages as follows. The hardware electrical circuit C8051F410 mixed signal MCU is adopted as the control core, which decreases the power loss and volume of the system, but enhances the ability of real-time control. The switch machine distributed monitor system model based on RS-485 bus line technique is provided, which can enhance the efficiency of the management work in the railway equipment.

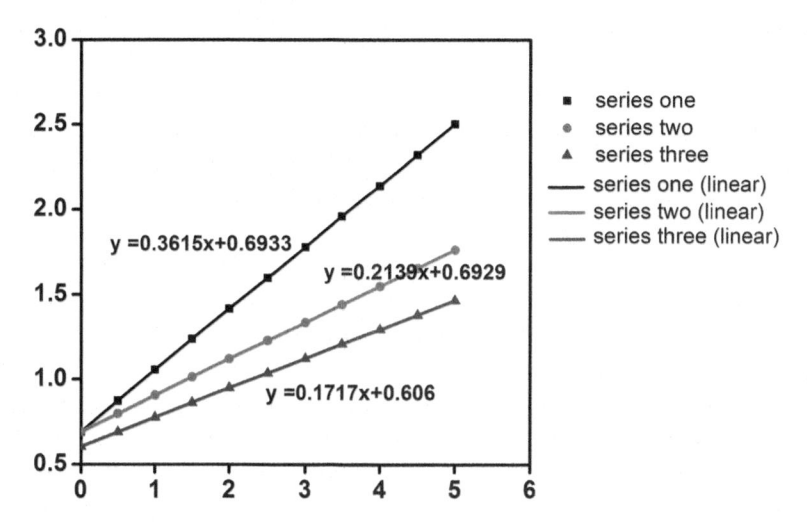

Fig. 4 The linearity of test results

According to the above design, the power switch machine was produced successfully. After adjusting of software, the output DC simulation signal achieves 4–20 mA. This system possesses advantages of high reliability, high anti-interference and can be applied to the railway line, which can supply technique guarantees for the steady and safety of train running.

References

1. Igarashi Y (2006) Development of monitoring system for electric switch machine. QR RTRI 47:78–82
2. Li G, Lin L (2002) Compatible with the 8051 high-performance, high-speed microcontroller the C8051FXXX. Beijing University of Aeronautics and Astronautics press, Beijing, pp 10–50, In Chinese
3. Xia L, Huang L, Jiang M (2006) Application of high speed system-on-chip MCU C8051F in experiment and leaching. Chin J Electron Device 4:1354–1358, In Chinese
4. Wang Z, Sun Y (2010) A new type AC/DC/AC converter for contactless power transfer system. Trans China Electrotech Soc 25:84–89, In Chinese
5. Qin P, Lin Y (2007) Simulation study of cargo oil handling system based on RS485 standard. J Syst Simul 19:5827–5831, In Chinese
6. Tian S (2001) Construction machinery condition monitoring and fault diagnosis. Constr Mach 32:26–30

Data Model Research of Subdivision Cell Template Based on EMD Model

Delan Xiong and Qingzhou Xu

Abstract Aiming at solving the problems of vast data, complex operation, and time consuming processing for remote sensing image, subdivision cell template was proposed based on Global Subdivision Grid (GSG). Subdivision cell template can realize a high level abstraction and generalization for remote sensing image. Based on a kind of GSG model of the Extended Model Based on Mapping Division (EMD), this paper discussed the model and structure of subdivision cell template, and put forward some new ideas for remote sensing image template processing. The relate experiment shows, this study might help improve remote sensing image processing speed, reduce repeated handling of huge amounts of image data, and expand practical application of remote sensing.

Keywords GSG • EMD • Remote sensing image • Subdivision cell template

1 Introduction

With the rapid development of modern science and technology, the acquisition and application of spatial information resources get more and more attention. As a real-time, rich-information and covering a wide range of spatial information resources, remote sensing image has become an important foundation data in many military and civilian areas [1]. In recent years, much kind of remote sensing image data have appeared and expanded rapidly. But in many practical fields, we need to think of the compatibility, timeliness, accuracy, and reliability of these data. At the same time, processing speed has become the bottleneck of remote sensing applications [2]. Therefore, it needs to develop a more efficient theory and technology system to organize and manage global spatial information.

D. Xiong (✉) • Q. Xu
International School of Education, Xuchang University, Xuchang, China
e-mail: xiongdelan@aliyun.com

Global Subdivision Grid (GSG) provides a new way to build a global, multi-resolution level of open geospatial data organization framework [3]. Based on space partition organizational framework, the earth is divided into levels of regular, hierarchical, discrete cells. It will achieve rapidly storage, extraction and analysis of a worldwide mass data. In this paper, a kind of GSG model of EMD is introduced [4]. Then we propose the concept of subdivision cell template, and deeply discuss its application in remote sensing image processing. Some useful experiments and application demonstration will be done for testing.

2 Aerospace Information Subdivision Framework of EMD

2.1 EMD Method of Subdivision

In support of national 973 projects, Aerospace Information Engineering Research Center of Peking University proposed a set of multi-level geospatial framework, namely the Extended Model Based on Mapping Division (EMD) [5, 6]. The EMD mode can integrate the advantages of various global subdivision models in the world. The main idea of the EMD model is: using regular polyhedron to divide in the high-latitude regions, while using equal latitude and longitude grid to divide in low-latitude regions. The specific subdivision methods are as follows:

- First level subdivision. One divided cell is from south latitude 88° to the South Pole, marked as by 0. The other cell is from north latitude 88° to the North Pole, marked as 21. There are two latitude belts if we divide the surface of the earth from north latitude 88° to south latitude 88° through the equator. Each latitude belt can divide into ten longitude belts with width of 36° initially from west longitude 180°. So there are 20 subdivision cells, each one covers a range of 88° × 36°. These cells are orderly marked as 1, 2,..., 20. Therefore, the first level subdivision can get 22 cells, as shown in Fig. 1a.
- Second and Third level subdivision. Each first subdivision is divided into four equal second level subdivision cells. Each one has latitude difference of 44° and longitude difference of 18°. These cells are coded by 0, 1, 2 and 3. Then, each second subdivision cell is further divided into 33 third subdivision cells. Each one covers a range of 11° × 6°. They are shown in Fig. 1b.
- More than Fourth level subdivision. Subdivisions from the fourth to sixth level are divided according to mapping division rules. They are corresponding to 1:50000, 1:25000 and 1:10000 map respectively. From seventh level, subdivisions are recursively carried on by quadtree method with equal latitude difference and longitude difference.
- Pole Subdivision. The South Pole and North Pole were divided regular polyhedron division. Detail subdivision method was introduced in Ref. [5].

Data Model Research of Subdivision Cell Template Based on EMD Model 819

a First Subdivision

b Second and Third Subdivision

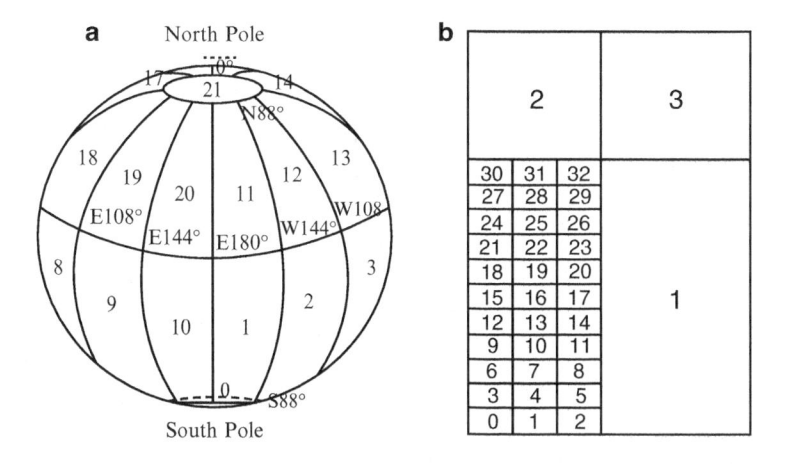

Fig. 1 Subdivision method of EMD model

2.2 The Advantages of EMD Model

Compared with the other method of subdivision, the EMD model has many advantages [7].

- The EMD model can provide a unified geospatial framework for various types of spatial data. Remote sensing image can be described in structure style using EMD model, which will improve expression ability of target.
- The EMD model can realize multi-scale representation of spatial data. All levels of subdivision cells form a strict hierarchy system, which can meet data requirements of different scales. Certain level subdivision cell has a unique correspondence with the cells in its upper level and lower level cells.
- The EMD model provides a unified spatial information integration services. Practice has proved that index of location is the basic approach to organize and use spatial data. EMD model can make spatial data of different levels and different regions have spatial logical relationships, just like a bridge.
- The practicality of EMD model. This model has a simple and clear correspondence with existing spatial data and coordinate expression Existing massive spatial data can be organized by this system easily.

3 Subdivision Cell Template Based on EMD Model

3.1 Subdivision Cell Template

GSG is a larger-scale hierarchical open spatial data management framework. It researches on how to subdivide the Earth into a series of cells with same area and similar shape. Subdivision cell is a multi-scale, discrete segmentation unit. It can be

Type	Item	Description
Cell Information	ID	Coding of subdivision cell
	Level	Subdivision level
	Area	the area of cell
	Length	the length of cell
	Location	Geographical region of cell
	Curvature	the average curvature of cell
Template Type	Name	the type name of subdivision template
	Application	Application goal of template
	Data Format	Metadata format for template
	Data Type	Data type and requirement of template
	File Interface	Data file and data processing algorithm interface
Image Data	Resolution	the resolution of remote sensing image
	Projection-type	Remote sensing plat and projection type
	Coordinate Information	Main coordinate value in image
	Entity Feature	Geometrical feature of main entity in image

Fig. 2 Main structure of subdivision cell template

expressed by an only digital or character encoding in the world [4]. It corresponds to a real region in the earth surface through subdivision. It has advantages of global, multi-resolution, well proportioned spatial location. Subdivision cell template is a standard data sample of remote sensing image corresponding to appointed region of certain subdivision cell. It contains data sets of spatial characteristics, geographic features and control points information of subdivision cell. Subdivision cell template inherits the advantages of subdivision cell, and establishes a correlation between the abstract model and specific remote sensing image. It will provide a good condition to organize, manage, process remote sensing image.

Subdivision cell template is composed of cell information, template type and image data. The main items are as shown in Fig. 2. Here, cell information contains detail subdivision cell information, such as coding, level, shape, size, spatial position, projection transformation and so on. Template type contains a variety of the template metadata, which is corresponding to specific subdivision processing algorithm. The template can be created by the template management module. Template data contains standard remote sensing image information according to certain subdivision cell. It not only includes resolution, coordinate information of remote sensing image, but also includes the color, texture, shape feature of spatial entity.

3.2 Modeling Process for Subdivision Cell Template

Constructing model for subdivision cell template is the key of remote sensing image processing. Hierarchical strategy and partition method are the main idea in construct data for subdivision cell template [8]. Modeling process can be described as following.

- Determine the number of layers for original remote sensing images. Firstly, make the original highest resolution images as the bottom resource. Then, determine relate layer number of other resolution image data based on multiplying power relationship.
- There are two kind methods to construct template. One is using existing image data correspond to appropriate layer. The other is circularly resample original data and generates other layer data.
- When meet a termination condition, it will stop creating new layer. Otherwise, creating new layer until the data amount of image is less than a image block.

Hierarchical strategy can be adjusted flexibly according to the situation of the data source. There is a greater flexibility in establishing image pyramid for multi-source data. The goal is as far as possible to ensure the accuracy of the image, reduce the data calculation, and improve processing speed.

Partition method is to improve I/O access efficiency of image data. Generally, the block of $2^n \times 2^n$ pixels is selected as the standard image data size [9].

4 Subdivision Cell Template Applications in Remote Sensing Image Processing

4.1 Subdivision Cell Template Processing Idea

Subdivision cell template can be fit for massive remote sensing image information, and improve processing speed of data. According to specific subdivision model, the subdivision cell has outstanding feature in size and shape. We can select several continuous subdivision cells as research object. Thinking of different application requirements, the high resolution and orthogonal projected remote sensing image of typical hot area can be selected. Then, subdivision cell template should be established through standardized processing. Subdivision cell template processing task includes image segmentation, feature extraction, manual annotation. The processing flow can be described by Fig. 3.

4.2 Application Demo for Remote sensing image

Based on EMD model, we have research on shape, feature and coding of subdivision cell in certain level. Then, we constructed the corresponding subdivision template of remote sensing image. An initial prototype system for subdivision template of remote sensing image processing system was developed [10]. The simple demonstration system was developed in Windows Server 2003. At present,

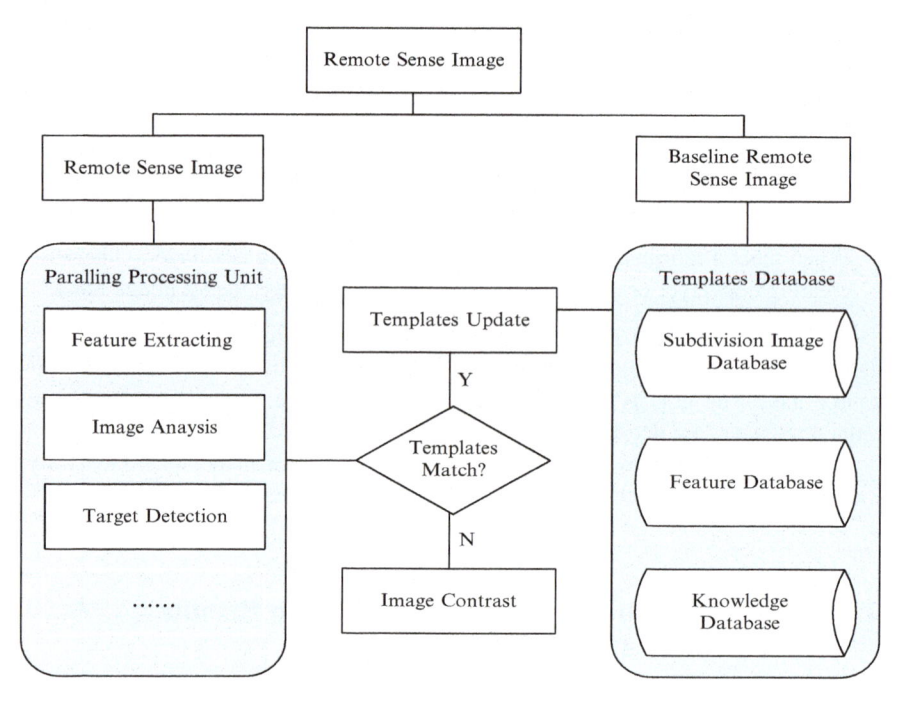

Fig. 3 Subdivision cell template processing framework

it used SQL Server 2005 to manage remote sensing data. And basic interface was designed and developed by VC++.

Preliminarily, we selected several high-resolution remote sensing images of tourist attractions, and create subdivision template manually. At present, basic functions such as image browsing, template view and feature retrieval according to the specified conditions were completed. A kind of operation is shown as Fig. 4. You can select retrieve mode (such as image subject), and input the keywords (such as Henan AND tourism AND spa). Then, the search results will be shown. You can get more information of image and template by clicking the related buttons.

At present, the system has completed of basic functions testing in single computer environment. Preliminary tests showed that the system have highly-targeted, high practicability, high retrieval efficiency and good expandability. More functions would be further developed.

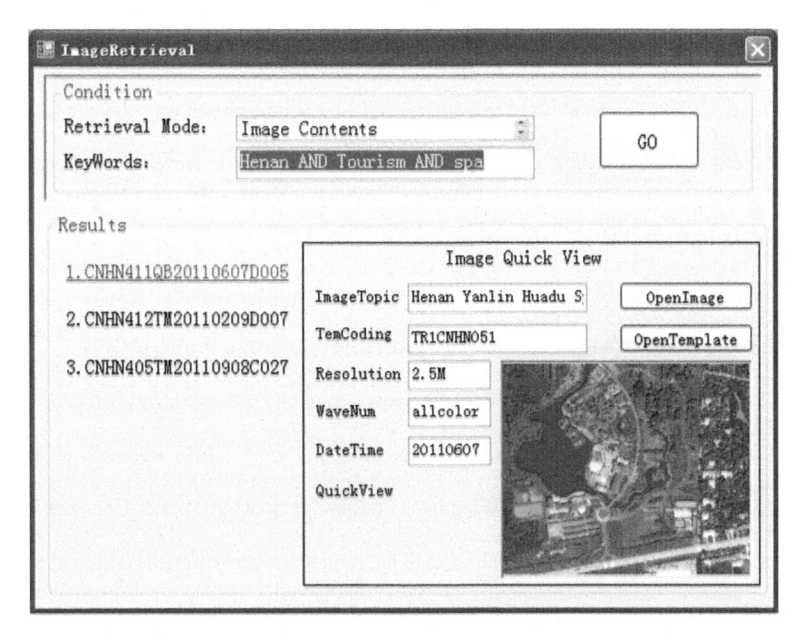

Fig. 4 Subdivision cell template and image retrieve interface

5 Conclusion

Remote sensing images have been widely used in various fields such as military reconnaissance, disaster forecasting, and environmental monitoring. Now the sensing images data is increasing, and its application requirement is expanding. The collection, organization, management and sharing of remote sensing data have become the most prominent problem to be solved for data producers and users.

The global subdivision model of EMD divides the Earth into cells at levels. Combining with the features and advantages of subdivision cell, this paper put forward the concept of subdivision template and proposed to construct subdivision cell template of remote sensing image. This method can be done gradually by level, by batches, and by cell region. With the increase of subdivision cell template, parallel processing technique would be adopted to improve processing efficiency and sharing service. This study might have significance for extending application fields of remote sensing image. It will open many practical applications of GSG, and enhance the value of spatial data.

Acknowledgements This work is supported by the Science and Technology Research Project of Henan Province under Grant No. 112102210079, and Third Young Backbone Teacher Support Program of Xuchang University.

References

1. LI De-ren (2005) On generalized and specialized spatial information grid. J Remote Sens 9 (5):513–520
2. Goodchild M (2000) Discrete global grids for digital earth. In: International conference on discrete global grids, Santa Barbara Press, California, pp 1271–1288
3. Sahr K, White D, Kimmerling A (2003) Geodesic discrete global grid systems. Cartogr Geogr Info Sci 30(2):121–134
4. CHENG Chengqi, GUAN Li (2010) The global subdivision grid based on extended mapping division and its address coding. Acta Geodaeticaet Cartographica Sinica 39(3):295–30
5. SONG Shu-hua, CHENG Cheng-qi, GUAN Li et al (2008) Analysis on global geodata partitioning models. Geogr Geo-Info Sci 24(5):11–15
6. GUAN Li, LU Xuefeng (2012) Properties analysis of geospatial subdivision grid framework for spatial data organization. Acta Scientiarum Naturalium Universitatis Pekinensis 48 (1):123–132
7. CHENG Cheng-qi, ZHANG En-dong, WAN Yuan-wei et al (2010) Research on remote sensing image subdivision pyramid. Geogr Geo-Info Sci 26(1):19–23
8. CHENG Qi-min (2011) Remote sensing image retrieve technology. Wuhan University Press, Wuhan, pp 123–127
9. Plaza A, Plaza J, Valencia D (2007) Impact of platform heterogeneity on the design of parallel algorithms for morphological processing of high-dimensional image data. J Supercomput 40 (1):87–107
10. DU Gen-yuan, MIAO Fang, GUO Xi-rong (2010) A novel network service mode of spatial information and its prototype system. Adv Mater Res 12(5):108–111, 319

A Spatial Architecture Model of Internet of Things Based on Triangular Pyramid

Weidong Fang, Lianhai Shan, Zhidong Shi, Guoqing Jia, and Xin Wang

Abstract In This paper, a novel spatial architecture model is proposed for the Internet of Things (IoT) based on triangular pyramid. Under the support of generic technologies, this spatial architecture model is composed of "Sensing and Controlling", "Ubiquitous Transmission" and "Diversified Requirements and Applications". The proposed architecture model may not only achieve effective convergence of technology, but also simplify the heterogeneous network and facilitate the application design of IoT. So, the current application situation of IoT is introduced first, and the architecture of IoT is then analyzed. Secondly, the spatial architecture model is proposed and some relevant compositions are discussed. Finally, the application prospect is described.

Keywords Internet of things • Architecture • Triangular pyramid

---

W. Fang • Z. Shi
School of Communication and Information Engineering, Shanghai University, Shanghai, China

L. Shan (✉)
Shanghai Internet of Things Co., Ltd, Shanghai, China
e-mail: shanlianhai@163.com

G. Jia
College of Physics and Electronic Information Engineering, Qinghai University for Nationalities, Xining, China

X. Wang
Changchun Institute of Engineering Technology, Changchun, China

1 Introduction

The Internet of Things (IoT) is an advanced multi-disciplinary research field nowadays. It is considered to be one of the most important technologies which have enormous influence in the twenty-first century. In 2005, ITU defined IoT as [1], "The connectivity for anything by embedding mobile transceivers into a wide array of additional gadgets and everybody items, enabling new forms of communication between people and things, and between things themselves." Actually, the IoT could sense, collect and monitor all kinds of information in real-time through various integrated micro-sensor, processing information by embedded systems, and converging information through self-organizing multi-hop relay communication networks. The IoT would transfer this information to the user's terminal to complete various applications and achieve the "ubiquitous computing" goal.

Currently, the IoT is gradually applied in various fields. K.H. Su focused and summarized the relationship between "smart city" and "digital city" [2]. A municipal solid waste recycling management information platform was proposed [3]. The application of "smart community" and "community security management" was depicted respectively [4, 5]. In terms of people's lives, G.X. Yang investigated security and defense system for home based on IoT [6], Y.G. Li researched the reinforcement of communication security of the intelligent home [7]. Of course, there are other application fields, such as "Logistics Industry", "Electronic Commerce" and so on.

Although the applications have a bright future, the IoT is an ongoing research and development work. At present, there is no IoT system of ITU strict definition and objective architecture. In this paper, we will introduce the current research status of IoT architecture. For assuring the objectivity and rationality of assumption, we will propose our novel architecture based on the understanding to IoT, and then put forward key technologies involved with this architecture.

2 Related Works

It is very heterogeneous for each joint of IoT from sense, transmission to application. The future needs to be taken openly and expansive network system architecture for achieving interconnection, interaction and interoperability of information.

2.1 Architecture of IoT Based on Technology

At present, the mainly research describing the architecture of IoT is divided into two types. One is based on the high-level architecture of Ubiquitous Sensor Networks (USN) in Y.2221 of ITU [8]; the other refers to the OSI model.

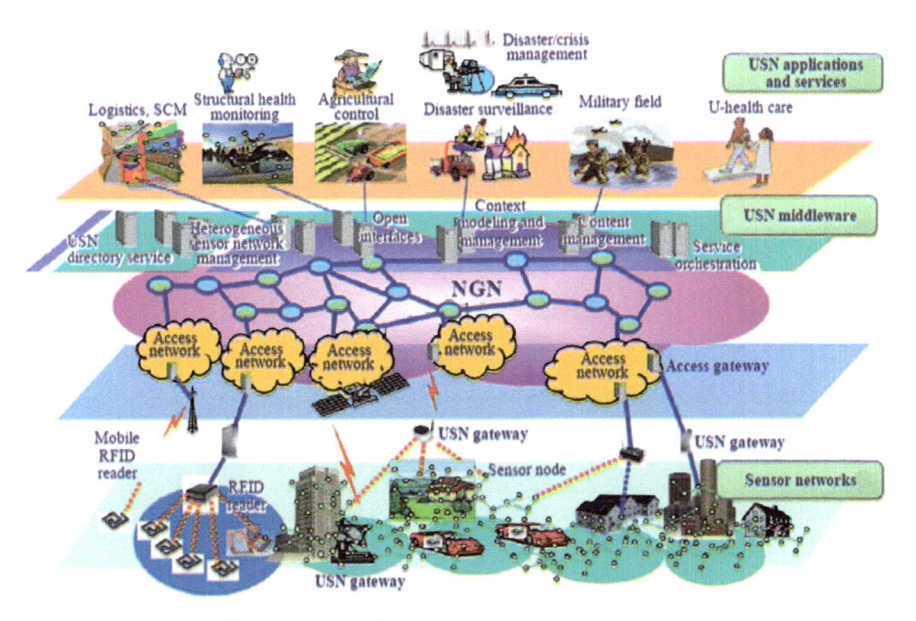

Fig. 1 An overview of USN with related technical areas

The architecture is divided into four layers based on oriented applications and services of the USN in Fig. 1. It includes the underlying sensor network, access networks, USN middleware, USN applications and services from bottom to up.

Another architecture that refers to OSI model is divided into three layers by using hierarchical design. It includes the Perception Layer, the Network layer and the Application Layer. Although the architecture ignores the network heterogeneity, it is cursory and flat. This architecture's layer is divided more from the application's point of view, mentioned less for the interface between the layers (data interface or physical interface) and scalability.

In fact, there is no specialized research for IoT in the ITU's technology roadmap, but take the communication between persons and things, things and things as the IoT's important function, which is adopted by ITU.

2.2 Architecture of IoT Based on Application

There are three types of IoT's architectures based on application, which include Radio Frequency IDentification (RFID), WSN and Machine-to-Machine (M2M).

1. Base on RFID
   The electronic tags that change "things" into intelligent objects may be the most flexible in the three types of systems architecture, tagging mobile and

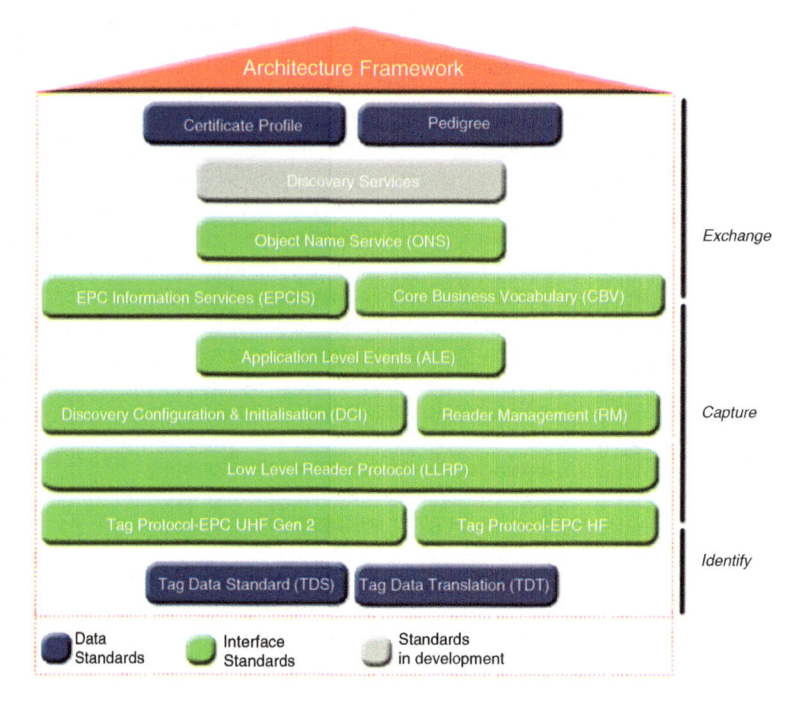

Fig. 2 EPCGlobal standard architectures

non-mobile assets are their major application for tracking and management. Prof. E. Fleisch believes that RFID, just like punch card, keyboard and barcode, is a data entry method, belongs to a scope of IoT.

In the coding, the Auto-ID Centre proposed EPCGlobal system for all electronic encoding; RFID is just a carrier of encoding. It proposed five technical components of the Auto-ID system in Fig. 2, including an EPC (electronic product code) tag, RFID reader, ALE Middleware for information filtering and gathering, EPCIS (EPC Information Service) and EPCIS Discovery Service.

2. Based on WSN

WSN is composed of a set of "autonomy" wireless sensor nodes in free space, which collaborate to complete the monitoring of specific environmental conditions, such as temperature, humidity, and so on. It is self-organizing or self-configuring network, including Mesh Networks and Mobile Ad-hoc Network.

Although WSN is still a hot spot in the research field, there is seldom a success case in the industrial field. This is because, the research on WSN focus on sub-layer of network mostly. According to its current development, we think the WSN has certain distance from the IoT of real sense, because it is inadequate for research on the issue of its level. On the other hand, some researchers of WSN are so high on wireless technology that they ignore the combination of using field bus in perception aspect and using long distance wireless communication in the transmission, which have achieved the goal of large scale application from the practicability and commerce popularization.

3. Based on M2M

   Generally speaking, the concepts and architecture of M2M cover a wide range, including the part of the contents of the EPCGlobal and WSN, also covering both wired and wireless communication.

   Similarly, the development of M2M lacks standardized specification and unified system architecture. Although there is some attempt (such as oBIX and oMIX), the unified specification has not yet formed, there is still a long way to go.

   In contrast, the technical architectures of WSN and M2M have not yet fully been raised to the IoT's ONS/PML technology system height. We believe that it should refer to ONS/PML technology system on the road towards the IoT.

3 Novel Architecture Profile

The architecture is the most important prerequisite for the guidance of the system design. In addition, the design of the architecture will determine the IoT's technical details, application patterns and trends.

We have followed the research methods used by ITU for architecture of IoT. Firstly, we abstracted its applications and scenarios, which are the basis of the design and verification of its architecture; then put forwards the general principles and requirements. At last, we proposed a common architecture and the functional structure models in Fig. 3.

We know that the objective of common architecture is the development of the functional methodology and general model. So, we propose the spatial architecture model of IoT based on triangular pyramid differs from the traditional form. This architecture consists of four parts: 'Sense and Control', 'Ubiquitous Transmission', 'Diversified Services and Applications' and 'Generic Technologies', in which 'Generic Technologies' are the bases for above mention.

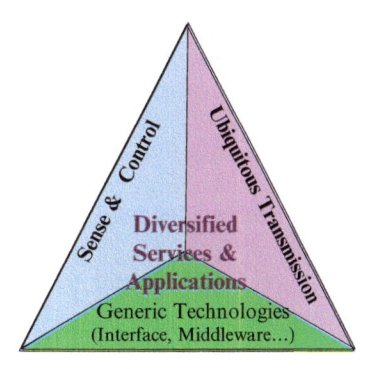

Fig. 3 A spatial architecture model of internet of things

1. Sense and Control

   The technology of things sensed and controlled is also called information things technology. It belongs to the interface technology between the physical world and network world. In a wide sense, it translates the analog into the digital, which would be used. According to the instruction, it could perform some operation.

   The connotation of this technology is very extensive. It includes the Cyber Physical System (CPS) [9]. Some researchers who study on CPS usually take the embedded systems as the basic technology between real-world and network systems in the U.S.A. Other researchers think that RFID, Near Field Communication and WSN constitute the basic technology to connect the real world and the digital world in Europe. This difference derived from the different research point of view.

2. Ubiquitous Transmission

   The ubiquitous transmission means information can be reached anywhere, not just the end to end. This transmission should have the high efficiency, low latency and QoS guarantee. The ubiquitous transmission relates to the network of self-management capability. On the other hand, it extended to "Any time, any place, to connect to any things, to transmit any type of information".

   From a macro perspective, Internet to carry transmission of thing's information is no longer Internet of traditional sense. It must ensure that the information about the query things has a time stamp and space markings, and instructions to manipulate things must be the instructions of the time and space semantics. All of these require the Internet to carry transmission of thing's information has the clock system and the coordinate system associated with the physical world,

3. Diversified Services and Applications

   The IoT is embracing the people and things of the modern society, so the applications and services involve various industries. Its applications can be divided into several categories. The transportation and logistics classes include the logistics management, automatic toll collection based on personal identification, and electronic navigation map. The applications of medical classes are composed of the tracking of medical objects, perception of physical symptoms, as well as a database system. The personal and social classes are composed of real-time interaction between people and networks, finding missing items. The application of the future class is not yet available in the current deployment of the conditions, including the robot taxi, smart city; and so on. These applications are various services, which include multimedia services and data services. Of course, these services could be divided into narrow-band services and wideband services from the aspect of bandwidth.

4. Generic Technologies

   In general, the generic technology is a class of technology that may be widely used in many areas in the future, It's R & D results can be shared, and profoundly impact one or more industries and enterprises. As a major part of the spatial architecture model of IoT based triangular pyramid, the generic technologies play an important role from cognizing and sensing the ambient world to diversified services and applications. These generic technologies could not only solve the heterogeneity of the network, but also promote the convergence of multiple technologies. The generic technologies of IoT include middleware, interface and information process, but not limited to these items.

4 Associated with Cloud Computing

In 2008, the concept of cloud computing was proposed by Google, caused widespread concern in the industry. Forester Research gave cloud computing a definition: A standardized IT capability (services, software, or infrastructure) delivered in pay-per-use, self-service way [10]. We believe that cloud computing is the combination of information processing, storage and operations based on cooperation. The main source of information is the massive data, which mostly comes from the ambient world.

In the traditional sense, the association of cloud computing and the IoT is vague. However, through the proposed architecture in this paper, they both could conveniently be associated by 'Generic Technologies'. This is due to that some technical features are not only owned by the two, and some events are but also driven jointly.

Many associated technologies combined with novel architecture of IoT and cloud computing, we focus on the pre-processing technology of massive data in IoT. At present, with the expansion of the scale of the IoT's applications, the amount of data that is generated by perceived information is growing exponentially. On the other hand, the perceptual characteristics of the sensor node, just as multi-point sensing a common target, as well as a single-point sensing multi-target, would cause the deviation and redundant of the information within a certain range. In addition, in the process of information transmission, especially the transmission process in the self-organizing multi-hop network, part of the data loss caused by dirty data – in the data of source system (data perception and transmission in IoT), the size of the data is not within the given range, or the data is meaningless for the actual services, or data format is illegal, as well as there are non-standard coding and vague logic of service. These dirty data would reduce the operating efficiency of the system, or even cause system collapse. In order to solve the above-mentioned issues, some measurements should be taken:

- Date filtering or data cleaning based on variable constraint for 'Sense and Control';
- Data aggregation based on distributed multi-source for 'Ubiquitous Transmission';
- Data reduction based on the feature of the perceived target for 'Diversified Services and Applications'.

5 Conclusion

There are some arguments about the research and development of IoT. One is whether it has its own technology architecture. Someone holds passive attitudes. They claim that it has just been the integration of existing technologies. However, others hold opposite opinions. In this paper, the novel architecture pro-posed ensures to provide guidance for how to define the functional structure of an IoT. The guidance includes how to divide into the appropriate feature set and how to define the reference point between the feature set corresponding to the physical realization and interface definition. Meanwhile, it is believed that although the IoT possesses features of computer, communication, network and control, the simple integration of these technologies cannot constitute a flexible and efficient IoT. Based on the convergence of above mentioned technology, the IoT forms own technical architecture through further research, development and application.

In accordance with the ITU point of view, "Interconnect Any Thing" is the expansion of capacity, services and applications of the NGN. Therefore, it is recommended that the IoT should be applied to NGN's research scope and be implemented in the NGN technology development roadmap. We should further study the functional architecture of the IoT and specific configuration model.

Acknowledgements This work is partially supported by the Science and Technology Innovation Program of Shanghai (12DZ2250200, 12511503300), and the Tianjin University – Qinghai University for nationalities Independent Innovation Fund Projects.

References

1. ITU internet reports: the internet of things (2005)
2. Su KH, Li J, Fu HB (2011) Smart city and the applications. In: Proceeding IEEE ICECC, Ningbo, China, pp 1028–1031
3. Chen T, Li X (2010) Municipal solid waste recycle management information platform based on internet of things technology. In: Proceeding IEEE MINES, Nanjing, China, pp 729–732
4. Li X, Lu RX, Liang XH, Shen XM (2011) Smart community: an internet of things application. IEEE Commun Mag 49:68–75
5. Liu JH, Yang L (2011) Application of internet of things in the community security management. In: Proceeding IEEE CICSyN, Bali, Indonesia, pp 314–318
6. Yang GX, Li FJ (2010) Investigation of security and defense system for home based on internet of things. In: Proceeding IEEE WISM, Nanjing, China, pp 8–12
7. Li YG, Jiang MF (2011) The reinforcement of communication security of the internet of things in the field of intelligent home through the use of middleware. In: Proceeding IEEE KAM, Sanya, China, pp 254–257
8. ITU. T. Recommendation Y.2221 (2010) Requirements for support of ubiquitous sensor network (USN) applications and services in the NGN environment
9. Atxori L, Iera A, Morabito G (2010) The internet of things: a survey. Comput Netw 54:2787–2805
10. http://www.forrester.com/Cloud-Computing

Design and Realization of Fault Monitor Module

Dongmei Zhao, Huanhuan Dong, and Xiang Xue

Abstract Fault monitor module is designed for power grid fault diagnosis and recovery system. It sets interface with dispatching automation platform to get real-time operating data. In order to solve the problem – "Data are flooding but useful fault information is lack and fuzzy", this paper analyzed and processed the upload of data in fault condition. It uses semantic analysis and logical analysis to delete interference, integrate some other messages and do some complement. With the more reliable and clear messages to identify the condition, it can avoid the fault diagnosis mistakenly start through fault identification functional module. It uses fast network topology analysis (using depth-first method search whole net and dynamic topology tracking the net changes) to monitor real-time data changes of the power grid running state. The whole program is divided into several separate threads to improve the efficiency. The application result shows that it improves the diagnosis and the recovery can be more accurate, more efficient and faster.

Keywords Monitor • Interface • Fast topology analysis • Multi-thread • Fault identify

1 Introduction

Modern grid becomes bigger and the harm caused by fault is more and more serious. At the same time, users' requirements are also rising about continuous power and its quality. Grid is developing toward the direction of building a strong and smart grid, trying to realize it's self-healing and self-adaptive. In other word it means when simple or complex fault happens to grid, the automation system can

---

D. Zhao • H. Dong (✉) • X. Xue
School of Electrical and Electronic Engineering, North China Electric Power University, Beijing, China
e-mail: dong.huanhuan@163.com

accurately and rapidly identify the fault zone to determine the true fault element, then to restore power to the users in blackout region with the faster speed and optimal path. Therefore, fast and accurate fault diagnosis and recovery system is significant to the stable and reliable operation of the power grid.

Fault diagnosis and recovery methods are all based on the gathered fault messages. Without the fault messages, the fault diagnosis and recovery system is just like bricks without straw, so fault messages analysis and processing is necessary and essential. We establish the monitor module to receive real-time information collected by the dispatching automation platform, and then to classify and select from the information to make network topology analysis in this paper. It can provide reliable and useful fault information and coordinate the execution order of the program.

2 The Overall Program Design

The monitor module has several sub-modules: data interface, alarm information identification, fast network topology analysis, the system self-test, multi-thread management. The whole system's function block diagram has shown in Fig. 1.

2.1 Data Interface

It connects SCADA/EMS to get the static and dynamic data for fault processing. The static data includes device model, topology information and some secondary device configuration information. We can get the static data from CIM. Dynamic data are real-time operation messages. It can provide grid operating information. We can get real-time operating codes and translate them into information that can be directly and easily used by following program based on the 104 rules.

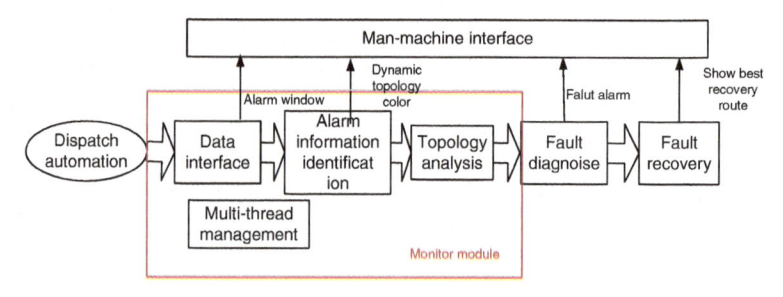

Fig. 1 Fault diagnosis and recovery system monitor module functional diagram

Fig. 2 Fault messages pre-treatments flow chart

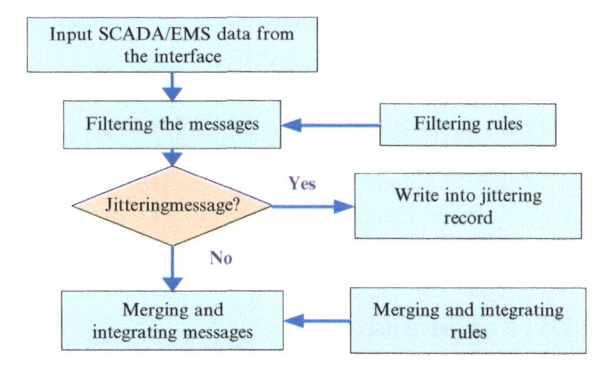

Alarm information window includes the new alarms fault diagnosis system getting data last time till it getting data this time(suppose the time interval flag is t). With testing many times, t equaling 10 sends is suitable for getting fault messages.

2.2 Fault Messages Pre-treatment

There are plenty of fault messages, but many of them are redundant and useless. The messages are incomplete inconsistency and possibly they are got not at the same time. All of these will lead the interface to miss some of the troubleshooting critical information, so the messages pre-treatments are necessary [1] (Fig. 2).

2.2.1 Filter

In order to highlight main information, we can filtering some alarm messages which contains some special keywords, for example, the alarm messages including words like "DC panel", "fault recorder" and so on. What is more, we can provide the editing functions such as add, delete, modify keywords to enhance the adaptability [2].

2.2.2 Anti-shaking

The jitter messages do great harm to fault diagnosis, so the rules of judge and deal with jitter messages are developed to ensure that jitter messages are written into jitter record table and being prevented from uploading.

2.2.3 Messages Merging

Based on message merging logic, the messages which have the same semantic content or extension meaning should be merged. E.g. some important components that have dual-microprocessor-based relay protection and the two sets of protection device may act and send messages at the same time. To merge the messages according to the logic, then choose the most suitable one for display.

2.2.4 Comprehensive

Most of the protection signals, such as over current protection, differential protection, and overload protection are all have a fixed protection tripping objects. We can integrate the related fault messages into one message according to the directly relationship between the relay signals and breaker tripping messages.

2.2.5 Dealing with the Mistake Messages and Omitted Messages

According to action logical relationship between the relay protection and circuit breakers, and considering some auxiliary secondary signals at the same time, it can do in-depth analysis of the protection signals to see the possibility of errors in logic or to see whether missing some circuit breaker tripping messages that should be uploaded. After these are done, it can suggest the possibly mistake messages and the omitted messages, and directly delete the insignificant ones [3, 4].

2.3 Fault Identification

The fault diagnosis and recovery program isn't running when grid is operating well. It starts only when fault happens, so fault identification is important to start diagnosis program. The fault criterion is made with the necessary fault information of relay protection, circuit breakers and so on.

$D = \{d_1, d_2, \ldots, d_i\}$ is a collection of all devices, including generators, transformers, buses, lines. $P = \{p1, p2, \ldots pj\}$ is a collection of all relay protections. $B = \{b_1, b_2, \ldots, b_k\}$ is a collection of all breakers.

Relationship $\Re$ is defined as follows:

$$\Re = \{(p, b) | p \text{ move contributes to } bmove, p \in P, b \in B\} \quad (1)$$

If p and b have the relationship as Eq. 1, we can say p and b are relations.
For $p \in P$, the related breakers are $B(p)$

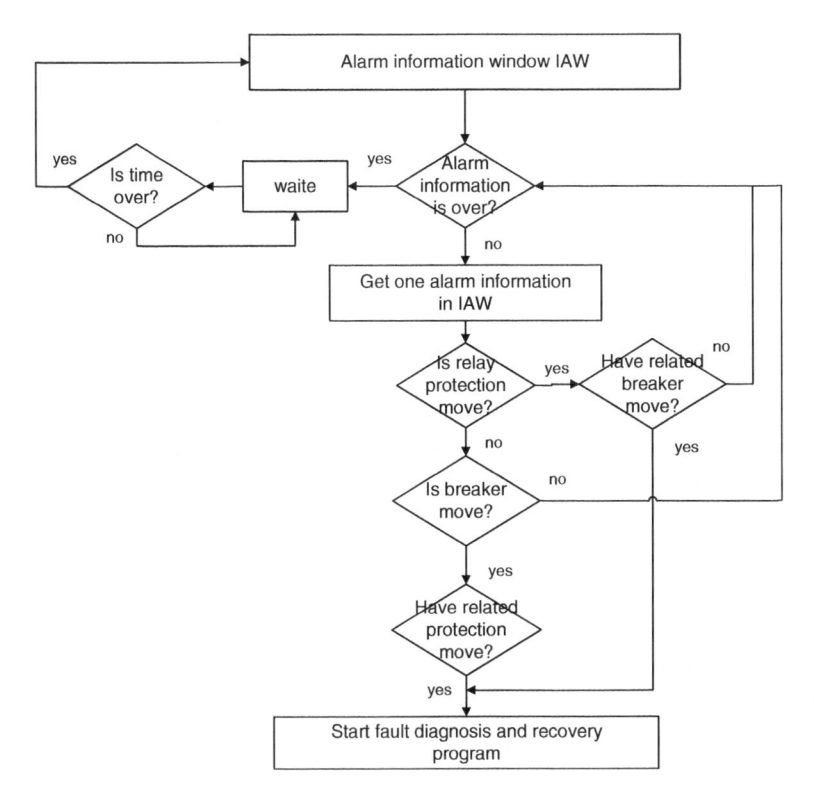

Design and Realization of Fault Monitor Module

Fig. 3 Fault identification flow chart

$$B(p) = \{b|p\Re b, b \in B\} \quad (2)$$

For $b \in B$, the related protections are $P(b)$

$$P(b) = \{p|p\Re b, p \in P\} \quad (3)$$

If there is protection $p \in P$ arising, and at the same time there is related breaker move-information $b \in B$, it tips fault happen [5]. The detailed identification process is shown in Fig. 3.

2.4 Fast Network Topology Analysis

2.4.1 Network Topology Analysis

The information of connecting relationship of the devices and the real-time status of switches which comes from the interface are important data for topology analysis.

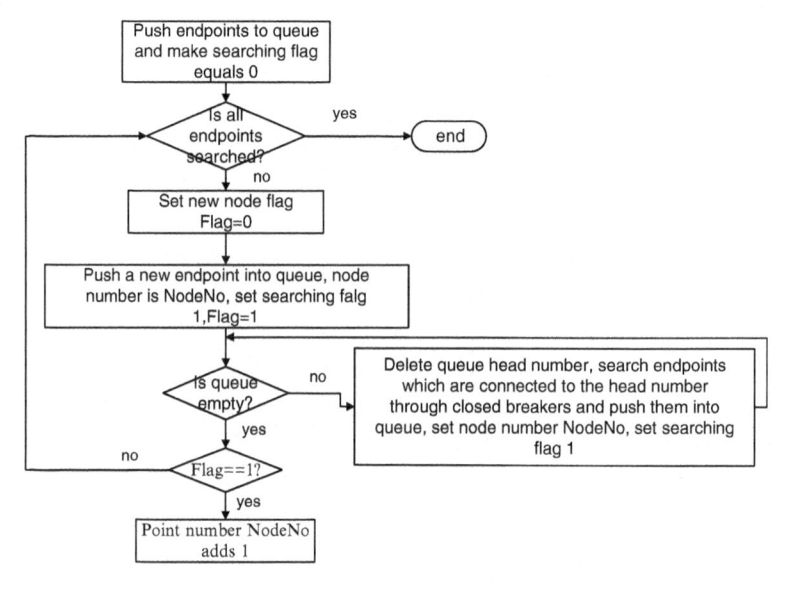

Fig. 4 Topology nodes analysis

If the switching device is closed then the two endpoints of it can integrate into one node. Check all the switching devices with this method to form the endpoints-nodetable. The detailed analysis process is shown in Fig. 4.

After that, integrating lines transformers and other impedance devices' two nodes into one subsystem. The method is similar with topology nodes analysis above.

The subsystem has power and load representing that it is operating well. Otherwise, it is an island that is isolated from the grid. In order to ensure the timeliness of the online application, it needs real-time information to update current grid dynamic topology and to provide the relations between devices for the whole program.

2.4.2 Dynamic Topology Tracking

Partial network topology tracking is essential for improving the speed. It directly focuses on the move zone and analyzes the partial zone based on the breakers movement [6]. Figure 5 is the flowing chart.

1. Are the breaker's two endpoints still belonging to one node? If the answer is yes, the track is over; If answer is no, to start searching from breaker's one endpoint based on closed neighbor breakers, set their node number MaxNodeNum+1 (MaxNodeNum is the max node number before the breaker opens).
2. Are the breaker's left and right nodes still belonging to one system? If answer is yes, the track is over; otherwise, to start searching from the new node based on impendence devices, set their system number MaxSysNum+1(MaxSysNum is the max system number before the breaker opens) (Fig. 6).

Design and Realization of Fault Monitor Module 839

Fig. 5 Dynamic topology tracking when breaker opens

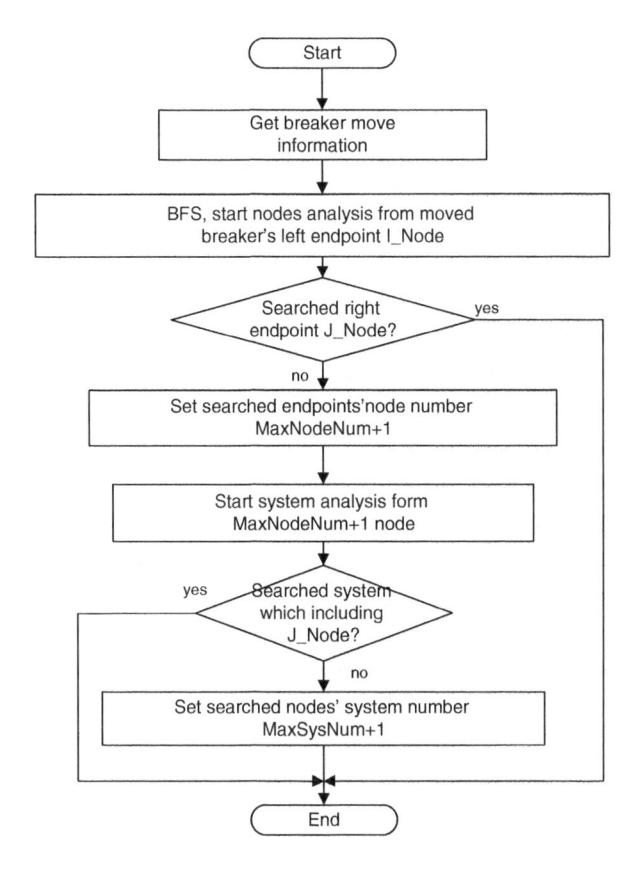

2.5 Multi-thread Management

Multi-thread management sub-module sets separate threads to improve the running efficiency as multi-thread technology aloud several threads running at the same time. It also sets their running prior to decide which one running first when crash happens.

Prior I: fault identification function starts the online fault diagnosis and recovery program has the first position, its running cannot be disturbed by other threads. Other threads are all hang.

Prior II: data interface is an always running thread, and it is listening for data transmission request all the time.

Prior III: users' request from man – machine interface such as electrical devices' parameter adding, deleting or changes and so on.

In a word, multi-thread management can solve threads' crash and order the sequence of their running. The thread which has the higher prior could obtain more CPU running time and recourse.

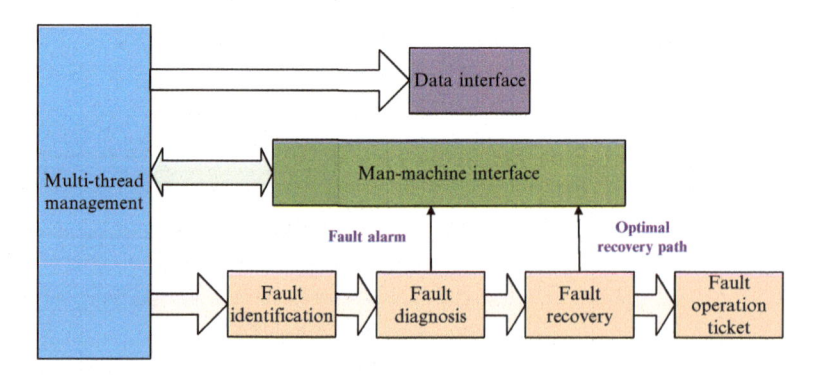

Fig. 6 Multi-thread management module functional block diagram

3 Software Testing

Due to the low probability of fault in the grid, it designs the study states fault simulation to test effectiveness and practicality of the software. Using the alarm information that read from the database to set fault condition to test the software, the results show that the software running smoothly and the diagnostic result and recovery path are correct and reasonable. We apply it to a regional power grid's fault diagnosis and recovery system to see that it could read real-time data, give warning tips when it accepts exceptional messages. It has high reliability and meets the development requirements of the system's function.

4 Conclusion

Monitor module is a data provider for fault diagnosis and recovery system, and it is also the entire system's manager and commander. It controls the priority order of the system program's implementation and trigger conditions, etc., keeps coordination of the various program modules and avoids conflicts. In advancing the process of fault diagnosis and recovery system utility, full attention must be paid to the collection and collation of the fault information, including the pretreatment of the fault information for building underlying data and extracting diagnostic knowledge for fault diagnosis.

References

1. Ding Jian, Bai Xiao-min, Zhao Wei (2007) Fault information analysis and diagnosis method of power system based on complex event processing technology. Proc CSEE 27(28):40–45, In Chinese
2. Guo Jin-zhi (2003) Study on the on-line intelligent fault diagnosis of power transmission network. North China Electric Power University, Beijing, pp 6–14, In Chinese
3. Liu Dao-bing, Gu Xue-ping, Zhao Jie-qiong (2010) Practical research of fault restoration for the regional power grid. Power Syst Prot Control 38(21):48–52, In Chinese
4. Zhou W (2008) Research and development of power grid intelligent alarm information processing system. North China Electric Power University, Beijing, pp 1–38, In Chinese
5. Liu Dao-bing (2012) Modeling and solving of a complete analytic model for fault diagnosis of power systems. North China Electric Power University, Beijing, pp 18–62, In Chinese
6. Chen Jiong (2008) Research on the method of automatically generate topology in the grid operator decision-making system. Nanjing Normal University, Nanjing, pp 30–40, In Chinese

Dynamical System Identification of Complex Nonlinear System Based on Phase Space Topological Features

Lei Nie, Zhaocheng Yang, Jin Yang, and Weidong Jiang

Abstract Dynamical system identification of complex nonlinear system can be regarded as parameter estimation task of uncovering the nature of the system, which is of much importance for many engineering applications. Sea clutter is one kind of complex nonlinear system, which has intrinsic complexity and high nonstationarity. Therefore, it is difficult to detect marine targets due to the low Signal-Noise-Ratio (SNR) and low Signal-Clutter-Ratio(SCR). To improve the performance of marine target detection in, a method to discover fundamental geometric structures from high-dimensional and multivariable data sets is proposed. The method can improve the target detection performance and its main novelty lies in phase space reconstruction and topological features extraction, based on the dynamical system identification scheme and topological data analysis (TDA) techniques. Experiments with measured sea-clutter data demonstrate the performance of the method proposed in this work.

Keywords System identification • Pattern discovery • Sea clutter

1 Introduction

Robust marine target detection [1] and adequate sea clutter modelling [2, 3] are significant capabilities in utilizing the modern maritime surveillance radar, which has high-resolution and broadband [4, 5]. The massive data of radar echo provide more possibilities as well as more difficulties in knowledge extraction.

L. Nie (✉) • Z. Yang • J. Yang • W. Jiang
School of Electronic Science and Engineering, National University of Defense Technology, Changsha, China
e-mail: Nielei@Nudt.Edu.cn

Recent years, Topological Data Analysis (TDA) methods [6] are being developed for data sets on which the natural structure is not a metric, but rather a partial order [7, 8]. By extracting global geometrical structures from lower dimension expression, this non-parametric approach is proved to be an effective tool for massive data analysis [9]. The primary processes applied in TDA are described as several common steps [10].

To implement TDA methods on the datasets of measured sea-clutter, the transformation from original data to point cloud data (PCD) must be made. The global geometrical features are extracted from PCD and are certain presentation for the original data. The hidden structures and properties consist in the sea-clutter represented as digital form time series.

In real world most systems are doubtless nonlinear and non-stationary systems. A prerequisite for the representation of nonlinear and non-stationary data is to find the adequate adaptive basis. It is advisable to apply TDA methods on sea-clutter, the typical complex system of practical radar clutter environment.

2 Nonlinear Dynamical System Identification

Devote to reconstruct dynamical systems from measured data, system identification can be regard as a general term to provide mathematical tools and algorithms representation. The field of system identification is well developed for linear systems. In fact, all physical systems are nonlinear to an extent and exhibit a variety of complex dynamic behavior, so nonlinear dynamical system identification is of great importance. To deal with the curse of dimensionality and insufficiency of priori knowledge, nonlinear dynamical system identification have to consider efficient method capable of dimension reduction and knowledge discovery, like TDA mentioned above. It is obvious that there is no particular representation of nonlinear systems which can be regarded as the best for all applications, however, it seems that system identification is in many aspects simpler and more convenient to implement.

As is shown in Fig. 1, A general implementation procedure to estimate a nonlinear model can be described as follows:

1. *Select the model structure*
2. *Split the observed data into two parts: an estimation data set and a validation dataset.*
3. *Select the input and output lag spaces nu and ny and the input time delayed.*
4. *Train a number of nonlinear models by using the estimation data set*
5. *For each of these models, use validation data to compute the value of the criterion.*
6. *Select the one which minimizes the fit on the validation data as the final model.*

Fig. 1 The common implementation procedure of system identification

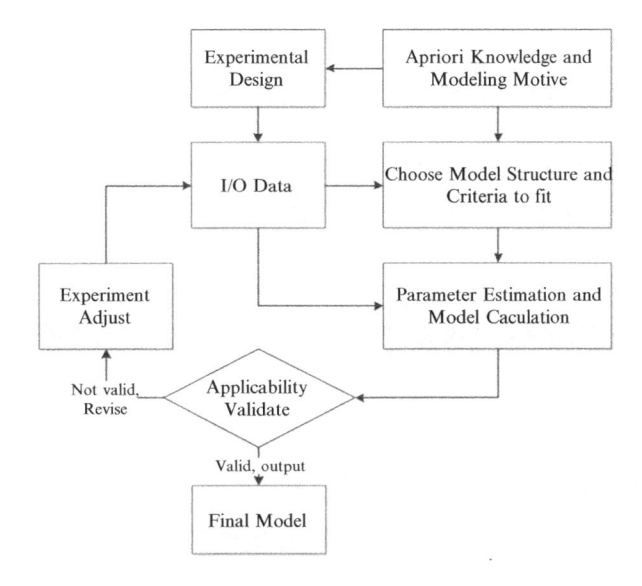

3 Topological Features Extraction

Discovering hidden patterns or structures is of significant importance to study complex systems. Actually, the description of pattern or structural knowledge can be classified into many kinds according to the research area and application objective. The topological structure and invariant are what we focused in this work.

Pattern recognition [4] and most related works is about methods of supervised learning and unsupervised learning. Usually, the patterns to be classified are groups of measurements or observations. Data or patterns classification take advantages of priori knowledge, or ulteriorly statistical information, whereas, pattern discovery takes advantages of using domain-specific knowledge, processes the data by reverse-engineering method, and aims to acquire the certain patterns which are able to character the complex system effectively. The patterns include topological structure, topological invariants, and manifold characteristics, et al. (Fig. 2)

As a novel approach of studying the inner structure and sub-structure of complex systems, pattern discovery use less theoretical assumption but provide more practically useful knowledge. It can be considered as the extension of pattern recognition and machine learning. Up to now, the idea of pattern discovery illustrates considerable superiorities through the integration with nonlinear dynamics, dimension reduction, statistical manifolds, and topological analysis.

The classical statistical and non-statistical models based on the amplitude statistics of sea clutter are extensively studied for many years. Topological methods under pattern discovery scheme, which are natural to be applied in this direction, is suitable to see through the instinct complexity by extract structural information and topological invariants, simply because geometry can be regarded as the study of distance functions.

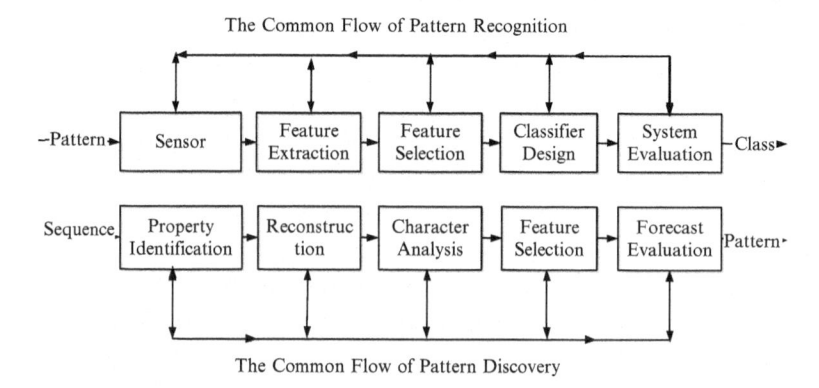

Fig. 2 Pattern discovery and pattern recognition comparison

Currently, the geometrical and topological methods for structural analysis of the data are being pursued in the three allied fields of computer science, mathematics and statistics.

Although each field has their own particular approach and questions of interest, there are still lots of similarities among all three fields. Taking advantages of geometry and topology methods, novel and effective methods could be proposed, among which TDA is a typical tool. In this work, the method based on TDA is implemented by the following steps:

1. Locate the critical points in the dataset
2. De-noise, convenient for the structural pattern extraction
3. According to proximity parameter, replace the data points with a family of indexed simplicial complexes
4. Through algebraic topology and especially persistent homology, analyze topological features
5. In the form of Betti number, get the representation of the persistent homology.

In order to utilize the topological structural patterns in the dynamical system identification, a distance function based on the coherence function could be defined to evaluate the "closeness" of two processes:

$$d(X_i, X_j) \triangleq \left[\frac{1}{T} \int_0^T \left(1 - C_{X_i X_j}(f)\right) \right]^{1/2} \quad (1)$$

Where $C(f)$ is coherence function of the two processes. The topological invariants then provide efficient parameters to evaluate if there exist significant differences between the reconstructed dynamical systems.

4 Experiments and Conclusion

Experimental sea-clutter data used in this work are provided with McMaster University IPIX radar. This radar is located at OHGR, Dartmouth, Canada, on a cliff facing the Atlantic Ocean. It is fully coherent X-band radar with several advanced features [2]: dual-polarized, coherent with an antenna beam width of 1″, built-in calibration equipment, and digital control system.

The antenna of this radar points along with a fixed direction, and the radar works at dwelling mode of operation. Only a piece of the ocean surface is illuminated, and the sea clutter features completely depends on the sea surface's motion modality.

The experimental results are given on the measured sea clutter data of the data set #17 and #18. Experiments on correlative datasets, for instance, #19, #28, #26, #283, #300, and most other datasets, bring similar results and conclusions.

The sea clutter time series can be reconstructed to get a set of vectors which are m-dimensional, and these vectors consist in the reconstructed phase space. To deal with effective structural features extraction in these vectors, Topological Data Analysis methods could be utilized (Fig. 3).

Figure 4 demonstrates the three dimensional phase space portrait of sea clutter datasets #17 and #18. It is notable that the distributions are severely imposed by the target. Based on these distribution patterns, the topological structures could be acquired by implementing dimension reduction methods and topological transformations.

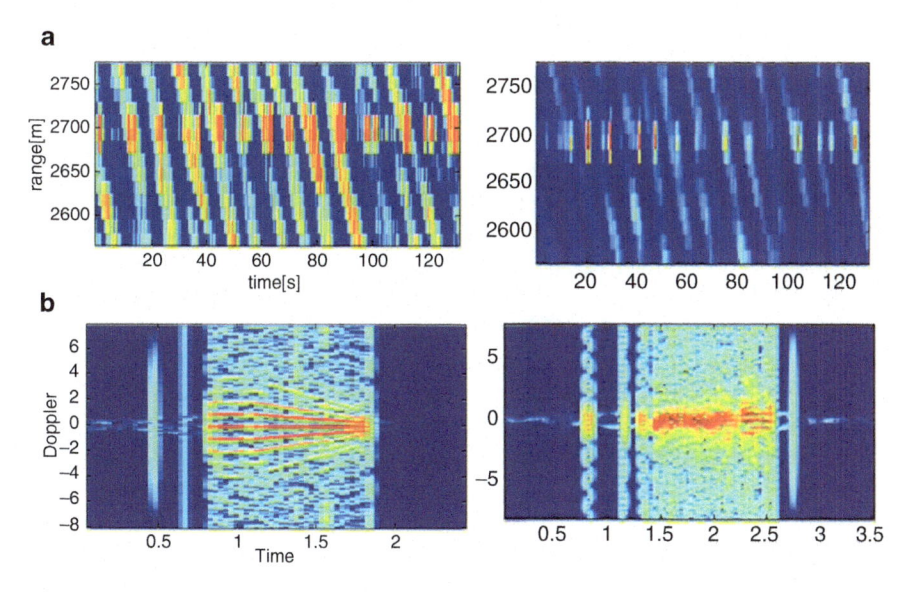

Fig. 3 The sea-clutter data of #17(*left*) and #18(*right*): (**a**) range profiles; (**b**) range bin 9 time-doppler imaging

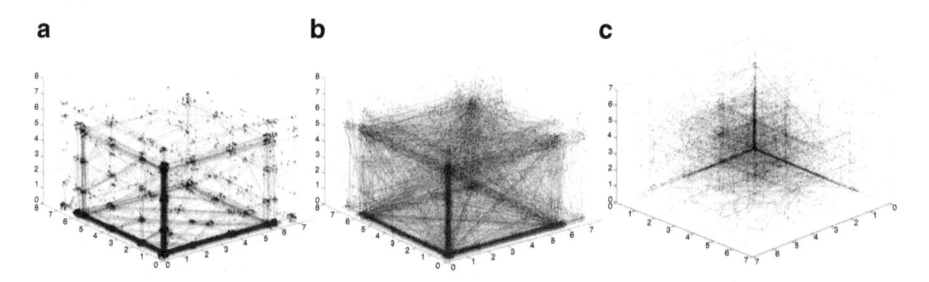

Fig. 4 (a) #17 No.14 bin 3-D phase space portrait of sea clutter data set; (b) #17 No.9 bin 3-D phase space portrait of sea clutter data set (c) #18 No.5 bin 3-D phase space portrait of sea clutter data set

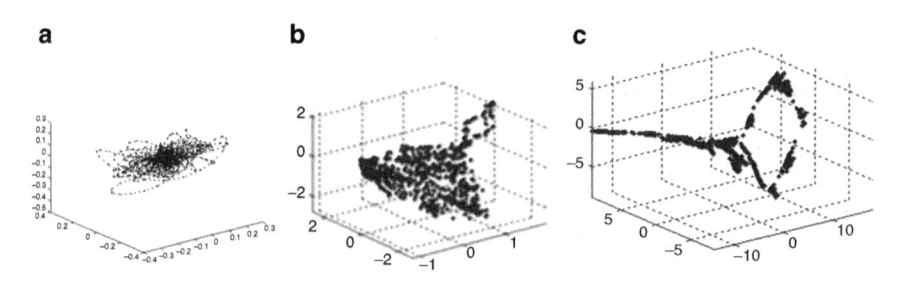

Fig 5 (a) #17 No.14 bin phase space topological structures; (b) #17 No.9 bin phase space topological structures; (c) phase space structural features simplified by TDA

As is shown in Fig. 5a, b and more obvious in Fig. 5c, the No. 9 range bin which contains the marine target illustrates a prominent topological structure compared with the other range bins. Furthermore, the topological structural patterns can be utilized in nonlinear dynamical system identification.

Knowledge discovery and system identification on sea clutter, which are typically massive, high-dimensional and multivariable, are undoubtedly valuable as well as full of challenge. A nonlinear dynamical system identification scheme based on the topological pattern discovery was proposed in this work, aiming to provide useful knowledge for corresponding complex sea clutter modeling and addressing analysis problem. Through the method proposed in this work, the target detection performance in the sea clutter can be improved significantly. Based on the dynamical system identification scheme and topological data analysis techniques, the method may help find ways through complex sea clutter data. Experiments with measured sea-clutter data demonstrate the performance of the method proposed in this work.

References

1. Xie N, Leung H, Chan H (2003) A multiple-model prediction approach for sea clutter modelling. IEEE Trans Geosci Remote Sensing 41:1491–1502
2. Haykin S, Bakker R, Currie BW (2002) Uncovering nonlinear dynamics-the case study of sea clutter. Proc IEEE 90:860–881
3. Totir F et al (2008) Advanced sea clutter models and their usefulness for target detection. MTA Rev Xviii(3):1–17
4. Duda RO, Hart PE, Stork DG (2001) Pattern classification, 2nd edn. Wiley, New York. ISBN 0-471-05669-3
5. Goodwin GC, Payne RL (1977) Dynamic system identification: experiment design and data analysis. Academic, New York, pp 171–172
6. Collins A, Zomorodian A, Carlsson G, Guibas L (2004) A barcode shape descriptor for curve point cloud data, eurographics symposium on point-based graphics, vol 28. Zürich, pp 881–894
7. Edelsbrunner H, Letscher D, Zomorodian A (2000) Topological persistence and simplification. FOCS '00 Proceedings of the 41st Annual Symposium on Foundations of Computer Science, IEEE Computer Society, Washington DC, USA, pp 454–463
8. Yao Y, Sun J, Huang X et al (2009) Topological methods for exploring low-density states in biomolecular folding pathways[J]. The Journal of chemical physics 130:144115
9. Robins V, Abernethy J, Rooney N (2003) Topology and intelligent data analysis. Springer, Berlin/Heidelberg, pp 111–122
10. Group S, Carlsson G (2009) Topology and data. Bull Am Math Soc 46(2):255–308

Field Programmable Gate Array Configuration Monitoring Technology for Space-Based Systems

Longxu Jin, Jin Li, and Yinan Wu

Abstract In order to solve the problem of field programmable gate array (FPGA) configuration failure after the Satellite TDI CCD camera power-up, a FPGA configuration process monitoring system was designed. First, this paper describes the core monitoring circuit in the FPGA configuration monitoring system. Then the design ideas of software are proposed for the monitoring system. Finally, the experiments were performed in the monitoring system. Experiment results showed that the proposed FPGA configuration monitoring system can timely monitor exception occurred in FPGA configuration process and can analyze the error origins and can configure successfully the FPGA applications. The FPGA configuration successful rate reaches 98.2 % after using this system. The monitor system can improve the reliability of FPGA configuration and solve the problem of FPGA configuration failure.

Keywords Space camera • Field programmable gate array configuration • Monitor state machine

1 Introduction

The Field Programmable Gate Array (FPGA) devices based on SRAM have been widely used in the Satellite TDI CCD camera these days [1, 2]. They are volatile devices- they do not retain their configuration when power is removed [3]. To ensure a FPGA system operating properly, people must reinitialize the CCLs inside the device each time power is cycled, that is the ICR (In-Circuit Re-configurability) [4, 5]. However, configuration failure often occurs in the current space and aviation FPGA system. Once FPGA configuration failure occurred, people have to make the

L. Jin (✉) • J. Li • Y. Wu
Changchun Institute of Optics, Fine Mechanics and Physics, Chinese Academy of Sciences, Changchun, China
e-mail: 1458312813@126.com

Satellite TDI CCD camera system re-power up. Under the satellite environment, the power-up operation is very trouble in the Satellite TDI CCD camera [6].

To ensure Satellite TDI CCD camera operating properly under the satellite environment, the traditional design method overcoming FPGA configuration failure is watchdog technology. When the power-off controller discovers a watchdog alarming, the power-off controller makes the camera power up again. This method need to design a dedicated power-off controller, watchdog monitor and a separate power supply system. And it has brought a lot of trouble in the Satellite TDI CCD camera design. So it is necessary that a FPGA configuration process monitoring system was designed in the FPGA module. On the one hand, the new design does not require additional power supply system, watchdog monitor and the controller that controls the Satellite CCD camera power-off; On the other hand, when the Satellite TDI CCD Camera after a one-time power-up, FPGA configuration is successful. And the FPGA system runs at normal state. Meanwhile, the monitoring system can make FPGA reconfigure successfully a bit-stream again when it is notified to reconfigure a bit-stream by FPGA. In addition, the monitoring system can analyze and monitor the FPGA configuration process. Once a FPGA configuration fault is detected by the monitoring system, the reconfiguration process begins during the Satellite TDI CCD Camera power up.

Take Xilinx's Virtex-IIPro family FPGA as an example, A FPGA configuration process monitoring system was designed in this paper. And the external system was designed to verify the feasibility and superiority of the new design method. The design can satisfy the Satellite TDI CCD camera requirements of high success rate and reliability in configuring the FPGA application, and improve the success rate of the FPGA configuration.

2 The Scheme of FPGA Configuration Monitoring System

Figure 1 illustrates the total block diagram of FPGA configuration monitoring system. The system contains a FPGA sub-system, power-off subsystem, communication module, and data transmission subsystem and host computer. The FPGA subsystem is mainly composed of two circuits which contain two same FPGA and FLASH devices. One has the FPGA configuration monitoring circuit, the other don't have the circuit. The two Flash devices separately load the configuration data into the corresponding SRAM-based FPGA devices after power-up when the Power-off subsystem has operated. If a FPGA configuration failure occurs, the FPGA sub-system sent configuration failure information to the communication module. And according to an instruction receiving, communication module issues a corresponding command to the power-off system, and the power-off system disconnects the power of the FPGA device configuration failure. In addition, the FPGA configuration information is transmitted by the communication module, and is stored in the host computer. After the FPGA configuration successful, application displays in the host computer through the data transmission module. If the two

Field Programmable Gate Array Configuration Monitoring Technology... 853

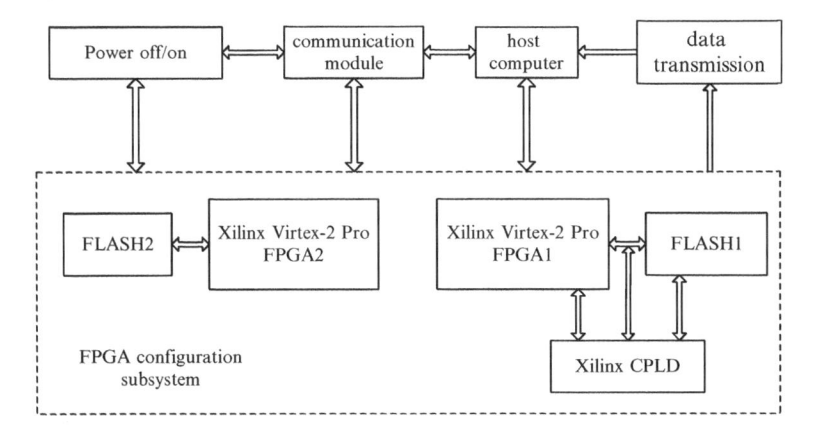

Fig. 1 Block diagram of configuration monitoring system

FPGA devices can successfully configure a bit-stream file, the power-off system is notified by the communication module, and the whole system starts the next configuration experiment. The host computer can inject the number of configuration experiments parameters into the communication module. The FPGA configuration process monitoring is completed by a CPLD device.

3 The Hardware Circuit and Soft Design of FPGA Configuration Monitoring System

The FPGA configuration monitoring system design is used for verifying feasibility and superiority of the monitoring circuit in FPGA configuration process. Therefore, the core of the FPGA configuration system is monitoring circuit. In order to monitor entire FPGA configuration process and real-timely reflect configuration information, data re-configuration is needed under power-up situation when configuration failure happens. As the result, circuit design requirements are:

1. The device used for monitoring FPGA configuration process should guarantee its configuration data won't lose when powered off. And it can quickly and reliably enter operation state at every power-up.
2. In order to real-timely monitor FPGA configuration process, FPGA, configuration memory and monitoring process need to work in the same clock.
3. FPGA, configuration memory and monitoring process should be in the same scan chain.

Figure 2 is a concise circuit diagram which is used for monitoring FPGA configuration process. XC2VP40-6FG676 chip produced by Xilinx has powerful processing property. It is selected as FPGA of this circuit. XCF16P-VQ44C produced by Xilinx is selected as configuration memory and XC2C64A chip in

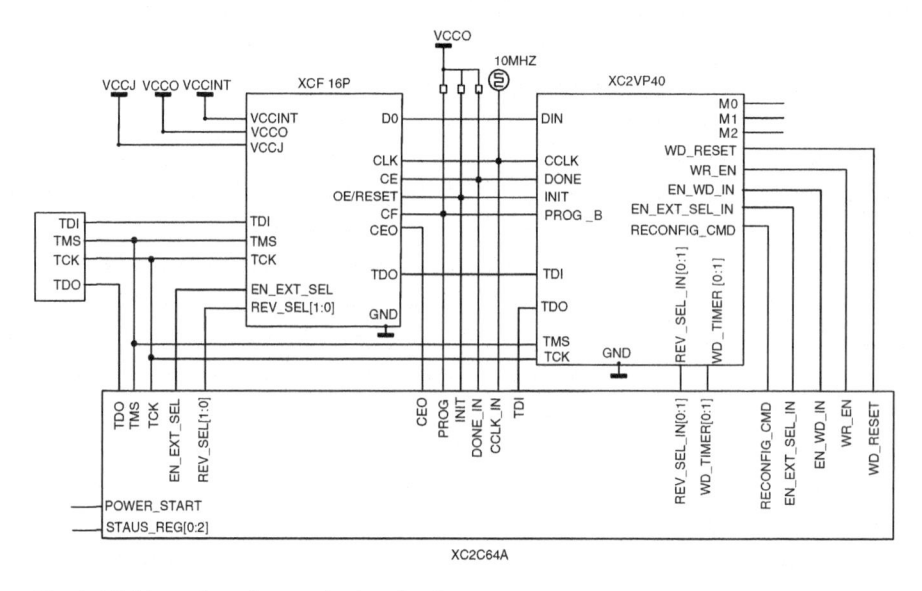

Fig. 2 FPGA configuration monitoring circuit

CoolRuner-II family produced by Xilinx is selected as monitor CPLD. XC2C64A chip is non-volatile device [7], so configuration memory is not needed and configuration data won't lose when powered off. The three devices in configuration monitor circuit are in the same scan chain and the configuration data are accordingly downloaded to chips by the same JTAG. In order that configuration clock and monitor operate in the same clock, slave serial configuration mode must be selected in FPGA and FLASH. As the processor which monitors FPGA configuration process, CPLD is used to control storage location of the FPGA bit-stream in FLASH, monitor configuration process, analyze failure reasons, re-configure configuration data when a configuration failure occurs, re-configure configuration data when FPGA re-configuration order is received and monitor whether the FPGA user program operating normally and so on. Furthermore, the method has been used for the FPGA configuration monitor soft [8].

4 Experiment Studying

Figure 3 illustrates tested simulation results which contain Configuration passing, CRC checksum failure, Missing synchronization word and watchdog alarm through Xilinx ISE 10.1 software. From the Fig. 3, a conclusion that the monitor state machine can handle various configuration failure problems during the FPGA configuration process can be drawn. When the state machine detecting configuration failure, it enables FPGA to reconfigure data until the configuration data is

Field Programmable Gate Array Configuration Monitoring Technology...

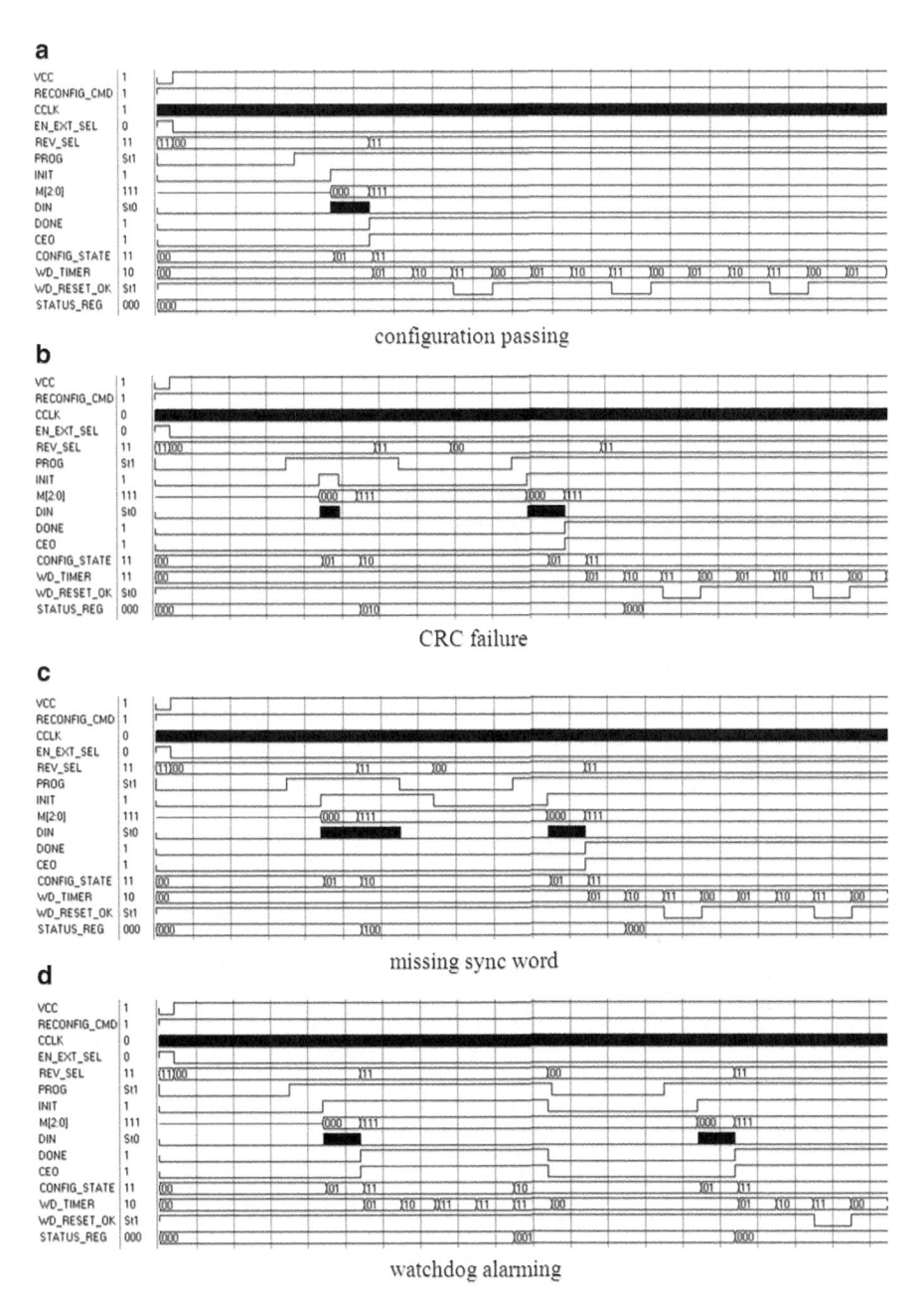

Fig. 3 Diagram of state machine simulation (**a**) configuration passing (**b**) CRC failure (**c**) missing sync word (**d**) watchdog alarming

Table 1 Statistics of configuration failure times per 100 times with different boards

Board number	Un-using monitor system	Using monitor system
1	3	1
2	7	0
3	0	0
4	9	2
5	2	0
6	7	3
7	4	1
8	8	2
9	5	2

successfully configured. At the same time, configuration registers can accurately reflect the results of the entire configuration process.

Up to now, the simulation result indicates that the design mentioned in this paper is correct. Then after FPGA configuration monitoring system designed power-up, PC computer through Xilinx ISE 10.1 software makes IMPACT software and circuit board communication. And the software scans the devices on circuit board, can find XC2VP40, XC2C64A, XCF16P in the same scan chain. In the same chain, different formats application can be downloaded into the corresponding device. Meanwhile, after the host computer injecting parameters into the power-off system, FPGA configuration experiment begin, and the configuration information is recorded. The experimental results show that the configuration causes from the actual device operation consistent with that shown Fig. 3. This shows the design mentioned in the paper is feasibility. We test the performance of configuration of nine blocks of circuit board. The test results are shown in Table 1 using the monitoring system and Table 1 un-using the monitoring system. The experimental data is analyzed by a binomial distribution formula. The formula is

$$f_X(x) = P[X = k] = \frac{n!}{k!(n-k)!} p^k (1-p)^{n-k} \qquad (1)$$

where n is the total number of configuration test, p is the probability of configuration passing, 1-p is the probability of configuration failure, x is the number of configuration passing in n configuration tests. From the Eq. 1, the experimental data were processed using MATLAB. The result processed shows that the configuration failure rate using traditional FPGA configuration method is 5 %, while 1.2 % using the monitoring system. This shows the new design greatly improves the configuration success rate, which reaches 98.8 %. This shows the new design greatly improves the configuration success rate, and fully is able to meet the requirements of Satellite TDI CCD cameras.

Table 2 Statistics of configuration failure average times per 100 times with different configuration rate

Configuration rate/MHz	Un-using monitor system	Using monitor system
1	0	0
5	1	0
10	1	0
15	3	0
20	4	0
25	7	1
30	7	1
35	9	2
40	10	2
45	12	2
50	13	3
55	16	3

In addition, the author tests the performance of configuration of nine blocks of circuit board at different configuration rate using the method proposed by this paper when working temperature is +20 °C. The test results are shown in Table 2. From Table 2, the probability of FPGA configuration success is 98 % under configuration rate low 40 MHz, which is the requirement of space camera. Therefore, the method proposed by this paper meets fully the application of space camera.

5 Conclusion

Based on the TDI CCD cameras requirements of the high success rate and reliability of FPGA configuration, the FPGA configuration monitoring system mentioned in the paper was designed. The feasibility and superiority of the system designed was verified through tens of thousands of test. The results show that, FPGA configuration monitoring system can overcome a variety of failure problems in the FPGA configuration process in practice; the FPGA configuration successful rate is up to 98.2 % after using this system. Compared with the traditional methods of FPGA configuration, the new design made configuration success rate to increase by 3.8 %. It can satisfy the satellite camera requirements of high success rate and reliability in configuring the FPGA application.

References

1. Li Jin, Jin Longxu et al (2012) Application of ADV212 to the large field of view multispectral TDICCD space camera. Spectrosc Spect Anal 32(6):1700–1707
2. Li Jin, Jin Longxu et al (2012) Reliability of space image recorder based on NAND flash memory. Opt Prec Eng 20(5):1090–1101

3. Cheng Ming, Bi Liheng, Yang Xiaoguang (2007) Using CPLD and flash memory to configure FPGA. Control Autom 23(20):171–173
4. Zhou Wei, He Jianying, Nie Jugeng (2006) Implementation of the programming and configuring of CPLD and FPGA. Comput Digit Eng 34(1):100–102
5. Mao Jianhui, Hei Yong Wu, Qiao Shushan (2008) A of new scheme of multi-mode configuring FPGA devices. Microcomput Info 24(2):179–181
6. Li Jin, Lv Zengming et al (2012) Error correction technology for CCD image using high speed optical fiber transmission. Optics Precis Eng 20(11):2549–2558
7. Li Yamchun (2008) Fast dynamic reconfiguratuion of FPGA based on CPLD. Telecommun Eng 48(7):87–89
8. Li Jin, Li Guoning et al (2010) Design of FPGA configuration monitoring system. Chin J Liq Cryst Disp 25(6):851–857

HVS-Inspired, Parallel, and Hierarchical Traffic Scene Classification Framework

Wengang Feng

Abstract In order to quickly, accurately, and efficiently classify traffic scene based on the image semantics, a bionic, parallel, and hierarchical scene classification framework is presented in this paper. Based on the perception as defined by the human visual system (HVS), the model is built. At first, an image pyramid was used to present both the global scene and local patches containing specific objects. Secondly, codebooks were built, which satisfy both long stare and short saccade similar to those of humans. Next, the visual words by generative and discriminative methods were trained respectively, which obtain the initial scene categories based on the potential semantics using the bag-of-words model. Then, a neural network was used to simulate a human decision process. This led to the final scene category. Experiments showed that the parallel, hierarchical image representation and classification model obtained superior results in terms of accuracy.

Keywords Image pyramid • Visual codebook • Generative method • Discriminative method • Neural network

1 Introduction

Scene classification [1–3] plays an important role in efficient image resource management and automatic image classification and is very important in computer vision and pattern recognition domain as well as content-based retrieval. The goal of a computer vision system is to build a model of the real world that allows a user

---

W. Feng (✉)
Police Intelligence Department, Public Security Intelligence Research Center, Chinese People's Public Security University, Beijing, China
e-mail: wengang.feng@gmail.com

to interact with it. The ultimate goal is to allow a computer to "see" the world in a manner similar to the biological visual system. First it must build the framework then realize cognition and inference. The goal of this research is to allow images to be quickly, accurately, and efficiently classified based on the image semantics.

In cognitive theory [4], the majority of research on scene classification has focused on the roles of either low level visual feature cues or high-level factors such as object-, semantic-level cues, or behavioral task demands. However, the latest research in cognitive neuroscience found the partial separation scanning the global scene relocate to new areas, as well as evidence to support the existence of small amplitude, the importance of "correct" image browser glanced natural cycle. So both global features and local features are good visual feature candidates for targets on which the eyes fixate. They should be used together for scene modeling, in a parallel fashion.

The perception in the human visual system follows a hierarchical model in which the information of scale space is not changed by altering the scale [5]. Humans can recognize one scene in a very short time because they have a large and complex scene dataset in the brain. The model needed to classify a scene has been trained sufficiently to fit most situations. Based on cognitive neuroscience, we propose a novel parallel, hierarchical scene classification framework that has superior results for scene recognition.

2 Traffic Scene Classification Framework

In the training step, there are five stages: First, each one image will be presented as 21 images (patches) in three layers using an image pyramid. Second, we will build codebooks, and third, each visual word extracted from one image patch is trained in a generative way using pLSA model. Each training patch is then represented by z-vectors ($p(z|d)$) where z is a topic and d is an image. Fourth, the z-vectors obtained from last step are subjected to discriminative training. Finally, we use neural network to get the final classification category on the 21 patches scene categories. The approach simulates human's decision process based on local and global visual perception.

The test step also proceeds in five stages. The difference is that the test image is projected onto the learning during the training period by the vector spanned simplex. This is used in learning the expectation maximization (EM) to run in a similar manner, but now, only in the z vector updating steps with other learned vector fixed in each of M step. As a result, the test image is represented by the z-vector. Then test the multi-class image classification discriminant classification (Fig. 1).

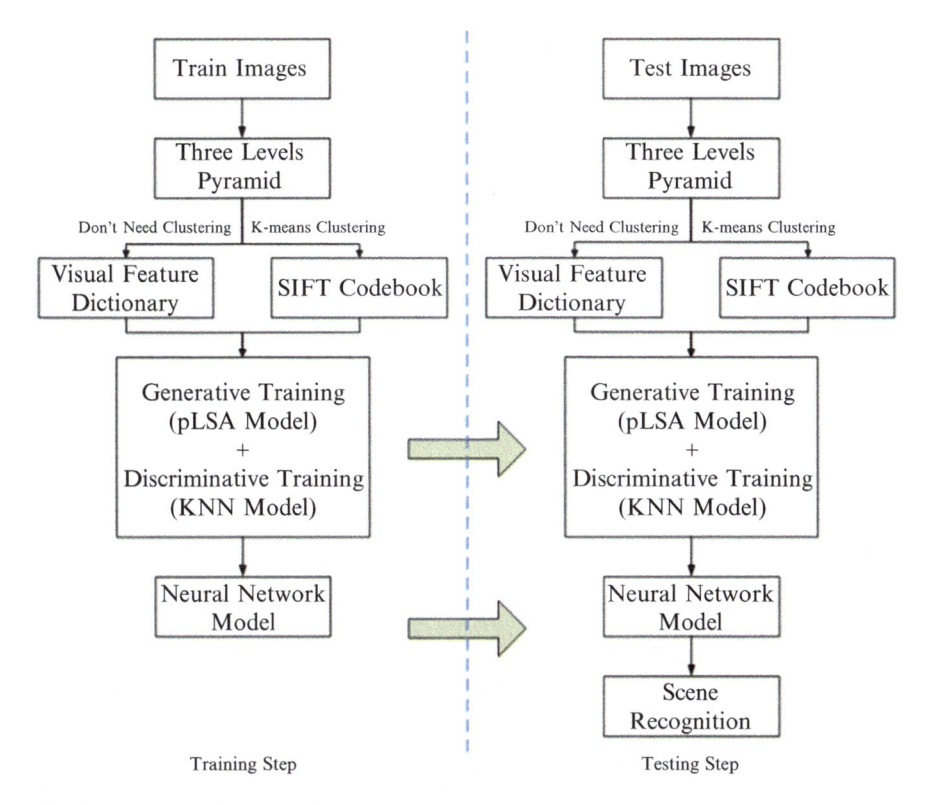

Fig. 1 Flowchart of the algorithm

2.1 Image Pyramid

On the basis, the image is further divided into a planar sequence, $I_k = \{I_i^k\}_{i=1}^{4^{k+1}}$, where I_i^k is the image content of the subscript area R_i^k in the image I.

$$I_i^k = \{I_{x,y}, (x,y) \in R_i^k\}_{i=1}^{4^{k+1}} \tag{1}$$

Where R is the subscript matrix of the image I, and R_i^k is the subscript set of the i-th patch of the k-th layer in the pyramid. Here we use three layers as shown in Fig. 2: (Where (X, Y) is the pixel resolution of image I.)

$$R = \{R_i^k\}_{i=1}^{4^{k+1}}, R = \{(x,y) | x = 1, \ldots, X, y = 1, \ldots, Y\} \tag{2}$$

Here we use a three-layer pyramid: The first layer is the original image; the second layer has four grids, which means each patch is a quarter of the original image; the last layer has 16 grids, which is one sixteenth of the original image.

Fig. 2 Image scale space

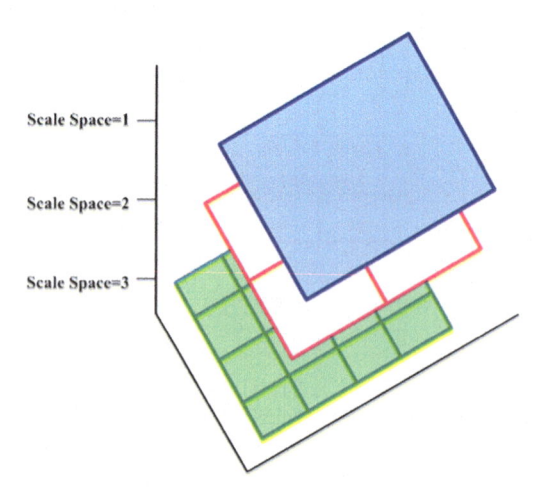

2.2 Feature Extraction

In the framework, first an input image is separated into three layers and 21 patches by an image pyramid. We treat each patch from the original image as one independent image. For each patch we extract two types of features to build codebooks used in the experiments of Sect. 4. First, we have the so-called "clustering codebook," which are SIFT [6] (scale- invariant feature transformer) features. The SIFT feature is highly distinctive, shown to provide robust property of affine distortion, image scale and rotation, in the three-dimensional views of the changes, noise, and lighting changes.

We also use a non-clustering codebook combination of color, texture, and fractal characteristics. The description of all colors can be extracted image or image region. The structure of the color distribution and spatial color structure histogram. Texture descriptors can extract the image or image area. There are five common texture descriptors: left, right, upper, lower, and the internal boundary of the texture histogram, which describes the spatial distribution of the image of the local structure. Complex and irregular fractal dimension can be represented by the image, we can use the histogram of the global and the local fractal dimension.

2.3 Generative Model Training

Probabilistic Latent Semantic Analysis (pLSA) is a generative model from statistical text literature [7]. The model is applied to images by using a visual analog of words, formed by clustering and non-clustering codebooks (as described in Sect. 2.2). pLSA is appropriate here because it provides a correct statistical model for clustering in the case of multiple object categories per image.

In detail, each image is modeled as a mixture of Z topics, and each topic is a distribution over the vocabulary of words. More formally, we have a set of images

$D = d_1, \ldots, d_N$, each modeled as a histogram of word counts over a vocabulary of $W = w_1, \ldots w_V$. The corpus of images is represented by a co-occurrence matrix of size $V \times N$, with entry $n(w, d)$ listing the number of times the words w occurred in an image d. Image d has N_d words in total.

The key values used or obtained from training via pLSA are probabilities. $p(d_j)$ denotes the probability of observing a particular image d_j. $p(w_i|z_k)$ signifies the probability of a specific word conditioned on the unobserved topic variable z_k. Finally, $p(z_k|d_j)$ expresses an image specific probability distribution over the latent variable space. Based on these, we can express the specific pLSA model for the problem. First, an image d_j is selected with the probability $p(d_j)$; second, a latent topic z_k is picked with the probability $p(z_k|d_j)$; finally, a word w_i is generated with the probability $p(w_i|z_k)$. It is the probability of the latent variable, $p(w_i|z_k)$, for which we must solve.

$$p(w,d) = \sum_{z=1}^{Z} p(w,d,z) = \sum_{z=1}^{Z} p(d)p(w|z)p(z|d) \quad (3)$$

2.4 Discriminative Model Training

For a discriminative model, one needs only the conditional probabilities of the target labels, $p(z|d)$. For each patch, we now have a vector for each image whose cardinality equals the number of the topics, $\langle p(z_1|d), p(z_2|d), \ldots, p(z_n|d) \rangle$. Two discriminative models are used, one to obtain the best estimate of the label (and its probability) for each patch and one to obtain the best label for the overall scene.

To obtain the single-estimate probability and label for one patch we use KNN over an R^2-space, in which the training set is embedded, KNN where $k = 7$ is used to select the best label. The probability is estimated as the average probability over neighbors having similar labels. The probability is used in the next step.

2.5 Neural Network Model Training

Radial basis function (RBF) networks [7] typically have three layers: an input layer, a hidden layer, and a linear output layer. The hidden layer consists embeds an RBF activation function, each having stronger responses when the N inputs are nearer the mean. The output, y, is a weighted sum of the hidden neurons,

$$y = \sum_{i=1}^{S} \alpha_i \rho(\|x - c_i\|) \quad (4)$$

Where S is the number of neurons in the hidden layer, c_i is the center vector of neuron i, α_i are the weights of the neuron, and β_i is the response of hidden neuron:

$$\rho(\|x - c_i\|) = \exp\left[-\beta_i \|x - c_i\|^2\right] \tag{5}$$

Here, we set the initial value of weights, i, but in the training, all weights are changed to optimum values. The norm is taken to be the Euclidean distance.

The number of inputs, N, is 21 – one for each patch. The value of the input is the value $p(z|w, d)$ instantiated by the KNN step described earlier. There is one output, y, from which a Z-way decision is extracted by $Z - 1$ thresholds.

3 Experiments

In this section, we present the traffic scene classification results of our approach, and comparing it with pLSA model [7]. Our set of experiments was divided into three scene categorical datasets. The first one involved day-time, night-time, and twilight-time. The second one included fog, rain, snow, and sunny traffic scenes. And the last one was to consisted of free-flow, light-congestion and heavy congestion traffic scenes. This database contained 200 images per category. The images were of medium resolution, i.e., about 500×330 pixels.

Why does the proposed framework obtain the more accurate results? As opposed contrast to pLSA model, the proposed framework uses the image pyramid to connect the global and local semantic information, and get better choice depending on the discriminative model. Simultaneously comparing with the image pyramid-SVM model, the proposed frame work has generative model training. Besides such advantages, choosing the different visual words and using neural network upgrade the performance of the proposed framework.

We investigated three aspects: the visual found (1) locally extracted from the image measurement functions, ranging from the number 40–100,000. Constant descriptor records, canceled unexpected lighting, viewpoint or scale differences. To capture the complexity of the world, we extracted descriptors including texture, spirit, SIFT and color, in of visual global and local, then extract the image and patch within each pyramid. (2) visual vocabulary of semantics – this is not verbal, but the gradient image of a general description of a detail, the patch every word every place. We only summarize the most significant points, wherein semantic scene category learned set of images corresponds to, and then using the determined object class, and (3) the model into a semantic concept of visual words. In fact, it is assigned a probability at the same time all the concepts. These possibilities are used to rank the concept of existing image (Fig. 3).

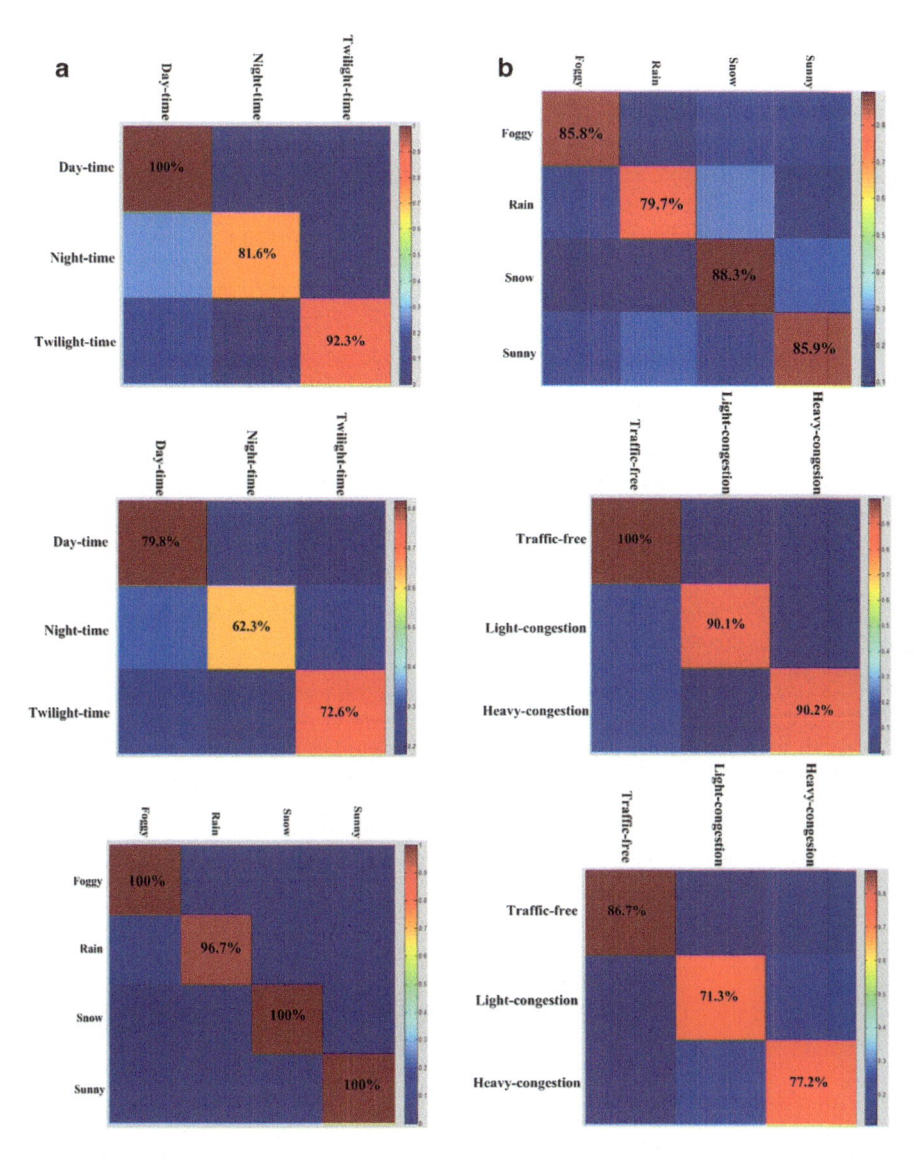

Fig. 3 Confusion matrix for traffic scene categories, (**a**) the result of our method; (**b**) the result of the baseline pLSA method

4 Conclusion

This paper proposed an HSV-inspired, parallel, hierarchical traffic scene classification framework, which can recognize the traffic scene category, enabling it to be possible that the most efficient vehicle tracking can be automatically selected and traffic surveillance algorithms for the conditions can be obtained. The model is able to obtain semantic information from global and local manners despite using coarse features. The image pyramid was used to acquire the information at scales to emulate human perception. This study also built different codebooks to simulate long stare and short saccade to imitate human vision. Besides, generative method, discriminative method, and neural network were used here to train the visual words to get the semantic information from images to simulate human beings visual system. Finally a traffic scene classification result depending on the generic framework could be obtained as well.

References

1. Lazebnik S, Schmid C, Ponce J (2006) Beyond bags of features: spatial pyramid matching for recognizing natural scene categories. In: Proceedings of the IEEE conference on computer vision and pattern recognition, New York, USA, pp 1–8
2. Rasiwasia N, Vasconcelos N (2009) Holistic context modeling using semantic co-occurrences. In: Proceedings of the IEEE conference on computer vision and pattern recognition, Miami, USA, pp 1–7
3. Malisiewicz T, Efros AA (2008) Recognition by association via learning per-exemplar distances. In: Proceeding of the IEEE conference on computer vision and pattern recognition, Anchorage, USA, pp 1–8
4. Torralba A, Sinha P (2001) Recognizing indoor scenes. Technical report, Boston, MIT AI lab
5. Siagian C, Itti L (2007) Rapid biologically-inspired scene classification using features shared with visual attention. IEEE Trans pattern anal mach learn 29(2):300–312
6. Agarwal A, Triggs B (2008) Multilevel image coding with hyperfeatures. Int J Comput Vis 78(2):15–27
7. Haykin S (1994) Neural networks. Prentice-Hall, Upper Saddle River

Analysis and Application on Reactive Power Compensation Online Monitoring System of 10 kV Power Distribution Network

Yan-jun Shen and Jin-liang Jiang

Abstract Data shows that the reactive power compensation devices break down frequently in 10 kV power distribution network. But the failure cannot be discovered in a short time, so lots of manpower and resources are wasted. This paper proposes one method which uses the JST-C20 series Low-voltage Intelligent Capacitors to solve the problem and formulates the reactive power optimization plan for several typical lines based on the reactive power compensation optimization computation. Finally, the article analyzed the validity of this method on voltage quality improvement, energy saving, on-line monitoring and fault diagnosis.

Keywords Low-voltage intelligent capacitor • Reactive power compensation • Online monitoring system • Fault diagnosis

1 Introduction

Lots of manpower and resources has been put into the routine line patrol work in Jiangmen Power Supply Bureau. They discovered the high failure rate of the dynamic reactive power compensation devices and controllers, such as the abnormal phenomenon when the capacitors are using or exiting, the burning events of the capacitors and so on. It leads to that the operational state of the device cannot be uploaded to the main station, and the flowing of information cannot be realized. Now, the reactive power compensation devices belong to Jiangmen Power Supply Bureau is lack of the function of online monitoring and fault diagnosing. Because of the large amount of the compensation points, the failure capacitors cannot be

---

Y.-j. Shen (✉)
School of Electric Power, South China University of Technology, Guangzhou, China
e-mail: syjzb.2007@163.com

J.-l. Jiang
School of Business Administration, South China University of Technology, Guangzhou, China

repaired timely. It is not only harmful to the user's normal using of electricity and the quality service of Jiangmen Power Supply Bureau, but also is a waste of the manpower and resources [1, 2].

Therefore, it is a useful method to use a new generation of intelligent low voltage reactive power compensation device which can achieve the following functions, including reactive power compensation, on-line monitoring, fault diagnosis, alarming and other functions to solve the problem discovered in Jiangme Power Supply Bureau.

2 Introduction of the Advanced Low-Voltage Intelligent Capacitor

The reactive power compensation device selected in the paper is a new generation of intelligent product which is based on power measuring technology, computer technology and modern mobile communication technology. It meets the need of the network users of the Power Supply Bureau at the county level. The Low-voltage Intelligent Capacitor consists of a CPU control unit, thyristor composite switch, protector, two (Δ type) or one (Y type) low self-healing power capacitors [3]. It uses the advanced GPRS/CDMA public wireless communication technology to achieve the function of reactive power compensation, on-line monitoring and fault diagnosis. The system of the low-voltage intelligent capacitor can be divided into three layers, including the layer of terminal equipment, layer of communication network, layer of the main station, shown in Fig. 1.

1. The terminal device layer which is installed in the distribution room includes intelligent capacitors and the communication management unit. When it is compensating the reactive power, it can also collect the information and monitor the device's operational state, record and analyze the operating data and movement time. Most importantly, it can send out alarming signal and store the running information when the equipment works in an abnormal condition.

   The communication management unit has the function of the communicating, managing and storing. It connects the intelligent capacitor with the GPRS network, and stores the operating information at the same time.
2. Communications network layer is not only the GPRS public network, but also is the bridge between the main station and terminal equipment. It transmits the information of the terminal equipment to the main station, and issues the control commands of the main station to the terminal equipment.
3. The main station is mainly responsible for dealing with the data and alarming information coming from the terminal device layer, and dealing with the historical data, issuing various types of control commands to the terminal device layer.

Following are the details of the three major functions of the low-voltage intelligent capacitor.

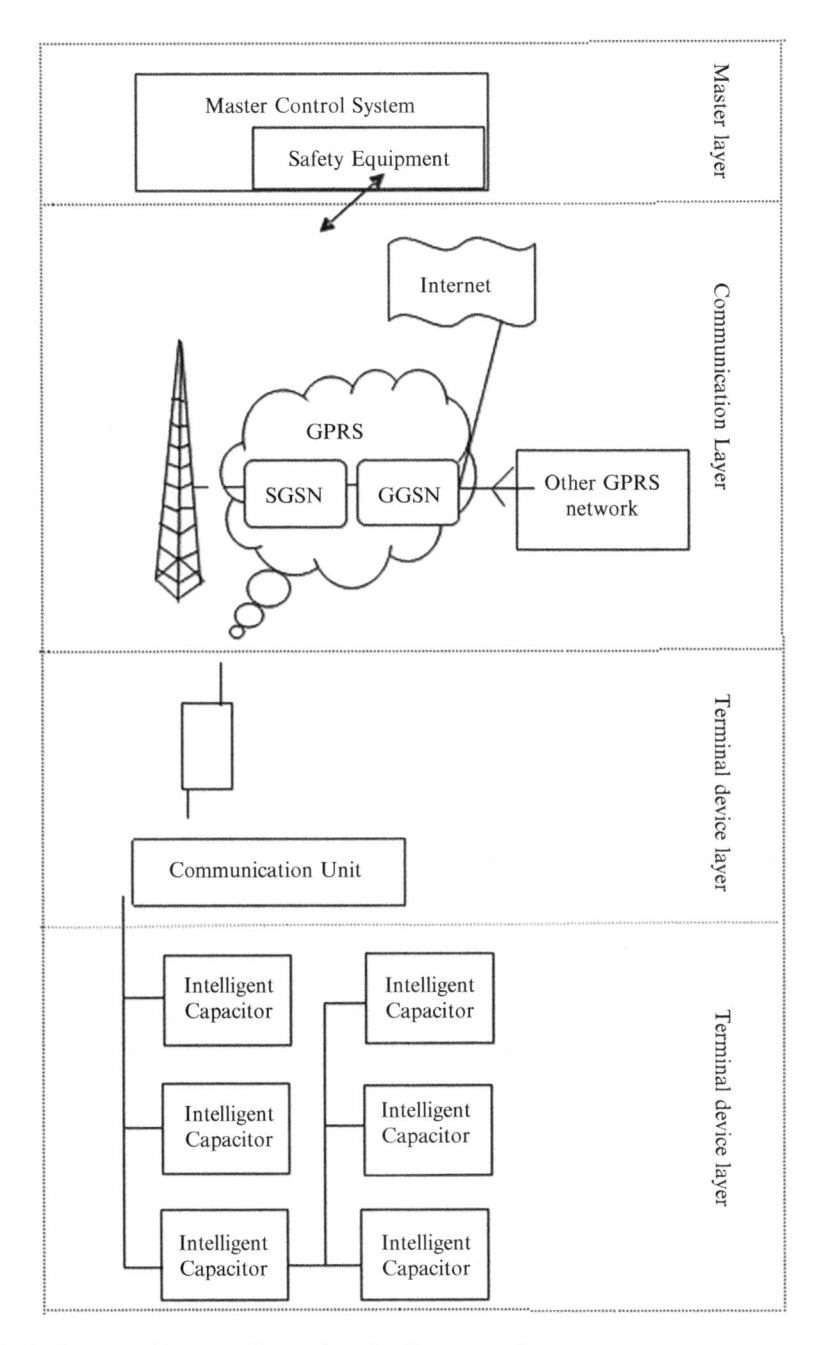

Fig. 1 System architecture of low-voltage intelligent capacitor

2.1 Reactive Power Compensation Function

The low-voltage intelligent capacitor calculates the system's voltage, current, active power, reactive power and power factor according to the three-phase voltage and three-phase current of the system. It can also make the decision to input or quit the capacitors so as to realize real-time reactive power compensation function based on the reactive power and power factor. Following are the characteristics [4]:

1. The switching switch is composed by relay, zero trigger turn-on circuit; thyristor protection circuit. It can reach the purpose of inputting capacitors when the voltage is zero and quitting the capacitors when the current is zero.
2. The power switch has the characteristics of responding quickly and operating frequently, micro-power, controlling Common and phase compensation easily, lower failure rate, longer working life.

2.2 Online Monitoring Function

The online monitoring system of the low-voltage intelligent capacitor chooses GPRS as the method to realize data communication and the public GSM network as the carrier. It is supplemented by the line of RS485 and infrared. The system regards the large users, public distribution transformer and household as the main controlled and management objects. It can realize the electric power online monitoring, controlling and managing form power lines to power users [5]. Following are the characteristics:

1. Input the accounting information of the distribution transformer intelligently.
2. The man–machine interface is simple, graphics and the charts are rich.
3. Because of the online monitoring function on voltage, it can monitor the running configuration state of the low-voltage intelligent capacitor in real time.
4. Provide the statistical information of operating conditions of the transformer and the comprehensive data for users to analysis the economic operation of the transformer.
5. Follow the standard database designing principles and provide strongly technical guaranty for the application integration system of the power supply enterprises.
6. The system has powerful data mining capabilities.
7. The system has self-diagnosis function, maintenance-free function and a variety of statistical analysis functions.
8. The network system is safe and reliable composed by routers, hardware firewall or dedicated network security isolation device.

2.3 Fault Diagnosis and Alarming Function

The low-voltage intelligent capacitor judges whether the device is over voltage, under-voltage or lack of phase according to monitoring the system's voltage. It can also discover abnormal operating conditions of the capacitors such as short-circuit, over-current or overload. Above is the stage of fault diagnosis. If it diagnosed the fault, the capacitor will send the information to communication unit. After the communication unit stores the electrical parameters before and after the accident, the alarming signal will be reported to the main station. With the acceptance of the main station, it can remind the low-voltage intelligent capacitor of the alarming, then save and record the information.

3 Practical Application and Analysis of Low-Voltage Capacitor

According to the operating data of Jiangmen power distribution network and the field investigation, this paper puts forward a plan to make the low-voltage intelligent capacitors into use in Jiangmen, and analyzes the effectiveness of the program on reactive power compensation, on-line monitoring and fault diagnosis. To some extent, it fills the blank of the practical application on online monitoring function of reactive power compensation equipment of power distribution network.

3.1 Determining the Point and Capacity of Reactive Power Compensation

This paper chooses two typical lines of 10 kV power distribution network in Jiangmen to complete reactive power optimization calculations based on the software of energy saving of power distribution network (PSAS). As follows: the total length of A line is 15.698 km, the number of its public transformer is 17, the total capacity is 2,600 kVA, the number of the dedicated transformer is18, the total capacity is 8610kVA; the total length of B line is 13.69 km, the number of the public transformer is 9, the total capacity is 2,364 kV.A, the number of the dedicated transformer is 59, the total capacity is 11,650 kVA.

According to the simulation results of PSAS and the operational characteristics of the grid of Jiangmen, the project selects three points for reactive power compensation named A, B and C in B line, the original capacity of the three points above are 0 kvar, 90 kvar and 210 kvar, and the added capacity are 200 kvar, 190 kvar and 210 kvar; The project also selects one point for reactive power compensation named D in A line, the original capacity of the D point is 210 kvar, the added capacity is 790 kvar.

Fig. 2 Voltage contrast before and after optimization on A line

Fig. 3 Voltage contrast before and after optimization on B line

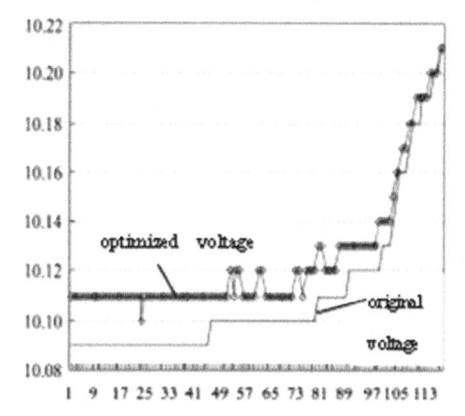

3.2 Realizing the Function of Real-Time Reactive Power Compensation Function

After inputting the corresponding capacity of reactive power compensation on the four points selected above, this paper compares the original voltage with the optimized voltage on A line and B line, shown in the Figs. 2 and 3.

According to the results above, this paper draws the following conclusions: the original node voltage on A line and B line is generally lower than the voltage before the reactive power compensation. After inputting the corresponding capacity of reactive power compensation on the four points selected above, not only the voltage quality has been improved significantly, but also the line losses and loss expenses are both cut down, shown in the Table 1

As is shown in the Table 1 above, the line loss and Loss costs of A line and B line are both reduced after applying the corresponding capacity of reactive power compensation.

Table 1 Comprehensive results of reactive power compensation on A and B line

	A line		B line	
	Original data	Optimal dat	Original data	Optimal data
Total active power of power supply/kW	1103.15	1104.02	1512.54	1506.35
The total supply of reactive power and reactive/kvar	692.36	585.13	1096.15	446.31
Total active power loss	36.49	37.52	71.68	66.45
Total reactive power loss	11.54	11.89	21.65	21.91
Loss costs/Million	3.39	3.42	4.76	4.56
Power factor	0.85	0.88	0.81	0.96

After installed the low-voltage intelligent capacitors in the actual compensation points, on one side the Reactive power compensation function has been realized, on the other side the power factor has been improved. The results proved the Practicality and economy of the low-voltage intelligent capacitor in reactive power compensation.

3.3 Realizing the On-line Monitoring Function

The low-voltage intelligent capacitors installed in A line and B line are equipped with a Four-in-one terminal system so as to realize its function, including collecting the voltage, current, active power and reactive power automatically, and also monitor harmonic wave. It can also upload the capacitors' operating condition parameters and real-time information to realize the on-line monitoring function accurately.

3.4 Realizing the Fault Diagnosis and Alarming Function

Because of the fault diagnosis function, the low-voltage intelligent reactive power compensation device can monitor the fault conditions including over voltage, under-voltage, short circuit, over-current, over harmonic wave, phases unbalance, over-temperature.

It can also carry on reactive power compensation control algorithm based on the picked parameters to make the decision to input or quit the single intelligent capacitor.

4 Conclusion

1. The advanced low-voltage intelligent reactive power compensation device used in Jiangmen Power Supply Bureau can carry on a variety of settings based on the actual requirements. It can complete the task of picking and controlling data, on-line monitoring the related electricity information so as to realize the functions including monitoring, controlling, managing the electricity consumption from the line to the user. It can also monitor the distribution transformer, measure remotely, monitor voltage and reactive power compensation and so on.
2. The fact proved that the advanced low-voltage intelligent reactive power compensation device installed in Jiangmen Power Supply Bureau has brought good results. On one hand, because of the real-time reactive power compensation function not only the voltage quality has been improved, but also the users' satisfaction has been enhanced greatly. On the other hand, the devices online monitoring function and its fault diagnosis and alarming function can help the workers to find out and repair the failure capacitors without delay. In one word, the advanced reactive power compensation device help guarantee the grid running safely and reliably, saving a lot of manpower and material resources.

References

1. ZHANG Yong-jun, REN Zhen (2005) Readjusting cost of dynamic optimal reactive power dispatch of power systems. Autom Electr Power Syst 29(2):34–38
2. ZHANG Yong-jun, YU Yue, REN Zhen, LI Bang-feng (2004) Research on dynamic modeling for reactive power optimization under real-time circumstance. Power Syst Technol 28(12):12–15
3. Wang Pei, Raghuveer MR, Mcdermid W et al (2001) A digital technique for the on-line measurement of dissipation factor and capacitance. IEEE Trans Dielectr Electr Insul 8(2):228–232
4. Allan D, Blundell M, Boyd K et al (1992) New techniques for monitoring the insulation quality of in-service HV apparatus. IEEE Trans Electr Insul 27(3):578–581
5. Moore PJ, Carranza RD, Tohns AT (1996) Model system test on a new numeric method of power system frequency measurement. IEEE Trans Power Deliv 11(2):696–701

Histogram Modification Data Hiding Using Chaotic Sequence

Xiaobo Li and Quan Zhou

Abstract In order to solve the security problem of data hiding with obvious detectable traces to attacker, a histogram modification data hiding scheme using chaotic sequence is presented. This algorithm modifies a cover image histogram by using a chaotic sequence to conceal the embedding process, and then secret information can be embedded into the modified image with encrypted trace. Experimental results show that the algorithm can achieve sound invisibility and large embedding capacity while guaranteeing high security. With the proposed scheme, the hiding trace of secret data is concealed. Even though the attacker detects the existence of hidden message under stego-image, the secret message cannot be extracted without private keys.

Keywords Data hiding • Histogram modification • Chaotic sequence • Embedding capacity

1 Introduction

With the development of internet and demand of users, image, video and multimedia data transmission in the internet increased rapidly over the last few years. Therefore, data security becomes more and more important. Data hiding [1] is a data security technique to undetectably insert secret message into a cover media to create a stego medium such that the existence of hidden information will not be detected by attacker. At the receive end, the hidden information can be extracted correctly by legal users. i.e., it provides a safer and securer data communication manner.

X. Li (✉) • Q. Zhou
Key Laboratory of Space Microwave Technology, China Academy of Space Technology, Xi'an, China
e-mail: lxb619@126.com

A number of data hiding methods were proposed. There are three general types: spatial domain methods [2], frequency domain methods [3] and compression domain methods [4]. Spatial domain data hiding manner mixes secret data into the distributed pixels directly. A commonly used method, called the least significant bit (LSB) [5] method, is a simple spatial domain data hiding method by replacing the least significant bit of cover image pixels to embed secret bits. For the frequency domain manner, firstly, the cover image pixels must be transformed into frequency coefficients by using a frequency transform method such as the discrete cosine transformation (DCT) [6] and discrete wavelet transformation (DWT) [7]. Later, the secret data are embedded by modifying the relative coefficients in the frequency-form image. Finally, stego image can be obtained by utilizing corresponding inverse transform. The compression domain method means that the secret message are embedded into the compression codes, such as block truncation coding (BTC)-based [8] scheme and side match vector quantization (SMVQ)-based [9] scheme, and so on.

The desires of good data hiding schemes are high security and low image distortion as well as large payload. However, despite claiming good imperceptibility in previous schemes mentioned above, it leaves obvious detectable traces to attacker inevitably. In other words, once the attacker realizes there is a hidden information under stego-image, the secret data would be extracted possibly. To further reinforce the security of data hiding, we develop a novel histogram modification data hiding scheme by using chaotic sequence to control the embedding procedure. With this scheme, the hiding trace of secret message is concealed. Even though the attacker detects the existence of hidden message under stego image, the secret message cannot be extracted without private keys. Furthermore, the proposed scheme has nice image quality and high payload.

The rest of this paper is organized as follows. The proposed data hiding method is presented in Sect. 2. In Sect. 3, the experimental results of our method are demonstrated. Finally, the conclusions of paper are presented in Sect. 4.

2 Proposed Scheme

In this section, the proposed data hiding scheme will be presented. The proposed algorithm has the merits of high security and low image distortion as well as large embedding capacity. To enhance the security, we utilize chaotic sequence to conceal the histogram modification trace. Figure 1 shows an example of the embedding process. Figure 1a shows an image block of size 4 × 4 and histogram of the image block. Consider a binary chaotic sequence as "0110001111010110". There is a one-to-one relationship between pixels in image block and bits of chaotic sequence. Before embedding, the image block is scanned by raster-scan order. Once an even pixel value is encountered, if the corresponding bit of chaotic sequence is "1" the pixel is added by "1", else it is kept intact. On the contrary, once an odd pixel value is encountered, if the corresponding bit of chaotic sequence is "0" the

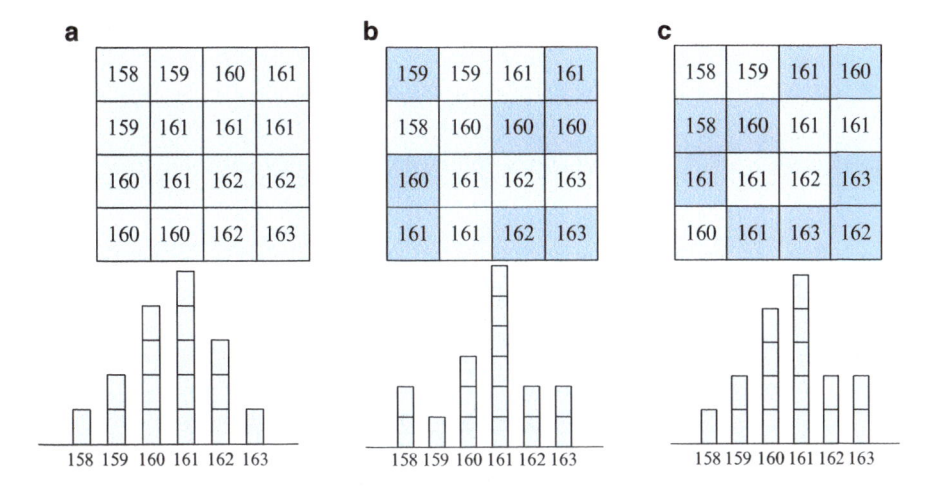

Fig. 1 An example of the embedding process : (**a**) original image block and histogram;(**b**) modified image block and histogram; (**c**) embedded image block and histogram

pixel is decremented by "1", else it is kept intact. The result of modifying operation by using chaotic sequence as shown in Fig. 1b. Suppose that the payload data is the stream of "10010011100101l". To embed the stream of data, the modified image block is scanned in the same scan order once again. Whenever the same chaotic sequence '0' is encountered, we sequentially check the bit of payload data. If the payload data bit is "1" the pixel is added by "1", else it is kept intact. When the chaotic sequence "1" is encountered, if the corresponding payload data bit is "1" the pixel is decremented by "1", otherwise it is not changed. As a result, the embedded image block is obtained as shown in Fig. 1c. Data extraction is actually the reverse of the embedding process. Note that the number of payload data bits that can be hidden into an image and the number of pixels associated with the image are equal. The detail algorithm is shown as below.

2.1 Embedding Process

Assume that the cover image X of size $M \times N$ is an 8-bit grayscale image. Denote the pixel value as $X(i,j)$, where (i,j) indicates the location in the original image. We select logistic map [10] to generate chaotic sequence, as shown in formula (1):

$$l_{n+1} = \alpha \cdot l_n(1 - l_n) \tag{1}$$

Where α is bifurcation parameter, and $l_n \in (0,1)$, $n = 0, 1, 2 \cdots$. When $3.5699456 \cdots < \mu \leq 4$, the logistic map is in chaotic state. In this paper, the parameter α and initial value l_0 are used as private keys.

Step1: Select parameter α and initialize value l_0, chaotic sequences S_k can be obtained by

$$S_k = \begin{cases} 0 & if \ 0 < l_n < 0.5 \\ 1 & if \ 0.5 \leq l_n < 1 \end{cases} \quad (2)$$

Where $k = 1, 2, \ldots, M \times N$, $n = 0, 1, \ldots, (M \times N) - 1$.
Step2: Scan the whole image in a given order, modify the pixel value $X(i,j)$ by

$$Y(i,j) = \begin{cases} X(i,j) & if \ S_k = 0 \& R_{i,j} = 0 \\ X(i,j) - 1 & if \ S_k = 0 \& R_{i,j} = 1 \\ X(i,j) & if \ S_k = 1 \& R_{i,j} = 1 \\ X(i,j) + 1 & if \ S_k = 1 \& R_{i,j} = 0 \end{cases} \quad (3)$$

Where $R_{i,j} = rem(X(i,j), 2)$ is the remainder of pixel value $X(i,j)$ and integer 2, $Y(i,j)$ is modified value of pixel $X(i,j)$, $1 \leq i \leq M$, $1 \leq j \leq N$, and $k = 1, 2, \ldots, M \times N$.
Step3: Scan the whole image in the same order in step2, and modify $Y(i,j)$ according to the secret message B.

$$Z(i,j) = \begin{cases} Y(i,j) + B & if \ S_k = 0 \\ Y(i,j) - B & if \ S_k = 1 \end{cases} \quad (4)$$

Where S_k is the same chaotic sequences in step2, $k = 1, 2, \ldots, M \times N$. B is a binary sequence, and $B \in \{0, 1\}$.
Finally, the final stego image Z with the embedded secret information is constructed.

2.2 Extraction Process

The secret information extraction process is similar with the embedding process exception the image is stego image. The extraction process is described as follows:

Step1: Generate chaotic sequences S_k by utilizing private keys α and l_0.
Step2: Scan the whole stego image Z in the same order as during the embedding. Extract message B by

$$B = \begin{cases} 0 & if \ S_k = 0 \& R_{i,j} = 0 \\ 1 & if \ S_k = 0 \& R_{i,j} = 1 \\ 0 & if \ S_k = 1 \& R_{i,j} = 1 \\ 1 & if \ S_k = 1 \& R_{i,j} = 0 \end{cases} \quad (5)$$

Where $R_{i,j} = rem(Z(i,j), 2)$ is the remainder of pixel value $Z(i,j)$ and integer 2, $1 \leq i \leq M$, $1 \leq j \leq N$, and $k = 1, 2, \ldots, M \times N$.

Thus, the secret message hidden in the stego image are obtained.

3 Results and Discussions

We select four 8-bits grayscale images of size 512×512 to evaluate the performance of the proposed scheme, as shown in Fig. 2. These natural images, "Lena", "Baboon", "Boat", and "Pepper", are selected from CVG-UGR image database [11]. To evaluate the visual quality of stego image, we use the function of peak-signal-to-noise-ratio (PSNR), which is defined as in

$$PSNR(dB) = 10 \times \log_{10} \left(\frac{255^2 \times M \times N}{\sum_{i=1}^{M}\sum_{j=1}^{N}(X(i,j) - Z(i,j))^2} \right) \quad (6)$$

Where, X and Z denote the original cover image and stego image, M and N denote the width and height of the cover image, respectively.

The results of four test images for embedding are listed in Table 1, which shows the embedded capacity and PSNR of the stego image. From the Table 1 it is seen that the values of PSNR of four test images are all greater than 51 dB. The embedding capacity of four test images are 262,144 bits, i.e. 1 bpp (bits per pixel). This demonstrates that our proposed algorithm can achieve nice invisibility and large embedding capacity.

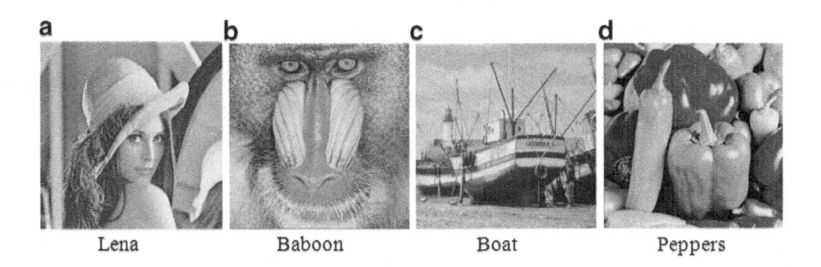

Fig. 2 Original cover images (a) Lena (b) Baboon (c) Boat (d) Peppers

Table 1 Embedding capacity and distortion for test images

Test images	Capacity (bits)	PSNR (dB)	Key
Lena	262,144	51.13	Yes
Baboon	262,144	51.13	Yes
Boat	262,144	51.14	Yes
Peppers	262,144	51.13	Yes

Stego Lena   Stego Baboon   Stego Boa   Stego Peppers

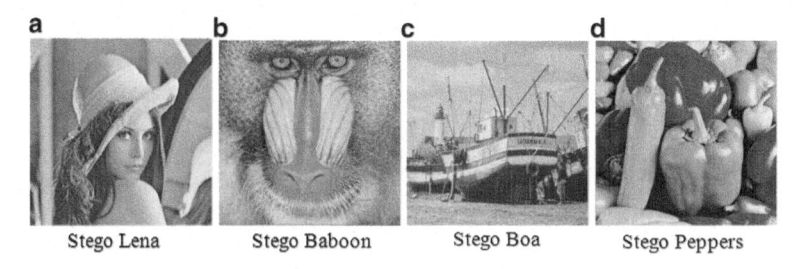

Fig. 3 The stego images (**a**) Stego Lena (**b**) Stego Baboon (**c**) Stego Boa (**d**) Stego Peppers

Figure 3 shows the visual quality of stego images when embedding same secret data with length of 262,144 bits. The visual differences cannot be detected by the Human Vision System (HVS) between the stego images and the corresponding original cover images. This is the most important desire for data hiding application.

Figure 4 shows the histogram distribution of Lena image after embedding data by using LSB method and our proposed method, respectively. Figure 4a is the histogram of original Lena image. Figure 4b is the histogram of stego image obtained from applying the classical LSB data hiding approach. From Fig. 4b, the histogram of stego image has changed obviously. it leaves detectable traces to attacker inevitably. In other words, once the attacker realizes there is a hidden information under stego image, the secret data would be extracted possibly by analysing the distribution of histogram of the stego image. Figure 4c is the histogram of stego image obtained from applying our proposed data hiding method. With reference to Fig. 4c we can clearly see that there is almost no change on histogram between stego image and original image. i.e., the attacker cannot detect the changes done for data hiding.

The result of security test for our proposed method is showed in Fig. 5. Figure 5a is a secret binary image with size 512×512. We embed this secret binary image into cover image by applying our proposed embedding process with private keys, and then extract the hidden information under stego image by applying our proposed extraction process with wrong keys. Finally, the extracted image is shown in the Fig. 5b. From the Fig. 5b it is seen that the extracted image is similar to white noise image, which demonstrates the extracted image varies sensitively with the variances of the private keys.

4 Conclusion

This paper put up a histogram modification data hiding method using chaotic sequences for increasing the security. The secret message is embedded at different histogram modification traces, which are encrypted by private keys. Even though the attacker detects the existence of hidden message under stego image, the secret message cannot be extracted without private keys. Experimental results

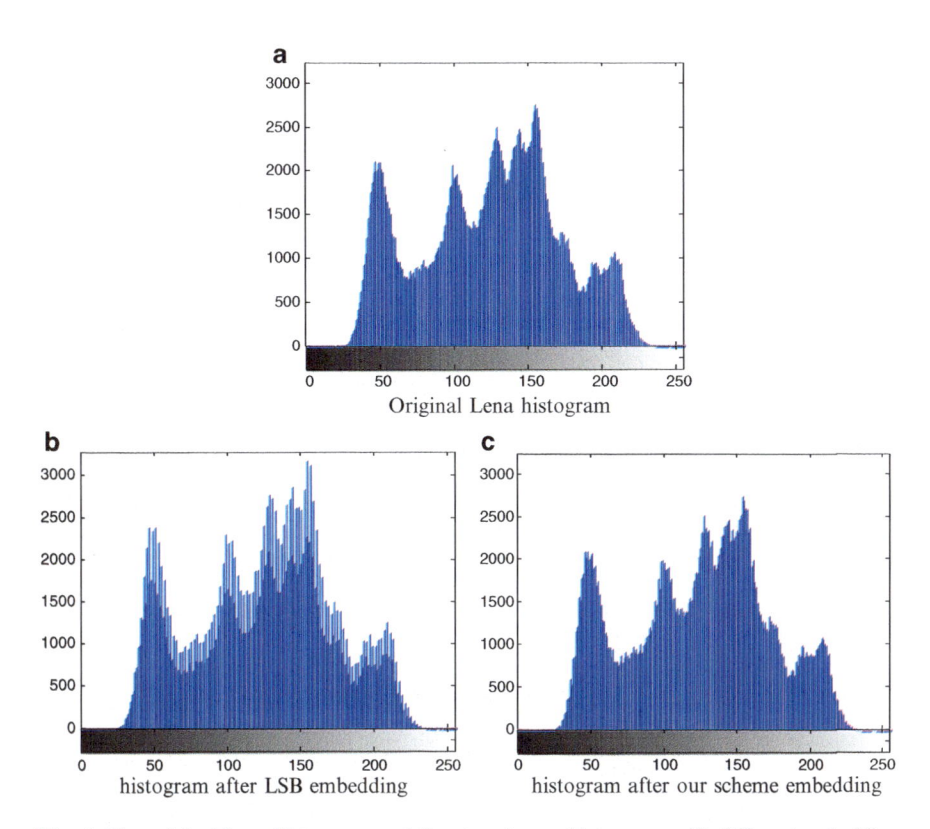

Fig. 4 The original Lena histograms and the stego image histograms with different embedding methods (**a**) original Lena histogram, (**b**) histogram after LSB embedding (**c**) histogram after our scheme embedding

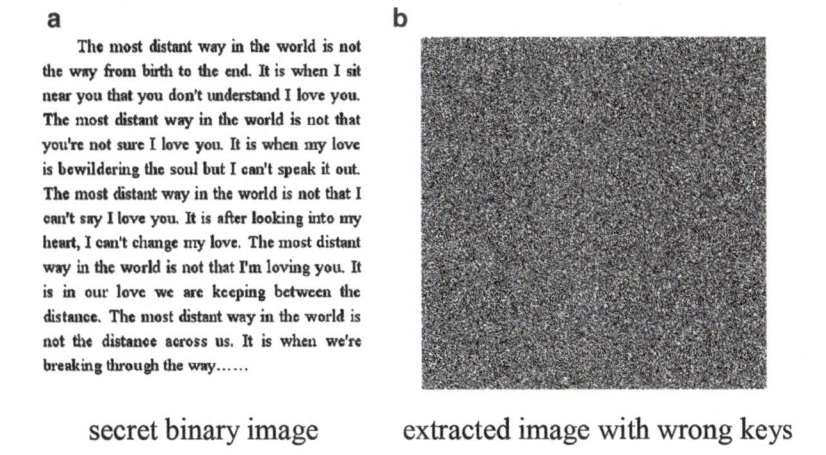

Fig. 5 Security test (**a**) secret binary image (**b**) extracted image with wrong keys

demonstrated that the proposed data hiding scheme can provide higher security while keeping low distortion and large embedding payload. Besides, the embedding process of the proposed scheme just deals with the scanning, adding, and subtracting operations that the computation complexity of this scheme is also very small. It is expected that the proposed scheme having high security can be deployed for extensive application fields.

References

1. Petitcolas FAP, Anderson RJ, Kuhn MG (1999) Information hiding-a survey. Proc IEEE 87(7):1062–1078
2. Wu DC, Tsai WH (2000) Spatial-domain image hiding using image differencing. IEE Proc Vis Image Signal Process 147(1):29–37
3. Bao P, Ma X (2005) Image adaptive watermarking using wavelet domain singular value decomposition. IEEE Trans Circuits Syst Video Technol 15(1):96–102
4. Chuang JC, Chang CC (2006) Using a simple and fast image compression algorithm to hide secret information. Int J Comput Appl 28(4):329–333
5. Chan CK, Cheng LM (2004) Hiding data in images by simple LSB substitution. Pattern Recog 37(3):469–474
6. Singh S, Siddiqui TJ, Singh R et al (2011) DCT-domain robust data hiding using chaotic sequence. In: International conference on multimedia, signal processing and communication technologies, IEEE Press, Aligarh, India, pp 300–303
7. Huang HY, Chang SH (2010) A lossless data hiding based on discrete Haar wavelet transform. In: IEEE 10th international conference on Computer and Information Technology (CIT), IEEE Press, Bradford, England, pp 1554–1559
8. Chang CC, Lin CY, Fan YH (2008) Lossless data hiding for color images based on block truncation coding. Pattern Recog 41(7):2347–2357
9. Lee JD, Chiou YH, Guo JM (2010) Reversible data hiding based on histogram modification of SMVQ indices. IEEE Trans Info Forensics Secur 5(4): 638–648
10. Sun Y, Wang GY (2011) An image encryption scheme based on modified logistic map. International workshop on Chaos-Fractals Theories and Applications(IWCFTA), IEEE Press, Hangzhou, China, pp 179–182
11. CVG-UGR Image Database: http://decsai.ugr.es/cvg/dbimagenes/index.php

A Family of Functions for Generating Colorful Patterns with Mixed Symmetries from Dynamical Systems

Jian Lu, Yuru Zou, Guangyi Tu, and Haiyan Wu

Abstract This paper investigates generating artistic patterns with mixed symmetry by means of dynamical systems. First, a new family of functions is introduced to produce colorful patterns with cyclic/dihedral symmetries and some crystallographic symmetries. Successively, two functions with distinct symmetries are applied to create patterns having mixed symmetries in different planar regions with the aid of a mixing technique. The method might provide a novel approach for automatically generating a great variety of artistic patterns.

Keywords Crystallographic groups • Cyclic group • Dihedral group • Mixed symmetry

1 Introduction

The fact that the attractors arising from the iteration of functions in the plane that are designed to have symmetry has been the subject of much recent study. Carter et al. [1] explored the construction of the families of functions which can be used to create chaotic attractors with symmetries of the frieze and wallpaper groups. Field and Golubitsky [2] investigated how to generate chaotic attractors with cyclic, dihedral and some of the planar crystallographic symmetries. In addition, Carter [3] illustrated two functions of the plane that have distinct symmetries intertwined using a sinew function to produce a function of three variable having different symmetries in different regions of space. Dumont et al [4] studied the evolution of families of attractors that change from one symmetry type to another. Dumont and Reiter [5] further discussed the chaotic attractors with symmetries that are close to forbidden and their associated functions. The methods of coloring the chaotic

---

J. Lu • Y. Zou (✉) • G. Tu • H. Wu
College of Mathematics and Computational Science, Shenzhen University, Shenzhen, China
e-mail: yrzou@163.com

attractors mentioned above are based on the point-hitting frequency. And always suppose that the attractor has a Sinai-Ruelle-Bowen (SRB) measure [6, 7].

In this paper, different from Dumont et al.'s work constructing attractors with evolving symmetry or forbidden symmetry, we first present an effective method to generate artistic patterns with cyclic/dihedral symmetry or some wallpaper symmetries from dynamical systems. Furthermore, the colorful patterns with mixed symmetry are investigated by using a mixing technique from dynamical systems. The presented method is based on our previous work [8, 9] and can be used to create a great variety of artistic patterns.

2 Planar Symmetry Groups

A mapping $g : \mathbf{R}^2 \to \mathbf{R}^2$ is said to be *equivariant* with respect to a symmetry group if $g \circ \gamma = \gamma \circ g$ for all elements γ of the group. The discrete planar symmetry groups are well known to consist of the identity, cyclic and dihedral groups, 7 frieze groups and 17 crystallographic groups. Cyclic and dihedral groups are subgroups of permutation groups. The cyclic group, C_n, is generated by n-fold rotations about a single point. A mapping with C_n symmetry should be equivariant with respect to

$$S_n = \begin{pmatrix} \cos 2\pi/n & -\sin 2\pi/n \\ \sin 2\pi/n & \cos 2\pi/n \end{pmatrix}$$

The dihedral group, D_n, contains n-fold rotations as well as a reflection through a line passing through the point of rotation. A mapping with D_n symmetry should be equivariant with respect to S_n and

$$M = \begin{pmatrix} 1 & 0 \\ 0 & -1 \end{pmatrix}$$

The crystallographic groups are characterized by two independent translations, which give rise to a lattice. In addition, the crystallographic groups may contain rotations of order two, three, four and six. Together with reflections and glide reflections, there are a total of 17 crystallographic groups. For instance, the $p31m$ symmetry group contains third-fold rotations and mirrors about y-axis as well as two independent translations. Since a lattice does not map to itself by third turns, we can take $u_0 = 2\pi(1, 0)$ and $u_1 = 2\pi(-\frac{1}{2}, \frac{\sqrt{3}}{2})$ as the generators of the lattice L. Connecting vertices in this lattice produces a regular tiling of the plane that is called a hexagonal lattice. The vectors $v_0 = (1, -\frac{1}{\sqrt{3}})$ and $v_1 = (0, -\frac{2}{\sqrt{3}})$ form a basis for the dual lattice L^*. The generators of $p31m$ include $\delta(x, y) = (x, -y)$ and $\delta(x, y) = (-\frac{1}{2}x - \frac{\sqrt{3}}{2}y, \frac{\sqrt{3}}{2}x - \frac{1}{2}y)$.

The theory of which polynomials have cyclic and dihedral symmetries is discussed in Field and Golubitsky [1]. Carter et al. [2] investigated a family of crystallographic functions using trigonometric functions. In traditional methods [1, 2], all these equivariant mappings are applied to create chaotic attractors with different planar symmetries. In next section, we will present another new family of functions for creating colorful patterns with cyclic/dihedral symmetries and some wallpaper symmetries. The construction of these equivariant functions is based on the theory introduced by Dumont and Reiter [5] for creating attractors with forbidden symmetry.

3 New Symmetric Functions

Suppose $g : \mathbf{R}^2 \to \mathbf{R}^2$ being an arbitrary function and G being a finite group on $\mathbf{R}^2$, then [10],

$$F(\vec{x}) = \sum_{\gamma \in G} \gamma^{-1}(g(\gamma(\vec{x}))) \qquad (1)$$

is equivariant with respect to G, where $\vec{x} = (x, y) \in \mathbf{R}^2$. In fact, it is easy to verify that for any $\sigma \in G$, $F(\sigma(\vec{x})) = \sum_{\gamma \in G} \sigma(\gamma\sigma)^{-1}(g(\gamma\sigma(\vec{x}))) = \sigma(F(\vec{x}))$. So, various functions can be used for creating cyclic or dihedral symmetry provided that the group G is C_n or D_n [10]. In the method here, according to lattice theory, we will discuss a family of functions that can be used to generate cyclic and dihedral patterns as well as some specific crystallographic patterns, such as $p3$, $p31m$, $p4$, $p4m$, $p6$, $p6m$ etc. We first consider a lattice L in $\mathbf{R}^2$ along with a dual lattice L^*, which implies that $\vec{\alpha} \bullet \vec{\beta}$ is an integer for any vectors $\vec{\alpha} \in L$ and $\vec{\beta} \in L^*$. Thus, $\sin(\vec{\beta} \cdot (\vec{x} + \vec{\alpha})) = \sin(\vec{\beta} \cdot \vec{x})$, if $\vec{\beta} \in 2\pi L^*$, and likewise for cosine. This means that $\vec{a}\sin(\vec{\beta} \cdot \vec{x}) + \vec{b}\cos(\vec{\beta} \cdot \vec{x}) + \vec{c}\sin(\vec{\beta} \cdot \vec{x})\cos(\vec{\beta} \cdot \vec{x}) \bmod L$ is equivariant with respect to the translations L, where $\vec{a}, \vec{b}$ and $\vec{c}$ are constant vectors and $\vec{a}, \vec{b}, \vec{c} \in \mathbf{R}^2$. In practical use, $\vec{a}, \vec{b}, \vec{c}$ are produced randomly. Based on this analysis, similar to Dumont and Reiter [5], we can construct a new equivariant function g with respect to the translational symmetries of L and

$$g(\vec{x}) = \sum_{\vec{\beta} \in \Omega} (\vec{a}\sin(\vec{\beta} \cdot \vec{x}) + \vec{b}\cos(\vec{\beta} \cdot \vec{x}) + \vec{c}\sin(\vec{\beta} \cdot \vec{x})\cos(\vec{\beta} \cdot \vec{x})) \bmod L \qquad (2)$$

Note that even if we take a function of the form $F_g(\vec{x})$ in Eq. 1 with g having the form in Eq. 2, $F_g(\vec{x})$ has the symmetry of G and may not take the symmetry of the lattice. Naturally, if we define the function

Fig. 1 A pattern with D_7 symmetry

Fig. 2 A pattern with C_6 symmetry

$$\tilde{F}_g(\vec{x}) = F_g(\vec{x}) \bmod L \qquad (3)$$

and the group G can map the lattice L back onto itself, then the desired symmetries of $F_g(\vec{x}) \bmod L$ are obtained [5].

Combining with the above equivariant functions $F_g(\vec{x})$ or $\tilde{F}_g(\vec{x})$, colorful pattern with specific planar symmetries can be created by using invariant mapping from dynamical systems. For instance, Figs. 1, 2, and 3 are generated by using Eq. 1. Figure 4 is created by using Eq. 3.

Fig. 3 A pattern with D_{13} symmetry

Fig. 4 A pattern with $p31m$ symmetry

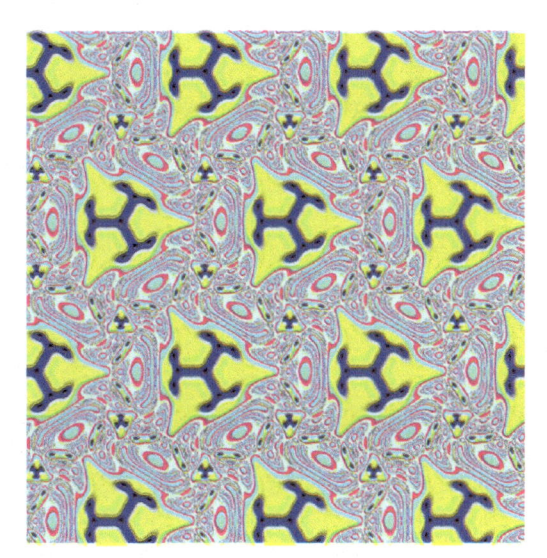

4 Colorful Patterns with Mixed Symmetry

Our construction of mixed symmetric patterns utilizes two functions with symmetry in the plane intertwined and connected via stochastic scheme. Let $F_1(\vec{x})$ and $F_2(\vec{x})$ be two desired symmetric functions. In order to mix the two symmetries visually, we need to consider a smooth transition by following function

$$F(\vec{x}) = \lambda(\vec{x})F_1(\vec{x}) + (1 - \lambda(\vec{x}))F_2(\vec{x}) \quad (4)$$

Fig. 5 A mixed symmetric pattern with dominated D_7 outer symmetry and dominated D_3 inner symmetry

Note that, when $\lambda(\vec{x}) \equiv 1$, the function is $F_1(z)$, and when $\lambda(\vec{x}) \equiv 0$, the function is $F_2(z)$. The values of $\lambda(\vec{x})$ between 0 and 1 show the intermediate steps for intertwining between one symmetry and the other symmetry. Suppose a disc region Ω is defined by $\Omega = \{(x,y)^T \in R^2 | \sqrt{x^2 + y^2} \leq r\}$ where r is a constant and $r \in R$. Here, we proposed two methods for constructing mixed symmetry. One simple method is introduced by defining $\lambda(x,y)$ as

$$\lambda(x,y) = \begin{cases} \lambda_1 & (x,y) \in \Omega \\ \lambda_2 & otherwise. \end{cases} \quad (5)$$

where λ_1 and λ_2 is constant and $\lambda_1, \lambda_2 \in [0,1]$. Equation 5 implies that if $\lambda_1 \gg 0.5$, the symmetries in region Ω is dominated by $F_1(x,y)$; conversely, the symmetries in region Ω is dominated by $F_2(x,y)$ if $\lambda_2 \gg 0.5$. For example, we take $\lambda_1 = 0.1$ and $\lambda_2 = 0.9$ to create the mixed symmetric pattern with dominated D_7 outer symmetry and dominated D_3 inner symmetry, as shown in Fig. 5.

The other method is implemented by defining $\lambda(x,y)$ as

$$\lambda(x,y) = \begin{cases} \left(\dfrac{\sqrt{x^2+y^2}}{r}\right)^5 & (x,y) \in \Omega \\ 1 & otherwise \end{cases} \quad (6)$$

Equation 6 means that the mixed symmetry is gradually transformed from $F_2(x,y)$ to $F_1(x,y)$ in the region Ω and the symmetry is dominated solely by $F_1(x,y)$. For instance, the mixed symmetric pattern with dominated D_9 outer

Fig. 6 A mixed symmetric pattern with dominated D_9 outer symmetry and dominated D_5 inner symmetry

symmetry and dominated D_3 inner symmetry shown in Fig. 5 is created by using Eq. 6.

Let $\psi(X)$ a symmetry group of a certain subset of the plane, namely X. A mapping $f : \mathbb{R}^2 \to \mathbb{R}$ is said to be *invariant* with respect to a symmetry group if $f \circ \gamma = f$ for all elements γ of the group. Since the crystallographic symmetries contain two translations in nonparallel directions, we only need to design the invariant functions on a period parallelogram. In other words, f need not be invariant with respect to $\gamma(x,y) = (x+T, y+T')$, where T, T' is the translation period. So, It can directly verify that $f(x, y)$ is invariant for C_n/D_n and crystallographic symmetries if $f(x, y) = h \circ v(x, y)$, where $v(x, y) = x^2 + y^2$ and $h : \mathbb{R} \to \mathbb{R}$ is any real function [10]. For example, we take $f(x,y) = v(x,y) + |\cos(v(x,y))| + 0.15v^3(x,y) + 12v(x,y)$ to generate the images in Figs. 1, 2, and 5; and take $f(x,y) = |\sin(v(x,y)) + \cos(h^3(x,y))| + 10v(x,y)$ and $f(x,y) = |6\sin(v^3(x,y)) + 2\tan(v^3(x,y))|$ to create the image in Figs. 3, 4, and 6. Here, Figs. 1, 2, 3, and 4 are patterns with D_7, C_7, D_{13} and $p31m$ symmetries, respectively. It is worth noting that Fig. 4 is also a D_3 symmetric pattern. Figure 5 is a mixed symmetric pattern with dominated D_7 outer symmetry and dominated D_3 inner symmetry; Fig. 6 is a mixed symmetric pattern with dominated D_9 outer symmetry and dominated D_5 inner symmetry.

5 Color Scheme

Suppose that the scope for generating patterns is denoted by U; f is a invariant function and $f(x, y) \geq 0$; F is equivariant with respect to planar symmetry groups as described above; M is a given positive integer and τ is a positive real number.

For each point $p(x, y)$ in displaying area U, the equivariant mapping F is determined by the position of p. Calculate the f-values $\{f(F^j(p))\}_{j=1}^{J}$ of the orbit $\{F^j(p)\}_{j=1}^{J}$ of p under F where F^j s the j-th iteration of F. Once $f(F^j(p)) < \tau$ for some $j \leq J$, the iteration exits and then we take $\hat{F}(p) = \frac{F(f^j(p))}{\tau}$, which is used to color the pixel p. Otherwise, $\hat{F}(p)$ is defined as $\hat{F}(p) = 0$ and the pixel p is colored black. Since $\hat{F}$ is a symmetric density function on X, the group symmetrically placed pixels can get the same color with the same density. Outstanding features of this color method appear in Lu et al.'s work [9].

6 Conclusion

This paper introduced a new family of functions to create artistic images with mixed symmetry. Specially, two functions with different symmetries were investigated and integrated for constructing mixing symmetric images in displaying regions by using a mixing technique. The generated symmetric patterns verified the validity of this method.

Acknowledgements The authors were supported by NSFC #61003178, #10926142, #11201312, #61272252, #11071150, Shenzhen MSTP #JC2011051 70615A and #JC201005280508A.

References

1. Carter NC, Eagles RL, Hahn AC, Grimes CA (1998) Chaotic attractors with discrete planar symmetries. Chaos Soliton Fract 12:2031–2054
2. Field M, Golubitsky M (1992) Symmetry in chaos. Oxford University Press, New York, pp 89–101
3. Carter NC (1997) Attractors with duelling symmetry. Comput Graph 21:263–271
4. Dumont JP, Heiss FJ, Jones KC et al (1999) Chaotic attractors and evolving planar symmetry. Comput Graph 23:613–619
5. Dumont JP, Reiter CA (1998) Chaotic attractors near forbidden symmetry. Chaos Soliton Fract 11:1287–1298
6. Field M (2001) Designer chaos. Comput Aided Des 33:349–365
7. Robinson C (1994) Dynamical systems. CRC Press, Boca Raton, pp 351–356
8. Lu J, Ye ZX, Zou YR (2007) Automatic generation of colorful patterns with wallpaper symmetries for dynamics. Vis Comput 23:445–449
9. Lu J, Zou YR, Li WX (2011) Colorful patterns with discrete planar symmetries from dynamical systems. Fractals 18:35–43
10. Jones KC, Reiter CA (2000) Chaotic attractors with cyclic symmetry revisited. Comput Graph 24:271–282

Automatic Calibration Research of the Modulation Parameters for the Digital Communications

Kai Wang, Zhi Li, Ming-zhao Li, Chong-quan Zhu, Qiang Ye, and Juan Lu

Abstract Calibration of the modulation quality as an important parameter of wireless communication, it played a vital role in the performance of wireless com-munications test. We proposed a auto-calibration method of digital modulation parameters of a digital mobile communication tester, against the unstable modulation parameters and high error rate of manually test, what's more, analyzed the uncertainty of the measurement results. The experiment results showed that the automatic test method was precision, high stability, and practical.

Keywords Digital mobile communications integrated tester • Automatic calibration • Digital modulation

1 Introduction

TD-SCDMA (Time Division-Synchronous Code Division Multiple Access), the first complete mobile communication technology standards of the Chinese telecommunications industry for centuries, and got the full support of the China Communications Standards Association (CWTS) and 3GPP international organizations, was one of the technical specifications by ITU formally released the third generation mobile communications space interface to of the modulation quality of the main parameters of the digital communications tester, also an important part of the conformance testing of TD-SCDMA terminal, the main measure of the modulation error generated by the modulator and RF devices quality of the modulated signal.

K. Wang • Z. Li • M.-z. Li (✉) • C.-q. Zhu • J. Lu
Shenzhen Academy of Metrology and Quality Inspection, Shenzhen, China
e-mail: smqlmz@163.com

Q. Ye
College of Information and Engineering, China Jiliang University, Hangzhou, China

The GPIB (General-Purpose Interface Bus)-based automated test system was a product of the combination of computer technology and automated testing techniques and now widely used in many fields [1]. E5515C wireless communications test instrument of Angilent under the Visual Basic environment called the GPIB DLL send SCPI commands to control the E5515C, and accomplished TD-SCDMA standard digital modulation quality parameters of the automatic calibration, and introduced an automatic calibration method of a wireless communication tester.

2 System Model

In this paper, the system model was divided into two parts of the hardware and software systems to introduce:

First, the automatic measurement system was consist of a E4445A spectrum analyzer with vector signal analysis function, computers, and GPIB interface card and Angilent 8960 digital communications tester, system hardware block diagram shown in Fig. 1.

The automatic measurement system was consist of Angilent 8960 Series E5515C digital communications, E4445A spectrum analyzer with a vector signal

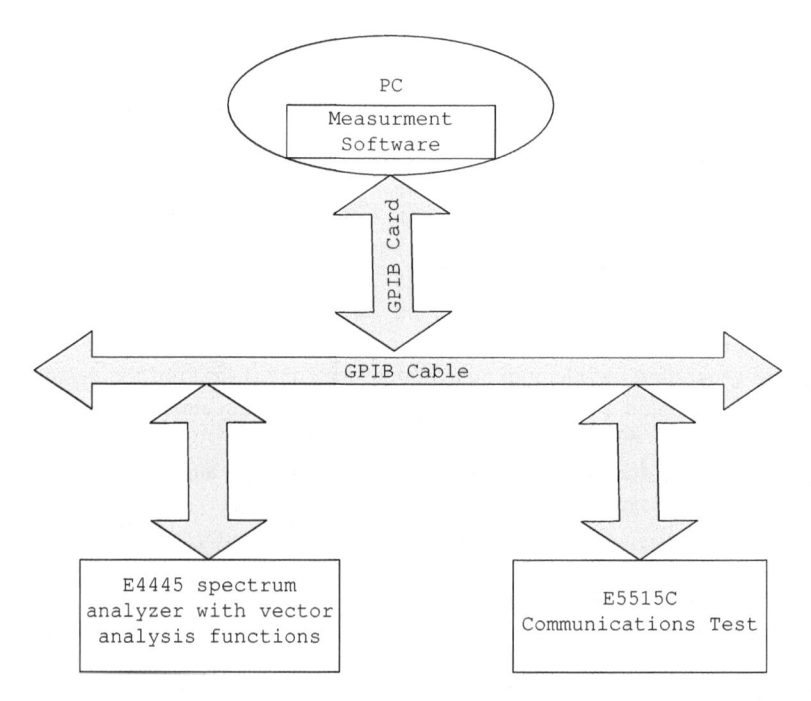

Fig. 1 System hardware framework

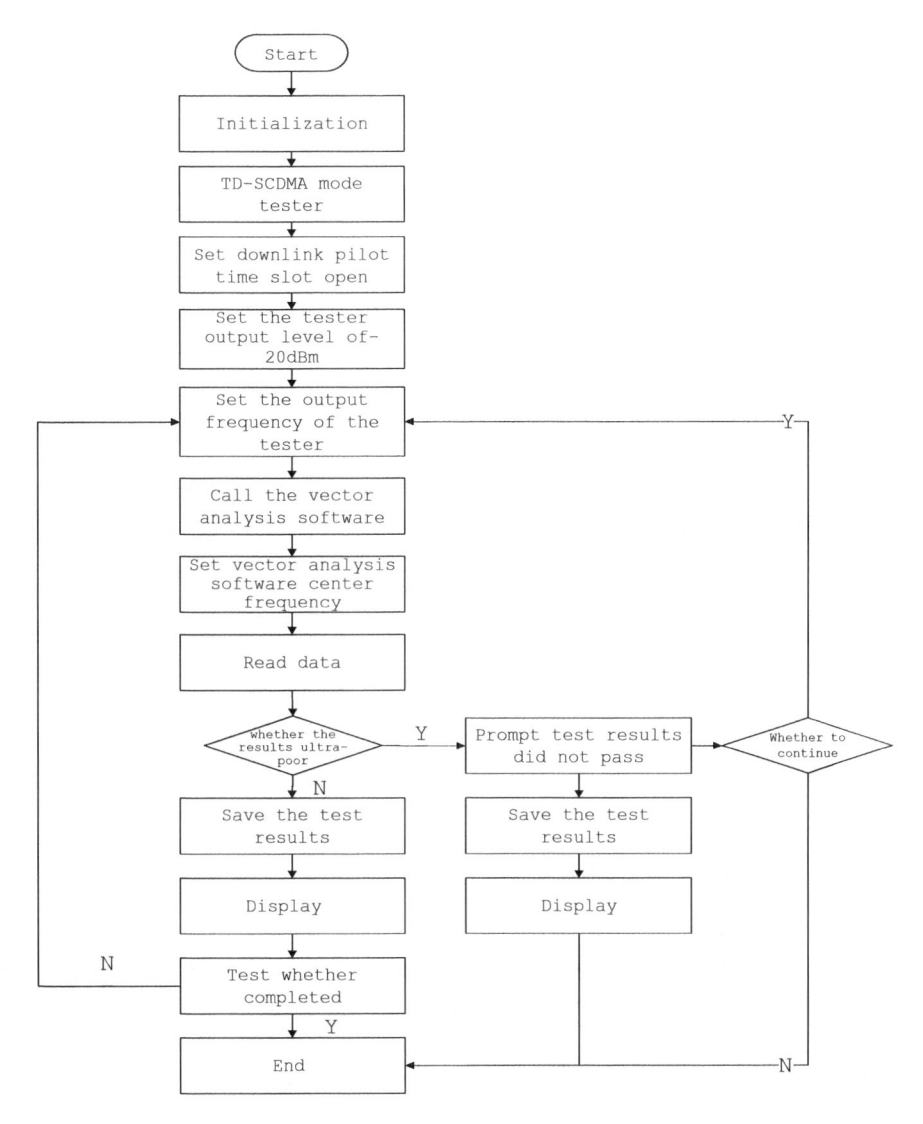

Fig. 2 Software flow chart

analyzer function, computers, GPIB interface card and cable. The system took computer as the core, connected the GPIB interface card, the standard and the measured instrument with GPIB cables. Computer sent SCPI command to control and management standards and the measured instrument, and accomplished data transmission and processing [2].

Secondly, the software system was the core part of the whole system, the realization of the process shown in Fig. 2.

Software sent commands to initialize the instrument, the tester was set to TD-SCDMA standard, and set the modulation signal of the tester for the downlink, at the same time, vector analysis software was also needed to set to a downlink demodulation, then, the tester would be able to output the TD-SCDMA modulated signals. Tester output level too large or too small would lead vector analysis software would not be able to demodulate, so the modulated signal was typically set to -20dBm. The vector analysis of the EVM software demodulation fluctuate, in order to reduce its error, returned average results of 20 groups.

The software system was the core part of the whole system, in order to achieve the data acquisition and analysis, under VB environment, Communication between the standard device and the gpib32 of the dll library in GPIB interface card the gpib32 of the dll library were connected through the IEEE-488. The E5515C tester issued TD-SCDMA modulated signals, relying on the modulation signal Angilent 89600 vector analysis software to analyze, re-used testing software to make data processing and judgment, the final test results showed in the test software interface and deposited to ACCCESS database [3].

3 System Performance Analysis

ATS (Automatic Testing Systems) was a system, which could automatically measure, data process, and display or output the test results in an appropriate manner with people rarely participate or not participate in. Compared with manual testing, automated testing could save time, improve labor productivity and product quality, which had an important role in production, research and defense [4].

TD-SCDMA combine code division multiple access, time division multiple access frequency division multiple access as one, had a class of noise and autocorrelation characteristics. There were two general methods of measurement such as these signals: the appropriate amount of margin of error by using the CDMA waveform quality RHO and WCDMA EVM.RHO and EVM reflected the overall quality of the modulation was a composite indicator for CDMA [5].

TD-SCDMA system used QPSK modulation, which was a fly constant envelope modulation, there were errors in the amplitude, phase error and frequency difference did not adequately reflect the raised accuracy, therefore, needed a indicator comprehensive measured the error of the signal amplitude and phase. The error vector could clearly reflect the degree of damage of the signal, so the error vector magnitude parameters proposed [6]. EVM measurements of the vector difference between the actual signal waveforms with the theoretical reference signal, was the ratio of RMS error vector magnitude and the reference signal amplitude. EVM was a measure of the overall quality of the modulated signal, if you want to further analyze modulation quality factors were also needed to measure the code domain error [7].

Basis JJF1204-2008 TD-SCDMA digital mobile communications test instrument calibration specification, took direct measurement.

According to the method used, the measured output tester modulation error was expressed as the formula (1):

$$\Delta R = R_X - R_N \tag{1}$$

According to the formula (2):

$$u_c^2(\Delta R) = \sum (\partial f/\partial R_i)^2 u^2(R_i) \tag{2}$$

Variance as formula (3):

$$u_c^2 = c^2(R_X) \cdot u^2(R_X) + c^2(R_N) \cdot u^2(R_N) \tag{3}$$

Propagation coefficient as formula (4) and (5):

$$c(R_X) = \partial f/\partial R_X = 1 \tag{4}$$

$$c(R_N) = \partial f/\partial R_N = -1 \tag{5}$$

Select spectrum analyzer E4445A with vector analysis software 89601A spectrum as standard instrument, 8960 Mobile Phone Tester as measured instrument to do uncertainty evaluation [8].

Under reference conditions, use E4445A spectrum analysis instrument with TD-SCDMA Digital demodulation functions on the 8960 Mobile Phone Tester TD-SCDMA output digitally modulated signals of the EVM measurement repeated 10 times.

Standard deviation expressed as formula (6):

$$S = \sqrt{\frac{\sum_{i=1}^{n}(R_i - \overline{R})^2}{(n-1)}} \tag{6}$$

Calculated by the data in the experimental, we can get standard uncertainty of measurement results from (6):

$$u_1 = S = 0.02\%$$

Standard uncertainty components expressed as formula (7):

$$u_2 = a_2/k_2 \tag{7}$$

Check technical manuals that the E4445A measurement EVM value of the maximum allowable error was $\pm 1.00\ \%$, $a_2 = 1.00\ \%$. Assumed that the components were uniformly distributed, $k_2 = \sqrt{3}$. we could get the standard uncertainty components from formula (7):

$$u_2 = a_2/k_2 = \frac{1.00\%}{\sqrt{3}} = 0.58\%$$

The above component was independent and unrelated, we could get the combined standard uncertainty from formula (3):

$$u_c = \sqrt{u_1^2 + u_2^2} = 0.58\%$$

Taken to contain factor k = 2, the expanded uncertainty was:

$$U = k \cdot u_c = 1.2\%$$

4 Test Results Analysis

After testing, draw curve of the test results automatically at same frequency and different temperatures, and the contrast curve of the same frequency of manual testing and automated testing at same temperature. Shown in Fig. 3, at different temperatures, the same frequency point of the automatic measurement of EVM value fluctuations, but very smooth, showed that the stability of the system under different circumstances.

Shown in Fig. 4, at the same temperature, the contrast curve of the same frequency, manual testing and automated test showed that the accuracy of the automated testing. Thus, the automatic test method was reliable, high stability, and practical.

5 Conclusion

In this paper, the automatic calibration method for TD-SCDMA standard digital modulation quality parameters compared to manual testing, greatly improved the speed of measurement and saved the time of measurement. Meanwhile, repeated measurements after averaging that significantly improved the accuracy of measurement. The research of TD-SCDMA standard digital modulation quality parameters auto-calibration method, provided a good reference for automatic calibration of other standard digital modulation quality parameters, with the rise of 4G networks, the next step would research LTE standard calibration method, so as to realizing the automatic calibration of the LTE standard digital modulation quality parameters.

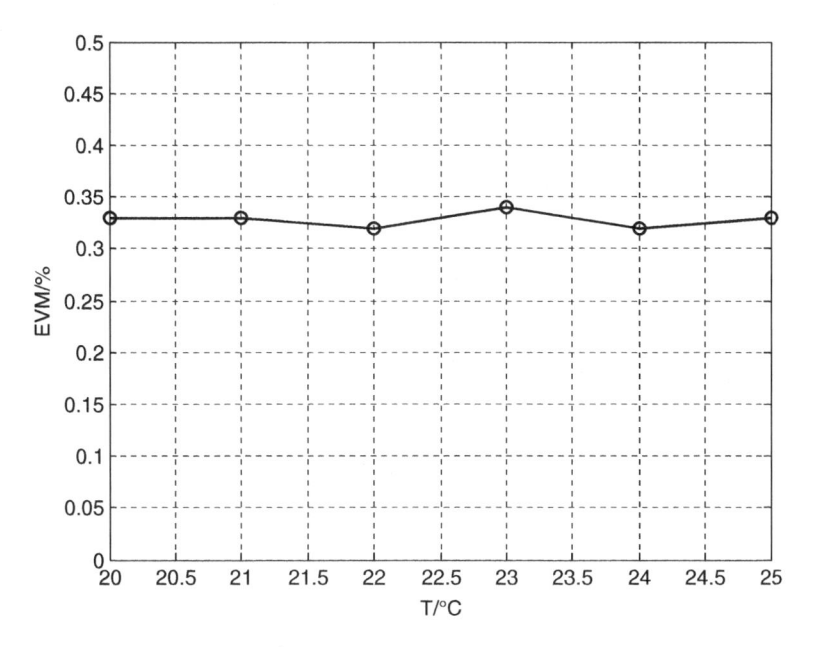

Fig. 3 EVM curve of the automatic measurement at different temperature

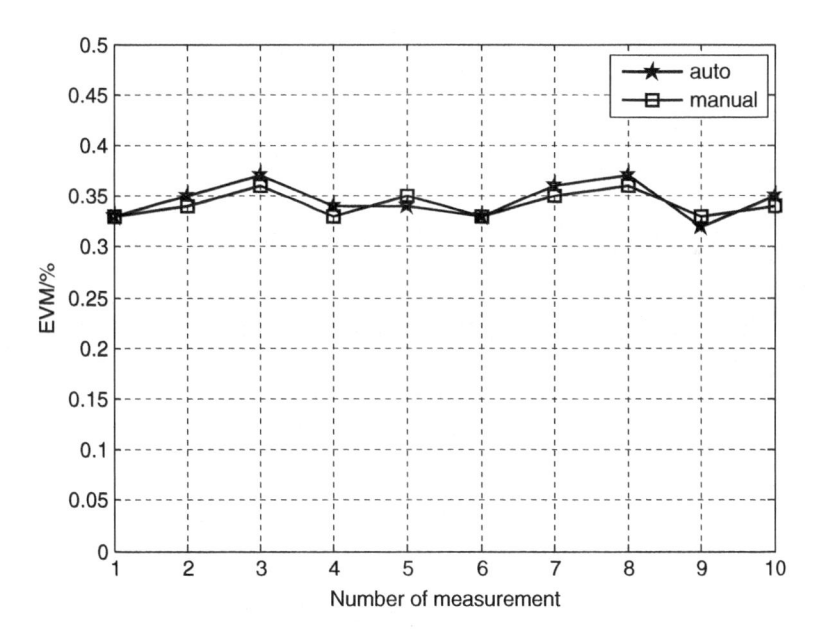

Fig. 4 The comparison of the manual and automatic measurement, under the same conditions

References

1. Qing JI (2007) Modern metering communications. Realize the GPIB-based automated test system with VB. 13(5):19–21
2. Qiu Hai Bo, Zhan Zhi Qiang, Zhao Ji Xiang et al (2011) China Institute of Metrology. The E5515C tester 1xEV-DO standard digital modulation parameter calibration. 22(2):139–143
3. Fan Xiu Li, Zheng Jian Hong, Chen Li et al (2005) TD-SCDMA, EVM testing and analysis. Journal of Chongqing College of Post and Telecom (Natural Science Edition) 17(2):156–159
4. Nassery A, Ozev S, Verhelst M et al (2011) Extraction of EVM from Transmitter system parameters. European Test Symposium (ETS), IEEE 33(5):75–80
5. Indrawati, Murugesan S, Raman M (2010) 3G mobile multimedia services (MMS) utilization in Indonesia: an exploratory research. Technol Soc (ISTAS) 21(5):145–155
6. Zhu Z, Zhang H, Shen L (2010) The application of structure arrays and files in the SCPI parsing system. Intell Comput Technol Autom (ICICTA) 15(6):710–713
7. Dasnurkar S (2009) Hybrid BiST solution for analog to digital converters with low-cost automatic test equipment compatibility. Circuit Syst ISCAS 14(7):9–12
8. Ross WA (2003) The impact of next generation test technology on aviation maintenance. IEEE Autotestcon 22(5):2–9

Adaptive Matching Interface Technology Based on Field: Programmable Gate Array

Xuejiao Zhao, Xiao Yan, and Kaiyu Qin

Abstract In order to handle the error happened when FPGA (Field-Programmable Gate Array) reads the data from the ultra-fast ADC. We studied and developed an adaptive matching interface based on the internal resources of FPGA. It can solve the problems occurred in high-speed reading of data between the FPGA and ADC. Based on the Xilinx Virtex-5 chip, guided by the phase-locked loop theory, and take advantage of the internal IP CORE of the Virtex-5, this study tried to construct a model which is able to predict and revise the phase of datas between the FPGA and ADC as well as to align the phase and read datas accurately. The model was designed to correct the error of phase matching which is caused by the delay time after the data entering into the internal FPGA chips and it can't be predicted while it was programming. Tests showed that the adaptive matching interface can totally apply to the adaptive reading of signal whose input rate is 0–250 MHz, and it can solve the problems happening in data reading between FGPA and ADC effectively by same peripheral devices.

Keywords FPGA • Adaptive • ADC interface matching • ISEDERS • Deserializer

1 Introduction

In recent years, Digital signal processing speed improved fast, also including the digital conversion chip. However, with the development of the ADC module speed, there are some new problems on using FPGA to read the ADC output data.

X. Zhao (✉) • X. Yan • K. Qin
Institute of Astronautics & Aeronautics, University of Electronic Science and technology of China, Chengdu, China
e-mail: 106906195@qq.com

Especially in each FPGA synthesis process, ADC module output data matches to FPGA internal D flip-flop in the correct phase difficultly after entering different routes with its clock, because the internal wiring is different, device delay, input pin delay [1]. Thus it often leads to large amounts of data read errors. Even in the case that the clock frequency is variable, the relative phase relationship will still change with the frequency [2].

The traditional method is to constraint wiring before comprehension [2], or to calculate the wiring delay with graph theory and to modify the wiring considering the device delay after comprehension [3]. Each FPGA integrated wiring is different, so this method must modify delay values each time comprehension. But with the increase of the chip number, operability seriously decline. So it urgently needs adaptive matched interface that can be apply to most of the FPGA chips.

2 The Analysis of Data Read Principle and Traditional Method

2.1 Data Read Principle

For example, the 14-bit analog-digital converted chip ADC9642 of ADI company, shown in Fig. 1. There are seven output pins in the rising edge of DCO clock with seven even-numbered bit outputted. In the falling edge of DCO clock with seven odd-numbered bit outputted, and the sampling rate are 250 MSPS.

The 14-bit parallel ADC standard path of reading is shown in Fig. 2. Now take read of lowest bit D0/D1 in the seven-channel data for instance. Signal output from the ADC, with the arrival to the internal FPGA via pins.

$$DCO\_cycle = \frac{1}{Sample\_rate} \qquad (1)$$

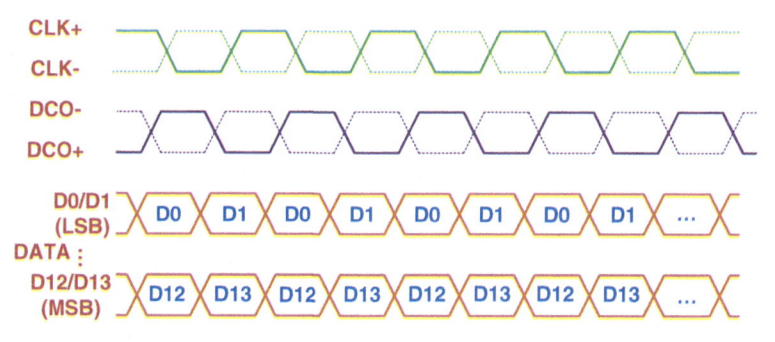

Fig. 1 The timing diagram of 14-bit high speed ADC

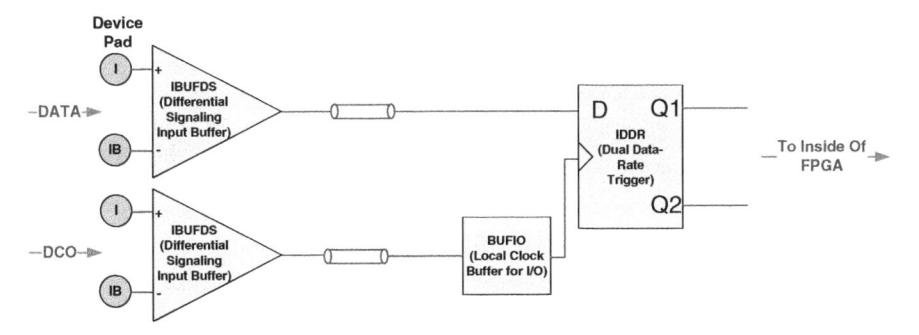

Fig. 2 The internal transmission path of DATA and DCO of FPGA

Known by the result of formula (1), the cycle of the DCO-outputting is 4ns, the delay of one device in the FPGA is about 4.5ns [4]. So, when the DCO and DATA get through the two paths with difference in devices and transmission line length to reach Dual D flip-flop within the FPGA, the initial phase relationship between DCO and DATA signals has lost [4]. But dual d flip-flop is still reading D0 and output to Q1 in accordance with the rising edge, falling edge to read the D1 and output to Q2, leading to a large number of errors in reading data. What is worse, wiring is different in each integration. When the routing path delay is just an integer multiple of the cycle after wiring, if its phase can align, then it is read correctly, otherwise it is wrong.

2.2 Traditional Method

2.2.1 Wiring Constrained

With proposition by some methods, as wiring delay can be calculated based on graph theory, we can adjust the wiring in accordance with its delay in after routing and adjust its delay to an integer multiple of the cycle [3].

Though this method has high accuracy, the actual operation is quite difficult in the actual work. For the difference in each wiring, calculation and adjustment to the wiring delay is necessary each time. Not only increase the workload in debug, but also wiring adjustment is required for each product while in mass production. Obviously, the feasibility of this method is not very high.

2.2.2 Constant Delay Time

It is suggested that after the DCO is read via the input buffer [5], the DCO get through the clock buffer and the module which can generate fixed delay, then get into the double D flip-flop to use a fixed waveform to help its "training".

According to the effect of the output to modify the delay value of the fixed – delay module, until a delay value which make the output relatively stable is get.

In the same chip, each wiring is not very different, which can make the chip work in a safer range. But different FPGA chips differ much, it is impossible to use the same delay value to help the delay compensation [4]. Although this method is simple and feasible, but its portability is poor and the delay value is only an estimate with low accuracy.

2.2.3 Data Training

A way to send data in the training is proposed in the literature [6], solutions are proposed which apply to this problem according to the principle, whose internal link structure is similar to the fixed delay program. What the difference is that the ADC chip send a series of fixed data to the FPGA within a fixed period of time, the ASCII code sequence order of ABCD for instance. Within this period of time, FPGA compare the received data with local data, adjust the fixed delay if some differences occur, and will not stop the adjustment until the same training data is consistent with local data.

This method is accurate, but which is based on the cooperation between ADC and FPGA with each other. For the stabilization time of FPGA is not definite, the time of ADC sending a fixed sequence is difficult to accurately control, and will increase the turn-on time of the system. And the portability of this method is poor.

3 Implementation of Technology of Adaptive Interface Matching

3.1 Clock Interface of DCO

Again, take the ADC9642 LSB D0/D1 port and DCO as an example. As shown in Fig. 3, DATA and DCO pass through pad and input buffer, and then to ISEDERS front-end. The production process of the FPGA can guarantee that the delay time in this distance of DATA as well as DCO, so before entering ISEDERS input interface, the DATA and DCO are strictly in phase matching.

3.1.1 Description of Matching Interface of DCO

After entering the ISEDERS, DCO passes through the delay module of ISEDERS, then to BUFIO (clock-driver). Sample the input DCO with the output clock of BUFIO as the sampling clock of the ISEDERS. BUFR (clock frequency divider)

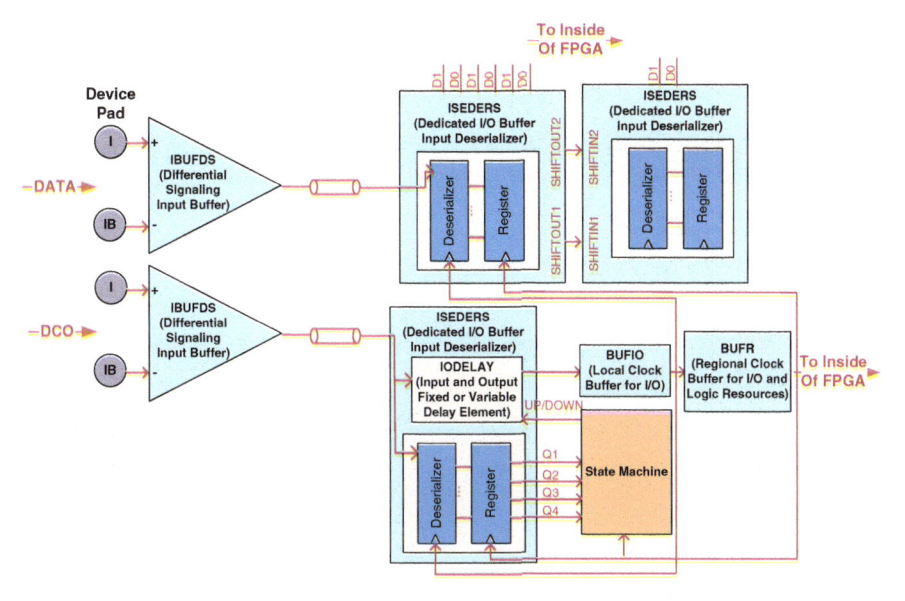

Fig. 3 Adaptive interface matching of ADC in FPGA

divide the input signal of BUFIO into four, then use to ISEDERS register as output clock and trigger clock of state machine [4]. ISEDERS works in SDR state, which is used to make sure that the sampling only happens at the positive edge of clock.

3.1.2 Operation of State Machine

The state machine is divided into five states.

STATE1: When use the output clock of BUFIO to sample the input clock of ISEDERS, if taken to a low level, the output sampling values of Q1, Q2, Q3, Q4 in Fig. 4 are all 0 s, we should increase the IODELAY TAP at that time, after that the sampling point will move to the right. Repeat testing until the output sampling values of Q1, Q2, Q3, Q4 are not 0 s, then stop increasing the IODELAY TAP and reduce 2 TAPS, so that the two clock edges align more accurately.

STATE2: When use the output clock of BUFIO to sample the input clock of ISEDERS, if taken to a high level, the output sampling values of Q1, Q2, Q3, Q4 in Fig. 5 are all 1 s. We should reduce the IODELAY TAP at that time, after that the sampling point will move to the left. Repeat testing until the output sampling values of Q1,Q2,Q3,Q4 are not 1 s, then stop reducing the IODELAY TAP and increase 2 TAPS, so that the two clock edges align more accurately.

STATE3: When use the output clock of BUFIO to sample the input clock of ISEDERS, if taken to a edge, the output sampling values of Q1, Q2, Q3, Q4 in Fig. 6 are neither all 0 s nor all 1 s we should increase or reduce the IODELAY TAP

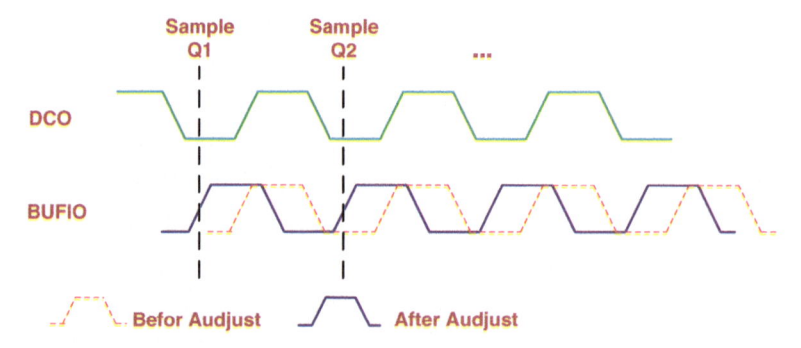

Fig. 4 Sampling to low level

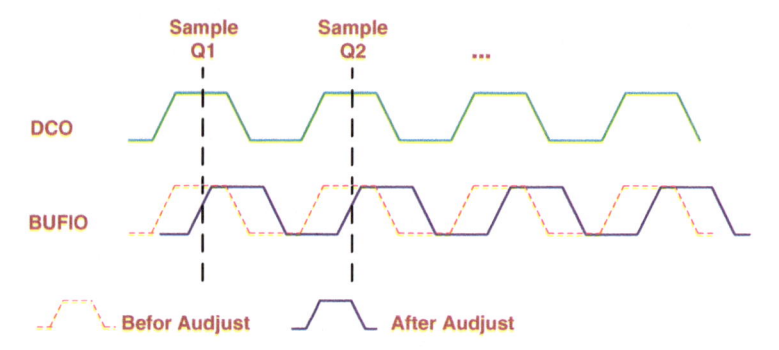

Fig. 5 Sampling to high level

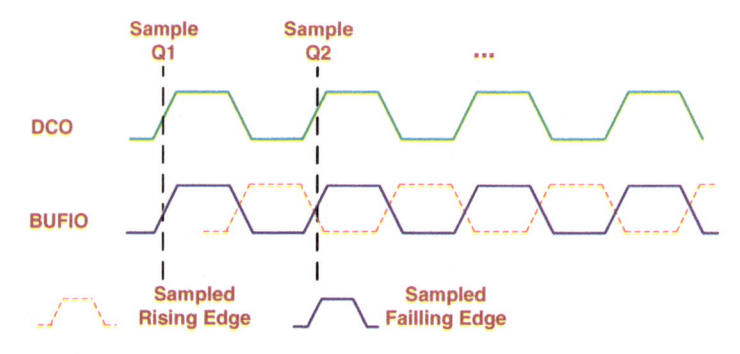

Fig. 6 Sampling to edge

at that time continuously. Repeat testing until the output sampling values of Q1, Q2, Q3, Q4 are all 0 s or all 1 s, then stop to go into the STATE1 or STATE1 for further operation.

STATE4: When the cycle is greater than 4.9 ns and less than 10 ns, it would not yet encountered electrical level conversion of Q1, Q2, Q3, Q4 although the 32 TAPS is full filled, the TAP will go back to 0 to re-start to accumulate automatically, and the

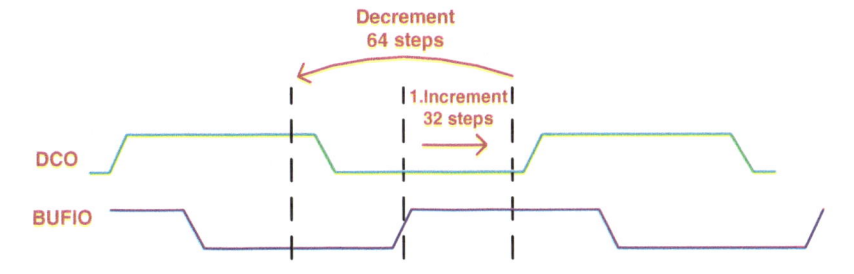

Fig. 7 The level converse when 32 TAP is full filled and went back to 0

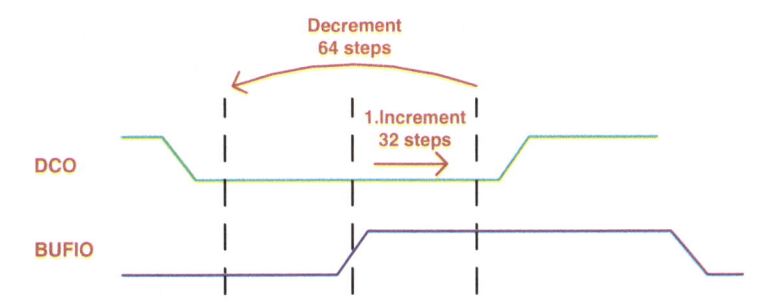

Fig. 8 The level do not converse when 32 TAP is full filled and went back to 0

electrical level will converse at that time. We can accumulate the TAPS which we increase at first. When the level conversion occurs and the value of accumulated register is 32, we can't see this situation as matching to edge acquiescently, instead we should go into STATE1 or STATE2 for a further operation according to the new level (Fig. 7).

STATE5: When the cycle is greater than 10 ns, the electrical level conversion of Q1, Q2, Q3, Q4 will not happen not only when the 32 TAPS is full filled, but also when the TAP go back to 0 to re-start to accumulate automatically. If you start to experience high level, increase the delay TAPS to the maximum. If you start to experience low level, reduce the delay TAPS to the minimum. Because when the cycle is greater than 10 ns, the impact of delay time of lines and devices to the matching of data and DCO is limited, so the approximate treatment is enough (Fig. 8).

3.2 DATA Interface of DCO

After the step 3.1, the sampling clock of ISEDERS of DATA route can use the clock which was adjusted. Because ISEDERS work in the DDR (sampling happen at both positive edge and negative edge) pattern [6], data wide should be set to 6, so to the output wide. Then we should use two ISEDERS cascade. As shown in Fig. 9, it's an

Fig. 9 The combination of output data

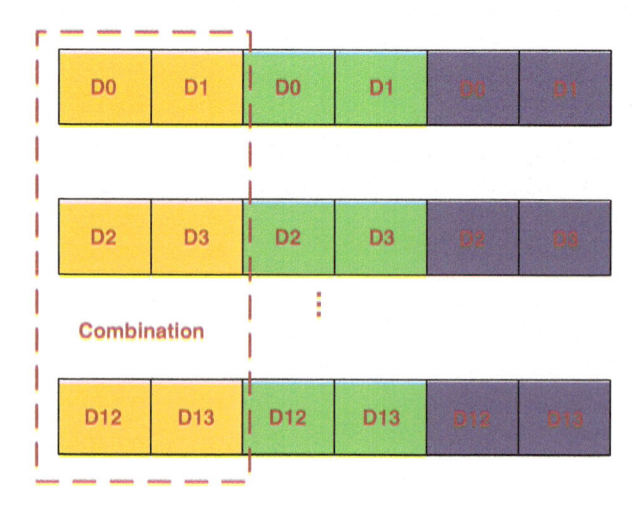

output data of a quarter of the frequency of the cycle, each line include six output data of a DATA route [7]. To combine each line of the two data which are shown in Fig. 9, we will gain a 14-bit data.

4 Output Effect of Experimental Platform

As shown in Fig. 10, before using the matching techniques described in this paper, the output effect is very poor and the wave contains too much ripples almost lose the original appearance of sine wave when putting the normal sine wave to the input port of the ADC. The phase of the DATA and DCO do not match, so the odd and even bit of DATA have reversed. After using adaptive ADC interface matching technique, the output effect is very perfect, as shown in Fig. 11 distinctly.

5 Conclusion

This technical structure and related software were confirmed completely in the real-time spectrum analyzer of Institute of Astronautics & Aeronautics UESTC (University of Electronic Science and Technology of China), which can work stably. The output effect is sound, and the error rate is reduced greatly after the simulation data is converted by the ADC and continued to input into FPGA. And it can run for quite a long time without any mistakes. This method can be used for other general engineering practices, and also it has certain reference value for other types of optimization of FPGA system.

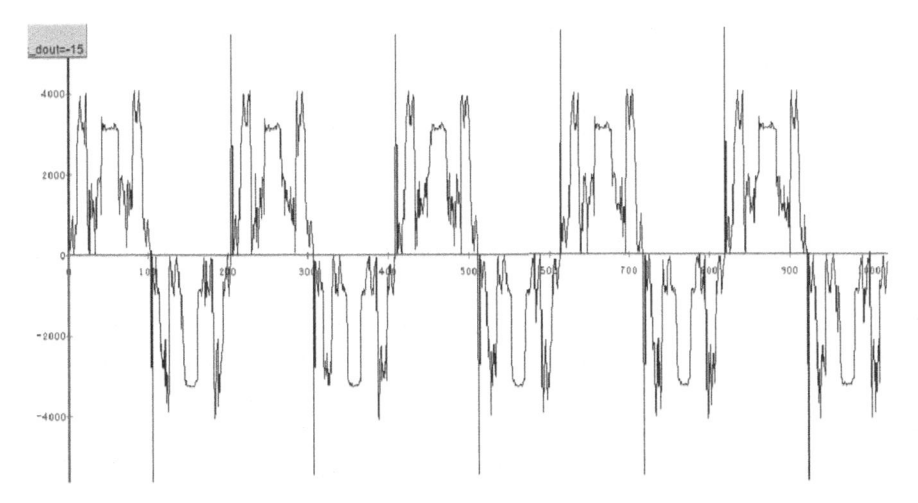

Fig. 10 The output wave before the using of matching techniques

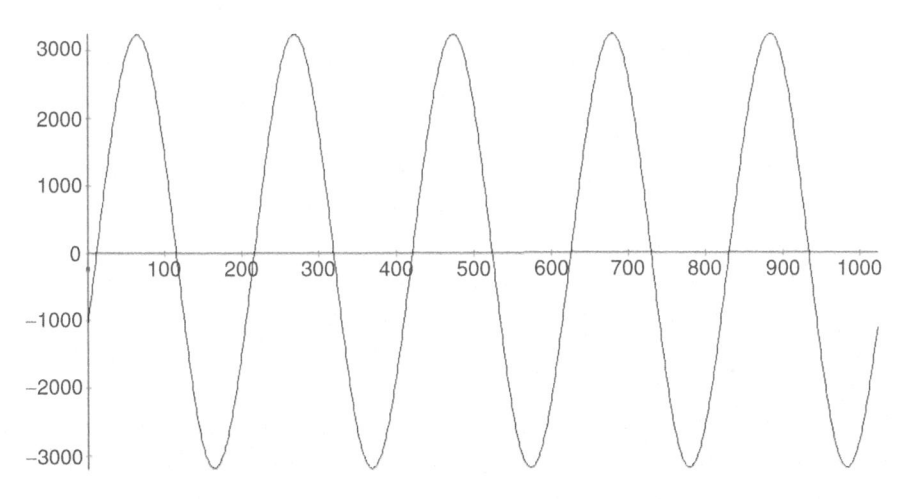

Fig. 11 The output wave after the using of matching techniques

References

1. Betz V (1998) Architecture and CAD for speed and area optimization of FPGAs[M]. Ph.D. thesis, National Library of Canada, University of Toronto
2. Guoxing Zhang, Junfeng Gao (2007) Adaptive synchronizer based on FPGA. Appl Integr Circ 348(06):63–65
3. Wang Chun-zao, Zhang Han-fu, Shi Liang (2007) The optimized design of routing switch and wire segment of FPGA. Electron Pack 52:27–29
4. Defossez M (2008) An interface for Texas instruments analog-to-digital converters with serial LVDS outputs, XAPP866 (v3.0). Available for download from http://www.xilinx.com. 7th Apr 2008

5. Betz V, Rose J (1999) FPGA routing architecture: segmentation and buffering to optimize speed and density[R]. In: FPGA '99 Proceedings of the 1999 ACM/SIGDA seventh international symposium on Field programmable gate arrays, Ann Arbor, Michigan, USA, pp 59–68
6. Luo Zhi nian, Li Si, Zhang Wenjun (2007) Improved frame synchronization algorithm for IEEE 802.11a. Comput Eng Appl 43(32):31–32
7. Wu GuangMing, Chang YaoWen (1999) Quasi-universal switch matrices for FDP design. IEEE Trans Comput 48(10):1107–1109

Power Grid Fundamental Signal Detection Based on an Adaptive Notch Filter

Zhi-xia Zhang, Xin-yu Zhang, Chang-liang Liu, and Tsuyoshi Funaki

Abstract In order to detect the power grid fundamental signal in the polluted environment, a new adaptive notch filter (ANF) algorithm is used in this paper. A prominent advantage of the algorithm is that it does not require a phase-locked loop (PLL) for the synchronization, but it can track the change of fundamental component automatically by adjusting its own parameters, γ and ζ. Empirical results show that the algorithm is simple and can offer a high degree of immunity and insensitivity to voltage mutations, harmonics, three-phase unbalanced and other types of pollution that exist in the grid signal. The ANF algorithm can accurately extract the real-time power grid fundamental signal in the polluted environment.

Keywords ANF • Fundamental signal estimation • Distributed generation (DG)

1 Introduction

With the development of distributed generation technology, the distributed generation technology is one of the important directions of energy technology development in the future. However, some problems of the grid voltage, such as, voltage swells/sags, phase shifts, frequency shifts and three-phase voltage unbalance etc. are often happen for grid-connected. These problems affect the detection of frequency, phase and amplitude of fundamental positive sequence voltage. Therefore, it is of an important significance to implement steady, accurate and real-time extraction of the fundamental component in grid [1–3].

At present, a variety of algorithms to detect the power grid fundamental signal were proposed by many domestic and foreign scholars. These algorithms can be

Z.-x. Zhang (✉) • X.-y. Zhang • C.-l. Liu • T. Funaki
Shenyang Agricultural University, Shenyang, China
e-mail: syzzx7@163.com

mainly summarized into the following three stages. The first detection stage is based on a simple sinusoidal signal model. The primary methods in this stage are voltage zero-crossing method and maximum algorithm. The structural simplicity of this algorithm makes it desirable from the standpoint of implementation in both software and hardware environments. However, these methods ignore all harmonics and noises. Therefore, these two methods cannot be applied to detect the power grid fundamental signal in the polluted environment. The second detection stage is based on a cycle signal model. The primary method in this stage is the discrete Fourier transform (DFT). In fact, grid is in a dynamic state, and sometimes the fundamental frequency is changing. Therefore, there are spectral leakage and fence effect. The last detection stage is based on a complex signal model. The primary methods in this stage are wavelet transform, neural network, PLL based on d-q transformation and Kalman filtering. The characteristic of wavelet transform and neural network methods are large calculative volume and poor real-time. The cutoff frequency of low-pass filter in PLL is set relatively low in order to achieve good filtering, however, the low cutoff frequency will affect the dynamic characteristics of detection system. The Kalman filtering belongs to IIR filter and has unsure group delay, thus it is difficult to accurately calculate the phase angle of phasor [4–6].

This paper presents an ANF algorithm in order to settle this problems which are mentioned above. A prominent advantage of the algorithm is that it does not require a phase-locked loop (PLL) for the synchronization, but it can track the change of fundamental component automatically by adjusting its own parameters, γ and ζ. Therefore, ANF algorithm is real-time and accurate.

2 Synchronization Technique Based on ANF

2.1 ANF

An ideal notch filter whose frequency response is characterized by a unit gain at all frequencies except at a particular frequency, at which its gain is zero. This structure cannot make the particular frequency of signal pass, while the rest of frequency of signal can be passed completely. This particular frequency is called notch frequency. ANF can automatically adjust the notch frequency by tracking the change of signal frequency, which makes it possible that ANF can estimate fundamental component of voltage signal.

Very often, when a sinusoidal signal is contaminated, it can be modeled by [7]

$$y(t) = \sum_{k=1}^{\infty} A_k \sin \phi_k(t) + A_0 + n(t) \qquad (1)$$

Where $f_k = \omega_k t + \varphi_k$ is phase angle of the k, A_k is amplitude of the k (k = 1, 2, ...), A_0 is the dc component, $n(t)$ is the noise. ϕ_k and A_k is typically unknown parameters. Estimating the unknown parameters is the key in this study.

The dynamic behavior of the ANF is characterized by the following set of differential equations:

$$\begin{cases} \ddot{x} + \theta^2 x = 2\zeta\theta e(t) \\ \dot{\theta} = -\gamma x \theta e(t) \\ e(t) = y(t) - \dot{x} \end{cases} \quad (2)$$

Where x is the state variable, θ is the estimated frequency, ζ and γ are adjustable positive parameters that determine the estimation accuracy and the convergence speed of ANF.

For a single sinusoid, $y(t) = A_1 \sin(\omega_1 t + \varphi_1)$. This ANF has a unique periodic orbit located at [2, 8, 9]

$$O = \begin{pmatrix} \bar{x} \\ \dot{\bar{x}} \\ \bar{\theta} \end{pmatrix} = \begin{pmatrix} -A_1 \cos(\omega_1 t + \varphi_1) \\ A_1 \omega_1 \sin(\omega_1 t + \varphi_1) \\ \omega_1 \end{pmatrix} \quad (3)$$

The third entry is the estimated frequency, ω_1. Dynamic equations in (2) are stable, which means that the proposed ANF is stable. We proceed with a discussion on the stability of the Eq. 2, and it can be rewritten as

$$\dot{\theta} \approx -\frac{\gamma}{2\zeta} x^2 (\theta^2 - \omega_1^2) \quad (4)$$

That is, it is stable in the process of approaching periodic orbit O.

2.2 ANF Block Diagram

Figure 1 shows the schematic structure of ANF. At Eqs. 2 and 3, reveals that ANF is composed of adders, multipliers and integrators. The structure has two independent design parameters, γ and ζ. Parameters γ determines the adaptation speed. Parameters ζ determines the depth of the notch. However, increasing one of the parameters will affect the other factor. Therefore, A tradeoff between the accuracy and convergence speed can be carried out by adjusting design parameters, γ and ζ.

Fig. 1 Block diagram of ANF

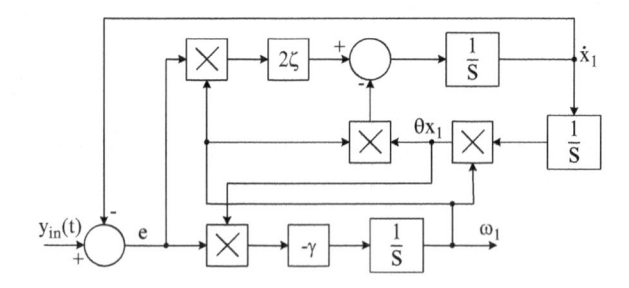

3 Simulation Analyses

Under the Matlab7.6.0/Simulink, that uses the proposed algorithm to establish system simulation model and make an analysis. The parameters of ANF can set to $\zeta = 0.75$ and $\gamma = 800$. The initial condition for the integrator that outputs the frequency, ω, is set to $2\pi 50$ rad/s. The initial conditions for all other integrators are set to zero.

3.1 ANF Performance Evaluation

Figure 2 shows the tracking performance of ANF (there only gives the tracking characteristics of ANF on amplitude). Where, (a), (b), (c) are the input signal, the amplitude of input signal, and the error between input signal and extraction signal, respectively. At $t = 2\ s$, the amplitude of input signal jumps from *1pu* to *0.75pu* [10]. The fast response and accurate performance of ANF can be seen from the Fig. 2.

3.2 Positive and Negative Sequence Extraction

In the abc frame, a three-phase voltage signal $u(t)$ can be decomposed into $u(t) = u^+(t) + u^-(t) + u^0(t)$. Where, $u^+(t)$, $u^-(t)$ and $u^0(t)$ are positive, negative, and zero-sequence components, respectively. Its sequence components can be decomposed as follows [11, 12]:

$$\begin{cases} u^+(t) = T_2 X_1(t) + T_1 X_2(t) \\ u^-(t) = T_2 X_1(t) - T_1 X_2(t) \\ u^0(t) = (I - 2T_2) X_1(t) \end{cases} \quad (5)$$

Fig. 2 Tracking performance of ANF

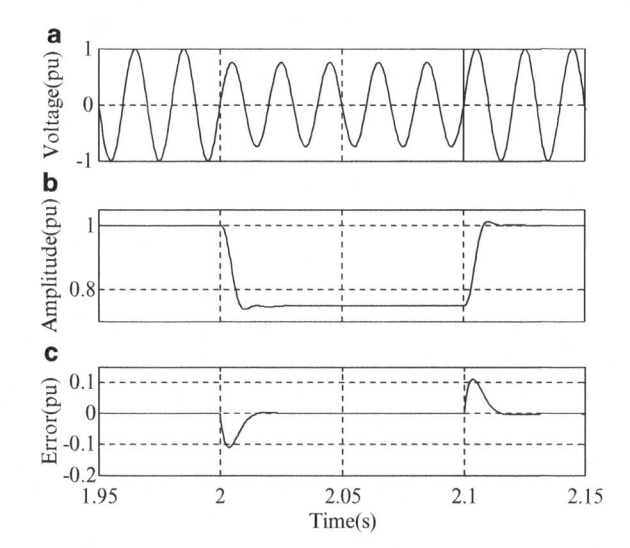

Where, $X_1(t)$ and $X_2(t)$ are the fundamental components of the input signal and its 90° phase shift, respectively. I is a 3×3 identity matrix, T_1 and T_2 are 3×3 matrixes. As follows:

$$T_1 = \frac{1}{2\sqrt{3}} \begin{pmatrix} 0 & 1 & -1 \\ -1 & 0 & 1 \\ 1 & -1 & 0 \end{pmatrix} \quad (6)$$

$$T_2 = \frac{1}{3} \begin{pmatrix} 1 & -0.5 & -0.5 \\ -0.5 & 1 & -0.5 \\ -0.5 & -0.5 & 1 \end{pmatrix} \quad (7)$$

From the above, the positive-sequence and negative sequence extractor unit can be comprised of three ANFs and simple arithmetic operators. ANFs adaptively extract the fundamental voltages and their 90° phase shift. That is, $X_1(t)$ and $X_2(t)$.

The simulation results are shown in Fig. 3.

Under Matlab/Simulink, that uses electronic components to set up a signal model which can produce polluted signal (Fig. 3a). The positive, negative components of fundamental signal can be extracted after those treatments. It can be seen from the Fig. 3 that the system will be able to accurately extract the positive sequence and negative sequence components of fundamental signal after two cycles (Fig. 3b, c).

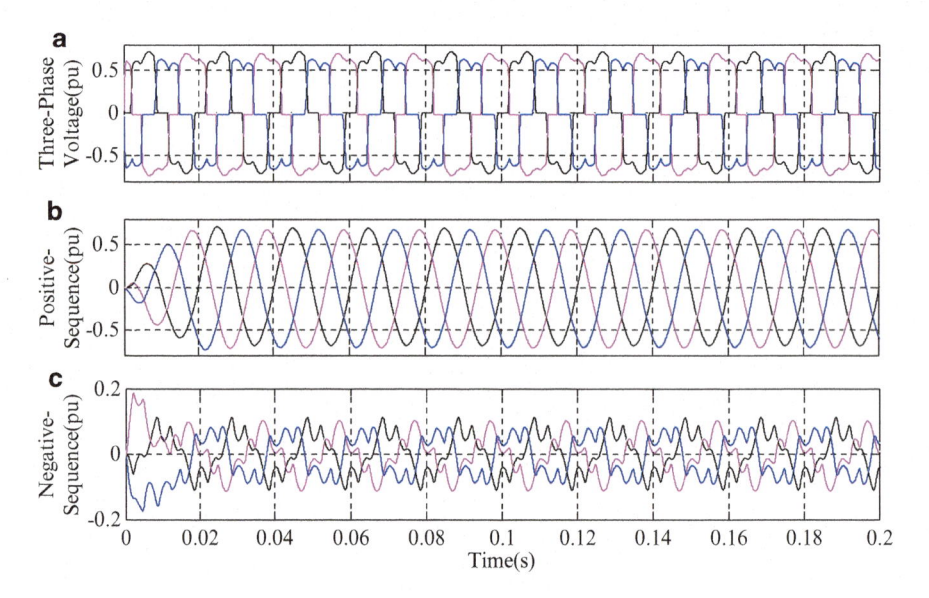

Fig. 3 Extract the fundamental positive, negative sequence component

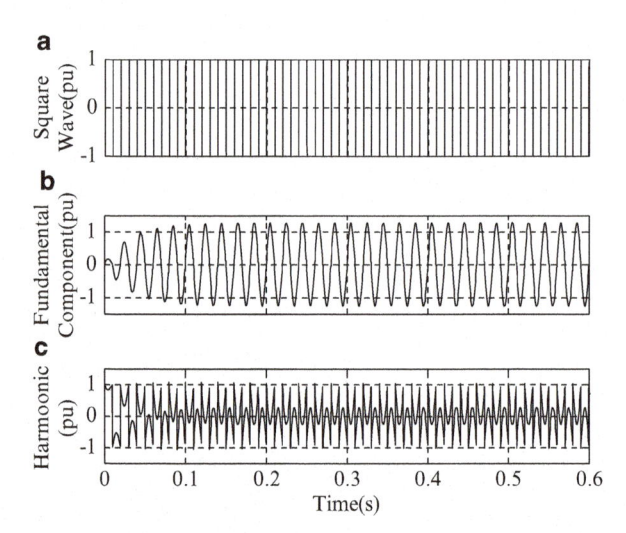

Fig. 4 Extraction of the square wave

3.3 Harmonic Extraction

In this section, a square wave as an example. The fundamental signal can be extracted by ANF, the simulation results are shown in Fig. 4.

Under Matlab/Simulink, a square wave signal can be generated by M-File. Then, a simulation model is established to extract the fundamental signal. It can be seen from the Fig. 4 that the system can effectively isolate from its fundamental and total harmonic distortion components, and output their values after five cycles. The M-File of square wave signal can be rewritten as

```
clear all;
t=(0:0.0001:10)';
for c=1:500
for m=(200*(c-1)+1):(200*c-100); u1(m)=1; end;
for m=(200*(c-1)+1+100):(200*c); u1(m)=-1; end; end;
u1(100001)=1;u=u1';plot(t',u');axis([0 10 -1.5 1.5]);
grid on;
```

4 Conclusion

In this paper, a series of simulation examples illustrated that the rapidity and accuracy of ANF in extracting the fundamental component in the polluted environment. A prominent advantage of the approach is that it does not require a PLL for the synchronization, but it can accurately extract the fundamental component in real-time grid by adjusting its own parameters. In the three-phase unbalanced system, the positive sequence component and negative sequence component can be extracted by ANF, and in the harmonic pollution environment, the fundamental and harmonic distortion components can be extracted by ANF. The principle of this algorithm is reliable and simple, so it has a high practical value.

References

1. Chu Zhao-bi (2009) Adaptive notch filter-based time-frequency analysis of signals in power system[D]. Hefei University of Technology, Hefei
2. Chu Zhao-bi, Zhang Chong-wei, Feng Xiao-ying (2010) Adaptive notch filter-based frequency and amplitude estimation. Acta Automat Sin 36(1):60–66
3. Zou San-hong, Pei Wei, Qi Zhi-ping (2010) Common interface for interconnection between distributed generation and microgrid. Autom Elec Power Syst 34(3):91–95
4. Mai Rui-kun, He Zheng-you, He Wen et al (2010) Adaptive frequency tracking algorithm for power systems. Proc CSEE 30(16):73–78
5. Zhang Shi-ping, Zhao Yong-ping, Zhang Shao-qing et al (2003) A novel approach to measurement of power system frequency using adaptive notch filter. Proc CSEE 23(7):81–83
6. Zhang Zhi-xia, Piao Zai-lin, Guo Dan, Xie Ying (2012) A kind of phase-locked loop for power system. Trans China Electrotech Soc 27(2):250–254, 222
7. Mojiri M, Karimi-Ghartemani M, Bakhshai A (2007) Estimation of power system frequency using an adaptive notch filter. IEEE Trans Instrum Meas 56(6):2470–2477
8. Mojiri M, Bakhshai AR (2007) Estimation of n frequencies using adaptive notch filter. IEEE Trans Circ Syst 54(4):338–342
9. Mojiri M, Bakhshai AR (2004) An adaptive notch filter for frequency estimation of a periodic signal. IEEE Trans Autom Control 49(2):314–318

10. Zhang Bin, Zhang Dong-lai (2011) Adaptive fundamental component extraction and frequency tracking algorithm for power systems. Proc CSEE 31(25):81–89
11. Karimi-Ghartemani M, Iravani MR (2004) A method for synchronization of power electronic converters in polluted and variable-frequency environments. IEEE Trans Power Syst 19(3):1263–1270
12. Du Xiong, Wang Li-ping, Li Shan-hu et al (2012) A voltage synchronization method with adaptive frequency response. Proc CSEE 32(4):115–121

The Precise Electric Energy Measurement Method Based on Modified Compound Simpson Integration

Min Zhang, Huayong Wei, and Weimin Feng

Abstract In time domain the asynchronous sampling is easy realized in hardware, but it leads to measurement error in energy measurement. Modified compound Simpson integral method is proposed to reduce the error. The computing formula is deduced. Ideal signals and distortion signals are simulated. Simulating results show the proposed integral method has the highest precision, compared with previous method, the precision is improved dramatically. The results also show the sampling rate has little influence on the measurement accuracy. So the new algorithm is practical and is easy realized by DSP.

Keywords Asynchronous sampling • Leakage error • Modified compound Simpson integral method • Electric energy measurement

1 Introduction

With the far-ranging use of non-linear components in electric and electronic devices, the electrical harmonic pollution has deteriorated the power quality in electrical power networks significantly. The harmonic distortion of voltage and current greatly affects on the accuracy of electric energy measurement [1]. In order to improve the measurement accuracy, many methods have been proposed. These methods measure the energy in time domain or in frequency domain. In time domain the precise measurement methods are used such as Newton-Cotes integration algorithm [2] and Compound Trapezoidal [3]; the error caused by window

---

M. Zhang (✉) • H. Wei
Xinyang Power Supply Company, Henan, China
e-mail: peidiandianlan@yahoo.cn

W. Feng
College of Electric and Information Engineering, Zhenzhou University of Light Industry, Zhengzhou, China

function is corrected in time domain [4]. The time domain methods are low processing time and easy applicability for digital signal processors, but are very sensitive to the presence of harmonic and inter-harmonic components [4]. The frequency domain methods use DFT to obtain the harmonics amplitude and phase angle, then the power is calculated according to harmonics respectively. The data is smoothed by window function, then reduce the leakage effects [5]; an overview of active power measurement has been made and a new method is proposed infrequency domain [6]; a new decomposition of voltage, current and power is presented to improve the accuracy [7]. In order to improve the accuracy of electric energy measurements of harmonic in the complex industrial environment, the wavelet packets decomposition and reconstruction algorithm are employed [8].

At present, the electric energy measurements are mainly based on the synchronized sampling of signals. The synchronized sampling is realized by using PLL (Phase Locked Loop) [4]. However, synchronized sampling is not normally provided for in practical applications. So the electric energy measurement based on synchronous sampling has larger measurement error. In order to improve the measurement precision, some error compensation algorithms are presented according to the asynchronous sampling [3, 5, 9]. In this paper, Modified Compound Simpson Integration algorithm is proposed to improve the energy metering accuracy.

2 Algorithm Principle

A periodic signal $f(t)$, T is the fundamental cycle time and the fundamental frequency is $f = 1/T$. According to the Nyquist sampling theorem, if the maximum frequency component of the signal is f_{max}, the sampling rate f_s must be higher than $2f_{max}$. N points are sampled in one fundamental cycle, so the interval of ideal synchronous sampling $\Delta T = T/N$. But the synchronous samplings are difficult to realize because of harmonics and noise pollution. If the sampling rate f_s is fixed, so sampling time is $T_s = 1/f_s$. The ratio of the fundamental cycle T and T_s can be expressed as: $T = T_s(N + \Delta)$ ($0 \leq \Delta < 1$). Where N is integers and Δ is a fraction. In order to improve the measurement accuracy, Δ must be calculated precisely.

An asynchronous sampling is shown in Fig. 1. $f_0(N)$ and $f_1(N)$ are the last sampling values of the first and second cycle respectively; $f_1(0)$ and $f_2(0)$ are the first sampling values of the first and second cycle respectively. From the picture can be seen: $f_0(N)$ and $f_1(0)$ are opposite sign, so $f_0(N)^*f_1(0) \leq 0$. There is a zero between $f_0(N)$ and $f_1(0)$. Likewise, $f_2(0)^*f_1(N) \leq 0$, there is always a zero between $f_1(N)$ and $f_2(0)$. According to the value of several points, the value of Δ can be calculated.

If the sampling frequency f_s is high, the sampling time-interval T_s between $f_0(N)$ and $f_1(0)$ is very small, the connection of the two points can be approximated as a straight line. In the second cycle, according to linear interpolation, the time delay Y_s can be calculated which is the time interval between the starting point of this cycle and $f_1(0)$. Y_s can be expressed as:

Fig. 1 Schematic of sampling error of asynchronous

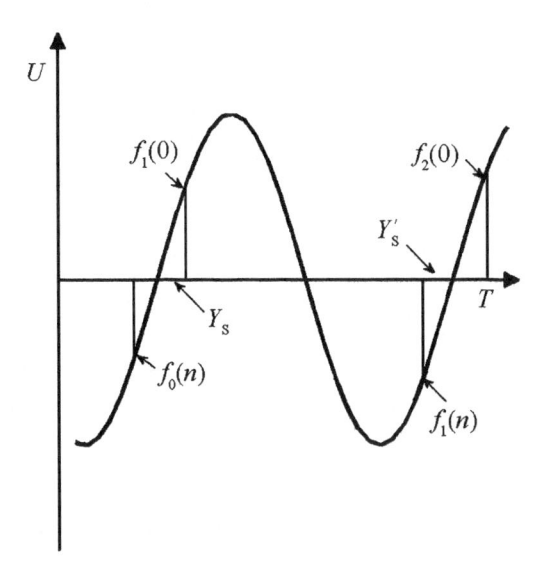

$$Y_s = \Delta_1 {}^*T_s = \frac{f_1(0)}{f_1(0) - f_0(N)} {}^*T_s \tag{1}$$

Similarly, the time interval Y_s' between $f_1(N)$ and the ending point of this cycle can be expressed as:

$$Y_s' = \Delta_2 {}^*T_s = \frac{-f_1(N)}{f_2(0) - f_1(N)} {}^*T_s \tag{2}$$

So: $\Delta = \Delta_1 + \Delta_2$.

In order to improve the accuracy, The Compound Trapezoidal is used in [3], the formula has been given in the paper, and we do not write it here.

The compound Simpson integration is also a useful method in improving measurement accuracy, it can be expressed as:

$$S_n = \frac{1}{T}\int_0^T f(t)dt = \frac{1}{T}\sum_{k=0}^{n/2-1}\int_{2k^*\Delta T}^{(2k+2)\Delta T} f(t)dt = \frac{1}{T}\sum_{k=0}^{n/2-1}\frac{1}{3}(f(2k^*\Delta T) + 4f((2k+1)^*\Delta T)$$

$$+ f((2k+2)^*\Delta T))^*\Delta T \approx \frac{1}{3^*n}[f(0) + f(n^*\Delta T) + 2\sum_{k=1}^{n/2-1} f(2k^*\Delta T)$$

$$+ 4\sum_{k=0}^{n/2-1} f((2i+1)^*\Delta T)] \tag{3}$$

In practice, the frequency of power system is always fluctuated with time. If the sampling frequency is fixed, the signals are asynchronous sampling. Then the fraction Δ will bring leakage error. In order to reduce the leakage error, the modified compound Simpson integration is presented. It is shown as below:

$$\begin{aligned} S_{nm} &= \frac{1}{T}\int_0^T f(t)dt = \frac{1}{T}\left[\sum_{k=0}^{N/2-1}\int_{2k^*T_s}^{(2k+2)T_s} f(t)dt + \int_{N^*T_s}^{(N+\Delta)^*T_s} f(t)dt\right] \\ &\approx \frac{1}{T}\left[\frac{1}{6}\sum_{k=0}^{N/2-1}(f(2k^*T_s)+4^*f((2k+1)^*T_s)+f((2k+2)^*T_s))^*2^*T_s \right. \\ &\quad \left. +\frac{1}{2}{}^*\Delta^*(f(N^*T_s)+f((N+\Delta)^*T_s)^*T_s\right] \\ &= \frac{1}{6^*(N+\Delta)}\left[2^*(f(0)+f(N^*T_s)+2\sum_{k=1}^{N/2-1} f(2^*k^*T_s) \right. \\ &\quad \left. +4\sum_{k=0}^{N/2-1} f((2^*k+1)^*T_s))+3^*\Delta^*(f(N^*T_s)+f(0))\right] \end{aligned} \qquad (4)$$

3 Simulation and Result Analysis

The input sinusoidal voltage and current are:

$$u(t) = \sqrt{2}V_p \sin(2\pi ft) \qquad (5a)$$

$$i(t) = \sqrt{2}I_p \sin(2\pi ft + \alpha) \qquad (5b)$$

The average (active) power is:

$$P = \frac{1}{T}\int_0^T u(t)^*i(t)dt = \frac{V_p{}^*I_p}{T}\int_0^T [\cos\alpha - \cos(4\pi ft - \alpha)]dt \qquad (6)$$

The active power is calculated by the three methods (Compound Trapezoidal integration, Compound Simpson integration and Modified compound Simpson integration are aforementioned). All the simulating programs are written in C# language and are simulated in personal computer. The simulating time is 10 cycles.

Table 1 The simulating results of sinusoidal signal (f = 49.8 Hz)

Delay Y_s	Compound Trapezoidal integration (%)	Compound Simpson integration (%)	Modified compound Simpson integration (%)
0	0.418582	0.418916	8.806647×10^{-5}
$0.3\ T_s$	0.398320	0.398638	8.380354×10^{-5}
$0.5\ T_s$	0.384621	0.384928	8.092129×10^{-5}

3.1 Sinusoidal Signal

Given the input voltage $u(t) = 220^* \sin(2\pi ft)$.

The current waveform $i(t) = 20^* \sin(2\pi ft + \theta)$.

Where f is the fundamental frequency, $f = 49.8 Hz$, $\theta = \pi/3$. The sampling frequency is $f_s = 6400 Hz$. The input average power is 1,100 W per cycle.

Table 1 shows the results of the three different methods. From the table we can see Compound Trapezoidal integration and Compound Simpson integration almost have the same accuracy. But compared with the Modified compound Simpson integration, these two methods bring larger error, the error rate maintains at 0.385 %. While the results of proposed method show that the computation errors are very small and are all under 10^{-6}. The results illustrate the proposed method improve the precision dramatically. It means that the method can be used to design the high-precision power measurement equipment. From the table we can see also, if Ys are taken into account (Ys is the interval between the sampling start time and the origination time of the signals), the computation errors of these three methods have little changed and are at least of a similar order of magnitude. So, when discussing these three methods, Ys can be taken no account of.

3.2 Non-sinusoidal Signal

In the electrical power system, the input voltage and current generally include some harmonic waves. So the input voltage and current waveforms are non-sinusoidal and are same as [3]:

$$u(t) = 220^* \sin(2\pi ft) - 20^* \sin\left(2\pi f + \frac{\pi}{3}\right) + 2.5^* \sin\left(6\pi f + \frac{\pi}{3}\right)$$
$$+ 2.5^* \sin\left(10\pi f + \frac{\pi}{3}\right) + 2.5^* \sin\left(14\pi f + \frac{\pi}{6}\right) \quad (7a)$$

$$i(t) = 20^* \sin\left(2\pi f + \frac{\pi}{3}\right) + 2.5^* \sin\left(6\pi f + \frac{\pi}{6}\right)$$
$$+ 2.5^* \sin(10\pi f) + 2.5^* \sin\left(14\pi f + \frac{\pi}{6}\right) \quad (7b)$$

Table 2 The simulating results of non-sinusoidal signal (f = 49.8 Hz)

Delay Y_s	Compound Trapezoidal integration (%)	Compound Simpson integration (%)	Modified compound Simpson integration (%)
0	0.551523	0.553543	5.452108×10^{-4}
0.3 Ts	0.526144	0.528120	5.331358×10^{-4}
0.5 Ts	0.508808	0.510752	5.246681×10^{-4}

Table 3 The simulating results of non-sinusoidal signal (f = 50.2 Hz, Ys = 0.3 Ts)

Sampling rate	Compound Trapezoidal integration(%)	Compound Simpson integration(%)	Modified compound Simpson integration(%)
1.6 kHz	0.612635	0.684873	7.6663699×10^{-3}
3.2 kHz	0.580093	0.589462	7.7982671×10^{-4}
6.4 kHz	0.529072	0.524328	4.9546879×10^{-4}

Where f is the fundamental frequency and f_s is the sampling frequency, $f_s = 6400 Hz$. The real input average power is 909.375 w. The simulating results are shown in Table 2.

The simulation results of Table 2 are similar to Table 1. The Modified compound Simpson integration is more precise than other two methods, while Compound Trapezoidal integration and Compound Simpson integration have the same accuracy. But compared with the results of Table 1, the harmonics have great affluence on the measurement accuracy of the modified compound Simpson integration, an order-of-magnitude has been decreased. Because the Compound Trapezoidal integration and Compound Simpson integration have lower precision, the results of the two methods have been affected little. Similarly to ideal signals, the delay Ys has little affected on measurement.

The signals have been simulated with three different sampling rates, the signal frequency is $f = 50.2$ Hz. All the results are shown in Table 3. We can see that the bigger the sampling rate is, the higher the precision of all integration methods is. Among these integrations, the precision of Modified Compound Simpson integration is highest. We can also see that at sampling rate 3.2 and 6.4 kHz, the error of the proposed method is below 10^{-6}. The results illustrate this method is fitted in lower sampling rate.

In the last, the active power of non-ideal signals is test in a DSP Development Board, all the sampling date are input the DSP and calculate by the Modified Compound Simpson Integration. The experimental results illustrate the proposed method is effective.

4 Conclusion

The modified compound Simpson integration has been proposed to improve measurement accuracy under asynchronous sampling. The computing formulas are deduced. Ideal signals and distortion signals are simulated and the results are compared with other methods. The results show the proposed method has the highest precision in different frequency of signals; the delay sampling time has little influence on measurement accuracy. Different sampling rate has also been simulated; the results show when the sampling rate is high to some extent, increasing the sampling rate has little influence on the accuracy. Simulation and experimental results show that the modified compound Simpson integration improves the accuracy of energy measurement dramatically. So the proposed method has the characteristics of generality, less calculation, high velocity and high efficiency and can be used in DSP easily.

References

1. Kusˇljevic´ MD, Tomic´ JJ, Marcˇetic DP (2009) Active power measurement algorithm for power system signals under non-sinusoidal conditions and wide-range frequency deviations. IET Gener Transm Distrib 3(1):57–65
2. Yi LQ, Song HN (2010) The application of compound Newton-cotes integral algorithm to energy measurement. Electr Meas Instrum 47(10):15–19 (In Chinese)
3. Feng WM, Shi BS, Wu ZJ (2011) The electric energy measurement method based on modified trapezoidal integral. In: International conference on electronics, communications and control, Ningbo, China, pp 2914–2917
4. Novotny M, Sedlacek M (2009) Measurement of active power by time-domain signal processing. Measurement 42(8):1139–1152
5. Agrez D (2008) Power measurement in the non-coherent sampling. Measurement 41:230–235
6. Sedlacek M, Stoudek Z (2011) Active power measurements–an overview and a comparison of DSP algorithms by noncoherent sampling. Metrol Meas Syst XVIII(2):173–184
7. Ortiz A, Ma˜nana M, Renedo CJ (2011) A new approach to frequency domain power measurement based on distortion responsibility. Electr Power Syst Res 81:202–208
8. Teng Zhaosheng, Luo Zhikun, Sun Chuanqi et al (2010) Harmonic energy measurement based on wavelet packet decomposition and reconstruction algorithm. Trans China Electrotech Soc 25 (8):200–206 (In Chinese)
9. Nyarko D, Stromsmoe KA (1991) Analysis of truncation errors in asynchronous sampling of periodic signals. In: Proceedings of the 34th Midwest symposium on circuits and systems, Monterey, vol 2, pp 1117–1120

A Frequency Reconfigurable Microstrip Patch Antenna

Yong Cheng, ZhenYa Wang, XuWen Liu, and HongBo Zhu

Abstract A novel microstrip antenna with the frequency reconfiguration characteristic is proposed in this article. A U-slot has been introduced in the square patch. The slot is switched on and off by using three PIN diodes, which realizes the frequency reconfiguration characteristics. The antenna has been studied with Zeland's IE3D, simulated return loss and radiation results are proposed in this article. As the results show, from 3.85 to 7.05 GHz, the antenna can operate at six different frequencies with similar radiation patterns. The antenna has 2 dB gain flatness with the maximum gain being 5 dBi over the whole range. The conclusion that radiation performance within the operating bandwidth of the antenna is stable can be obtained.

Keywords Frequency • Reconfigurable • Antenna microstrip antenna • PIN diode

1 Introduction

Frequency reconfigurable antennas have the advantages of compact size, multiband capability. As the same time, the shape of the radiation patterns of the antennas must maintain unchanged when the operating frequencies are switched from one band to the other. Various methods have been reported in the literature to achieve reconfigurable antennas such as using PIN switches tuning the antenna [1], varying the ground plane electrical length [2] supporting a patch antenna or changing the

---

Y. Cheng (✉) • Z. Wang • X. Liu • H. Zhu
College of Electronics Science and Engineering, Jiangsu Key Laboratory of Wireless Communications, Nanjing University of Posts and Telecommunications, Nanjing, China
e-mail: chengy@njupt.edu.cn

induced electric current distribution [3]. The reconfigurable operating frequencies [4] obtained by switching different feeding location. Several reconfigurable antenna designs [5, 6] has been reported for the flexibility of integrating electronic switches.

In this article, a frequency reconfigurable microstrip patch antenna is investigated. The reconfiguration has been carried by etched a U-slot into the patch. The frequency characteristic has been realized by switching three PIN diodes on the slot on and off configurations. The geometrics of the antenna and the results will be explained in next sections. The full-wave electromagnetic simulation and analysis for the proposed antenna has been performed using Zeland's IE3D, which is based on the Method of Moments (MoM). Return loss, radiation pattern and the gain of the antenna are simulated and the results are presented. The antenna shows similar pattern and flat gain at different operating frequencies.

2 Design of the Antenna

The geometry and coordination of the antenna is shown in Fig. 1. It is a microstrip patch antenna with a U-slot. The antenna has been designed on a PCB with the relative permittivity of the dielectric substrate is 2.2, loss tanδ is 0.001, thickness h = 3 mm and surface area is A × B. To obtain the configurable frequency characteristic, Three PIN diodes has been used as switches. PIN diode1, 2 and

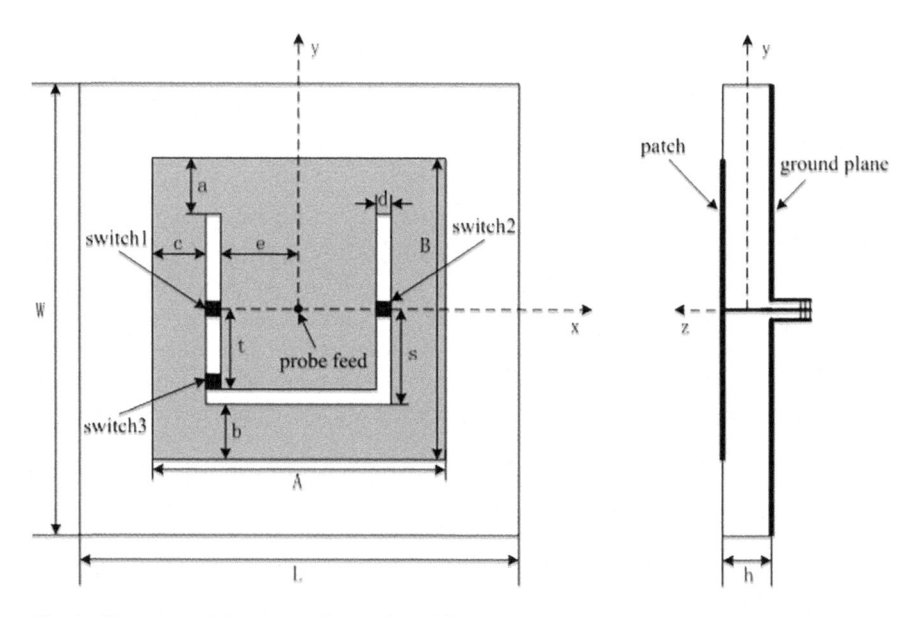

Fig. 1 Geometry of the proposed reconfigurable antenna

Table 1 Dimension parameters (mm)

L	30	c	4
W	30	d	0.5
A	20	e	5.5
B	20	t	5.5
a	4	s	6
b	4	h	3

PIN diode3 have been inserted the gap between the side and the center part of the patch. By switching the diodes on and off, the resonating frequencies of the antenna have been controlled. A coaxial probe is used to feed the antenna at the center of the patch. Detail dimension parameters of the antenna are shown in Table 1.

3 Results of Simulation

The return loss of the proposed antenna has been simulated with IE3D at five switch configurations as shown in Table 2. The simulation results for different cases of switches combinations is shown in Fig. 2

From Fig. 2, the operating frequency have been controlled by the switches configurations. At case1, the three switches are all on, the resonating frequency of the antenna is 5.05 GHz; When switch1 is on while the others are off, the resonating frequency of the antenna is 5.5 GHz; When switch1 and switch2 are both on, switch3 is off, the antenna resonates at 3.85 and 6.1 GHz;When switch1 and switch3 are both on, switch2 is off, the antenna resonates at 5.95 GHz; When the three switches are all off, the antenna operates at 7.05 GHz. In the above six bands, the return loss of the antenna is less 15 dB as shown in Table 2.

Then, the E-plane and H-plane radiation patterns of the antenna at six frequencies have been simulated and shown in Figs. 3, 4, 5, 6, 7 and 8, respectively. As the results show, the radiation patterns of the antenna are similar at the six frequencies. That performance is very important for a frequency reconfigurable antenna.

Finally, the maximum gain of the proposed antenna also has been simulated from 3.85 to 7.05 GHz, and the result is shown in Fig. 9. It is seen that the average maximum gain of the antenna is about 5 dBi, and the antenna has 2 dB gain flatness from 3.85 to 7.05 GHz. From the simulated results of the radiation patterns and gain, it can be obtained the conclusion that the proposed antenna has stable performance within the whole operating frequency range.

Table 2 Operating frequency bands and bandwidth for different cases

Case	Switch1	Switch2	Switch3	Center frequency (GHz)	Bandwidth
Case1	Off	Off	Off	5.05	4.952–5.115(3.24 %)
Case2	On	Off	Off	5.5	5.405–5.572(3.04 %)
Case3	On	On	Off	3.85/6.1	3.844–3.895(1.32 %)/ 6.06–6.147(1.43 %)
Case4	On	Off	On	5.95	5.882–6.024(2.39 %)
Case5	On	On	On	7.05	6.954–7.115(2.29 %)

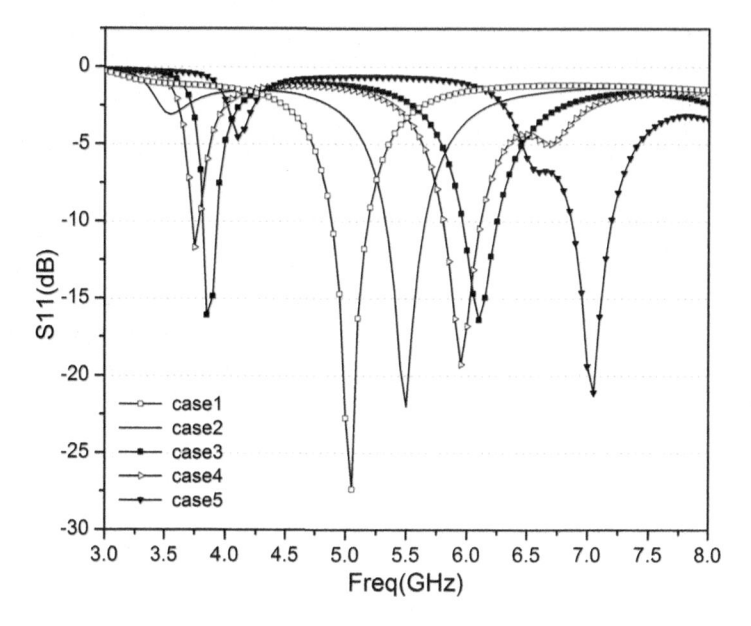

Fig. 2 S_{11} for different cases

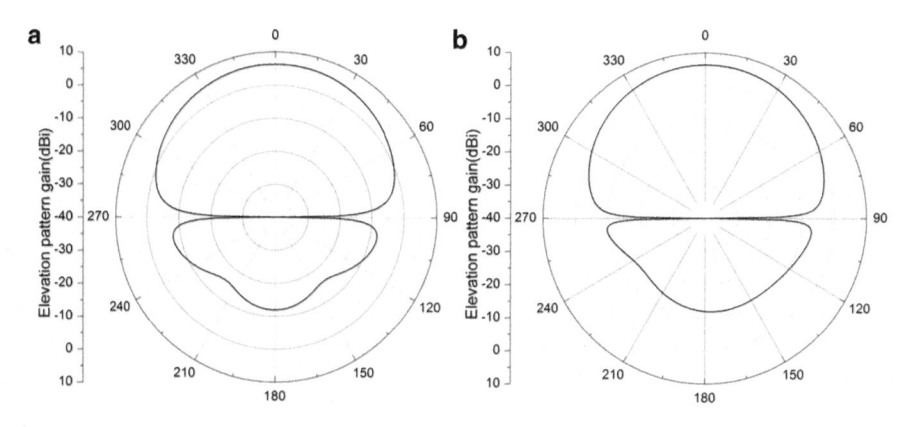

Fig. 3 Radiation pattern at 3.85 GHz (**a**) H plane (**b**) E plane

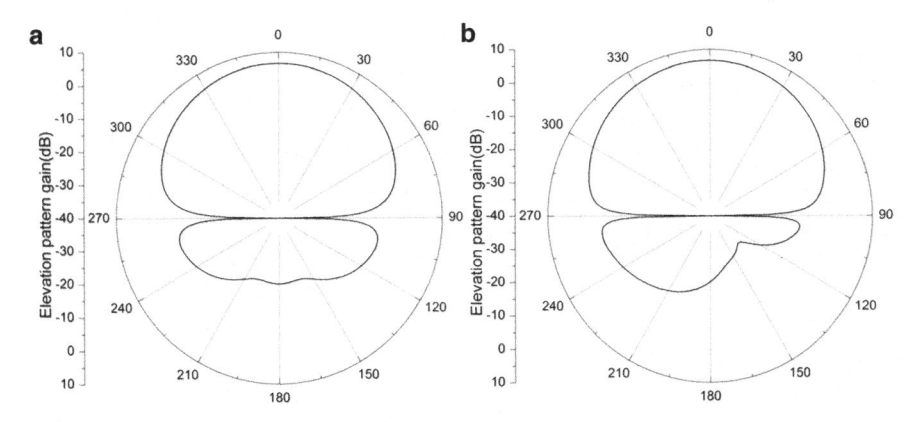

Fig. 4 Radiation pattern at 5.05 GHz (**a**) H plane (**b**) E plane

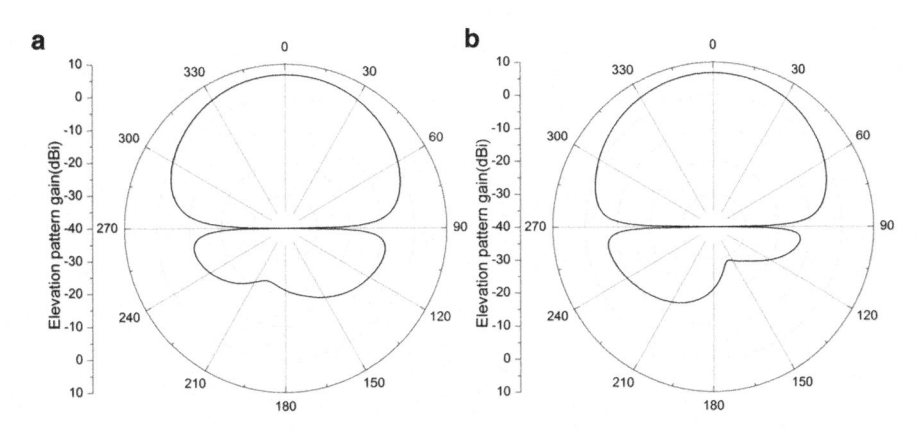

Fig. 5 Radiation pattern at 5.5 GHz (**a**) H plane (**b**) E plane

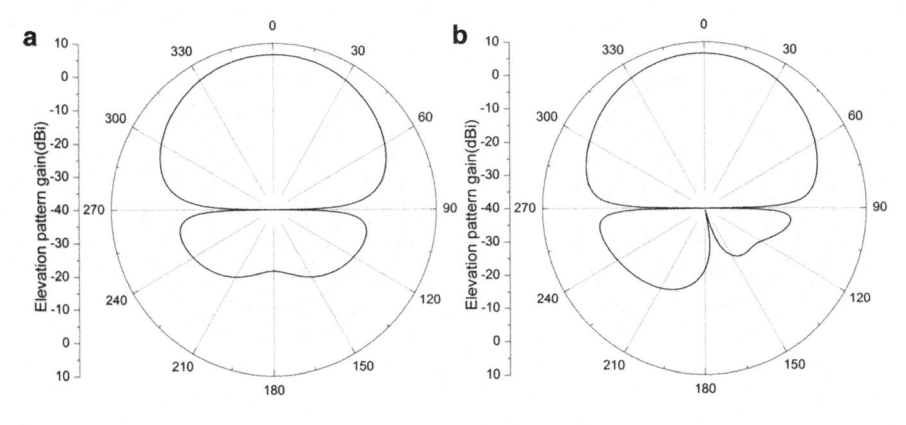

Fig. 6 Radiation pattern at 5.95 GHz (**a**) H plane (**b**) E plane

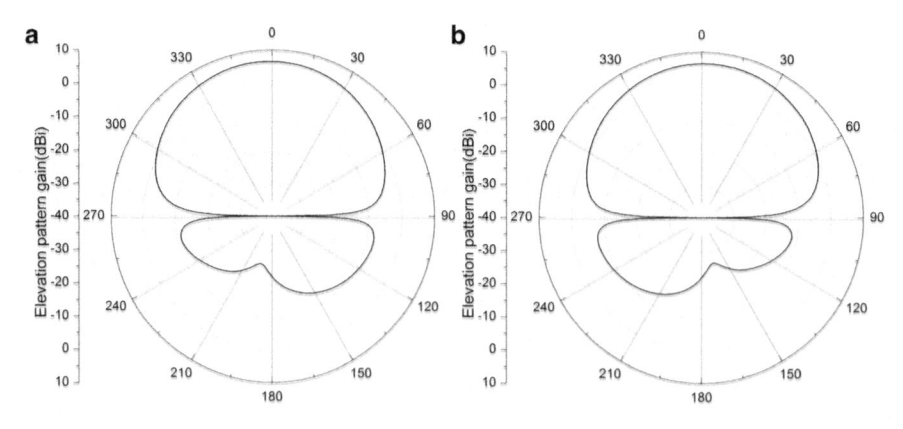

Fig. 7 Radiation pattern at 6.1 GHz (**a**) H plane (**b**) E plane

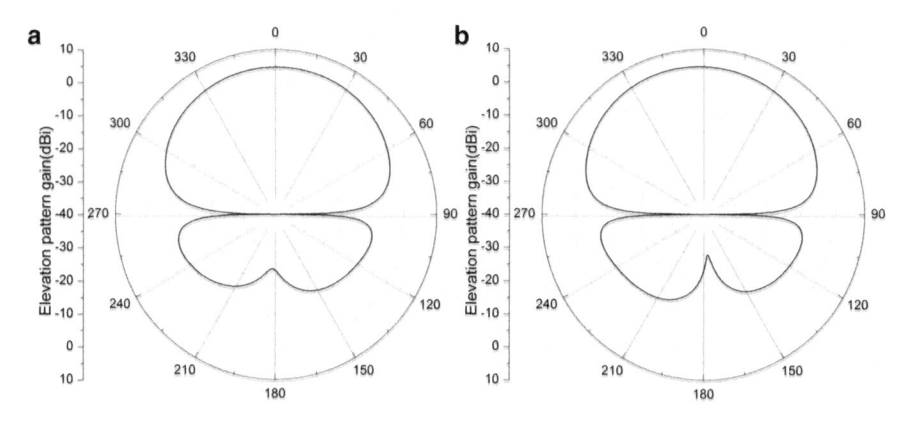

Fig. 8 Radiation pattern at 7.05 GHz (**a**) H plane (**b**) E plane

4 Conclusion

Design concept of microstrip patch etched U- slot antenna with three PIN diodes switches has been proposed in this article. The radiation performances of the antenna have been studied by using EM software and presented in this paper. As the results show that the proposed antenna has advantages such as reconfigurable frequency, stable radiation patterns, flat gain, compact size, etc., so it can be used for other wireless communications system.

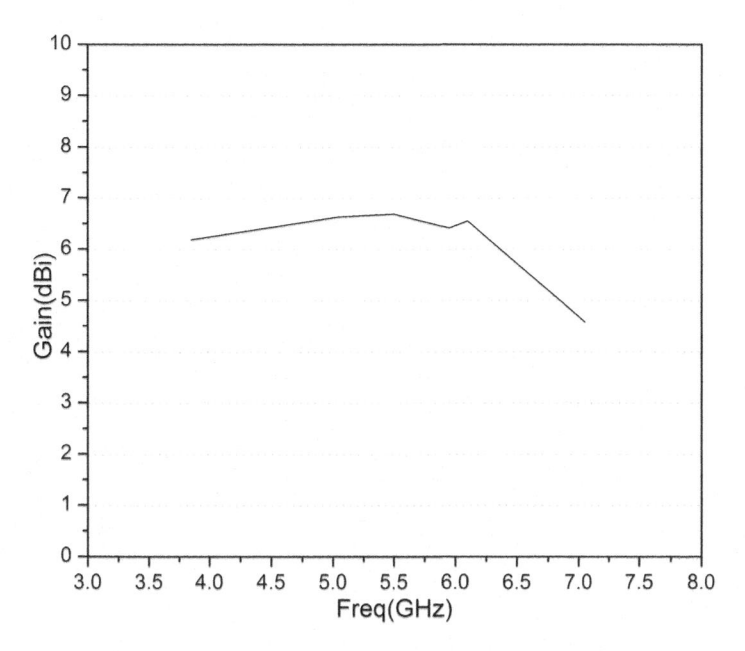

Fig. 9 Gain of proposed antenna

Acknowledgements This work was supported by National Science and Technology Major Project under granted no. 2012ZX03001028-005 and Research Project of Nanjing University of Posts and Telecommunications (208035)

References

1. Sheta AF, Mahmoud SF (2008) A widely tunable compact patch antenna. IEEE Anten Wirel Propag Lett 7:40–42
2. Byun SB, Lee JA, Lim JH, Yun TY (2007) Reconfigurable ground-slotted patch antenna using PIN diode switching. ETRI J 29(6):832–834
3. Lai ML, Wu TY, Hsieh JC, Wang CH, Jeng SK (2009) Design of reconfigurable antennas based on an L-shape slot and PIN diodes for compact wireless devices. IET Microw Anten Propag 3(1):47–54
4. Mak ACK, Rowel CR, Murch RD, Mak CL (2007) Reconfigurable multiband antenna design for wireless communication devices. IEEE Trans Anten Propag 55(7):1919–1928
5. Peroulis D, Sarabandi K, Katehi LPB (2005) Design of reconfigurable slot antennas. IEEE Trans Anten Propag 53(2):645–654
6. Li H, Xiong J, Yu Y, He S (2010) A simple compact reconfigurable slot antenna with a very wide tuning range. IEEE Trans Anten Propag 58(11):3725–3728

Electricity Consumption Prediction Based on Non-stationary Time Series GM (1, 1) Model and Its Application in Power Engineering

Xiaojia Wang

Abstract The construction of smart grid will comprehensively enhance the intelligent level of every step in the power grid of our country. The data prediction ability determines the quality of smart grid. This paper addresses situations in which the prediction accuracy of the Grey Model (GM (1, 1) model) is high for non-negative smooth monotonic sequences but inadequately low for non-stationary sequences, and isolates the trending sequence from the non-stationary time series using a numerical filtering algorithm, which is then used to make predictions. Numerical examples demonstrate that this method can improve the prediction accuracy of the GM (1, 1) model.

Keywords Non-stationary time series • Numerical filtering algorithm • GM (1, 1) model • Prediction

1 Introduction

With the continuous progress of social economy, power industry has been developing rapidly, and the network scale of power grid system is expanding constantly. On the basis of conventional power grid, the concept of smart grid is presented. Smart grid is based on physical power grid and implements optimal allocation of electric power by applying advanced modern technology in various fields.

Data prediction ability determines the quality of smart grid. As for electricity consumption, low prediction will cause power cut due to lack of allocated electricity, while high prediction will bring unnecessary generation cost and energy waste.

X. Wang (✉)
China Key Lab of Process Optimization and Intelligent Decision-making, Hefei University of Technology, Hefei, China
e-mail: tonysun800@sina.com

Therefore, it is essential to predict the electricity consumption accurately. One of the commonest electricity consumption prediction models is grey model- GM (1, 1) [1]. GM (1, 1) model can play a greater role in data forecast of smart grid. By use of its theory, prediction data accuracy can be improved to meet the requirement for data with high quality in smart grid.

After 20 years of development, grey system theory has been widely applied in many areas, including social science and economics. Deng [1]created the grey system theory and system described the principle of grey system, the applications of grey system in many different fields, such as science and economy; Wang, Yang and Wang [2]used cubic spline formula to improve the background value, and constructed a novel grey forecasting model, they used this new model to forecast electricity consumption and obtained high predict accuracy; Hsu and Wang [3, 4] applied the grey prediction model to the global integrated circuit industry and obtained a good prediction effect; Shen, Chung and Chen [5] introduced a novel application of grey system theory to information security, expanded the application field of grey system; Chang and Tsai [6] used neural network adaptation to support vector regression grey model, obtained effect forecasting results, the grey system and artificial intelligence method are combined well; Chen [7] combined the grey system with Bernoulli model, constructed a new grey forecasting model NGBM (1, 1), use this NGBM (1, 1) model to forecast the foreign exchange rates of Taiwan's major trading partners, and receive good effect; Huang [8] use a hybrid grey model to forecast the stock market also achieved good results. But the GM (1, 1) model is ineffective in predicting non-stationary time series, however, mainly because the sequence computed by the GM (1, 1) model is monotonic and because the inverted sequence is also monotonic. Based on these previous studies, this paper aims to find the appropriate treatment to turn a non-stationary sequence into a sequence suitable for the GM (1, 1) model and then to create the GM (1, 1) model to improve modelling and prediction accuracy.

The remainder of the paper is organized as follows: Sect. 2 present some preliminaries of this work, and in Sect. 3 we give the main results of construction process of the prediction model, the data simulation and accuracy comparison are in Sect. 4, finally, we make a conclusion for this paper in Sect. 5.

2 Preliminaries

In this section, we firstly give some preliminaries.

Definition 2.1 Given the time series $\{x_t\}$, $t = 1, 2, \cdots, n$, if

$$Bx_t \triangleq x_{t-1} \qquad (1)$$

then B is denoted as the backward shift operator of the time series $\{x_t\}$. B^{-1} is denoted as forward shift operator.

Definition 2.2 Compute the d-order differential for non-stationary time series $\{x_t\}$, $(t = 1, 2, \cdots, n)$, which results in the following differential sequence:

$$\nabla^d x_t = (1-B)^d x_t = \sum_{i=0}^{d}(-1)^i C_d^i x_{t-i}, \ t = 1, 2, \cdots, n, \ t \geq d+1 \quad (2)$$

C_d^i denotes the coefficients of the binomial $(1-B)^d$, and there exist i combinations of set d.

$$C_d^i = \frac{d!}{i!(d-i)!} \quad (3)$$

Definition 2.3 If B is the backward shift operator of time series $\{x_t\}$, then the following expression is the binomial of B:

$$F(B) = \sum_{j=0}^{n}(-1)^j C_d^j (B^j + B^{-j}), \ j = 0, 1, \cdots n \quad (4)$$

3 Main Results

If $\{x_t\}$ is a known non-stationary sequence according to the following numerical filtering algorithm:

$$x_t = T_t + S_t + \varepsilon_t, \ t = 1, 2, \cdots, n \quad (5)$$

S_t represents the periodic sequence (i.e., the seasonal component, where the specified interval is 12) denoted as $\{S_t\}$. T_t represents the aggregation of all trends (i.e., the trend component) other than $\{S_t\}$, which is denoted as $\{T_t\}$. ε_t represents the stationary random sequence (i.e., the irregular component), which is denoted as $\{\varepsilon_t\}$.

Rewriting Eq. 6 results in the following:

$$\varepsilon_t = x_t - T_t + S_t, \ t = 1, 2, \cdots, n \quad (6)$$

If the periodic sequence $\{S_t\}$ and the trend aggregation sequence $\{T_t\}$ are decomposed, we can eventually compute the trending stationary random sequence $\{\varepsilon_t\}$ from the non-stationary sequence $\{x_t\}$. The following steps demonstrate the computation:

Step 1: Multiply the binomial $F_1(B)$ for B with $\{x_t\}$, resulting in sequence $\{\tilde{T}_t\}$, which roughly estimates sequence $\{T_t\}$:

$$\tilde{T}_t = F_1(B)x_t, \quad t = 1, 2, \cdots, n \tag{7}$$

Step 2: Multiply the binomial $F_2(B)$ for B with sequence $\{x_t - \tilde{T}_t\}$, resulting in sequence $\{\tilde{S}_t\}$, which roughly estimates sequence $\{S_t\}$:

$$\tilde{S}_t = F_2(B)[x_t - \tilde{T}_t], \quad t = 1, 2, \cdots, n \tag{8}$$

Step 3: Multiply the binomial $F_3(B)$ for B with sequence $\{x_t - \tilde{S}_t\}$, resulting in T_t:

$$T_t = F_3(B)[x_t - \tilde{S}_t], \quad t = 1, 2, \cdots, n \tag{9}$$

Step 4: Multiply the binomial $F_4(B)$ for B with sequence $\{x_t - T_t\}$, resulting in S_t:

$$S_t = F_4(B)[x_t - T_t], \quad t = 1, 2, \cdots, n \tag{10}$$

Step 5: Compute $\varepsilon_t = x_t - T_t + S_t$, resulting in the trending stationary random sequence $\{\varepsilon_t\}$, $t = 1, 2, \cdots, n$.

From definition 2.3, we can obtain the binomials for B shown in Eqs. 7 through 10:

$$F_1(B) = \frac{1}{24}\left[\sum_{j=-6}^{5} B^j + \sum_{j=-5}^{6} B^j\right] \tag{11}$$

$$F_2(B) = \frac{1}{9}\left[\sum_{j=0}^{2} B^{12j} + \sum_{j=-1}^{1} B^{12j} + \sum_{j=-2}^{0} B^{12j}\right] \tag{12}$$

$$F_3(B) = 0.24 + 0.214(B + B^{-1}) + 0.147(B^2 + B^{-2})$$
$$+ 0.066(B^3 + B^{-3}) - 0.028(B^5 + B^{-5}) - 0.019(B^6 + B^{-6}) \tag{13}$$

$$F_4(B) = 0.2 + 0.2(B^{12} + B^{-12}) + 0.13(B^{24} + B^{-24}) + 0.07(B^{36} + B^{-36}) \tag{14}$$

If $\{x_t\}$ is a non-stationary sequence, it can be transformed into a trending stationary random sequence using numerical filtering. The transformed sequence is denoted as:

$$f : x_t \to y_t, \quad t = 1, 2, \cdots, n \tag{15}$$

$\{y_t\}$ represents the transformed sequence; f represents the rules of relevant numerical filtering transformation, as shown in Step 1 through Step 4. This paper use GM (1, 1) prediction model forecasting series $\{y_t\}$, can effectively suppress the influence of both high and low frequency noise in the sequence information from the non-stationary sequence, effectively improve the prediction accuracy

Next, we introduce the modelling mechanism of GM (1, 1) model.

Assume that $X^{(0)} = \{x^{(0)}(1), x^{(0)}(2), \cdots, x^{(0)}(n)\}$ is the original series. Applying accumulated generating operation, it can be getting that:

$$X^{(1)} = \{x^{(1)}(1), x^{(1)}(2), \cdots, x^{(1)}(n)\}$$

where $X^{(1)}(k) = \sum_{i=1}^{k} x^{(0)}(i)$ $(k = 1, 2, \cdots, n)$. $X^{(1)}(k)$ is called accumulated generating operation of $X^{(0)}(k)$ denoted as 1-AGO.

The first order linear ordinary differential equation expressed as

$$\frac{dx^{(1)}}{dt} + ax^{(1)} = b \tag{16}$$

is called whitened differential equation of GM(1, 1), of which the difference form is:

$$x^{(0)}(k) + az^{(1)}(k) = b \tag{17}$$

Where a, b are parameters to be identified. a is called developing coefficient, and b is called grey input. Solve it using least square method and obtain:

$$[a, b]^T = (B^T B)^{-1} B^T Y_n \tag{18}$$

Where

$$Y_n = [x^{(0)}(2), x^{(0)}(3), \cdots, x^{(0)}(n)]^T, \quad B = \begin{bmatrix} -z^{(1)}(2) & 1 \\ -z^{(1)}(3) & 1 \\ \vdots & \vdots \\ -z^{(1)}(n) & 1 \end{bmatrix} \tag{19}$$

In Eq. 19, the background value is formulated as

$$z^{(1)}(k+1) = \frac{1}{2}[x^{(1)}(k) + x^{(1)}(k+1)] \quad k = 1, 2, \ldots, n-1 \tag{20}$$

The discrete solution of Eq. 16 is:

$$\hat{x}^{(1)}(k+1) = \left(x^{(0)}(1) - \frac{b}{a}\right) \cdot e^{-ak} + \frac{b}{a} \quad (21)$$

The reduction value is:

$$\hat{x}^{(0)}(k+1) = \hat{x}^{(1)}(k+1) - \hat{x}^{(1)}(k) = (1 - e^a)\left(x^{(0)}(1) - \frac{b}{a}\right) \cdot e^{-ak} \quad (22)$$

Where $k = 1, 2, \cdots, n$.

4 Data Simulation and Accuracy Comparison

Now using the GM (1, 1) prediction model based on non-stationary time series treatment process to forecasting the electricity consumption and C# program to realize the forecast data. Figure 1 shows the smoothing processing results and Table 1 shows the calculated results. (Unit: 10^7 kwh) 1 (Fig. 2).

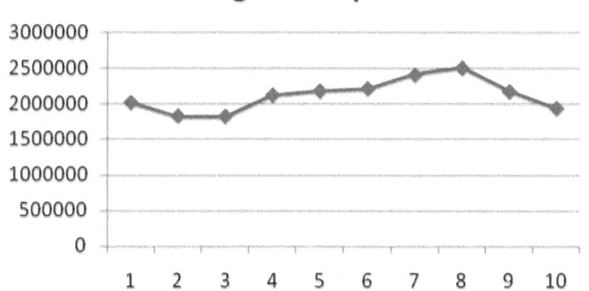

Fig. 1 Smoothing processing result

Table 1 Prediction results of smoothing data

Smoothing data	Forecast value	RE (%)
2,015,064	2,165,983	7.48
1,822,061	1,977,702	8.54
1,818,564	2,014,071	10.75
2,119,474	2,051,108	3.22
2,178,475	2,088,827	4.11
2,212,318	2,127,239	3.84
2,406,550	2,166,357	9.98
2,503,051	2,206,195	11.85
2,170,697	2,246,765	3.50
1,932,446	2,188,082	13.22

Fig. 2 Comparison chart of prediction accuracy

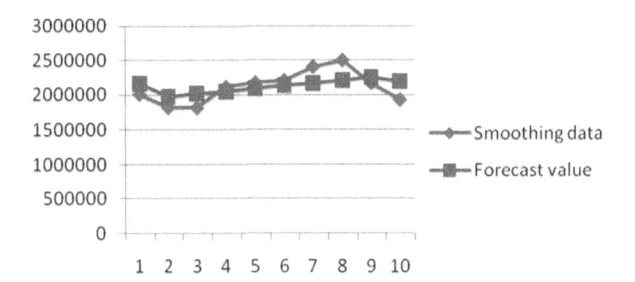

5 Conclusion and Future Work

This paper introduced a new thought to make electricity consumption prediction. The simulation example showed a stronger applicability that the smoothing processing for non-stationary time series effectively suppress the influence of both high and low frequency noise in the sequence information from the non-stationary sequence. And then by using the GM (1, 1) model to forecast the series obtained above, which is close to the real situation of power data, the forecasting accuracy can be really improved, and it will work more effectively in the smart grid and other practical applications. After exploring the GM (1, 1) model, this paper further extended the other new prediction model, such as the use of the artificial intelligent method.

Acknowledgements This research was partially supported by National Natural Science Foundation of China, grant No.71101041, National 863 Project, grant No. 2011AA05A116, Foundation of Higher School Outstanding Talents Grant No. 2012SQRL009 and National Innovative Experiment Program No.111035954.

References

1. Deng julong (1985) Gray system: social and economics[M]. National Defence Industry Press, Beijing, pp 24–76
2. Wang xiaojia, Yang shanlin, Wang haijiang et al (2010) Dynamic GM(1,1) model based on cubic spline for electricity consumption prediction in smart grid. China Commun 7(4):83–88
3. Hsu liChang (2003) Applying the gray prediction model to the global integrated circuit industry. Technol Forecast Soc Change 70(6):563–574
4. Hsu lichang, Wang chaohung (2009) Forecasting integrated circuit output using multivariate gray model and gray relational analysis. Expert Syst Appl 36(5):1403–1409
5. Shen VRL, Chung yufang, Chen teshong (2009) A novel application of gray system theory to information security (Part I). Comput Stand Inter 31(1):277–281

6. Chang baorong, Hsiu Fen Tsai (2008) Forecast approach using neural network adaptation to support vector regression gray model and generalized auto-regressive conditional heteroscedasticity. Expert Syst Appl 34(2):925–934
7. Chenchun, Chen honglong, Chen shuopei (2008) Forecasting of foreign exchange rates of Taiwan's major trading partners by novel nonlinear gray Bernoulli model NGBM(1, 1). Commun Nonlinear Sci Numer Simul 13:1194–1204
8. Huang kuangYu, Jane CJ (2009) A hybrid model for stock market forecasting and portfolio selection based on ARX, gray system and RS theories. Expert Syst Appl 36(5):5387–5392

Electricity Load Forecasting in Smart Grid Based on Residual GM (1, 1) Model

Jianxin Shen, Haijiang Wang, and Shanlin Yang

Abstract The construction of smart grid has put forward higher requirements on deployment accuracy of the energy. Power generation and electricity sectors have carried out more accurate data analysis and forecasting. In this context, a residual GM (1, 1) model is proposed. This model can overcome the lack of traditional grey model and make accurate forecasting of electricity consumption in smart grid. Finally, numerical examples demonstrate that this method can efficiently improve the prediction accuracy.

Keywords Smart grid • Electricity load forecasting • Residual GM (1, 1) model

1 Introduction

Compared with the traditional power grid, smart grid is characterized by environmental protection, safety, efficiency, etc. Particularly, it assists decision-making, which is useful to the optimal allocation of power resource. Data of every terminal in power grid can be controlled real timely by advanced communication facilities in smart grid. According to the forecasting of electricity consumption, smart grid allocates electricity to make a balance between supply and demand of electricity, which realizes optimizing the usage of power energy. Prediction of too much electricity consumption will cause excess power generation and distribution. While prediction of too little electricity consumption will bring power cut due to lack of power supply, then arouse economic losses and social unrest. Therefore, it is of great significance to predict the power consumption accurately. That is to say, the prediction accuracy of electricity consumption determines the quality of smart grid.

J. Shen • H. Wang (✉) • S. Yang
Key Lab of Process Optimization and Intelligent Decision-making,
Hefei University of Technology, Hefei, China
e-mail: tonysun800@sina.com

Data prediction ability determines the quality of smart grid. As for electricity consumption, low prediction will cause power cut due to lack of allocated electricity, while high prediction will bring unnecessary generation cost and energy waste. Therefore, it is essential to predict the electricity consumption accurately. One of the commonest electricity consumption prediction models is grey model- GM (1, 1) [1]. GM (1, 1) model can play a greater role in data forecast of smart grid. By use of its theory, prediction data accuracy can be improved to meet the requirement for data with high quality in smart grid.

After 20 years of development, grey system theory has been widely applied in many areas, including social science and economics. Deng [1]created the grey system theory and system described the principle of grey system, the applications of grey system in many different fields, such as science and economy; Wang, Yang and Wang [2]used cubic spline formula to improve the background value, and constructed a novel grey forecasting model, they used this new model to forecast electricity consumption and obtained high predict accuracy; Hsu and Wang [3, 4] applied the grey prediction model to the global integrated circuit industry and obtained a good prediction effect; Shen, Chung and Chen [5] introduced a novel application of grey system theory to information security, expanded the application field of grey system; Chang and Tsai [6] used neural network adaptation to support vector regression grey model, obtained effect forecasting results, the grey system and artificial intelligence method are combined well; Chen [7] combined the grey system with Bernoulli model, constructed a new grey forecasting model NGBM (1, 1), use this NGBM (1, 1) model to forecast the foreign exchange rates of Taiwan's major trading partners, and receive good effect; Huang [8] use a hybrid grey model to forecast the stock market also achieved good results. But the GM (1, 1) model is ineffective in predicting non-stationary time series, however, mainly because the sequence computed by the GM (1, 1) model is monotonic and because the inverted sequence is also monotonic. Based on these previous studies, this paper aims to find the appropriate treatment to turn a non-stationary sequence into a sequence suitable for the GM (1, 1) model and then to create the GM (1, 1) model to improve modelling and prediction accuracy.

The remainder of the paper is organized as follows: Sect. 1 present some preliminaries of this work, and in Sect. 2 we give the main results of construction process of the prediction model, the data simulation and accuracy comparison are in Sect. 3, finally, we make a conclusion for this paper in Sect. 4.

2 Main Results

In this section, we firstly introduce the modeling mechanism of traditional GM (1, 1) model, then, introduce the building process of residual GM (1, 1) forecasting model [9].

Assume that $X^{(0)} = \{x^{(0)}(1), x^{(0)}(2), \cdots, x^{(0)}(n)\}$ is the original series. Applying accumulated generating operation, it can be get that:

$$X^{(1)} = \{x^{(1)}(1), x^{(1)}(2), \cdots, x^{(1)}(n)\} \quad (1)$$

where $X^{(1)}(k) = \sum_{i=1}^{k} x^{(0)}(i)$ $(k = 1, 2, \cdots, n)$. $X^{(1)}(k)$ is called accumulated generating operation of $X^{(0)}(k)$ denoted as 1-AGO.

The first order linear ordinary differential equation expressed as

$$\frac{dx^{(1)}}{dt} + ax^{(1)} = b \quad (2)$$

is called whitened differential equation of GM (1, 1), of which the difference form is:

$$x^{(0)}(k) + az^{(1)}(k) = b \quad (3)$$

where a, b are parameters to be identified. a is called developing coefficient, and b is called grey input. Solve it using least square method and obtain:

$$[a, b]^T = (B^T B)^{-1} B^T Y_n \quad (4)$$

$$Y_n = [x^{(0)}(2), x^{(0)}(3), \cdots, x^{(0)}(n)]^T, \quad B = \begin{bmatrix} -z^{(1)}(2) & 1 \\ -z^{(1)}(3) & 1 \\ \vdots & \vdots \\ -z^{(1)}(n) & 1 \end{bmatrix} \quad (5)$$

In Eq. 3, the background value is formulated as

$$z^{(1)}(k+1) = \frac{1}{2}[x^{(1)}(k) + x^{(1)}(k+1)] \; k = 1, 2, \ldots, n-1 \quad (6)$$

The discrete solution of Eq. 2 is:

$$\hat{x}^{(1)}(k+1) = (x^{(0)}(1) - \frac{b}{a}) \cdot e^{-ak} + \frac{b}{a} \quad (7)$$

The reduction value is:

$$\hat{x}^{(0)}(k+1) = \hat{x}^{(1)}(k+1) - \hat{x}^{(1)}(k) = (1 - e^a)(x^{(0)}(1) - \frac{b}{a}) \cdot e^{-ak} \quad (8)$$

Where $k = 1, 2, \cdots, n$.

To further improve the accuracy of the model, we establish the GM (1, 1) model using the residual sequence

$$q^{(0)}(k) = \{z^{(0)}(k) - \hat{z}^{(0)}(k)\} \tag{9}$$

Considering most residual sequences include both positive and negative elements, we needed to process the data by dividing the sequence by the aggregate of the negative elements and thus yielding new sequences $q_1^{(0)}, q_2^{(0)}$. We can then establish the GM (1, 1) models separately with new residual sequences. The improved prediction model is shown as follows [10–12]:

$$\hat{x}^{(1)}(k+1) = (x_1^{(0)}(1) - \frac{b}{a})e^{-ak} + \frac{b}{a} - \delta(k - i_1)(q_1^{(0)}(1) - \frac{b_1}{a_1})e^{-a_1 k}$$
$$+ \frac{b_1}{a_1} + \delta(k - i_2)(q_2^{(0)}(1) - \frac{b_2}{a_2})e^{-a_2 k} + \frac{b_2}{a_2} \tag{10}$$

Where

$$\delta(k - i) = \begin{cases} 1, & k > i \\ 0, & k < i \end{cases} \tag{11}$$

3 Data Simulation and Accuracy Comparison

Now using the residual GM (1, 1) model forecasting the electricity consumption and $C^{\#}$ program to realize the forecast data. Next, we also apply the traditional GM (1, 1) model for comparison purposes. First, we give the trends of original data in Fig. 1. Secondly, the comparison results are show in Table 1 and Fig. 2.

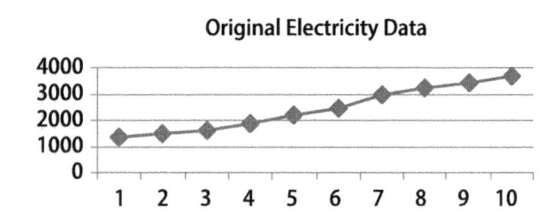

Fig. 1 Trends of original electricity load data

Table 1 Prediction and relative error comparison

Original data	Research model	RE(%)	Traditional GM model	RE(%)
1,346.62	1,283.549	4.68	1,435.28	6.58
1,468.25	1,404.6689	4.33	1,553.87	5.83
1,638.62	1,560.8704	4.74	1,740.65	6.22
1,889.12	1,877.8124	0.59	1,949.87	3.21
2,176.13	2,263.1869	4.00	2,184.24	0.37
2,468.85	2,660.8946	7.77	2,446.78	0.89
2,936.82	3,072.6476	4.62	2,740.87	6.67
3,245.84	3,192.2110	1.65	3,070.32	5.40
3,426.85	3,247.8194	5.22	3,439.37	0.36
3,659.53	3,611.2760	1.31	3,852.77	5.28

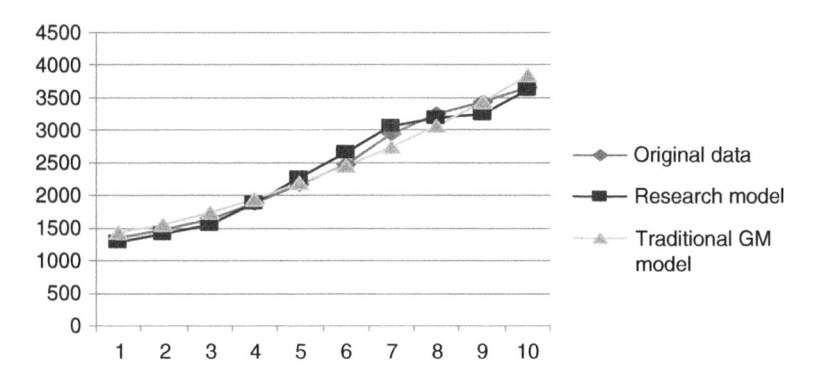

Fig. 2 Comparison chart of prediction accuracy

4 Conclusion and Future Work

This paper introduced a residual GM (1, 1) model to forecast the electricity load. Simulation example shows that the residual GM (1, 1) model can more accurately make prediction than the traditional GM (1, 1) method. The result showed that the proposed method can be effective for electricity load forecasting in smart grid.

Acknowledgements This research was partially supported by National Natural Science Foundation of China, grant No. 71101041, National 863 Project, grant No. 2011AA05A116, Foundation of Higher School Outstanding Talents Grant No. 2012SQRL009 and National Innovative Experiment Program No. 111035954.

References

1. Deng julong (1985) Gray system: social and economics[M]. National Defence Industry Press, Beijing, pp 24–76
2. Wang xiaojia, Yang shanlin, Wang haijiang (2010) Dynamic GM(1,1) model based on cubic spline for electricity consumption prediction in smart grid. China Commun 7(4):83–88
3. Hsu liChang (2003) Applying the gray prediction model to the global integrated circuit industry. Technol Forecast Soc Change 70(6):563–574
4. Hsu lichang, Wang chaohung (2009) Forecasting integrated circuit output using multivariate gray model and gray relational analysis. Expert Syst Appl 36(5):1403–1409
5. Shen VRL, Chung yufang, Chen teshong (2009) A novel application of gray system theory to information security(Part I). Comput Stand Inter 31(1):277–281
6. Chang baorong, Hsiu Fen Tsai (2008) Forecast approach using neural network adaptation to support vector regression gray model and generalized auto-regressive conditional heteroscedasticity. Expert Syst Appl 34(2):925–934
7. Chenchun, Chen honglong, Chen shuopei (2008) Forecasting of foreign exchange rates of Taiwan's major trading partners by novel nonlinear gray Bernoulli model NGBM(1,1). Commun Nonlinear Sci Numer Simul 13:1194–1204
8. Huang kuangYu, Jane CJ (2009) A hybrid model for stock market forecasting and portfolio selection based on ARX, gray system and RS theories. Expert Syst Appl 36(5):5387–5392
9. Mao mingzi, Chirwa EC (2006) Application of gray model GM(1, 1) to vehicle fatality risk estimation. Technol Forecast Soc Change 73(5):588–605
10. Li DC, Chang CJ, Chen WC, Chen CC (2011) An extended gray forecasting model for omnidirectional forecasting considering data gap difference. Appl Math Model 35:5051–5058
11. Tien TL (2009) The deterministic gray dynamic model with convolution integral DGDMC (1, n). Math Pract Theory 33:3498–3510
12. Xie naiming, Liu sifeng (2009) Discrete gray forecasting model and its optimization. Appl Math Model 33:1173–1186

Electricity Consumption Forecasting Based on a Class of New GM (1, 1) Model

Mei Yao and Xiaojia Wang

Abstract In recent years, with the development of society and the progress of science and technology, the GM (1, 1) model has been widely applied in economy, management, industry, control, social development plan, etc. But the GM (1, 1) model in the application process has encountered a low accuracy of predictions. Therefore, the improvement and optimization research of GM (1, 1) model is an important issue. In order to improve the prediction accuracy of GM (1, 1) model, it is necessary to consider reconstructing the background value of the model, which can improve the quality of the forecasting model. This paper discusses the improved GM (1, 1) model. The improved model has better prediction accuracy. It is used to forecast the society electricity consumption in eastern China, and the specific numerical examples demonstrate that this method can improve the simulation and the prediction accuracy.

Keywords GM (1,1) model · Background value · Electricity consumption prediction

1 Introduction

Since professor Deng JuLong in the early 1980s puts forward the gray system theory [1], after three decades of development, this theory has been widely applied in economy, management, industry, control, social development plan, etc., and has achieved many good results. GM (1, 1) model is the one core contents of gray

M. Yao (✉)
School of Mathematics, Hefei University of Technology, Hefei, China
e-mail: ymwalzn@163.com

X. Wang
Key Lab of Process Optimization and Intelligent Decision-making,
Hefei University of Technology, Hefei, China

system theory [2, 3]. The value of its application is to be reflected in an increasing number of areas [4–6].

GM (1, 1) model in the application process has encountered a low accuracy of predictions, therefore, in recent years, the improvement and optimization research of GM (1, 1) model has been the concern of many scholars [7–9]. Through the observation, the author found that these improvements can be divided into the following three types: (1) the improvement of the model parameter estimation methods; (2) the reconstruction of the model background value; (3) the improvement of the model initial conditions. Because of the iterative nature of the model GM (1,1), type (1) and type (3) eventually can be attributed to the reconstruction of the background value, so the type (2) that is the reconstruction of the model background value, has great significance. Therefore, the background value construction method will directly affect the accuracy and applicability of the model. Based on the comprehensive analysis of existing research literature, we make use of the improved method of optimized background value in GM (1, 1) model to forecast the society electricity consumption in eastern China. The results show that compared with the traditional gray GM (1, 1) model, the improved GM (1, 1) model has higher precision of the simulation.

2 Establishment of Mathematical Model

2.1 Modeling Ideas of the Traditional GM (1, 1) Model

Suppose that $X^{(0)} = \{x^{(0)}(1), x^{(0)}(2), \cdots, x^{(0)}(n)\}$ is the original sequence, and it can get through once cumulation:

$$X^{(1)} = \{x^{(1)}(1), x^{(1)}(2), \cdots, x^{(1)}(n)\}$$

Where $x^{(1)}(k) = \sum_{i=1}^{k} x^{(0)}(i)$ $(k = 1, 2, \cdots, n)$, we define $X^{(1)}(k)$ as a cumulative sequence of $X^{(0)}(k)$, then $X^{(1)}(k)$ is called $1 - AGO$.

An order linear differential equations

$$\frac{dx^{(1)}}{dt} + ax^{(1)} = b \tag{1}$$

is called as the winterization differential equation of the grey GM (1, 1) model (Gray differential equation)

$$x^{(0)}(k) + az^{(1)}(k) = b \tag{2}$$

Where a, b are parameters to be identified, and a is called the development coefficient, b is called the grey action. By means of the least square method, We have

$$[a, b]^T = (B^T B)^{-1} B^T Y_n \tag{3}$$

Where: $Y_n = [x^{(0)}(2), x^{(0)}(3), \cdots, x^{(0)}(n)]^T$

$$B = \begin{bmatrix} -z^{(1)}(2) & 1 \\ -z^{(1)}(3) & 1 \\ \vdots & \vdots \\ -z^{(1)}(n) & 1 \end{bmatrix} \tag{4}$$

In formula (4), the background value

$$z^{(1)}(k+1) = \frac{1}{2}[x^{(1)}(k) + x^{(1)}(k+1)] \quad k = 1, 2, \cdots, n-1. \tag{5}$$

The discrete solution of the Eq. 1 as follows:

$$\hat{x}^{(1)}(k+1) = (x^{(0)}(1) - \frac{b}{a}) \cdot e^{-ak} + \frac{b}{a} \tag{6}$$

Then the original value as follows:

$$\hat{x}^{(0)}(k+1) = \hat{x}^{(1)}(k+1) - \hat{x}^{(1)}(k) = (1 - e^a)(x^{(0)}(1) - \frac{b}{a}) \cdot e^{-ak} \tag{7}$$

Where $k = 0, 1, 2, \cdots, n-1$.

2.2 Establishment of Improved GM (1, 1) Model

The improved GM (1, 1) model is based on the reconstruction of the model background value. By formula (6), we find the simulation and prediction precision of the GM (1, 1) model depends on the constants a and b, but a and b depend on the value of the original sequence and structure form of the background value. Therefore the structure formula of background value $Z^{(1)}(k)$ is one of the key factors which lead to the simulation error $\varepsilon^{(0)}(k) = \hat{x}^{(1)}(k) - x^{(1)}(k)$ and the suitability of the GM (1, 1) model.

We can have integral calculation on both sides of the system (1):

$$\int_{k-1}^{k} \frac{dx^{(1)}}{dt} dt + a \int_{k-1}^{k} x^{(1)} dt = b$$

Thus, we get

$$x^{(1)}(k) - x^{(1)}(k-1) + a \int_{k-1}^{k} x^{(1)}(t) dt = b$$

$$x^{(0)}(k) + a \int_{k-1}^{k} x^{(1)}(t) dt = b \qquad (8)$$

By formula (2), the background value as follows:

$$z^{(1)}(k) = \int_{k-1}^{k} x^{(1)}(t) dt \, (k = 2, 3, \cdots n) \qquad (9)$$

Set $x^{(1)}(t) = Be^{At}$,
Where A and B are constants to be determined with

$$x^{(1)}(k) = Be^{Ak}, (k = 1, 2, \cdots, n) \qquad (10)$$

Generate $x^{(1)}(t) = Be^{At}$ into formula (9), we get

$$z^{(1)}(k) = \int_{k-1}^{k} Be^{At} dt = \frac{1}{A}(Be^{Ak} - Be^{A(k-1)}) = \frac{1}{A}(x^{(1)}(k) - x^{(1)}(k-1)) \qquad (11)$$

Since

$$\frac{x^{(1)}(k)}{x^{(1)}(k-1)} = \frac{Be^{Ak}}{Be^{A(k-1)}} = e^A$$

We have

$$A = \ln x^{(1)}(k) - \ln x^{(1)}(k-1) \qquad (12)$$

Generate formula (12) into formula (11), we have

$$z^{(1)}(k) = \frac{x^{(1)}(k) - x^{(1)}(k-1)}{\ln x^{(1)}(k) - \ln x^{(1)}(k-1)}, (k = 2, 3, \cdots, n) \qquad (13)$$

Generate formula (13) into formula (4), we get forecasting formulas of improved GM (1, 1) model

$$\hat{x}^{*(0)}(k+1) = \hat{x}^{(1)}(k+1) - \hat{x}^{(1)}(k) = (1-e^a)(x^{(0)}(1) - \frac{b}{a}) \cdot e^{-ak} \quad (14)$$

Thus we get the background value formula after optimized reconstruction.

3 Application Examples

Electricity as a special commodity has two main characteristics: the instantaneous characteristics of production, transmission and consumption.

The characteristics of electricity cannot be stored, which determines the demand for electricity is closely related to economic development. Because of the long construction period of the electricity, it is important to carry out its early warning systems in order to avoid its impact on social stability and the investment environment. The forecast methods of electricity consumption are GM (1, 1) model, multiple regression analysis, the exponential regression ARMA model, etc.

We select the monthly society electricity consumption of East China and record them as original data. We use the optimization method of background value and the traditional GM (1, 1) model to predict and comparatively analyze. The forecast scheme uses the data of the first 8 months to predict the data after the following 2 months. The calculated results show in Table 1.

From Table 1, using the optimization method of background value to carry out the forecast, the average relative error is 7.88 %, the forecasting accuracy has

Table 1 Contrast of the optimum one to the traditional GM (1, 1) about the prediction

Monthly	Original data	The optimum GM (1,1) model		Traditional GM (1,1) model	
		Model data	Relative error (%)	Model data	Relative error
1	2,140,773	2,312,850	8.0	149,475	30.2
2	1,503,399	1,909,487	27.0	1,991,297	32.5
3	2,133,739	2,069,753	2.9	2,013,600	5.6
4	2,105,210	2,078,968	1.2	2,036,153	3.3
5	2,251,740	2,145,116	4.7	2,058,958	8.6
6	2,172,995	2,092,433	3.7	2,082,019	4.2
7	2,640,204	2,240,846	15.1	2,105,338	20.3
8	2,365,997	2,290,378	3.2	2,128,918	10.1
9	1,975,497	2,141,057	8.3	2,152,763	8.9
10	1,999,394	2,092,908	4.7	2,176,874	15.2
Average relative			7.88		13.89

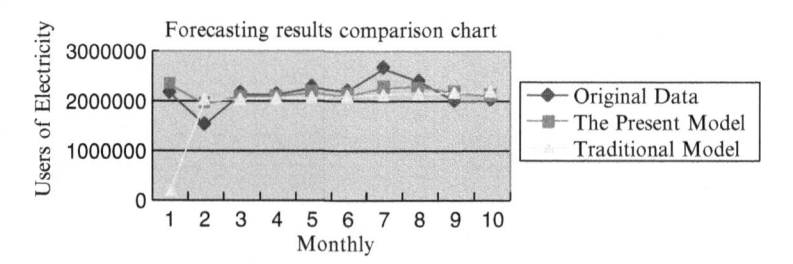

Fig. 1 Forecasting results comparison chart

greatly improved compared to the traditional GM (1, 1) method. It proves that the background value reconstruction is a key factor to affect the prediction accuracy and applicability and it also shows that the proposed method is effective.

The following is the figure of the Table 1 (Fig. 1).

4 Conclusion

By using gray system theory model, the GM (1, 1) model can overcome the weakness of related data and avoid the influence of the artificial factor. In this paper, the simulation results of the improved gray prediction model were closer to the practical value, prediction error was small and the model accuracy was better. This suggested prediction accuracy of improved gray prediction model was higher than the traditional gray prediction model. It verified the reliability and validity of the method presented in this paper.

Acknowledgements This research was partially supported by National Natural Science Foundation of China, grant No. 71101041, National 863 Project, grant No. 2011AA05A116, Foundation of Higher School Outstanding Talents Grant No. 2012SQRL009, the Youth Development Foundation of Hefei University of Technology, Grant No. 2013HGXJ0227 and National Innovative Experiment Program No. 111035954.

References

1. Deng Julong (1985) Gray system[M]. National Defence Industry Press, Beijing, pp 18–40
2. Liu Sifeng, Guo Tianbang, Dang Yaoguo (1999) Gray system theory and its application [M]. Science Press, Beijing, pp 60–72
3. Luo Dang, Liu Sifeng, Dang Yaoguo (2003) The optimization of Grey Model GM (1, 1). Eng Sci 5(8):50–53
4. Wang Xiaojia, Yang Shanlin, Haijiang Wang et al (2010) Dynamic GM (1, 1) model based on cubic spline for electricity consumption prediction in smart grid. China Commun 7(4):83–88

5. Xiaojia Wang, Shanlin Yang (2009) On delay-dependent stability for a class of neutral systems. In: Proceedings-5th international conference on wireless communications, networking and mobile computing, WiCOM 2009. IEEE Computer Society, Piscataway, pp 1–3
6. Xiaojia Wang, Shanlin Yang (2010) Electricity demand forecasting based on threepoint Gaussian quadrature and its application in smart grid. In: Proceedings-6th international conference on wireless communications, networking and mobile computing, WiCOM 2010. IEEE Computer Society, Piscataway, pp 1–3
7. Wang Xiaojia, Yang Shanlin, Hou Liqiang et al (2010) Simulation of orthogonalization prediction based on Grey Markov chain for electricity consumption. J Syst Simul 22(10):2253–2256
8. Xiaojia Wang (2011) Application research on electricity demand forecasting based on Gaussian quadrature formula. Procedia Eng 15:5574–5578
9. Wang Xiaojia, Shen Jianxin, Yang Shanlin (2010) Application research on Gaussian orthogonal interpolation method for electricity consumption forecasting of smart grid. Power Syst Prot Control 38(21):141–145

Spectral Visibility of High-Altitude Balloon by the Ground-Based Detection

Xiaoping Du, Yu Zhang, and Dexian Zeng

Abstract In order to determine the visible spectral region of the target, the paper made the analysis of the high-altitude balloon by the ground-based detection. By using the atmospheric transporting model (MODTRAN), the paper calculated the background noise intensity at the probe point and the brightness of the background of the target in the different time. And by using the reverse Monte-Carlo method, the paper calculated the scattering radiance of the high-altitude balloon the ground-based probe point received. After calculating the ratio of the signal to the noise, the paper made the analysis of the visibility of the target. The results show that, during the ground-based detection, with the smaller detecting angle or the greater solar zenith angle of the target, the signal to noise ratio can be higher when the target's location and the detecting altitude are both fixed. This study is beneficial to the determination of the threshold of detection equipment and the choice of spectral bands.

Keywords High-altitude balloon • Ground-based detection • Reverse Monte-Carlo method • Scattering characteristics • Signal to noise ratio • Spectrum • MODTRAN

1 Introduction

The high-altitude balloon is a non-powered aircraft which works in the stratosphere. With 30 years' development, average flying height of the high-altitude balloon can be 40–50 km. Having the features such that the experimental cost is low, the flight organizing is convenient and the testing cycle is short, the high-altitude balloon

---

X. Du (✉) • Y. Zhang • D. Zeng
The Academy of Equipment, Beijing, China
e-mail: dxp8600@163.com; xlrui522@163.com

has been used in the ground-based remote sensing, the atmospheric physics and the military. Thus carrying out the research of the recognition technology on the detection of high-altitude balloon should bring great value.

Due to the small radar scattering cross section the high-altitude balloon has, spectral detection technology is widely used as a sophisticated means of detection [1]. The paper used the ground-based detecting means to conduct the research of the high-altitude balloon. After completing the geometric modeling and determination of the spectral reflectance of the coating material, the paper calculated the scattering properties of the target on the visible spectrum by using the reverse Monte-Carlo method, and made the analysis of the spectral visibility of the target.

2 Analysis of the Radiance of the Background Lights

When the time to start observing, such condition must be met that the scattering intensity should be higher than the noise intensity entering the detector on the period of spectrum being used [2].

Generally speaking, the target-scattering lights can be divided into the direct sunlight and varieties of lights which have multiple scattering effects. To make the illustration conveniently, the upward scattering lights are all named as ground-lights, and the downward scattering lights, excepting the direct sunlight, are all named as sky-lights. Usually, people use the background radiation whose incident zenith angle is 45° instead of the average radiance of the sky-lights, and use the background radiation whose incident zenith angle is 135° instead of the average radiance of the ground-lights when conducting the engineering calculation [3, 4]. In the form of noise, the radiance of the sky-lights of the detector will enter the pupil surface if the detector is below the target in the altitude. Throughout the environment of exploration, the solar zenith angle and the solar azimuth of the target or the detector will both change as long as the time changes, which is mean that the radiance of the scattered light of the target and the intensity of the noise of the detector will change correspondingly.

For the target located in the atmosphere, apparently, the ground-lights and the sky-lights both exist. But for the reception of the detector which is below the target, the sky-lights of the target can be ignored because that although the under surface of the target can receive the sky-lights when the solar zenith angle of the target takes some angles, but comparing with the ground-lights, the reflected energy is too small to reach the detector. Although the direct sunlight is downward, but the reflected energy is so high that cannot be ignored.

Generally, the scattered lights of the target include the ground-lights and the direct sunlight at the location of the target. The ground-lights include the upward scattering sunlight, the upward atmospheric path radiation, the black-body radiation of the earth's surface and the reflected lights by the earth's surface for various atmospheric background lights. The sky-lights include the downward scattering sunlight, the downward atmospheric path radiation.

The energy of every kind of radiation is related with the latitude, season, time and atmospheric physical state. By using the atmospheric transporting model (MODTRAN), the paper calculated the background noise intensity at the different situation.

3 The Modeling of the Scattering Property of the Target

This section calculated the radiance, scattered from the target, the detector received by using the reverse Monte-Carlo method (RMC). RMC assumes that the detector emits lights in the field of view of the target. Only the light which can reach the target is useful for the follow-up recursive calculation on the light's energy. Obviously, this process is the reverse one of the real ray-casting. In details, RMC assumes that the energy of the light is from the black-body radiation of the target if the light is been absorbed by the target, and assumes that the energy of the light is from the background light scattered by the target if the light is scattered into the environment by the target [5–7]. Ultimately, the statistics for each light can be obtained.

The paper named the lights having no contact with the target as the invalid ones, and named the lights having intersection with the target as the effective ones. Ultimately some lights are absorbed by the target, the others are scattered into the environment. The light's energy will be gradually decreasing if it is multiply scattered. Given the actual situation of the spectral reflectance of the target coating material, the scattering times can be set to downsize the simulation scale. Figure 1 shows the transmission of energy between two surface elements in the vacuum.

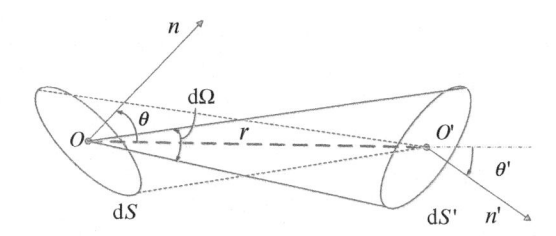

In Fig. 1, dS is the area of the surface element of the target, dS' is the area of the surface element of the detector's pupil surface, $d\Omega$ is the solid angle of the detector to the target, r is the length the line of centers of two surface elements, n is the normal line of the surface element of the target, n' is the normal line of the surface element of the detector, θ' is the included angle between n' and the line of centers of two surface elements.

The paper assumed that the change of the energy of the light being scattered is only relevant to the material's spectral reflectivity named as $\rho(\lambda)$. The spectral radiant flux named as $d\Phi'(\lambda)$ received by dS' can be expressed as

Fig. 1 The transmission of energy between two surface elements

$$\mathrm{d}\Phi'(\lambda) = \rho(\lambda) \cdot L_{\text{back}}(\lambda)\mathrm{d}\Omega\mathrm{d}S\cos\theta = \rho(\lambda) \cdot L_{\text{back}}(\lambda) \cdot \frac{\mathrm{d}S'\cos\theta'}{r^2}\mathrm{d}S\cos\theta \quad (1)$$

Here λ is the wavelength in microns and $L_{\text{back}}(\lambda)$ is the spectral radiance of the light scattered by the target. Thus the spectral irradiance named as $\mathrm{d}E(\lambda)$ received by $\mathrm{d}S'$ can be deduced as

$$\mathrm{d}E(\lambda) = \frac{\mathrm{d}\Phi'(\lambda)}{\mathrm{d}S'} = \frac{\rho(\lambda) \cdot L_{\text{back}}(\lambda)\mathrm{d}S\cos\theta\cos\theta'}{r^2} \quad (2)$$

Avoiding calculating complicated angle factor, that the multiplication of $\cos\theta$ can be replaced of the ratio of the amount of different lights is one of the advantages of RMC's. So it means that

$$\sum \mathrm{d}S\cos\theta = \frac{1}{N} \cdot [\sum_{a_1=1}^{S_{\text{sca\_1}}} S_{a_1} + \sum_{a_2=1}^{S_{\text{sca\_2}}} S_{a_2} + \cdots + \sum_{a_n=1}^{S_{\text{sca\_n}}} S_{a_n}] \quad (3)$$

where n is the scattering times, S_{a_n} is the area of the surface element from which the light numbered a_n is scattered into the environment ultimately, $S_{\text{sca\_n}}$ is the quantity of the lights whose scattering times are n.

Due to the presence of the scattering phenomenon, the energy of the light will change in different degree, so the above equation should be amended as

$$\sum \mathrm{d}S\cos\theta = \frac{1}{N} \cdot [\sum_{a_1=1}^{S_{\text{sca\_1}}} S_{a_1} + \rho(\lambda)\sum_{a_2=1}^{S_{\text{sca\_2}}} S_{a_2} + \cdots + \rho(\lambda)^n \sum_{a_n=1}^{S_{\text{sca\_n}}} S_{a_n}] \quad (4)$$

Considering the presence of the invalid light, the above equation should be amended again, which is showed as

$$\sum \mathrm{d}S\cos\theta = \frac{1}{N - S_{\text{sca\_0}}} \cdot [\sum_{a_1=1}^{S_{\text{sca\_1}}} S_{a_1} + \rho(\lambda)\sum_{a_2=1}^{S_{\text{sca\_2}}} S_{a_2} + \cdots + \rho(\lambda)^n \sum_{a_n=1}^{S_{\text{sca\_n}}} S_{a_n}] \quad (5)$$

Thus for the entire target, the radiance named as $L(\lambda)$ received by the detector can be deduced as

$$L(\lambda) = \frac{\tau(\lambda) \cdot \rho(\lambda) \cdot L_{\text{back}}(\lambda) \cdot \cos\theta'}{N - S_{\text{sca\_0}}} \cdot [\sum_{a_1=1}^{S_{\text{sca\_1}}} S_{a_1} + \rho(\lambda)\sum_{a_2=1}^{S_{\text{sca\_2}}} S_{a_2} + \cdots$$

$$+ \rho(\lambda)^n \sum_{a_n=1}^{S_{\text{sca\_n}}} S_{a_n}] \quad (6)$$

Where $\tau(\lambda)$ is the atmospheric spectral transmittance of the optical path from the target to the detector, N is the quantity of the lights the detector emits, and θ', also named as the detecting angle, here is the included angle between the normal line of the detector's pupil surface and the line from the target to the detector [8–10].

4 Simulations and Analysis

4.1 The Geometrical Model and the Coating Material of the Target

Taking the background of the research into account, the paper set the typical high-altitude balloon as the research target. Based on the software (3ds MAX), the paper completed its finite element modeling and exported its triangular mesh data. Figure 2 shows the Geometric model of the target.

The coating material made of the high-pressure polyethylene after its hot pressing welding has good properties of low temperature and tensile strength. Using the spectrometer, the paper made the determination of the spectral reflectance of the material being set as PVC. The bands range is set from 0.38 to 0.78 in microns. Figure 3 shows the curves of the spectral reflectance of the material after calibration by the whiteboard.

4.2 The Process and the Results of the Simulation

The location of the target is set to the urban area of Beijing (E 116.6° N 39.9°) and the vertical height from the surface is set to 40 km. The value of detecting angle determined the locations of the ground-based detectors. After analyzing the latitude

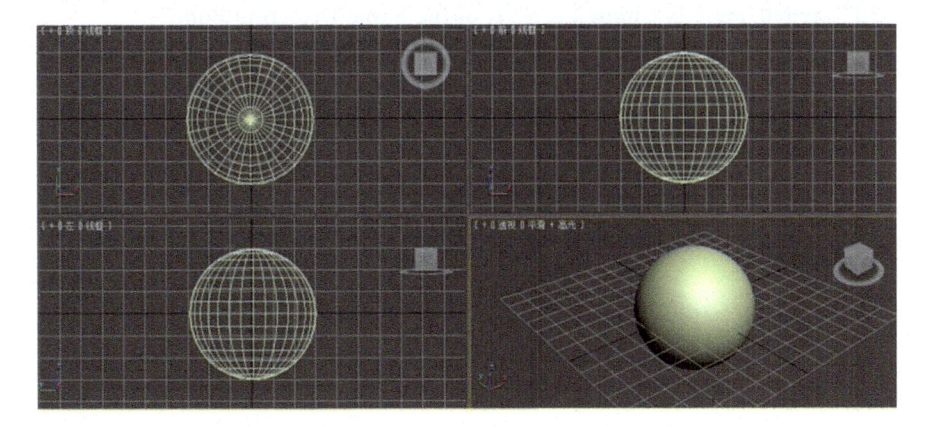

Fig. 2 The geometric model of the target

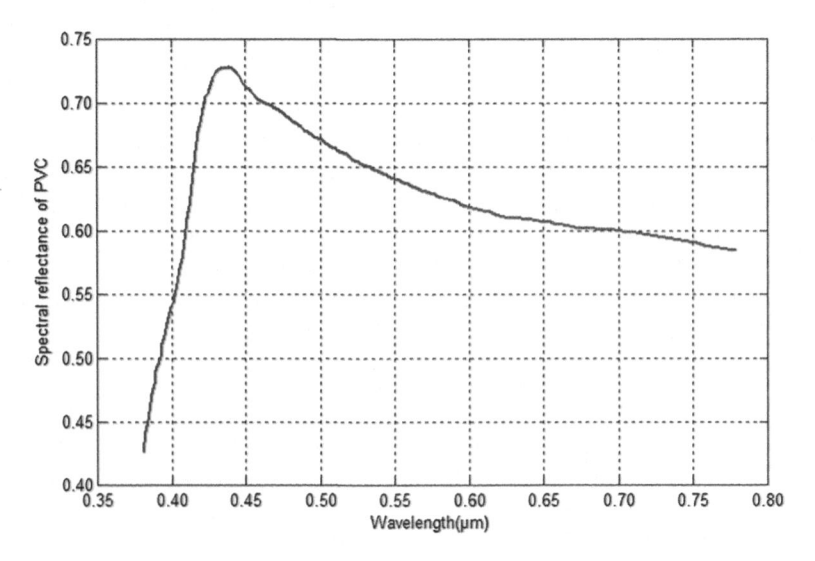

Fig. 3 The curves of the spectral reflectance of the material

and longitude of various districts and counties of China, the paper determined three typical positions. These three locations respectively were Tongzhou (E 116.65°, N 39.92°, $\theta' = 28.38°$), Changping (E 116.23°, N 40.22°, $\theta' = 44.13°$) and Pinggu (E 117.12°, N 40.13°, $\theta' = 59.35°$).

During the observation, the main difficulty lies in the low contrast and the SNR caused by the strong radiance of the background lights. Here SNR is expressed as the ratio of the radiance of the lights scattered by the target to the brightness of the background noise of the detector [11]. For simulation, time was set to May 4, 2012. The model of atmosphere was set as U.S. 1976 standard, the model of the aerosol was rural, and the visibility was 23 km [12].

4.2.1 When Time Is Changing, SNR of the Same Detecting Point

Setting the detecting point as Tongzhou District, Fig. 4 shows the curves of the SNR of Tongzhou at different time when the solar zenith angle of the target were 35°, 45°, 55° and 75° respectively.

Figure 4 shows that, with the adding of the solar zenith angle, the curve shows an increasing tendency. If the SNR is fixed, the start point of the visible area on the spectrum moves forward with the increasing of the solar zenith angle.

4.2.2 When Time Is Fixed, SNR of Different Detecting Point

When the solar zenith angle of the target is 75°, the solar zenith angle of each detector were 74.82° (Tongzhou), 75.16° (Changping) and 74.47° (Pinggu). Figure 5 shows the curves of the SNR of different detecting points at the same time.

Spectral Visibility of High-Altitude Balloon by the Ground-Based Detection 961

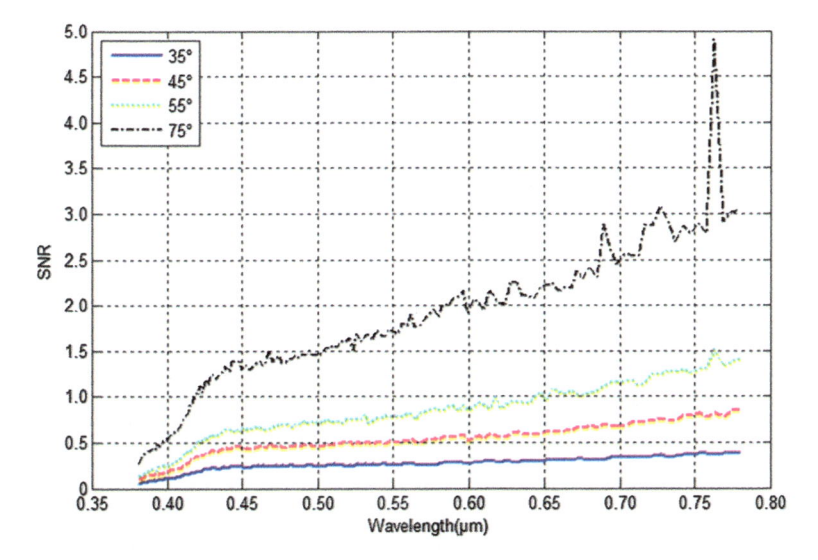

Fig. 4 The curves of the SNR of Tongzhou at different time

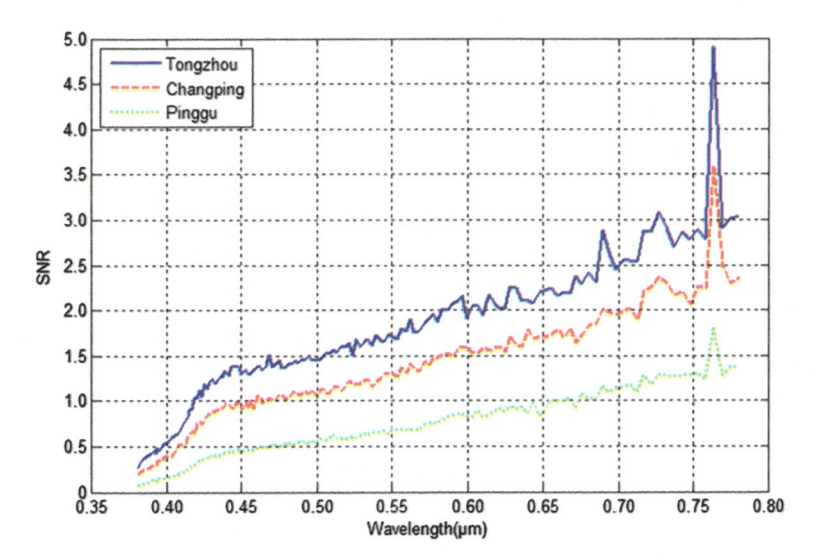

Fig. 5 The curves of the SNR of different detecting points at the same time

Figure 5 shows that with the adding of the detecting angle, the curve shows a decreasing tendency. If the SNR is fixed, the start point of the visible area on the spectrum moves backward with the increasing of the detecting angle.

5 Conclusion

By using the reverse Monte-Carlo method, the paper calculated the scattering radiance of the high-altitude balloon the ground-based probe point received and made the comparison with the background noise intensity at the probe point in different time and the brightness of the background of the target at different times. The following conclusions can be inferred:

(a) When the detecting point is fixed, with the increasing of the solar zenith angle of the target, SNR showed an upward trend on the whole. And when the SNR value is the same, with the increasing of the solar zenith angle of the target, the visible area of the spectrum will move forward.
(b) When the detecting time is fixed, with the increasing of the detecting angle, SNR showed a downward trend on the whole. And when the SNR value is the same, with the increasing of the solar zenith angle of the target, the visible area of the spectrum will move backward.

References

1. Zhou Linghui (2008) Simulation and identification of the spectral properties of space targets [D]. Master dissertation of the Huazhong University of Science and Technology, Wuhan, (In Chinese)
2. Li Daoyong, Wang Zengyu, Zhang Yanqun (2005) The visual conditions of properties of the light scattered by the space targets. Opt Optoelectron Technol 3(5):5–8 (In Chinese)
3. Wu Zhensen, Liu Anan (2002) Scattering of solar and atmospheric background radiation from a target. Int J IR Mill Waves 23(6):907–917
4. Li Daoyong, Wang Yunqiang, Gong Yanjun (2004) The properties of the light scattered by the space targets. J Yantai Univ 17(3):183–187 (In Chinese)
5. Shan Yong, Zhang Jingzhou, Guo Rongwei (2008) Numerical computation and analysis of the infrared radiation characteristic of missile scarfskin. J Aerosp Power 23(2):251–255 (In Chinese)
6. Lv Jianwei, Wang Qiang (2009) Numerical calculation and analysis of infrared radiation characteristics from aircraft skin by using RMC method. Infrared Laser Eng 38(2):232–237 (In Chinese)
7. Liu Gang, Riqitai E, Zhu Xijuan (2011) Simulation of infrared radiation characteristics of cruise missile skin by backward Monte Carlo method. Sci Technol Eng 11(5):1012–1017
8. Li Liangchao, Niu Wubin, Wu Zhensen (2011) Parallel calculation for scattering of background infrared irradiation from aerial complex targets. Syst Eng Electron 33(12):2573–2576 (In Chinese)
9. Yang Yufeng, Wu Zhensen, Cao Yunhua (2011) Scattering characteristics of complex background infrared radiation from a non-lambertian target. Infrared Laser Eng 40(5):800–804 (In Chinese)
10. Shu Rui, Zhou Yanping, Tao Kunyu (2006) The study of infrared spectrum of spacetarget. Opt Tech 32(2):196–199 (In Chinese)

11. Chen Yuheng, Zhou Jiankang, Chen Xinhua (2009) Calculation of SNR of a satellite CCD camera based on MODTRAN. Infrared Laser Eng 38(5):910–914 (In Chinese)
12. Mao Kebiao, Tan Zhihao (2004) The transmission model of atmospheric radiation and the computation of transmittance of MODTRAN. Geomat Spat Inf Technol 27(4):1–3 (In Chinese)

Part VI
Sensing Control Theory

A Novel Demodulation Method for Fiber Optic Interferometer Sensor Using 3 × 3 Coupler

Haiyan Xu and Zhongde Qiao

Abstract Fiber optic inetrferometric sensors have advantages of high sensitivity and can be employed to measure very small disturbance. In this paper, a novel demodulation method of fiber optic interferometer sensor is presented, which is based on the 3 × 3 coupler to demodulate the phase shift. By employing two output signals of the 3 × 3 coupler, it can eliminate the insensitive areas. The optical structure and the demodulation algorithm are easy to deploy, compared with the PGC (phase generated carrier) technique and other 3 × 3 coupler based demodulators. This algorithm can effectively overcome the phase distortion caused by the instability of the light output power. The simulation and experiment demonstrate that the high demodulate precision has been achieved.

Keywords Fiber optic sensor • Interferometer • 3 × 3 coupler

1 Introduction

Recently, distributed fiber-optic sensor is available for many applications, such as intruder detection, moving vehicle location, and oil pipeline leakage monitoring. The most distinguished techniques for distributed sensing involve the use of the optical time domain reflectometer (OTDR) [1, 2] and fiber-optic interferometer sensors [3, 4]. The optical fiber interferometer has been widely used in the physical measurement because of its high sensitivity, such as the sound of water, strain, magnetic field, current and acceleration. Double-beam interferometer is the main development direction for the interferometer sensors [5].

H. Xu (✉)
Jiangsu Key Laboratory of Power Transmission and Distribution Equipment Technology, Hohai University, Changzhou 213002, China
e-mail: xhyjstu@hotmail.com

Z. Qiao
Changzhou Ruize Microelectronics Co. Ltd, Changzhou, China

However, the photo detector can only respond to changes in light intensity, after the interference of light received by the photo detector of interference items, we need to use different methods to demodulate the phase signal. The demodulator methods are generally divided into phase generated carrier (PGC) [6] and the passive homodyne demodulation technique with a 3 × 3 coupler.

The main drawback of the PGC technology is the use of low-pass filter which ignore the higher order Bessel function, and results in the serious distortion of the demodulation. Dynamic range of the phase generated carrier has been restricted with the limited carrier frequency. When using the externally modulated carrier, the optical path is more complicated. A demodulation scheme utilizing a 3 × 3 coupler has the advantage of passive detection and low cost as it requires no phase or frequency modulation in the reference arm or of the laser source, and so there are no active components in the optical domain. There are basically three methods for demodulation with a 3 × 3 coupler: in the first one two of the three outputs are used to obtain the required 90° phase difference from their sum and difference and these form the inputs to a differentiating cross-multiplying demodulator [7], the method needs a dc offset in the circuit to eliminate the dc component resulting from summing up the two inputs, therefore when the light power in the interferometer changes, the circuit will become unbalanced; whereas in the second one, all three outputs are utilized symmetrically in an analog processing algorithm to obtain the demodulated time-changing phase difference [8], However, if the 3 × 3 coupler is not symmetric, the symmetric method will face rather serious problems; Whereas in the third one, an asymmetric 3 × 3 coupler demodulate method is proposed, but the demodulation process is more complex because of the gain control factor [9],so it is difficult to apply to the practical engineering.

In this paper, we present a new demodulation method which requires only two interference signals with a certain initial phase. The method can effectively overcome the phase distortion caused by the instability of the light output power. As compared to the PGC technique and past 3 × 3 coupler based demodulator, the method has no phase carrier, no filters, no calibration, no need for uniform splitting ratio of coupler, the optical structure and the demodulation algorithm are all simple. The demodulator method we proposed here can meet the requirements of practical application, which has a certain practical significance.

2 Theoretical Analysis

2.1 Signal Acquisition

We can achieve two AC output signals with constant phase difference when using a 3 × 3 coupler to construct the interferometer. The two outputs of the photodiodes are given by:

$$\begin{cases} I_1(t) = A'(t) + A(t)\cos(\Delta\varphi(t) + \phi) \\ I_2(t) = B'(t) + B(t)\cos(\Delta\varphi(t) - \phi) \end{cases} \quad (1)$$

Where $\Delta\varphi(t)$ is the phase difference of the interferometer at its output, which is a function of time containing the information on the quantity to be measured; and A, B, A', B' are dependent on the optical power of system input; ϕ is the constant, for a symmetric 3×3 coupler, we have $\phi = 2\pi/3$, and for an asymmetric 3×3 coupler, we can also achieve the value of ϕ. Use the two outputs signals from the 3×3 coupler, the phase shift to be measured can be recovered.

After removing the DC components, the following equations are met: $\Delta\varphi(t) = 0, I_1(t) = I_2(t) = 0$. According to the symmetry character of the 3×3 coupler, there is $A'(t) = -A(t)\cos\phi, B'(t) = -B(t)\cos\phi$, then we can rewrite Eq. 1 as:

$$\begin{cases} I_1(t) = A(t)\cos(\Delta\varphi(t) + \phi) - A(t)\cos\phi \\ I_2(t) = B(t)\cos(\Delta\varphi(t) - \phi) - B(t)\cos\phi \end{cases} \quad (2)$$

Discretized the signals in Eq. 2, suppose the sampling time interval is Δ, then $t = n\Delta$ ($n = 0,1,2......$), therefore

$$\begin{cases} I_1(n) = A(n)\cos(\Delta\varphi(n) + \phi) - A(n)\cos\phi \\ I_2(n) = B(n)\cos(\Delta\varphi(n) - \phi) - B(n)\cos\phi \end{cases} \quad (3)$$

2.2 DC Components Calculation

For simplicity, we only analysis the signal of $I_1(n)$, because of the amplitude of the signals is big enough, that is, the dynamic range of $\varphi(n)$ exceeds one cycle, the maxim and minimum of Eq. 3 is:

$$Max(I_1(n)) = A(n)(1 - \cos\phi) \quad (4)$$

$$Min(I_1(n)) = -A(n)(1 + \cos\phi) \quad (5)$$

so

$$A(n)\cos\phi = -\frac{Max(I_1(n)) + Min(I_1(n))}{2} \quad (6)$$

Similarly

$$B(n)\cos\phi = -\frac{Max(I_2(n)) + Min(I_2(n))}{2} \quad (7)$$

Substituting Eqs. 6 and 7 into Eq. 3, we can have

$$I_1(n) - \frac{Max(I_1(n)) + Min(I_1(n))}{2} = A(n)\cos(\Delta\varphi(n) + \phi) \quad (8)$$

$$I_2(n) - \frac{Max(I_2(n)) + Min(I_2(n))}{2} = B(n)\cos(\Delta\varphi(n) - \phi) \quad (9)$$

2.3 Normalization

We need to normalize the signals considering the amplitude difference of the two signals. Equation 4 subtract Eq. 5, can be expressed as

$$A(n) = \frac{Max(I_1(n)) - Min(I_1(n))}{2} \quad (10)$$

Similarly, $B(n)$ can be expressed as

$$B(n) = \frac{Max(I_2(n)) - Min(I_2(n))}{2} \quad (11)$$

By substituting Eqs. 10 and 11 into Eqs. 8 and 9, we get the normalized signal

$$I_1'(n) = C(n)\cos[\Delta\varphi(n) + \phi] = 2\left(\frac{I_1(n) - Min(I_1(n))}{Max(I_1(n)) - Min(I_1(n))}\right) - 1 \quad (12)$$

$$I_2'(n) = C(n)\cos[\Delta\varphi(n) - \phi] = 2\left(\frac{I_2(n) - Min(I_2(n))}{Max(I_2(n)) - Min(I_2(n))}\right) - 1 \quad (13)$$

Where $C(n)$ is a normalization factor. Ideally, $C(n) = 1$. During the actual experimental tests, due to systematic errors, each point corresponding to the C is vary, we can obtain the corresponding C according each set of experimental data (see Eq. 16).

2.4 Demodulation Algorithm

For simplicity, by add and subtract the two signals of Eqs. 12 and 13, there is

$$I_+(n) = I'_1(n) + I'_2(n) = 2C(n)\cos\phi\cos\Delta\varphi(n) \qquad (14)$$

$$I_-(n) = I'_1(n) - I'_2(n) = 2C(n)\sin\phi\sin\Delta\varphi(n) \qquad (15)$$

According to the above two equations, $C(n)$ can be denoted as:

$$C(n) = \sqrt{\left(\frac{I'_1(n) + I'_2(n)}{2\cos\phi}\right)^2 + \left(\frac{I'_1(n) - I'_2(n)}{2\sin\phi}\right)^2} \qquad (16)$$

From the amplitude ratio of Eqs. 14 and 15, it can be shown that $\Delta\varphi(n)$ can also be expressed as in one cycle:

$$\Delta\varphi(n) = \tan^{-1}\left(\frac{I_-(n)}{I_+(n)}/\tan\phi\right) \qquad (17)$$

Then need to further expansion of the phase to complete the automatic accumulation function of the signal over one cycle. For an arbitrary n, subtract $\varphi(n)$ and $\varphi(n-1)$, k = k − 1 while the value more than $\pi/2$; k = k + 1 while the value less than − $\pi/2$. $\Delta\varphi(n)$ can be rewrite as

$$\Delta\varphi(n) = k\pi + \tan^{-1}\left(\frac{I_-(n)}{I_+(n)}/\tan\phi\right) \qquad (18)$$

The phase difference $\Delta\varphi(n)$ can be achieved through software technology with the above algorithm. For a symmetric 3 × 3 coupler, $\phi = 2\pi/3$, as the coupler is an asymmetric, according to Eqs. 14 and 15:

$$\tan\phi = \frac{I_{-A}(n)}{I_{+A}(n)} \qquad (19)$$

Where subscript A denotes the amplitude of $I_-(n)$ and $I_+(n)$, by substituting Eq. 19 into Eq. 18, the demodulation signal can be achieved.

3 Simulation and Experimental

3.1 Simulation

To verify the correctness of the demodulation algorithm, the single-frequency sinusoidal signal is used. Make the amplitude of $\Delta\varphi(t)$ is 2.8 V, the frequency of $\Delta\varphi(t)$ is 5 kHz, the sample frequency is 500 kHz, according to the Eq. 3, the

Fig. 1 Time sequences before demodulation

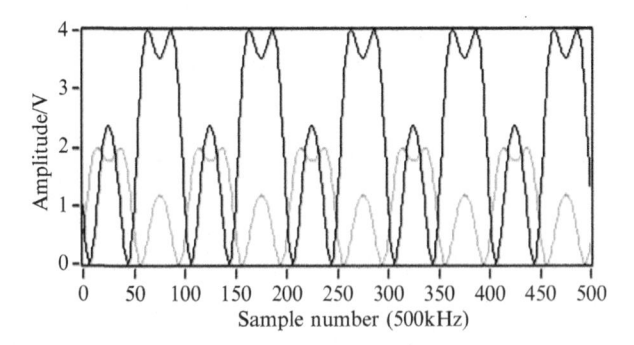

Fig. 2 Time sequences normalized

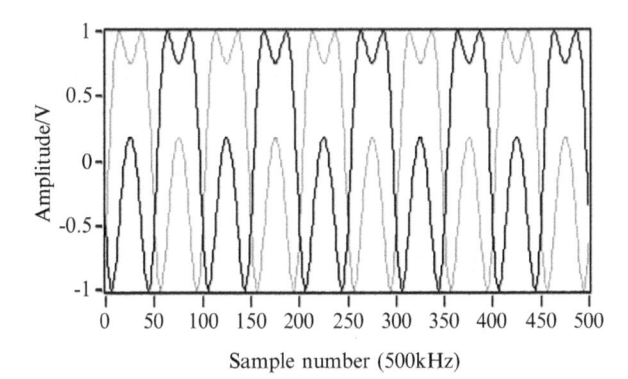

waveform of the two interference signals is shown in Fig. 1. The normalized signal which acquired according to the Eqs. 12 and 13 is shown in Fig. 2.

Figure 3 shows the demodulate signal which acquired by the demodulation algorithm that we proposed. The curve of the demodulation signal is relatively smooth. It reflects the original signal correctly. From the error curve (shown in Fig. 4), we can see that the error function presents a regular change along with the time, and the absolute error is less than ± 0.00017 V, the accuracy is relative higher. Construct the different analog signals with different frequencies and amplitudes, the signals can be demodulated correctly with the algorithm we proposed.

3.2 Experimental

To verify the feasibility of the demodulation algorithm, we utilized the fiber optic interferometer sensor to construct the experiment system. The schematic diagram of the distributed fiber-optic interferometer sensor is shown in Fig. 5. It includes a continuous wave super luminescent laser diode (SLD) source (40 nm bandwidth with a central wavelength of 1,310 nm and the output power is 5 mw), one isolator, one 3×3 fiber coupler, one symmetric 2×2 fiber coupler, one delay coil (Ld = 2 km), one Faraday rotation mirror, two photo-detectors (PD1 and PD2), one sensing fiber cable, D is the position of the disturbance acting at the sensing fiber.

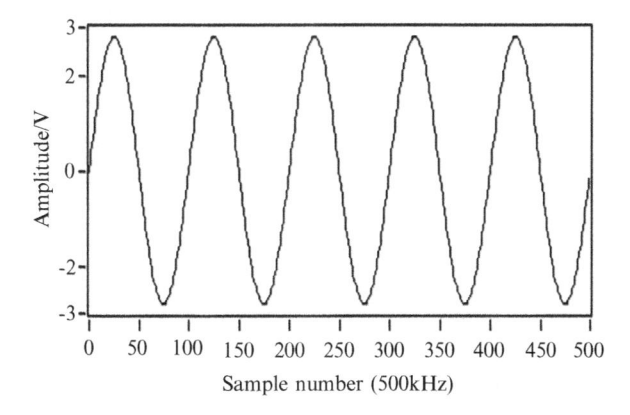

Fig. 3 Time sequences after demodulation

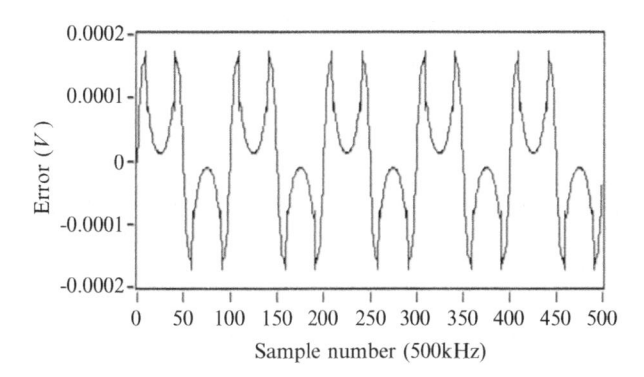

Fig. 4 Error of the demodulation signal

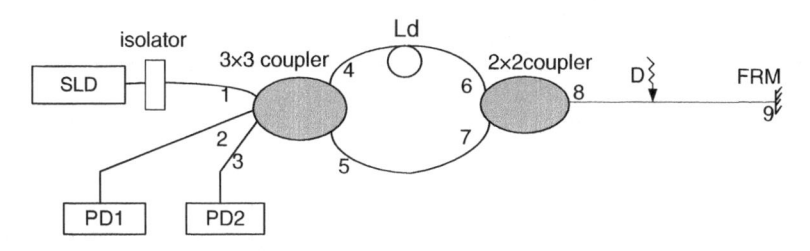

Fig. 5 The structure of optic fiber interferometer sensor

The two counter propagating beams can stably interfere at 3×3 coupler. The two beams have a delay 2nLd/c of the arrival time at the sensing section, which causes asymmetric phase shifts between the two counter propagating waves when a perturbation occurs. Intensity variation is converted to an electrical signal at the PD. The purpose of the demodulation is to acquire the external disturbance signals $\Delta\varphi(t)$ from the two phase modulation signals. The signals arrived at PD1 and PD2 can be expressed as:

Fig. 6 The two signals of PD1 and PD2

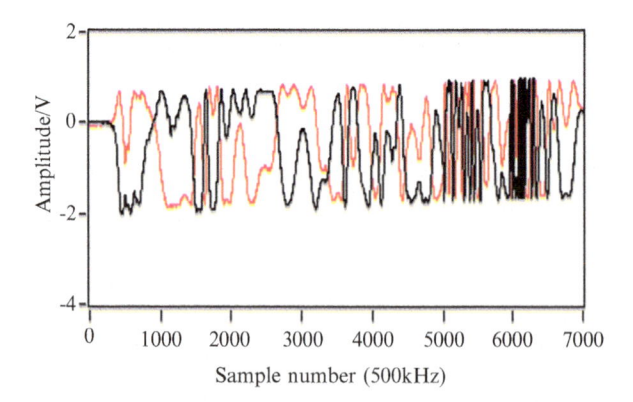

Fig. 7 The normalized signals of PD1 and PD2

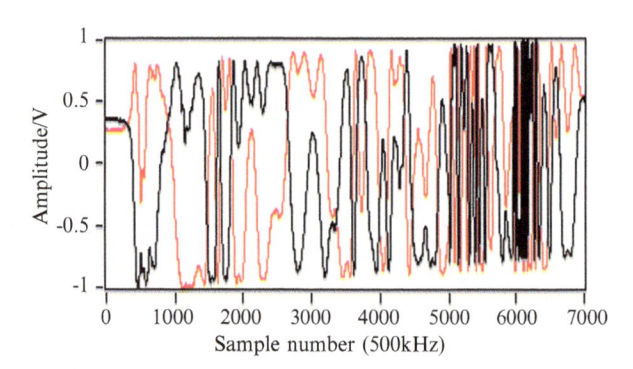

Fig. 8 The demodulation signal

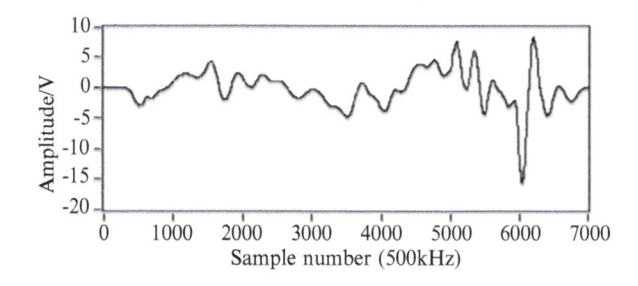

$$I_1(t) = A + B\cos(\Delta\varphi(t) + \phi) \qquad (20)$$

$$I_2(t) = A + B\cos(\Delta\varphi(t) - \phi) \qquad (21)$$

The two outputs of PDs acquired by a 16 bits acquisition card (the sampling rate is 500 kbps) according to Eqs. 3 and 5 is shown in Fig. 6. After normalization, signals of IP1 and IP2 are shown in Fig. 7. The demodulation signal of IP1 and IP2 is shown in Fig. 8. Figures 6, 7, and 8 shows that the disturbance signal can be demodulated according to the method we proposed.

4 Conclusion

From the above theoretical analysis, simulation and experiment, it is shown that the phase modulated signal can be demodulated with the two interferometer signals. The main features of the method are as follows : Firstly, the output phase difference of the two signals can be arbitrary(except for theπ); secondly, there is no need for uniform splitting ratio of 3×3 coupler (this overcome the drawback of the demodulation method based on the 3×3 coupler symmetry); thirdly, simple mathematical operations without complicated operations of the integral and the operations of the differential. The method we discussed overcomes the phase distortion caused by the instability of the light output power effectively and it has the high precision which can meet the practical applications.

Acknowledgements This paper is supported by "the Fundamental Research Funds for the Central Universities (2012B03814)" and "the National Natural Sciences Foundation of China (Grant No. 61273170)".

References

1. Parker TR, Farhadiroushan M, Handerek VA (1997) A fully distributed simultaneous strain and temperature sensor using spontaneous Brillouin backscatter. IEEE Photon Technol Lett 9:979–981
2. Fernandes N, Gossner K, Krisch H (2010) Low power signal processing for demodulation of wide dynamic range of interferometric optical fibre sensor signals. In: Proceeding SPIE 7653, Fourth European workshop on optical fibre sensors, Porto, Portugal, 765328
3. Hoffman PR, Kuzyk MG (2004) Position determination of an acoustic burst along a Sagnac interferometer. J Lightw Technol 22:494–498
4. Russell SJ, Brady KRC, Dakin JP (2001) Real-time location of multiple time-varying strain perturbations, acting over a 40-km fiber section, using a novel dual-Sagnac interferometer. J Lightw Technol 19:205–213
5. Xu HY, Xu Q, Xiao Q, Jia B (2010) Disturbance detection in distributed fiber-optic sensor using time delay estimation. Acta Opt Sinica 30:1603–1607
6. Danbridge A, Tveten AB, Giallorenzi TG (1982) Homodyne demodulation scheme for fiber optic sensors using phase generated carrier. IEEE Trans Microw Theory Tech 30:1635–1641
7. Koo KP, Tveten AB, Dandridge A (1982) Passive stabilization scheme for fiber interferometers using 3×3 fiber directional couplers. Appl Phy Lett 41:616–618
8. Jiang Y, Lou M, Wang HW (1998) Software demodulation for 3×3 coupler based fiber optic interferometer. Acta Photon Sinica 27:152–155
9. Xu Y, Li YQ, Jiang Y (2011) Application of 3×3 coupler based Mach-Zehnder interferometer in delamination patch detection in composite. NDT E Int 44:469–476

High Resolution Radar Target Recognition Based on Distributed Glint

Baoguo Li, Zongfeng Qi, Ying Zhou, and Jing Lei

Abstract To solve the problem of aspect angle sensitivity of range profile while used in radar target recognition, the concept "distributed glint" is presented for high resolution radar. The detailed deducing procedure of "distributed glint" is given through theoretical analysis. A new target recognition strategy based on range profile and distributed glint is proposed. Computer simulation proves that this method can greatly enhance the performance of recognition strategy based on range profile.

Keywords High resolution radar • Target recognition • Distributed glint

1 Introduction

Range profile, which is the projection of a target's backscattering on the radar line of sight, has been shown to be highly discriminative of target features. However, while used as the feature for radar target recognition, the main drawback of range profile is its aspect sensitivity, mainly caused by the interference of different

---

B. Li (✉)
The State Key Laboratory of Complex Electromagnetic Environmental Effects on Electronics & Information System of China, Luoyang, China

School of Electronic Science and Engineering, National University of Defense Technology, Changsha, China
e-mail: laglbg322@yahoo.com.cn

Z. Qi • Y. Zhou
The State Key Laboratory of Complex Electromagnetic Environmental Effects on Electronics & Information System of China, Luoyang, China

J. Lei
School of Electronic Science and Engineering, National University of Defense Technology, Changsha, China

scattering centers, which seriously degraded the performance of various kinds of target recognition methods exploiting range profile. To fully characterize the target, a larger number of range profiles at densely spaced aspect angles must be stored as templates [1]. The interference of different scattering centers is the main cause of target's angular scintillation (also called "glint"). Glint is harmful for tracking radar, and so engineers usually wish to suppress it [2]. Glint is a typical radar target characteristic signal. The most often used is linear glint offset, which means the deviation range, caused by glint. From target's glint, one can know some information about its physical structure. This leads us to consider using glint to enhance the target recognition performance. In this paper, the concept "distributed glint" is presented and a target recognition strategy using both distributed glint and range profile simultaneously is introduced. Computer simulation is performed to validate the efficiency of this method. The remainder of the paper is organized as follows. In section II, the concept of "distributed glint" is proposed and the detailed forming procedure is deduced through theoretical analysis. In section III, we give out a target recognition method using range profile and "distributed glint" simultaneously. Section IV gives out the simulation results of the above method. In section V, some conclusions and the forthcoming work are given.

2 Distributed Glint

For high resolution radar, range profile indicates the local RCS characteristic within each range cell. If the linear glint offset of each range cell is jointed together, the "distributed glint" is formed. Distributed glint can be induced through phase gradient method. Here we use a simplified multi-scatterers target model as Fig. 1 to illustrate the forming procedure of "distributed glint" [3]. In Fig. 1, target is represented by N statistically independent scattering centers; each scattering center is restricted within a cubic of L meters long. Radar transmits single carrier rectangular pulses as

$$s_0(t) = rect\left(\frac{t}{T_1}\right) \exp(-j2\pi f_0 t) \tag{1}$$

Where T_1 is the pulse width, f_0 is the signal's center frequency. Let R_k, a_k be each scattering centers' range from radar in radar's line of sight (LOS) direction and echo amplitude, respectively. Target's centroid is at the origin of the coordinates. So the echo signal can be expressed as

$$s(t) = \sum_{k=1}^{N} a_k rect\left(\frac{t - 2R_k/c}{T_1}\right) \exp(-j2\pi f_0(t - 2R_k/c)) \tag{2}$$

Fig. 1 Target model

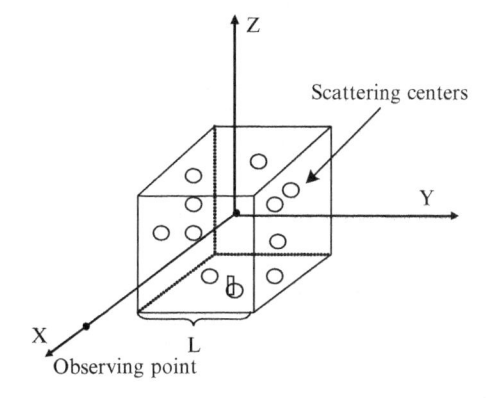

Remove the carrier, we get

$$s_1(t) = \sum_{k=1}^{N} a_k rect\left(\frac{t - 2R_k/c}{T_1}\right) \exp(j4\pi f_0 R_k/c) \qquad (3)$$

Sampling $s_1(t)$ inside the range gate, the result is

$$s_1(n) = \sum_{k=1}^{N} a_k rect\left(\frac{t_0 + nt_s - 2R_k/c}{T_1}\right) \exp(j4\pi f_0 R_k/c) \qquad (4)$$

Where $n = 1, 2, \ldots, M$, and M, t_s, t_0 denote the total number of sampling points, sampling interval and initial sampling instant, respectively. Letting $b(n,k) = rect\left(\frac{t_0 + nt_s - 2R_k/c}{T_1}\right)$ and $\varphi_k = (4\pi f_0/c)R_k$, we can get:

$$s_1(n) = |s_1(n)| \exp(j\varphi_E(n)) \qquad (5)$$

$$\varphi_E(n) = \tan^{-1}\left[\sum_{k=1}^{N} a_k b(n,k) \sin \varphi_k / \sum_{k=1}^{N} a_k b(n,k) \cos \varphi_k\right] \qquad (6)$$

The linear glint offset can be derived using phase gradient method [3]. The fluctuations of the normal to the phase front are determined by the gradient of φ_E, i.e.

$$\nabla \varphi_E = \frac{\partial \varphi_E}{\partial x} i_x + \frac{\partial \varphi_E}{\partial y} i_y + \frac{\partial \varphi_E}{\partial z} i_z \qquad (7)$$

Where i_x, i_y and i_z are the unit vector components in the LOS coordinates.

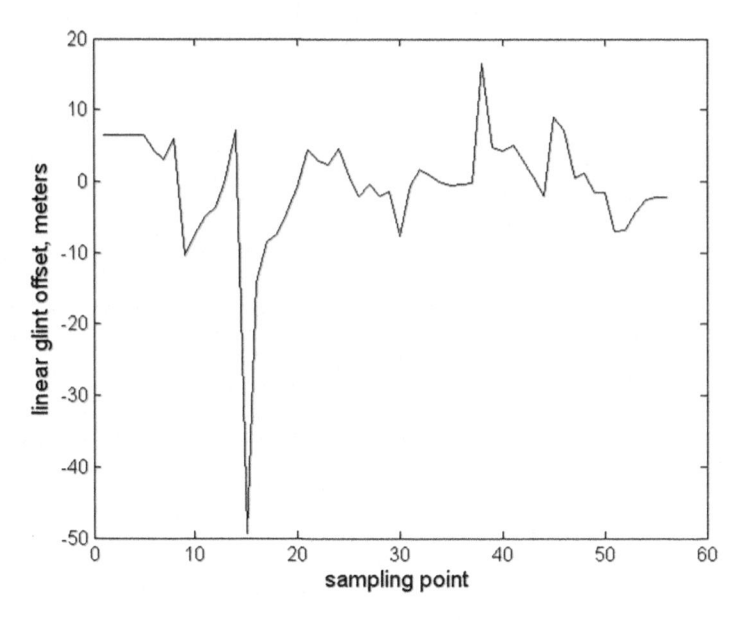

Fig. 2 Plot of a target's distributed glint

Take y direction as an example, the linear glint offset for each sampling point is as follows [2].

$$g_y(n) = \frac{\sum_{k=1}^{N}\sum_{l=1}^{N} a_k b(n,k) a_l b(n,l) y_l \cos(\varphi_k - \varphi_l)}{\sum_{k=1}^{N}\sum_{l=1}^{N} a_k b(n,k) a_l b(n,l) \cos(\varphi_k - \varphi_l)} \quad (8)$$

Here y_l is the scatter coordinate of y direction. Equation 8 shows how to compute the distributed glint of a target. Figure 2 is a plot of a target's distributed glint. In reality, it is impossible to know the accurate coordinate of each scatters.

In real battlefield environment, target is often noncooperative. For high resolution radar, distributed glint can be extracted from radar echoes as follows. Firstly, target can be separated into many range cells, for each range cell, the range and deviation angle can be computed, linear glint is the product of these two entries. From radar echoes, we can only get the absolute angle of each range cell, which subtracts the angle of range cell centroid is the deviation angle. So the key technologies lie in the estimation of range cell centroid. There are many methods to estimate the centroid, such as frequency diversity plus amplitude weighting [4], and it is not the emphasis of this paper. We lay the emphasis on the application of distributed glint.

3 Distributed Glint Based Target Recognition

It's well known that target glint (usually referred to linear glint offset) and target RCS (radar cross section) are not correlated, but the absolute value of target glint and target RCS are weakly negative-correlated [4]. The latter is the base of amplitude weighting glint suppressing algorithms and the former determines that glint can be used for target recognition. Since glint is not correlated with RCS, it must contain target structure information that RCS does not contain certainly. For high resolution radar, glint corresponds to distributed glint and RCS corresponds to range profile. So we propose a target recognition strategy based on distributed glint and range profile for high resolution radar.

In the recognition phase we have to select a suitable decision rule to compare the feature vector of an unknown target with feature vectors of known targets in the data base. Here we use the Matching-score rule introduced in Ref. [5]. The proposed recognition algorithm goes as follows:

Step 1: Construct the data base comprising the feature templates of M preselected targets at aspects of interest. These feature templates include range profiles, distributed glint in elevation direction and distributed glint in azimuth direction. Denote them by $g_i^r(\theta, \varphi), g_i^e(\theta, \varphi), g_i^a(\theta, \varphi)$, for each target i at aspect (θ, φ). The templates step angle obeys the criterion that the range motion of the scattering centers should not exceed a resolution cell if the target aspect changes an increment size.

Step 2: Calculate the normalized average correlation coefficients (ANCC) of three types of templates in the data base, respectively. Denote them by $\alpha_r, \alpha_e, \alpha_a$. These coefficients are stored as the fusion coefficients for future use.

Step 3: Fetch the input feature vector of an unknown target at a certain aspect, including range profiles, distributed glint in elevation direction and distributed glint in azimuth direction, and calculate the correlation coefficients with three types of feature templates at all aspects in the data base. Denote them by m_r, m_e, m_a, respectively.

Step 4: Use the fusion coefficients derived in step 2 to calculate the fusion matching-score of three types of features. The procedure can be expressed as $m_f = \alpha_r m_r + \alpha_e m_e + \alpha_a m_a$, where m_f is the fusion matching score.

Step 5: Identify the unknown target with one which has the maximum fusion matching score, m_f.

4 Simulations

Four simulated targets are chosen as the known targets. Target model is shown as Fig. 1. They each are combined with 80 scattering centers. All scattering centers are restricted within a cubic which is 20 m long. The coordinates of each scattering

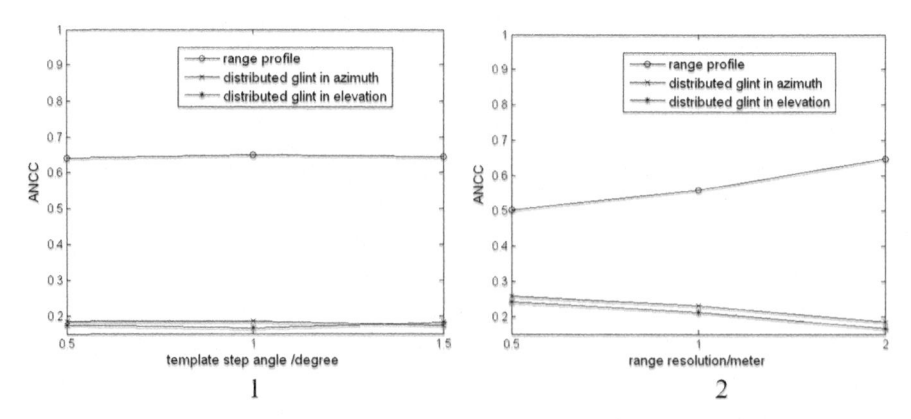

Fig. 3 ANCC of templates in the data base (1): constant range resolution 2 m, different step angle (2): constant step angle 1°, different range resolution

centers are uniformly randomly generated and then remain constant while aspect changes, the same is with each scattering centers' amplitude. The aspects region of interest is $(0, \pi/10)$. The transmitted radar signal's center frequency is 35 GHz. The range profile and distributed glint are supposed to have been aligned according to the center of the target already.

For range profile and distributed glint templates, ANCC indicate the similar degree of two consecutive templates. So the aspect sensitivity of each template can be characterized by ANCC. From Fig. 3 we can see that distributed glint is more aspect sensitive than range profile, so distributed glint can only be used to enhance the performance of target recognition based on range profile as an assistant feature. It is also shown from Fig. 3 that range resolution is the main factor that influences ANCC rather than the step angle.

Computer simulation is made to validate the performance of seven recognition methods based on three single feature and some combinations of each feature. The result is shown in Fig. 4. Seven recognition methods are symbolized by character "a" to "g" in the x-coordinate of Fig. 4, the relations are as follows:

a — only use range profile;
b — only use distributed glint in azimuth direction;
c — only use distributed glint in elevation direction;
d — combination of a and b (maximum fusion matching score);
e — combination of a and c (maximum fusion matching score);
f — combination of a, b and c (maximum fusion matching score);
g — combination of a, b and c (using vote rule);

It is seen from Fig. 4(1) that for all kinds of recognition methods, under constant range resolution, average recognition rate decreases when step angle increases, and vice versa. From Fig. 4(2) we can see that for all kinds of recognition methods, under constant step angle, average recognition rate decreases when range resolution

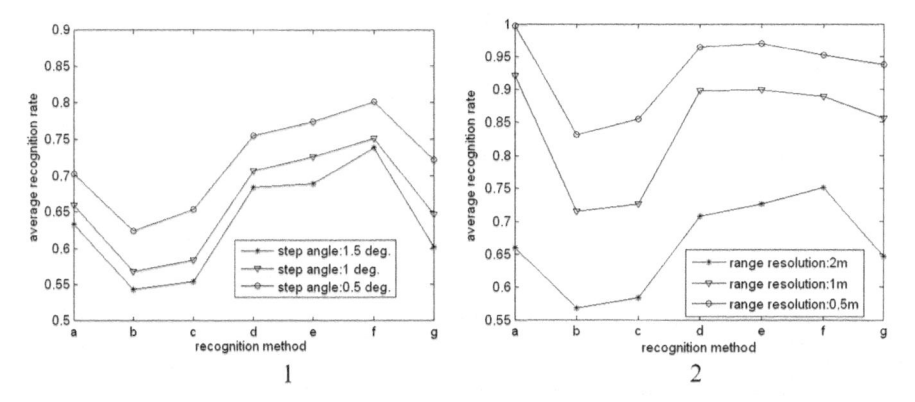

Fig. 4 Average recognition rate of each recognition method (1): constant range resolution 2 m, different step angle (2): constant step angle 1°,, different range resolution

decreases, and vice versa. The performance of method (a) is superior to method (b) and (c) under various conditions. This proves that distributed glint can only be used as an assistant feature of range profile for its much more serious aspect sensitivity than range profile. We can also see that the performance of fusion recognition methods (d, e, f) is far superior to that of recognition method using single range profile feature when range resolution is relatively low (such as 2 m in the simulation). When range resolution is relatively high (such as 1 and 0.5 m in the simulation), the fusion methods' recognition performance is nearly equivalent to that of the recognition method using single range profile feature. This phenomenon can be interpreted as: when range resolution is low, each range cell of the target contains many scattering centers, so distributed glint contains enough information about target's structure and it will enhance the performance of recognition method using single range profile. On the other hand, when range resolution is high, almost every scattering center is separated into different range cell, distributed glint contains many range cells of zero value and it does not contain enough information about target's structure.

5 Conclusion

In this paper, we proposed a new concept of "distributed glint" and discussed the application of this new target feature in radar target recognition. Through computer simulation, we see that target distributed glint can explicitly enhance the performance of target recognition based on target range profile. In real engineering applications, the accuracy of distributed glint directly affected the performance of target recognition method we proposed. How to improve the estimation accuracy of range cell centroid is the most critical problem, and this is the emphasis of our future work.

Acknowledgements The authors are with the State Key Laboratory of Complex Electromagnetic Environmental Effects on Electronics & Information System of China and the School of Electronic Science and Engineering, National University of Defense Technology. This work was supported by the fund of "The State Key Laboratory of Complex Electromagnetic Environmental Effects on Electronics & Information System, China (No.CEMEE2012K0303B)" and "National Natural Science Foundation of China (No.61101074, 61101097)".

References

1. Xuejun Liao, Zheng Bao (1998) Circularly integrated bispectra: novel shift invariant features for high-resolution radar target recognition. Electron Lett 34(19):1879–1880
2. Dongtao Zhao, Hao Wang (2010) Glint suppression based high resolution radar angle tracking. Electron Sci Tech 23(11):67–69
3. Sandhu GS, Saylor AV (1985) A real-time statistical radar target model. IEEE Trans A E S 21(4):490–507
4. Hong Cheng Yin, Pei Kang Huang (2008) Further comparison between two concepts of radar target angular glint. IEEE Trans Aerosp Electron Syst 44(1):372–380
5. Hsueh-Jyh Li, Sheng-Hui Yang (1993) Using range profiles as feature vectors to identify aerospace objects. IEEE Trans Antennas Propag 41(3):261–268

A Differential Capacitive Viscometric Sensor for Continuous Glucose Monitoring

Zhijun Yang, Meng Wang, Youdun Bai, and Xin Chen

Abstract The development of glucose micro sensors for clinical use is driven by the aim of automatic blood glucose normalization in diabetic patients. An improved affinity viscometric sensor for continuous glucose monitoring (CGM) is presented by using a micro-electro-mechanical system (MEMS) with a differential capacitor. A numerical model using Reynolds equation is used to simulate the dynamic response under different viscosities, and the relationship between capacitance and viscosity is revealed. Compared to the previous version presented by Columbia University, the sensor designed in this paper has enhanced the capacitor by introducing a differential capacitance, which also avoided volume changes of the air and polymer solution chambers during the vibration, increasing the linear range of the sensor. In addition, the simulation results show that the sensor can be driven by a Gaussian Pulse resulting in a significant power saving, when compared to a sinusoidal excitation.

Keywords Continuous glucose monitoring • Biosensor • Differential capacitance • Viscometer

1 Introduction

The development of glucose sensors for clinical use is driven by the aim of blood glucose normalization in diabetic patients [1]. Close monitoring of daily blood sugar levels reduces the risk of diabetes-related complications by allowing timely identification and correction of hyperglycemia as well as hypoglycemia, a condition that typically results from excessive insulin uptake or inadequate glucose intake.

Z. Yang (✉) • M. Wang • Y. Bai • X. Chen
School of Electromechanical Engineering, Guangdong University of Technology,
Guangzhou, China
e-mail: yangzj@gdut.edu.cn

This can be most effectively achieved by continuous glucose monitoring (CGM), which involves continuous measurements of physiological glucose levels.

There are two types of CGM devices that have been developed. The first one is an electrochemical method, including enzymatic or non-enzymatic reactions [2]. Electrochemical methods are capable of sensitive and specific glucose detection. However, the irreversible consumption of glucose in the electrochemical detection induces a potential change in the equilibrium glucose concentration in the tissue and thus affects the actual measured glucose level. In addition, the rate of glucose consumption is diffusion limited [3]. This lack of reliability has been severely hindering CGM applications in relation to practical diabetes management.

Another type of CGM is a physical method, in which the glucose is not consumed. More importantly, affinity sensing is considerably more stable, as the deposition of biological material on the implanted sensor surface results only in an increased equilibrium time, without any changes in the measurement accuracy. A widely used affinity sensing technique is based on concanavalin A (Con A), whose specific binding to glucose can be detected via methods such as fluorescence [4, 5] and viscosity [6]. The integration of capacitive detection within the MEMS affinity glucose sensor [7]represents a major improvement to the previously reported devices [6]. This viscosity change causes variations in the viscous damping on the magnetically driven vibration of a flexible diaphragm, whose deflection results in changes of the capacitance between two electrodes, one of which is embedded in the diaphragm. Thus, the capacitance change can be measured to detect the damped diaphragm vibration and further determine the interstitial fluid glucose concentration. However, the tension of the membrane changes with the compression of the trapped air, so the amplitude of vibration cannot be too large (less than 1 μm) in order to keep a linear relationship, such that the capacitance and its changes are very small [8].

In this paper, we introduce a relatively rigid silicon beam supported by a Parylene diaphragm in the middle, the magnetic force is applied to beam on the capacitor side, causing the beam to vibrate in a seesaw manner. The movement of the beam causes the diaphragm to deform in a seesaw manner "∽ ∾", avoiding any change in the overall volume of the air chamber and consequent change in the diaphragm tension. The damping forces caused by the movement of the membrane affect the beam's motion, further increasing the sensitivity of the sensor. In addition, the simulation results show that the sensor can be driven by a Gaussian Pulse resulting in a significant power saving when compared to a sinusoidal excitation, and the sensitivity can be adjusted by the clearance between the beam and walls.

2 Principle and Design

The affinity glucose sensor is based on a relatively rigid beam that is supported in the middle of a parylene membrane between two micro-chambers. The structure of the glucose sensor is shown in Fig. 1. In the figure, one end of the beam is situated inside a polymer solution micro-chamber, which is filled with the solution of a biocompatible polymer that binds specifically and reversibly with the glucose, and

Fig. 1 The structure of the designed glucose sensor

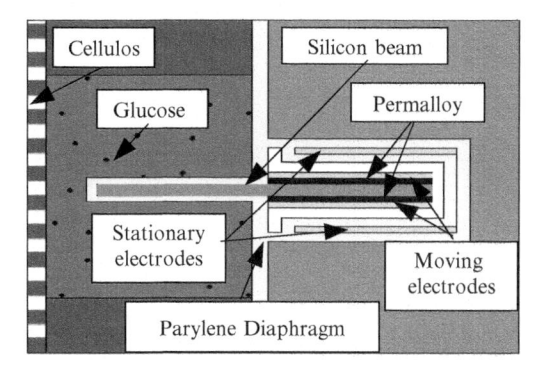

is sealed by a cellulose acetate semi-permeable membrane, which allows the glucose to permeate into and out of the chamber while keeping the glucose-sensitive polymer from escaping. The silicon beam is coated by a parylene film to avoid direct contact with the polymer solution.

The other end of the beam is located inside a micro air chamber and is coated with a Permalloy layer so that it can be excited by a magnetic force. The moving electrodes are situated on the beam, together with stationary electrodes on the chamber walls to form the differential capacitor. The differential capacitor is more sensitive to the vibration of the moving electrode of the capacitor [9].

The beam is excited by the microcoils that cause the Parylene membrane bending vibration. The electrodes on the beam are insulated by the Parylene to avoid direct contact with the polymer solution.

3 Modelling and Simulation

In order to investigate behavior of the presented glucose sensor, a multiphysics model has been developed using Reynolds equation for thin film lubrication, combining structural mechanics, fluid dynamics and electrostatic analysis, to analyze the vibration of the Parylene diaphragm and beam structure, and the capacitance of the differential capacitor. The parameters are shown is Fig. 2, and the detailed values are listed in Table 1.

The materials used are silicon, Parylene C, Permalloy (80 %Ni, 20 %Fe). The prestress of the Parylene diaphragm is set at $35 MPa$. The material properties for polymer solution with the mass density $1000 kg/m^3$.and the viscosity values studied are 10, 20, 30, 40 and 50 cP. The mass density and viscosity of air is $1.293 kg/m^3$, and $17.9 \times 10^{-6} Pa \cdot s$, respectively.

The fluid type can be determined by the relaxation time, where the relaxation time can be estimated according to the properties of the polymer solution. It is not known exactly what the fluid characteristic the polymer is, the possible characteristics are rigid rod, free draining and non free draining [10]. It has been proved that no matter what kind of molecular type it is, the relaxation frequency is

Fig. 2 The original parameters of the model

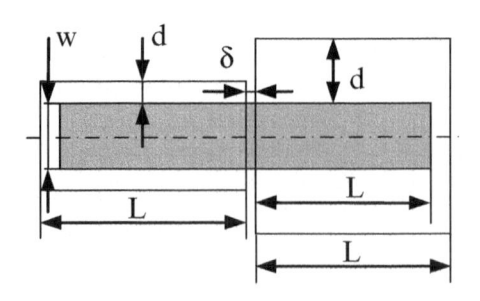

Table 1 Dimensions of the sensor

Parameters	Value (μm)	Parameters	Value (μm)
Thickness of Parylene diaphragm (δ)	10	The length of air chamber (L_a)	1,000
Width of silicon beam (w)	1,000	Thickness of silicon beam (δ_s)	50
The length of Silicon beam in both sides (L)	900	Clearance between silicon beam and air chamber (d_a)	10
The length of polymer solution chamber (L_s)	1,000	Clearance between beam and solution chamber (d_s)	10

Fig. 3 The vibration shape of the beam structure

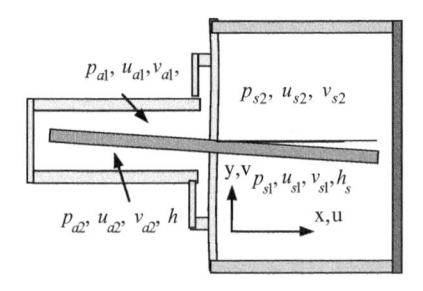

much higher than excitation frequency, so the fluid can be recognized as Newtonian fluid [8].

The velocity of the fluid

$$v < A\omega = 10^{-5}m \times 1000s^{-1} = 10^{-2}m/s \quad (1)$$

where ω and A are the circular frequency and amplitude of the response, respectively. The velocity of the fluid v is only $2.6 \times 10^{-5}Ma$, much less than $0.3Ma$, so the fluids are incompressible.

For the parylene diaphragm and silicon beam structure, the vibration (shown in Fig. 3) can be equivalent as

$$I\ddot{\theta} + c\dot{\theta} + k\theta = M_{ext} \quad (2)$$

Fig. 4 Different flow in the glucose sensor

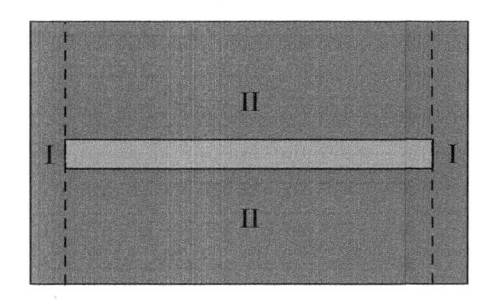

where I, c and k are the inertial, damping and stiffness, respectively. θ is the rotation of the membrane, and M_{ext} is the excitation moment caused by the magnetic force and interacted fluid flow pressure.

The length of beam is 900 μm, if the maximum displacement is 60 μm, the rotation of the parylene diaphragm is atan(60/900) = 3.81°, which is within the linear vibration of the diaphragm. The damping force of membrane is much smaller than that caused by polymer solution, can be neglected for simplicity.

Once the excitation moment M_{ext} is known, it is easy to solve Eq. 2. However, the flow inside micro-chambers are complicated, the pressure of the fluid flow is hard to get. Fortunately, if we divided the flow domain into sections I and II (shown in Fig. 4), and assume that the fluid is incompressible, the volume flow between two sections II all passes through the gap between the tip of beam and the wall of chamber as it is much wider than the gap between the sides of the beam and the chamber walls. The length of beams are much longer than the thickness, it can be seen that section I is oscillatory flow, and section II is squeezed film flow.

According to thin film lubrication theory, the fluid inside micro-chamber can be described by Reynolds equation [11].

$$\nabla \cdot \left[\left(\frac{\rho_i h_{ij}^3}{\mu} \right) \nabla p_{ij} \right] = 12 \frac{\partial (\rho_i h_{ij})}{\partial t} + 6 \nabla \cdot (\rho_i h_{ij} u_{ij}) \quad (3)$$

where ∇ represent the gradient operator, and ρ, h, μ, p and t are mass density, thickness, viscosity, pressure and time, respectively. The subscript $i = a, s; j = 1, 2$, a means air, s means polymer solution.

$$h_{ij}(x) = a_i \pm x_i \sin(\theta) \quad (4)$$

$$u_{ij}(x) = x_i \dot{\theta} \cos(\theta) \quad (5)$$

$$v_{ij}(x) = x_i \dot{\theta} \sin(\theta) \quad (6)$$

Where x is the distance starts from the membrane, u and v are the horizontal and vertical velocity, respectively. In addition, there still exists shear forces between the edges of the beam and the walls.

For the section I, the flow is a typical Stokes' second law problem [12], the shear force f_e is

$$f_e = \mu_i \frac{\partial v_i}{\partial d_i} \tag{7}$$

For the tip edge, assume that the fluid volume change all passes through the clearance between the tip and the wall. The volume change ΔV is

$$\Delta V_i = \pi L_i^2 w \cdot \frac{\theta}{2\pi} = \frac{1}{2} L_i^2 w \theta \tag{8}$$

The volume flow rate between the tip and the wall is

$$\dot{V}_i = \frac{\Delta \dot{V}_i}{w(L_i - L\cos(\theta))} = \frac{1}{2} \frac{L_i^2 \dot{\theta}}{L_i - L\cos(\theta)} \tag{9}$$

Therefore, the shear force between the tip and the wall ft is

$$f_t = \mu_i \frac{\partial v_i}{\partial r_i} = \mu_i \frac{L_i \dot{\theta} + \dot{V}_i}{L_i - L\cos(\theta)} \tag{10}$$

The electric force of the capacitor f_e is

$$F_e = \frac{\varepsilon A (\Delta V)^2}{2 h_a^2} \tag{11}$$

where ε is relative permittivity of air. Therefore, the total moment of the vibrating structure is

$$M_{ext} = \int_0^L (p_{i1} - p_{i2}) wx dx + 2 \int_0^L x f_e dx + w f_t L \cos\theta + \frac{1}{2} f_m L \tag{12}$$

where f_m is the total magnetic force applied to the beam.

The capacitance of the differential capacitor is

$$C = \int_0^L \frac{\varepsilon w}{4\pi k h_{a1}(x)} dx - \int_0^L \frac{\varepsilon w}{4\pi k h_{a2}(x)} dx \tag{13}$$

C is capacitance, k is electrostatic constant $k = 9.0 \times 10^9 N \cdot m^2 / coulomb^2$.

4 Results and Discussions

The vibration of the beam diaphragm structure is shown in Fig. 3. It can be seen that the main vibration is the bending of the diaphragm, which enable the volume of both air chamber and solution chamber to remain constant.

Assumed that the excitation force is sinusoidal function, the displacement response is shown in Fig. 5. The relationship between viscosity and capacitance is almost linear (Fig. 6), and the sensitivity is $-3 \times 10^{-8} pF/cp$.

However, the continuous excitation would cost too much energy. In order to save energy, the exciting force is a Gaussian pulse instead of sinusoidal function, the excitation magnetic forces with magnitude of $0.9\mu N$ in total.

The displacement response of a point at the tip of the silicon beam at air chamber side under different viscosities 10, 20, 30, 40 and 50 cP are shown in Fig. 7. It can be seen that the displacements are decreasing with viscosities.

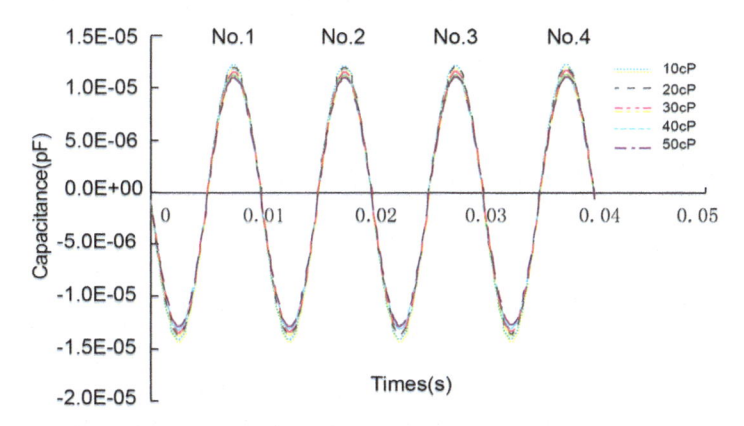

Fig. 5 The displacement response of sinusoidal excitation

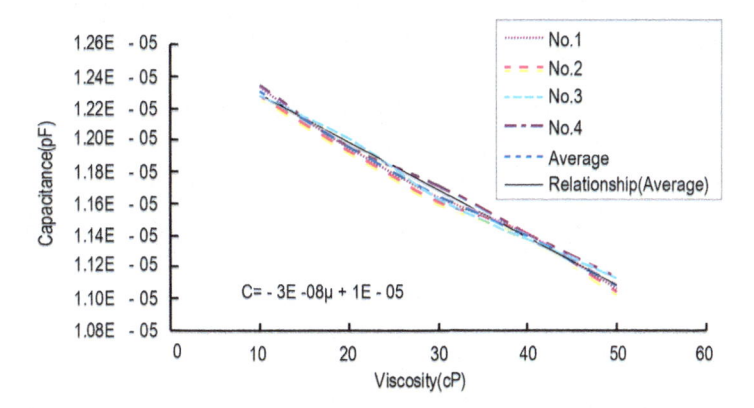

Fig. 6 The relationship between amplitude and viscosities

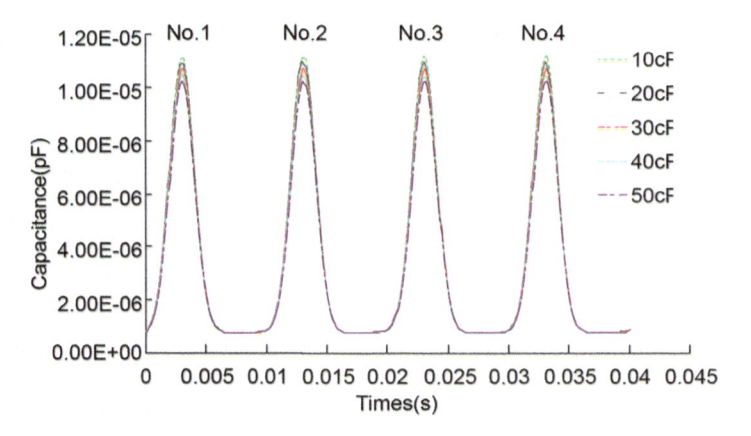

Fig. 7 The displacement response of Gaussian pulses with different viscosities (Original)

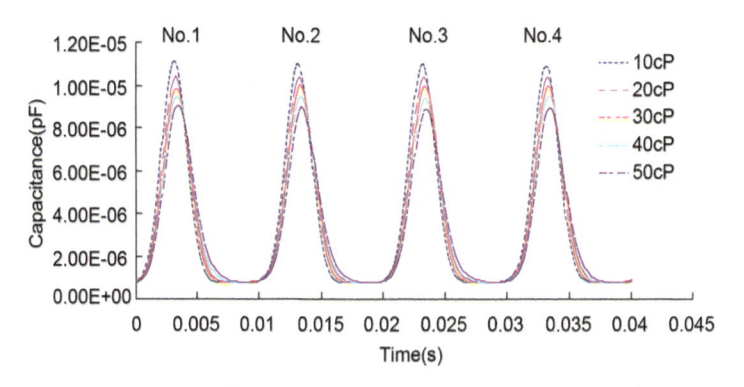

Fig. 8 The displacement response of Gaussian pulses with different viscosities (Improved)

The relationship between amplitude and viscosities is almost linear, the sensitivity is $-2 \times 10^{-8} pF/cp$, and the error is within 0.5 %.

From Eq. 7, one can see that the clearances d_a and d_s are very important to the damping forces. We increase the clearance in the air chamber to $d_a = 50\mu m$, while decreasing the one in the polymer solution chamber to $d_s = 10\mu m$. The results are shown in Fig. 8, the relationship between amplitude and viscosities is linear and the sensitivity is $-4 \times 10^{-8} pF/cp$. As a result, the sensitivity is doubled.

When the capacitance is measured, the dynamic viscosity of the polymer solution can be calculated by

$$\mu = -2.5 \times 10^7 C - 250 \qquad (14)$$

using the relationship between viscosity and glucose concentration [7], the glucose concentration can be easily detected.

The relationship has been estimated by Eq. 14, and the results of the original design and improved design show that the relative error of the improved design are within 0.341 %, which is much low than original design (0.945 %).

5 Conclusion

This paper presents an improved differential capacitance based viscometer glucose sensor. The operating principle is the vibration of a compound micro beam and pre-stressed membrane structure which is excited by Gaussian pulse. The vibration amplitude can be changed according to viscosity of the polymer solution. The sensitivity of the viscosity can be adjusted by the clearance between the beam and the walls. The relationship between vibration amplitude and viscosity is almost the same as sinusoidal excitation. Therefore a Gaussian pulse may be used as the exciting force, significantly reducing the power consumption of the sensor and enable a longer battery life.

Acknowledgements This work is supported by Natural Science Foundation of China (U1134004, 50905033), Guangdong Innovative Research Team Program (201001G0104781202), The National Basic Research Program of China (2011CB013100-G, National key technology support program (2012BAF12B10), Specialized Research Fund for the Doctoral Program of Higher Education of China (20094420120001)

References

1. Clark HR, Barbari TA (1999) Modeling the response time of an in vivo glucose affinity sensor. Biotechnol Prog 15:259–266
2. MiniMed Paradigm (2006) REAL-time insulin pump and continuous glucose monitoring system[P]. Medtronic MiniMed, San Antonio
3. Tracey Neithercott (2011) Continuous glucose monitoring system[J]. Diabetes Forecast 64:44–46
4. Schultz JS, Mansouri S et al (1982) Affinity sensor – a new technique for developing implantable sensors for glucose and other metabolites[J]. Diabetes Care 5(3):245–253
5. Schultz J, Sims G (1979) Affinity sensors for individual metabolites[J]. Biotechnol Bioeng Symp 9:65–71
6. Zhao yongjun, Li siqi et al (2007) A MEMS viscometric sensor device for continuous glucose monitoring[J]. J Micromech Microeng 17(12):2528–2537
7. Huang xian, Li siqi et al (2009) A capacitive MEMS viscometric sensor for affinity detection of glucose[J]. J Microelectromech Syst 18(6):1246–1254
8. Yang zhijun, Robert Kelley et al (2011) A numerical investigation of a capacitive viscometer with fluid–structure interaction using equivalent modeling[C]. Adv Mater Res Adv Mater Process, PTS 1–3(311–313):2423–2429
9. Lotters JC, Olthuis W et al (1999) A sensitive differential capacitance to voltage converter for sensor applications[J]. IEEE Trans Instrum Meas 48(1):89–96

10. Block H, North AM (1970) Dielectric relaxation in polymer solutions, advances in molecular relaxation processes. Elsevier Publishing Company, Amsterdam – Printed in the Netherlands, pp 309–374
11. Sherman FS (1990) Viscous flow[M]. McGraw-Hill Higher Education, New York, pp 746–747
12. Batchelor GK (1998) An introduction to fluid dynamics[M]. Cambridge University Press, New York, pp 302–303

An Improved Secure Routing Protocol Based on Clustering for Wireless Sensor Networks

Lin Chen and Long Chen

Abstract The security and efficiency are essential primary points for Wireless Sensor Networks (WSNs) when designing routing strategy. This paper proposes a secure routing solution based on LEACH protocol and clustering method in which system security is integrated into sensor node and clusters are changed dynamically and periodically according to node mobility, and it is different with the traditional encryption mechanism by using key clustering homemade management to reach the overall security. The simulation results show that the proposed secure routing protocol improve the survivability of node more efficiently in a harsh sensor network environment. Through clustering management dynamically, the immunity of WSNs has been enhanced, and it can be concluded that this routing protocol improved the network security effectively.

Keywords Wireless sensor network • Routing protocol • Security • LEACH

1 Introduction

A wireless Sensor Networks (WSNs) is an ad hoc wireless telecommunication network which embodies a number of tiny, low-powered sensor nodes densely or sparsely deployed in the area of interest to accomplish a particular mission like habitat monitoring, agricultural farming, battlefield surveillance etc. [1]. The extensive rise of using WSNs in diverse applications such as hostile, unattended, and inaccessible environments, have mandated the users to be more assured about the security and efficiency compared to the survivability. As long as security schemes

L. Chen (✉) • L. Chen
College of Computer Science, Yangtze University, Jinzhou, China
e-mail: chenlin@yangtzeu.edu.cn

provide confidentiality, authentication, and integrity, which are critical for such applications, a secure and survivable infrastructure is always desired. Network survivability has been defined as the ability of the network to fulfill its mission in the presence of attacks and/or failures in a timely mode [2]. In such hostile environments, the information exchanged between two communicating parties might include highly sensitive data that must be safeguarded, therefore security investigation and efficient routing strategy for WSNs have been a continuous research hot topic in the recent decades [3].

LEACH (Low Energy Adaptive Clustering Hierarchy) is a classical and hierarchical routing protocol which is based on the idea of clustering to lengthen the network lifetime and forward data reliably, but LEACH protocol is not secure enough to against all kinds of network threats, so it has no ability to resist some attacks in network layer. Therefore, a variety of improved routing protocols based on LEACH are investigated to improve security of WSNs, such as RLEACH protocol which is based on LEACH and embed encryption and authentication technologies to prevent unauthorized node to participate in route creating and data transmission works. However, it is vulnerable to a number of security attacks including jamming, spoofing and replay, the reason is that the LEACH protocol is a cluster-based protocol and rely on their Cluster Heads(CHs) for routing to a great extent, and some attacks involving CHs are most dangerous, if a attacker intrude networks successful and become a CH in cluster-based networks, it can stage attacks such as sink hole and data selective forwarding, to disrupt the WSNs network, the intruder may also leave the routing alone and try to inject bogus sensor data into the network to destroy the normal work [4].

A Homemade Low Energy Adaptive Clustering Hierarchy (HLEACH) protocol for WSNs is proposed to strengthen sensor node security in this paper. HLEACH is different with traditional encryption mechanism using KEY management which consumes considerable network and computing resources. In order to resolve the problem about KEY management, HLEACH protocol adopts distributed method between each clusters to reduce resource management consumption and improve routing efficiency, and during nodes migrating randomly, the KEY of the sensor nodes will change dynamically according to cluster, if one node is captured, the entire network security will not been affected.

2 The LEACH Protocol and Vulnerabilities

2.1 The LEACH Protocol Description

The LEACH protocol consists of three steps to create cluster and cluster membership.

In the first step, every node may decide probabilistically whether or not to become a CH for the current round or period. These candidate CH nodes will broadcast a message advertising this fact, and nodes at some extend can receive these broadcasting messages. In this step, the CSMA-MAC protocol is used to avoid channel collision.

In the cluster joining step, the remaining nodes select a cluster to join based on the largest received signal strength from a candidate CH, and send a join request message using CSMA-MAC to this candidate CH to show their intentions. At the end of this step, distributed clusters geographically will emerge, and membership between the CH and its receivers will also generate in a period.

Once a CH receive all the join requests, the confirmating step starts with the CH broadcasting a confirmation message that includes a time slot schedule to be used by their cluster members for communication during the steady state phase.

After the clusters are set up, the network will be a steady state phase, where actual communication between sensor nodes, CHs and the BS(base station) will take place. Each node knows when it is turn to transmit their sampling data in their environment according to the time slot schedule. The CHs collect messages from all their cluster members, and aggregate these data according some aggregating strategies, then send the aggregating results to the BS.

2.2 Security Vulnerabilities

Like most routing protocols for WSNs, LEACH does not take into account the security and is also vulnerable to a number of network attacks. In contrast to more conventional multi-hops transmitting schemes, member nodes around the CHs and BS are especially attractive for some attackers, but CHs in LEACH communicate directly with the BS, may be anywhere geographically in the network, and can change dynamically from round to round. All these characteristics make it harder for an adversary to identify CHs, so the dynamic CHs strategically are more important nodes to avoid network attack. Because LEACH is a cluster-based protocol, relying on the CHs fundamentally for data aggregation and routing, attacks involving CHs are the most dangerous.

LEACH routing protocol may extend the entire network life cycle, but not get a good improvement in terms of security. Because KEY management is an effective method to identify illegal nodes, and the dynamic CHs also make attackers harder to separate CHs and cluster members, the two ideas can be employed to improve WSNs' security. Therefore, in order to further improve the security, the proposed HLEACH uses dynamic changeable CHs and KEY encryption management to check the validity of each node in the identity authentication phase [5, 6].

3 The Proposed HLEACH Protocol

The setup of the proposed HLEACH protocol consists of four steps: the choice of Cluster Head nodes (CHs), CHs broadcasting, CHs identity authentication and cluster internal management.

1. Choice phase of CHs. The CHs choice of HLEACH protocol is same as that of the LEACH protocol. The choice of CHs is based on the node number in network, the total number of CHs and node distributed geographically. In this choice procedure, each candidate generates a value between 0 and 1 randomly, if this value is lower than a certain threshold, it will become a CHs and then enter the next phase.
2. CHs broadcasting message phase. After a candidate CH becomes a CH, it will broadcast this fact in the whole network through the radio, and other sensor nodes in a certain radius range in the network decide whether or not to join this clusters based on the signal strength of the received information, if the node decide to join in, then it will notify the appropriate CH, in a stable state a cluster will be established. Finally, the CHs allocate the interval of transmitting data for each node in the cluster.
3. CHs identity authentication phase. This step will ensure whether that CH is a legitimate node, and to make sure that the legitimacy of the CH has a crucial role in WSN security. Each node in the WSNs has an unique ID distributed by the base station, when a new node joins into in a cluster, it should get a ID form the BS, if a node become CH in the first phase, it should report its own ID to the BS, and the other nodes in the cluster can also receive this report messages and save the ID. Then the BS will validate this ID, if the ID verification is successful, the BS will send a broadcast message to all sensor nodes and notify that the CH is legitimate. In the same time the BS sends a public KEY to this CH, this public KEY will be used between CHs to ensure the security of information transmitting.
4. Cluster internal management phase. CHs notify its ID to every node in the cluster through the radio. The nodes use the CH's ID as the KEY to send messages, there are many encryption algorithms which may be used, in our simulation the DES encryption algorithm is adopted.

Figure 1 shows the relationship between KEYs and nodes, letter N delegate sensor node, A and B are the ID of CHs, C is the public KEY. The left side of Fig. 1 shows that when the nodes send the message to the CH which the KEY is CH's ID-A, the right side of Fig. 1 shows that when the nodes send the message to the CH which the key is CH's ID-B. When the CH sends a message to the BS or other CHs which the KEY is the public key-C.

In the stable phase, the nodes sample data and transmit them to the CH. The CHs collect data send by all their cluster members, and aggregate these data, and then send the result data to the BS. The steady state phase lasts much longer compared with the setup phase. The network will be restarted to establish the cluster and transmit sampling data in the next cycle repeatedly.

Fig. 1 Key distribution

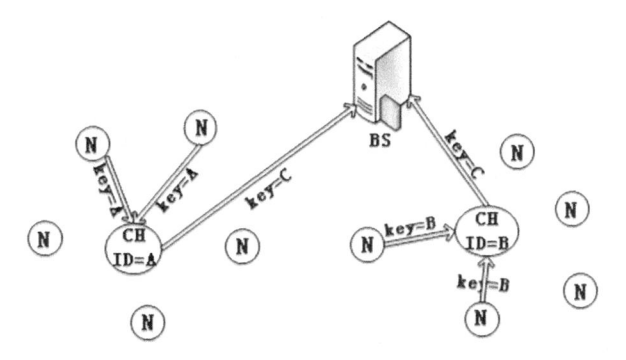

4 Performance Evaluation

In order to evaluate the performance of HLEACH protocol, NS2 simulation tools are used as experiment platform with 50 nodes and 2 malicious nodes in 1,200 × 1,200 range, simulation time is 1,000 s. In simulation procedure, two different parameters are adopted to measure performance.

Memory consumption: WSNs has some applications in memory, but sensor nodes have a tiny memory volume and can't hold a bigger memory, the lack of memory has always been a difficult problem in WSNs.

Network lifetime: Sensor node power supply is battery powered and is always a central point of the research. There are a lot of programs to save energy consumption. The life cycle is one of the most important indicators to measure the feasibility of a protocol.

The following figures show the performance curve which interference by malicious nodes.

Figure 2 shows the memory consumption of each node in the network, although HLEACH protocol embedded encryption mechanism, when the network was attacked, its memory consumption is significantly less than that of LEACH protocol. Figure 3 shows the node lifetime of HLEACH is reduced in the case of attack. The improved agreement to extend the life cycle of WSNs, it also proved that the improved protocol is feasible.

5 Conclusions and Future Work

WSNs is a low cost, fast ad hoc networks of anti-survivability characteristics and is applied strong and widely in many areas of military, industrial, etc. and needs a secure solution to solve many common challenging problems. The security is a prerequisite of a new technology application. In this paper, HLEACH is proposed to

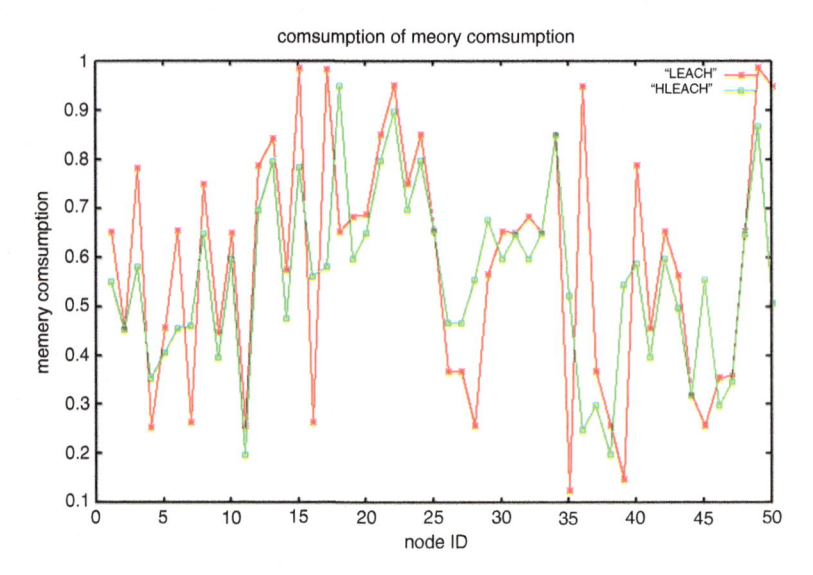

Fig. 2 Analysis of network memory

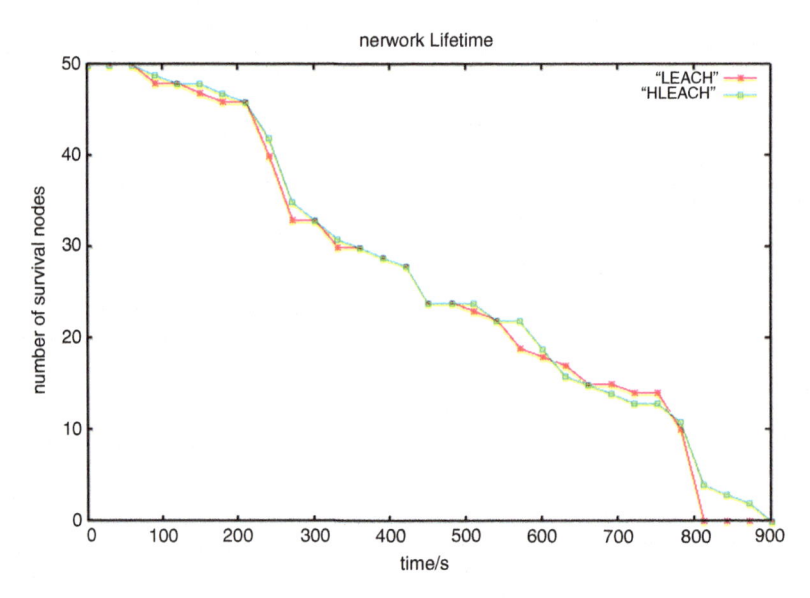

Fig. 3 Analysis of network lifetime

improve routing protocol on the basis of the LEACH protocol, which transmits the information through the sub-cluster autonomy and multi-channel, it can prevent some network attacks. But it adds a certain amount of memory and energy consumption, which is a gap of this paper, and it also needs further improvement.

Acknowledgements This work is supported by the Hubei Natural Science Foundation. I would like to show my deepest gratitude to this organization, without its support we will not finish this paper very smoothly.

References

1. Karlof C, Wagner D (2003) Secure routing in wireless sensor networks: attacks and countermeasures. Elsevier's Ad Hoc Networks Journal, Special Issue on Sensor Network Applications and Protocols 1:293–315
2. Jing Dengrichard, Shivakant, Mishra (2006) Decorelating wireless sensor network traffic to inhibit traffic analysis attacks. Special Issue of Elsevier Journal of Pervasive and Mobile Computing Journal (PMC)on Security in Wireless Mobile Computing Systems 2(2):159–186
3. Zhu S, Setia S, Jajodia S (2003) LEAP: efficient security mechanisms for large-scale distributed sensor networks. In: Proceedings of the 10th ACM conference on computer and communication security, ACM Press, Washington DC, USA, pp 62–72
4. Heinzelman WB, Chandrakasan A, Balakrishanan H (2002) An application-specific protocol architecture for wireless microsensor networks. IEEE Trans Wirel Commun 1(4):660–670
5. Akkaya K, Younis M (2004) Energy-aware routing of delay-constrained data in wireless sensor networks. J Commun Syst 17(6):663–687
6. Przydatek B, Song D, Perrig A (2003) SIA:Secure information aggregation in sensor networks. In: ACM SenSys 2003, Los Angeles, CA, USA, pp 175–192

Spatiotemporal Dynamics of Normalized Difference Vegetation Index in China Based on Remote Sensing Images

Yaping Zhang and Xu Chen

Abstract Based on 17 phases of Normalized Difference Vegetation Index and climate images obtained from 1982 to 1998, the paper analyzed normalized difference vegetation index trends, spatial distribution and their relationships with climate and human activity. The results indicate that at the national scale, the increases of monthly and seasonal NDVI correspond mainly to climate changes. However, corresponding mainly to human activities, the significant spatiotemporal heterogeneity of NDVI trends is found at the regional scale. Therefore, a set of policies must be established to ensure the ecological conservation and restoration, especially in ecological sensitivity areas.

Keywords NDVI • Spatiotemporal dynamics • Time-series

1 Introduction

Normalized Difference Vegetation Index (NDVI) is a general biophysical parameter that correlates with photosynthetic activity of vegetation. It is calculated as NDVI = (CH2 − CH1)/(CH2 + CH1), where CH1 and CH2 represent radiances from channels 1 (0.58–0.68 mm) and 2 (0.725–1.10 mm) of the AVHRR, respectively. Although NDVI does not provide land cover type directly, it provides an indication of the 'greenness' of the vegetation [1]. Therefore, a time series of NDVI values can separate different land cover types based on their phenology, or seasonal

---

Y. Zhang (✉)
School of Information Science and Technology, Yunnan Normal University, Kunming, China
e-mail: zhangyp.cs@gmail.com

X. Chen
Computer & Information Science Department, Southwest Forestry University, Kunming, China

signals [2], and have significantly improved our understanding of intra and interannual variations in vegetation from a regional to global scale [3]. Based on AVHRR data, numerous regional to global scale vegetation studies have been done. For example, using AVHRR NDVI time series data from regional to global scale, the changes of vegetation phenology could be analysed [4–7]. By intercomparing climatic variables such as rainfall and air temperature with long term time series analysis of AVHRR NDVI, the geo-biophysical causes of vegetation greenness or NPP changes could be revealed [8, 9]. Furthermore, combination with other geophysical parameters like albedo, continental scale trends of NDVI could be analyzed associated with environmental changes [10].

China has a large climate range because of its geographic size. This causes the country to have diverse and species-rich vegetation types, from the tropical to subarctic/alpine and from rain forest to desert. Since the late 1980s, land use patterns in China have been changed dramatically along with urbanization and loss of cultivated. In the late twentieth century and early twenty-first century, China has undergone a rapid socio-economic development, and a series of development strategies, including "Western Development", "Revitalization of Northeast", "Rising of Central China" and so on have been implemented across the nation. Meanwhile, the modification of industrial structure and acceleration of industrialization have resulted in remarkable changes and modifications in the spatial distribution of China's land use too [11]. The urbanization and large scale land use change over the last two decades in China have attracted international attention to analyze the impacts, driving forces, and future trends [12]. So, paying more attention to land cover monitoring and land use planning is necessary in china for protecting natural environment and ecosystem effectively [13]. In this paper, we used the NDVI and climate data from 1982 to 1998, together with information on land use to explore the variations of monthly, seasonal and annual NDVI and their relationships with climate and human activity.

2 Data and Method

2.1 Data

NDVI data used in this study were derived from the NOAA/AVHRR Land data set, produced by the Global Inventory Monitoring and Modeling Studies (GIMMS) group. Its spatial resolution was 8×8 km^2 with 15-day intervals, between January 1982 and December 1998 [14–16]. Annual mean air temperature and precipitation data at 1×1 km^2 resolution were compiled from the 1982–1998 temperature/precipitation database of China. DEM was derived from GTOPO30 that USGS distributes to the public through the Internet. And the 1:1,000,000 Land Use Map of China in 1980s was derived from sharing infrastructure of Earth system science.

2.2 Methodology

All data were aggregated to grid cells at 8×8 km^2 resolution, as done for the NDVI data sets. And then, the mask, geometric correction, coordination transformation, and other processes were carried out with ArcGIS 9.3. The mean-value iteration filter (MVI) [17], as a simple method, was used to reduce the noise and reconstruct high quality NDVI time-series. By using the maximum NDVI value in each month, season and year [18], we produced month, seasonal and annual NDVI data set to reduce residual atmospheric and bidirectional effect. Finally, the mean of NDVI, temperature and precipitation, the highest temperature and the lowest temperature, as well as aspect and slope were calculated for analysis.

3 Results

3.1 National Scale

From 1982 to 1998, annual NDVI did not appear the trend ($r^2 = 0.123$, $p = 0.169$), and the correlation between annual mean NDVI and climatic factors was not significant. But the significant increase trends of NDVI could be seen for spring and autumn. The largest NDVI increase ($r^2 = 0.533$, $p = 0.001$) was in spring, with a magnitude of 17.7 % over the 17 years and a trend of 0.00211 year^{-1} (the 17-year averaged NDVI is 0.3111). The other increase of NDVI ($r^2 = 0.355$, $p = 0.012$) for autumn was 7.21 % with a trend of 0.001126 year^{-1}. Despite the pronounced NDVI increases in two seasons, there were several large fluctuations in the NDVI trends. For example, seasonal NDVI was large in 1987 and 1990 but small in 1991 and 1995 for spring and autumn respectively (Fig. 1a).

As everyone knows, the magnitude of monthly NDVI and its change over time are important indicators of the contribution of vegetation activity in different months to annual plant growth total [16]. In China, except November and December, the positive values of monthly NDVI trends indicated that NDVI had increased throughout the year almost over the 17-year study period (Fig. 1b). The monthly NDVI trends for May and September increased significantly. The largest monthly NDVI increase ($r^2 = 0.531$, $p = 0.001$) was in May, with a magnitude of 19.96 % over the 17 years and a trend of 0.0022 year^{-1} (the 17-year averaged NDVI is 0.3047). The increase ($r^2 = 0.358$, $p = 0.011$) for September was 8.14 % with a trend of 0.001243 year^{-1}. But monthly NDVI got to maximum value in August and was rather small from December through March (Fig. 1b). The plant growth peak occurred in the middle of growing season (summer), while the largest NDVI increase appeared in the early growing season (spring). This means that the trends and patterns of monthly and seasonal NDVI are likely coupled with climate patterns and moisture availability [16].

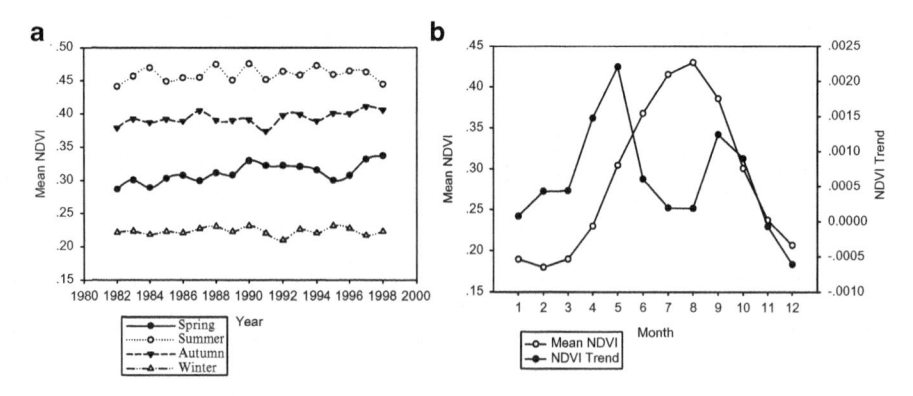

Fig. 1 Seasonal and monthly NDVI change in China (**a**) seasonal NDVI (**b**) monthly NDVI

Table 1 Land use types transformation matrix from 1980s to 2000 (10^4 km^2)

1980s	2000						
	Farm	Forest	Grass	Water	Urban	Unused	Total
Farm	167.69	4.03	3.80	0.34	0.66	0.62	177.15
Forest	2.62	245.88	6.20	0.09	0.08	0.53	255.40
Grass	3.81	6.52	275.48	0.51	0.026	18.15	304.48
Water	0.36	0.13	0.98	11.04	0.10	0.58	13.20
Urban	0.064	0.01	0	0.01	1.17	0.01	1.25
Unused	0.79	1.27	12.62	0.91	0.03	176.51	192.13
Total	175.34	257.84	299.08	12.89	2.07	196.40	943.62
I rate	−1.02 %	0.95 %	−1.77 %	−2.38 %	64.80 %	2.22 %	

3.2 Regional Scale

From 1982 to 1998, annual NDVI did not show the apparent trend at the national scale. But the spatial heterogeneity was found in the North China Plain, hilly and plain areas of Central China, Yangtze River deltas and Pearl River deltas, because of landuse changed significantly in these areas. Table 1 gave the main characteristics of the change in land cover types over the period 1980s–2000.

The areas of forestland, urban land and unused land increased; urban land had the greatest rate of increase, which was 64.8 % during the period. The areas of farmland, grassland and water area decreased; water area had the biggest rate of decrease, which was 2.38 % during the period. At the provincial scale, the provincial NDVI trends for Inner Mongolia ($r^2 = 0.235$, p = 0.049), Shanxi ($r^2 = 0.378$, p = 0.009), Xinjiang ($r^2 = 0.339$, p = 0.014) and Ningxia ($r^2 = 0.292$, p = 0.025) increased significantly. The largest provincial NDVI increase was in Ningxia, with a magnitude of 42.59 % over the 17 years and a trend of 0.002943 year^{-1} (the 17-year averaged NDVI was 0.323). The increase for Inner Mongolia, Shanxi and Xinjiang were 7.22 %, 10.76 %, and 17.61 % with a trend of

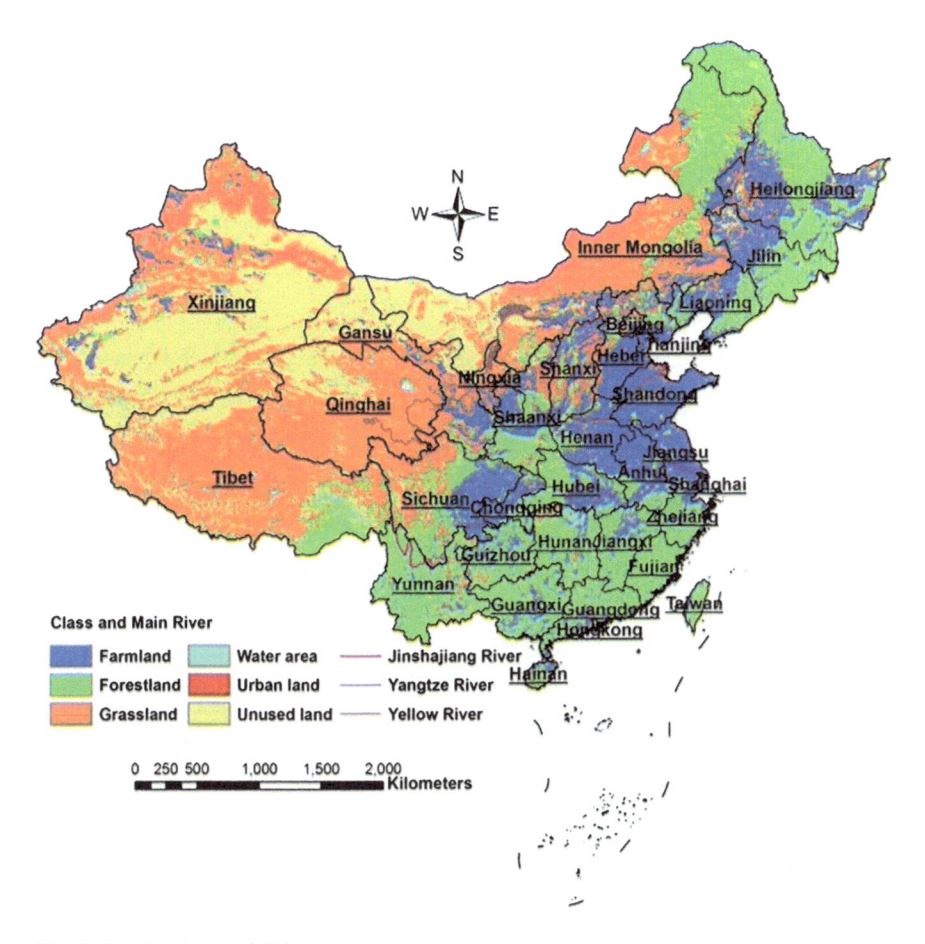

Fig. 2 Land use map of China

0.001908 year^{-1}, 0.002613 year^{-1}, and 0.001322 year^{-1}, respectively. But because a rapid urbanization had taken place over the past 20 years, there was significantly decrease in Guangdong ($r^2 = 0.27$, p = 0.032), with a trend of -0.001845 year^{-1}.

3.3 Land Use Scale

In this paper, the land use/cover types of Land Use Map of China were classified as: Farmland, Forestland, Grassland, Water Area, Urban Land and Unused Land, based on the land use/cover categories proposed by the Chinese Academy of Sciences (Fig. 2). If we assumed the locations of land covers were not change, annual NDVI and its trends showed spatial heterogeneity at land cover scale (Table 2). This is very similar to annual NDVI and its trend over the past 17 years at the provincial

Table 2 NDVI regression analysis from 1982 to 1998 in Landuse scale

Landuse	N	Reg coef	T	P	R	I rate
Farmland	17	0.001482	2.003	0.064	0.459	0.064469
Forestland	17	−0.000373	−0.671	0.512	0.171	−0.02506
Grassland	17	0.001004	1.807	0.091	0.423	0.053312
Water area	17	0.001295	2.259	0.039	0.504	0.121757
Urban land	17	−0.000344	−0.389	0.703	0.100	0.046945
Unused land	17	0.000606	1.720	0.106	0.406	0.105325

scale. Especially, the annual NDVI trends increased significantly for water area ($r^2 = 0.254$, $p = 0.039$) with a magnitude of 12.18 % over the 17 years and a trend of 0.001295 year^{-1}. This means that the trends and patterns of regional NDVI are likely associated with climate patterns and human activities.

4 Conclusion

By using the NDVI and climate data set from 1982 to 1998, together with information on land use, the variations of monthly, seasonal and annual NDVI and their relationships with climate and human activity were analyzed. The results indicated that both increase of monthly and seasonal NDVI at the national scale corresponded mainly to climate changes. It suggests that climate change is playing an important role at the national scale for the patterns of NDVI trends. But a large spatial and temporal heterogeneity of NDVI trends at the regional scale corresponded mainly to human activities. It suggests that human activities are playing an important role at the regional scale for the patterns of NDVI trends, besides some sensitive areas with climate changes.

Human activities have exerted a large effect on the spatiotemporal patterns of NDVI trends in some regions. So, it is necessary to establish a set of policies to ensure the ecological conservation and restoration, especially in ecological sensitivity areas.

Acknowledgements This work was financially supported by applied basic research projects of Yunnan Province (2010CD047) and (2011FZ140).

References

1. Sellers PJ (1985) Canopy reflectance, photosynthesis, and transpiration. Int J Remote Sens 6:1335–1372
2. Lenney MP, Woodcock CE, Collins JB (1996) The status of agricultural lands in Egypt: the use of multitemporal NDVI features derived from Landsat TM. Remote Sens Environ 56:8–20

3. Fensholt R, Rasmussen K, Nielsen TT, Mbow C (2009) Evaluation of earth observation based long term vegetation trends – intercomparing NDVI time series trend analysis consistency of Sahel from AVHRR GIMMS, Terra MODIS and SPOT VGT data. Remote Sens Environ 113:1886–1898
4. Anyamba A, Tucker CJ (2005) Analysis of Sahelian vegetation dynamics using NOAA AVHRR NDVI data from 1981 to 2003. J Arid Environ 63:596–614
5. Jeyaseelan AT, Roy PS, Young SS (2007) Persistent changes in NDVI between 1982 and 2003 over India using AVHRR GIMMS (Global Inventory Modeling and Mapping Studies) data. Int J Remote Sens 28(21):4927–4946
6. Stöckli R, Vidale PL (2004) European plant phenology and climate as seen in a 20- year AVHRR land surface parameters data set. Int J Remote Sen 25(17):3303–3330
7. Heumann BW, Seaquist JW, Eklundh L, Jonsson P (2007) AVHRR derived phenological change in the Sahel and Soudan, Africa, 1982–2005. Remote Sens Environ 108(4):385–392
8. Xiao J, Moody A (2005) Geographic distribution of global greening trends and their climatic correlates: 1982 to 1998. Int J Remote Sens 26(11):2371–2390
9. Hickler T, Eklundh L, Seaquist J, Smith B, Ardö J, Olsson L, Sykes MT, Sjöström M (2005) Precipitation controls Sahel greening trend. Geophys Res Lett 32:L21415
10. Govaerts YM, Lattanzio A (2008) Estimation of surface albedo increase during the eighties Sahel drought from Meteosat observations. Global Planet Change 64:139–145
11. Liu J, Zhang Z, Xu X (2010) Spatial patterns and driving forces of land use change in China during the early 21st century. J Geogr Sci 20:483–494
12. Zhang J, Zhang Y (2007) Remote sensing research issues of the National Land Use Change Program of China. ISPRS J Photogramm Remote Sens 62:461–472
13. Chen X, Tateishi R, Wang C (1999) Development of a 1-km landcover dataset of China using AVHRR data. ISPRS J Photogramm Remote Sens 54:305–316
14. Tucker CJ, Slayback DA, Pinzon JE (2001) Higher northern latitude NDVI and growing season trends from 1982 to 1999. Int J Biometeorol 45:184–190
15. Zhou LM, Tucker CJ, Kaufmann RK (2001) Variations in northern vegetation activity inferred from satellite data of vegetation index during 1981 to 1999. J Geophys Res 106:069–083
16. Piao S, Fang J, Zhou L, Guo Q, Henderson M, Ji W, Li Y, Tao S (2003) Interannual variations of monthly and seasonal normalized difference vegetation index (NDVI) in China from 1982 to 1999. J Geophys Res 108:4401
17. Ma M, Veroustraete F (2006) Reconstructing pathfinder AVHRR land NDVI timeseries data for the Northwest of China. Adv Space Res 37:835–840
18. Holben BN (1986) Characteristics of maximum value composite images from temporal AVHRR data. Int J Remote Sens 7:1417–1434

Part VII
Signal Processing and Control

The Application of Digital Filtering in Fault Diagnosis System for Large Blower

Changfei Sun, Yong Han, Zhishan Duan, and Yingge Xu

Abstract The analog signals that gathered from engineering test are often mixed with noises which will produce many adverse effects and greatly reduce the system operation speed. These useless and harmful signals can be got rid of by the digital filter. In the real-time monitoring and diagnosis system for the No. six blast furnace blower in The ChangZhi Steel Ltd, the Butterworth analog filter is converted into a digital filter, which can filter those useless and harmful signals. In order to confirm the validity of this digital filter, the high-pass digital filter in this real-time monitoring and diagnosis system is analyzed. The results show that the high-pass digital filter can effectively remove the unwanted signals under 5 Hz. Obviously, the digital filter can effectively remove the useless ingredients and harmful ingredients of the signals, and the system's operation speed can be greatly increased.

Keywords Fault diagnosis system • Filter • Digital filter

1 Introduction

For the influence of the factors of work environment and instrument, the analog signal that extracted from engineering test is often mixed with noise, even the signal is covered by noise. After A/D transformation, the quantization noise of the A/D converter is added into the discrete time signal besides the original noise. Many troubles for the subsequent judgment and diagnostic work are produced for the existing noise, so false-negatives, incorrect diagnosis are appeared in the system. In addition, not all of the signals have to be useful during fault diagnosis, which even reduce the speed of calculation. So the harmful signal must be deleted before data processing [1].

C. Sun (✉) • Y. Han • Z. Duan • Y. Xu
School of Mechanical and Electrical Engineering, Xi'an University
of Architecture and Technology, Xi'an, China
e-mail: sunchangfei27@163.com

Due to the higher vibration frequency of the blast furnace blower, some low frequency components which is under 5 Hz should be filtered out in order to improve the operation efficiency.

2 Digital Filter

Digital filter can filter out some unnecessary frequency components in signal by some numerical calculations. Digital filter is one of important contents in digital signal processing. Because of the operation way in digital filtering, it has the characteristics of high precision, high stability, using large scale integrated circuit, small volume, light weight, realizing flexible without requiring impedance matching [2]. In the real time monitoring and diagnosis system, the digital filtering method is taken.

The numerical calculation method is used in the digital filter to achieve the purpose of filter, so by following some certain algorithms, the software is established. In addition, at present many kinds of special digital signal processing chip are developed, which is easy to make the design of the digital filter [3].

3 The Design of Digital Filter

Digital filter is divided into infinity impulse response digital filter and finite impulse response digital filter, respectively called IIR filter and FIR filter. In this real-time monitoring and diagnosis system IIR filter is used.

Digital filter can be described by a N order difference equations

$$y(n) = \sum_{i=0}^{M} b_i x(n-i) - \sum_{k=1}^{N} a_k y(n-k) \quad (1)$$

its corresponding function is

$$H(z) = \frac{\sum_{i=0}^{M} b_i z^{-i}}{1 + \sum_{k=1}^{N} a_k z^{-k}} \quad (2)$$

The task of designing is finding a group of coefficients (a_k, b_i) according to the prescribed technical index, and making the function of the filter meet technical index.

There are many design methods for digital filter, the most commonly used one is designing digital filter from simulation filter, because the theory and design method of simulation filter have developed very mature. There are several kinds of typical simulation filter for choice, such as Butterworth filter, Chebyshev filter, Cauer filter, Bessel filter, etc. [4]. The Design method of digital filter is designing a simulation low-pass filter $H_a(s)$ according to the technical requirement, then the low-pass filter $H_a(s)$ is converted into a needed digital filters by converting the frequency. Frequency can be converted in the analog domain, also can be converted in the digital domain. so there are two conversion methods. The first method is to design low-pass filter in the simulation domain, then convert the frequency in the simulation domain, and convert it into a needed simulation filter. Then switch it from the s plane to the z plane to obtain the required digital filter. The second method is to switch the simulation low-pass filter from the s plane to the z plane to get digital low-pass filter, and then convert the frequency in the digital domain to get a desired digital filter [5].

4 The Application of Digital Filter

But the first method will cause the distortion of the frequency response, so it's not appropriate for the design of the high-pass filter and the band elimination filter. The second method is very trouble and inconvenient. So In this real-time monitoring and diagnosis system the high-pass filter is obtained by normalizing a simulation low-pass filter to a desired digital filter directly. The design method of the IIR digital filter is converting a Butterworth low-pass analog filter into a high-pass digital filter.

The technical parameters of this filter are: The cut-off frequency of the band pass filter is $f_p = 5$ Hz, the cut-off frequency of the stop band filter is $f_s = 1$ Hz, Sampling frequency is $f = 2500$ Hz, The maximum attenuation of passband is 3 dB, the minimum attenuation of stopband is 18 dB Resistance with inner minimum attenuation is 18 dB, Frequency characteristics as shown in Fig. 1 shows.

Firstly the boundary of frequency for equivalent high-pass digital filter is calculated

$$\omega_p = 2\pi f_p T = 2\pi \bullet 5/2500 = 0.004\pi \text{ rad} \tag{3}$$

$$\omega_s = 2\pi f_s T = 2\pi \bullet 1/2500 = 0.0008\pi \text{ rad} \tag{4}$$

From the above formula, the cut-off frequency of the band pass filter Ω_p and the cut-off frequency of the stop band filter Ω_s of the analog low-pass filter is calculated.

$$\Omega_p = \frac{1}{2}\Omega_c^2 T ctg \frac{1}{2}\omega_p \tag{5}$$

Fig. 1 Characteristics of the required digital filter's frequency

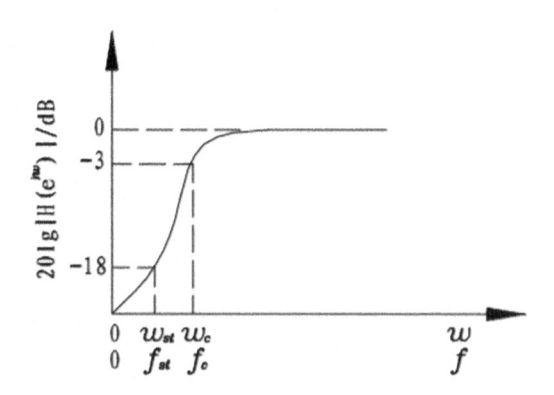

$$\Omega_s = \frac{1}{2}\Omega_c^2 T ctg \frac{1}{2}\omega_s \tag{6}$$

The requirement is Low-pass filter $|\Omega| < \Omega_p$, Attenuation is not more than 3 dB, $\Omega_s < |\Omega|$, Attenuation than 18 dB.

According to the above index, the simulation low-pass filter is designed.

$$\Omega_s/\Omega_p = \frac{\frac{1}{2}\Omega_c^2 T ctg \frac{\omega_s}{2}}{\frac{1}{2}\Omega_c^2 T ctg \frac{\omega_p}{2}} = \frac{ctg \frac{\omega_s}{2}}{ctg \frac{\omega_p}{2}} = \frac{ctg 0.0004\pi}{ctg 0.002\pi} = 5 \tag{7}$$

The basic parameters of Butterworth filter are:

$$N = -\frac{\lg k}{\lg \lambda} \tag{8}$$

$$k = \sqrt{\frac{10^{0.1a_p} - 1}{10^{0.1a_s} - 1}} = \sqrt{\frac{10^{0.3} - 1}{10^{1.8} - 1}} = 0.1257 \tag{9}$$

$$\lambda = \frac{\Omega_s}{\Omega_p} = 5 \tag{10}$$

$$N = -\frac{\lg 0.1257}{\lg 5} = 1.3 \tag{11}$$

so $N = 2$

According to $N = 2$, after looking up the table, transmission function for the second order normalized simulation low-pass filter is

$$A(p) = \frac{1}{p^2 + \sqrt{2}p + 1} \tag{12}$$

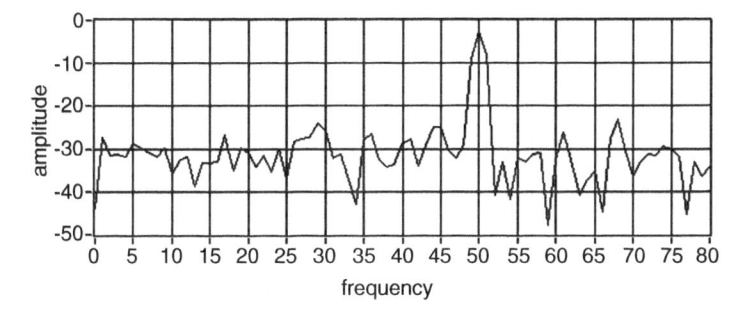

Fig. 2 The signal of a phase stator current of a electromotor

$$\because \Omega_p = \frac{1}{2}\Omega_c^2 T ctg\frac{1}{2}\omega_p = \Omega_c \tag{13}$$

$$\therefore \Omega_c = \Omega_p = \frac{2}{T} tg\frac{\omega_p}{2} \tag{14}$$

$$\therefore p = G(z^{-1}) = \frac{1}{2}\Omega_c T \frac{1+z^{-1}}{1-z^{-1}}$$
$$= \frac{T}{2} \cdot \frac{2}{T} tg\frac{\omega_p}{2} \frac{1+z^{-1}}{1-z^{-1}}$$
$$= tg\frac{\omega_p}{2} \frac{1+z^{-1}}{1-z^{-1}} = 0.0063\frac{1+z^{-1}}{1-z^{-1}} \tag{15}$$

$$H(z) = A(p)\Big|p = G(z^{-1}) = 0.0063\frac{1+z^{-1}}{1-z^{-1}}$$
$$= \frac{0.991 - 1.982z^{-1} + 0.991z^{-2}}{1 - 1.982z^{-1} + 0.982z^{-2}} \tag{16}$$

From the second order IIR digital filter system function, the second order difference equation for the designed digital filter is

$$y(n) = 0.991x(n) - 1.982x(n-1) + 0.991x(n-2) + 1.982y(n-1)$$
$$- 0.982y(n-2) \tag{17}$$

According to the above formulas program a filtering procedure, so the digital filter is achieved.

The signal shown in Fig. 2 is a phase stator current of an electromotor gathered by us, it can be filtered by this digital filter, the result is shown in Fig. 3. Compare Fig. 2 with Fig. 3, a conclusion can be drawn that the signal which frequency is below 5 Hz can be filtered out by this digital filter.

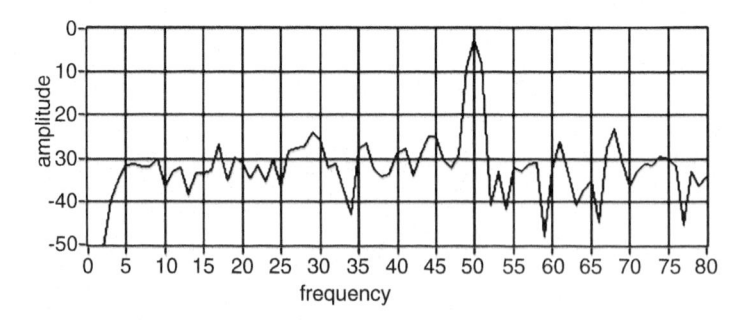

Fig. 3 The signal in Fig. 2 filtered by the digital filter

5 Conclusion

The results show that the high-pass digital filter can effectively remove the unwanted signals under 5 Hz. Obviously, the digital filter can effectively remove the useless ingredients and harmful ingredients of the signals, and the system's operation speed can be greatly increased.

As a consequence, a digital filtering is effective in the real-time monitoring and diagnosis system. The Butterworth analog filter can be converted into a digital filter, which can filter those useless and harmful signals and increase the operation speed of the diagnosis system.

References

1. Peiqing Cheng (2002) Digital signal processing tutorial. Tsinghua University Press, Beijing, pp 42–43
2. Mitchell JS (1990) The analysis and monitoring of mechanical fault. Machinery Industry Press, Beijing, pp 65–68
3. Jianhua Leng (2002) Digital signal processing. Defense industry press, Beijing, pp 52–55
4. Lichen Gu (2000) Mechanical signal processing and application. Shaanxi science and technology press, Xi'an, pp 62–65
5. Yong He (2003) The development of real-time monitoring and diagnosis system in rotating machine[D]. Xi'an University of Architecture & Technology, Xi'an

Performance Analysis on ST-ASLC with Wide-Band Interference

Xingcheng Li and Shouguo Yang

Abstract In order to find the relationship between the cancellation ratio with the wide interference and the delay lines, auxiliary channels and interference bandwidth. And we also would like to draw a quantification conclusion about the performance of ST-ASLC (Space Time Adaptive Side-lobe Canceller). Firstly, We will derive the ST-ASLC (Space Time Adaptive Side-lobe Canceller) processing models with wide-band interference in this article, and we also will derive the interference covariance and correlation matrix with amplitude and phase errors between the main and auxiliary channels; Secondly, based on the interference cancellation ratio, we will explain the performance influence of the auxiliary channels, delay time and delay lines number with simulation results. Finally, we will summarize the law about the ST-ASLC performance with wide interference as follows: the wider the interference bandwidth is, the lower the ASLC cancellation ratio will be; Auxiliary channels and delay lines increasing can improve the ASLC system's interference cancellation ratio, however, multi-auxiliary-channels system is better than the single channel system which has the same channels number of delay lines.

Keywords ST-ASLC · Wide-band interference · Space filter response · Cancellation ratio

---

X. Li (✉)
School of Air and Missile Defense, Air Force Engineering University, Xi'an, China
e-mail: lixingcheng2008@sina.com.cn

S. Yang
School of Air and Missile Defense, Air Force Engineering University, Xi'an, China

School of Electronic and Information, Northwestern Polytechnical University, Xi'an, China

1 Introduction

Under wide-band interference, different frequency component of interference has different delay in each antenna unit of radar side-lobe cancellation arrays, and the main and auxiliary channels have different responses to the interfering signals with different frequency components. Therefore, the wide-band interference (relative to the array) will lead to the mismatched amplitude and phase of ASLC, and it may result in the degradation of the interfering cancellation performance of ASLC system [1]. There is a widespread concern on improving the performance of ASLC system under wide-band interference. If the center frequency and band width of the wide-band interference are f_0 and Δf, and the incident angle is θ_0, which can be viewed as a group of sine-wave signals with different frequency, it can be equalized by the continuous-wave signals with f_0 which are incident simultaneously from several directions. The equivalent angle dispersion of the wide-band interference can be expressed as [2]

$$\frac{\Delta f}{f_0} = \frac{\Delta \sin \theta}{\sin \theta_0} \qquad (1)$$

As for the single-frequency incident wave, the optimum weights of ASLC depend on the incident angle. For one incident angle has the corresponding optimum weights, the optimum processing of the wide-band interference should have a group of optimum weights correspondingly. The wide-band interference will occupy ASLC system a large number of degrees of freedom (DOF) and lower its efficiency. To improve the interference cancellation performance of ASLC under wide-band interference, the most direct method is to reduce the bandwidth of the received interference, which can be fulfilled by using sub-band filter banks processing or fast Fourier transform(FFT) to obtain the multi-path signals divided according to the frequency [3, 4], and subsequently constructing independent adaptive array processing for each sub-band. Another method is to add delay lines in each auxiliary channel to add DOF, which can control the antenna's frequency response more profitably; this is space time adaptive side-lobe cancellation [5, 7]. It has been proved that these two methods have the same cancellation performance when the numbers of delay lines equal to the sub-band and delay time of each delay line equals to the sampling period of sub-band [6].

In this paper, the mathematical model of ST-ASLC with amplitude and phase error under wide-band interference is presented, and the performance is fully analyzed. Finally, the significant conclusions are presented.

2 ST-ASLC Processing Models

A ST-ASLC system with M auxiliary channels is shown in Fig. 1, each auxiliary channel has K delay lines. The signals of auxiliary channels can be expressed as

$$X = \begin{bmatrix} X_1 \\ \vdots \\ X_M \end{bmatrix} \quad (2)$$

Where $X_m = \begin{bmatrix} x_{m1} & x_{m2} & \cdots & x_{mK} \end{bmatrix}^T$, the system's weights are

$$W = \begin{bmatrix} W_1 \\ \vdots \\ W_M \end{bmatrix} \quad (3)$$

Where $W_m = \begin{bmatrix} w_{m1} & w_{m2} & \cdots & w_{mK} \end{bmatrix}^T$.

Suppose that the interference signal's spectrum response is flat in the entire bandwidth. The center frequency and bandwidth of the interference signal are f_0 and Δf respectively. Then when the interference signal comes from θ, and the delay time of delay lines $\Delta = T_d$, the Kth line's interference signal in the Mth auxiliary channel is

$$x_{mk} = G_a J[t - (k-1)T_d - (m-1)T] + n[t - (k-1)T_d] \quad (4)$$

Where $m = 1, \cdots, M, k = 1, \cdots, K$. T_d is the delay time of delay lines. G_a is the auxiliary antenna's gain (for all the auxiliary antennas).

$$T = \left(\frac{d}{c}\right)\sin\theta = \frac{\xi \sin\theta}{f_0} \quad \text{and} \quad \xi = \frac{d}{\lambda} \quad (5)$$

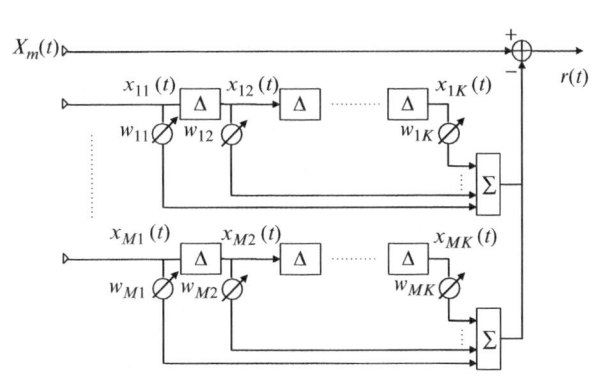

Fig. 1 Space-time ASLC processing system

T is the time delay between auxiliary antennas. The interference covariance matrix and correlation matrix are respectively

$$M = E(X^*X^T) = \begin{pmatrix} M_{11} & \cdots & M_{1M} \\ \vdots & \ddots & \vdots \\ M_{M1} & \cdots & M_{MM} \end{pmatrix} \quad R = E(X^*X_M) = \begin{bmatrix} R_1 \\ \vdots \\ R_M \end{bmatrix} \quad (6)$$

Where M_{mn} and R_m are respectively $K \times K$, $K \times 1$ dimension matrix. X_M represents the signal samplings of the main channel.

According to the expresses above, M_{mn} and R_m are given by

$$(M_{mn})_{pq} = G_a^2 P_j e^{j2\pi f_0[(p-q)T_d+(m-n)T]} \operatorname{sinc}[\Delta f((p-q)T_d + (m-n)T)] \quad (7)$$

$$(R_m)_q = G_m G_a P_j e^{j2\pi f_0[(q-1)T_d+mT]} \operatorname{sinc}[\Delta f((q-1)T_d + mT)] \quad (8)$$

Where $p, q = 1, \cdots, K$, P_j is the interference power. G_m is the main antenna gain in the interference direction.

Suppose that $T_d = rT_{90}$, $T_{90} = 1/(4f_0)$ is the required unit delay time, and

$$\Delta f T_d = \frac{rB}{4}, \quad B = \frac{\Delta f}{f_0}, \quad \Delta f T = B\xi \sin\theta \quad (9)$$

$$(M_{mn})_{pq} = G_a^2 P_j e^{j2\pi\left[(p-q)\frac{r}{4}+(m-n)\xi\sin\theta\right]} \operatorname{sinc}\left[(p-q)\frac{Br}{4} + (m-n)B\xi\sin\theta\right] \quad (10)$$

$$(R_m)_q = G_m G_a P_j e^{j2\pi\left[(q-1)\frac{r}{4}+m\xi\sin\theta\right]} \operatorname{sinc}\left[(q-1)\frac{Br}{4} + mB\xi\sin\theta\right] \quad (11)$$

Taking the amplitude and phase error into account, let b_m ($m = 1, \cdots, M$) be the Mth channel's maximum phase shift and a_m ($m = 1, \cdots, M$) be the Mth channel's maximum fluctuation amplitude. The interference covariance and correlation matrix can be derived as [8]

$$(M_{mn}^i)_{pq} = G_a^2 P_j e^{j2\pi\left[(p-q)\frac{r}{4}+(m-n)\xi\sin\theta\right]}$$
$$\times \frac{\operatorname{sinc}(\kappa)}{\sqrt{1+a_m^2/2}\sqrt{1+a_n^2/2}} \begin{pmatrix} 1 + \dfrac{a_m a_n}{2} + \dfrac{a_m a_n \kappa^2}{2(\kappa^2-4)} \\ - \dfrac{a_m \kappa^2 + a_n \kappa^2}{\kappa^2 - 1} \end{pmatrix} \quad (12)$$

$$(R_m^i)_q = G_m G_a P_j e^{j2\pi\left[(q-1)\frac{r}{4}+m\xi\sin\theta\right]} \frac{\operatorname{sinc}(\kappa')}{\sqrt{1+a_m^2/2}} \left(1 - \frac{a_m \kappa'^2}{\kappa'^2 - 1}\right) \quad (13)$$

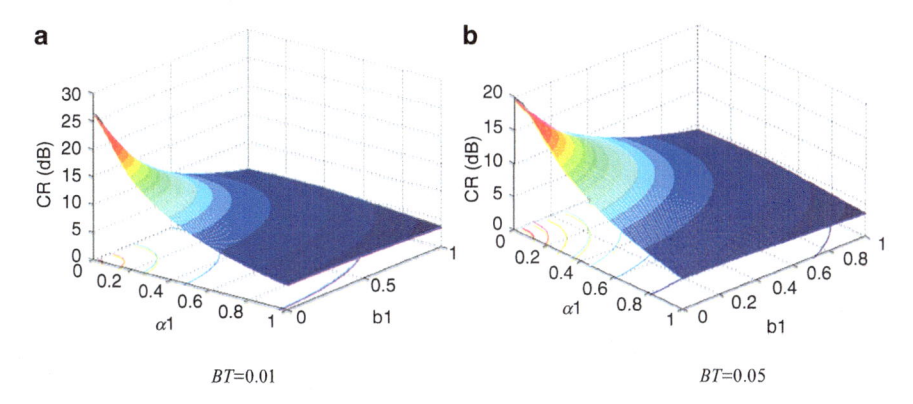

Fig. 2 ST-ASLC system CR with amplitude and phase error (**a**) $BT = 0.01$ (**b**) $BT = 0.05$

Where

$$\kappa = (p-q)\frac{Br}{4} + (m-n)B\xi\sin\theta + \frac{b_m - b_n}{\pi}, \quad \kappa' = (q-1)\frac{Br}{4} + mB\xi\sin\theta + \frac{b_m}{\pi} \quad (14)$$

In the following Fig. 2, suppose that the main channel is base channel, ASLC has only one auxiliary channel, and there are no delay lines in each channel, we named it channel 1. When the amplitude and phase error is considering, the cancellation ratio is descending when this error becomes larger. So the real equipment must have a good amplitude and phase characteristic corresponding, which must be strict to 0–0.12 with amplitude and 10^0 with phase, this is also equal to the engineering experiencing value in our exercise.

The optimum weights of ASLC system are $\boldsymbol{W}_{opt} = \boldsymbol{M}^{-1}\boldsymbol{R}$, the interference cancellation ratio (CR, Cancellation Ratio) is defined as the ratio between system output interference-to-noise power ratio in case of no auxiliary channels and auxiliary channels.

Without considering the expected signal, CR is

$$CR = \frac{E\{|X_M|^2\}}{E\{|X_M - \boldsymbol{W}^T\boldsymbol{X}|^2\}} = \frac{E\{|X_M|^2\}}{E\{|X_M|^2\} - \boldsymbol{W}^T\boldsymbol{R}^*} = \frac{E\{|X_M|^2\}}{E\{|X_M|^2\} - \boldsymbol{R}^H\boldsymbol{M}^{-1}\boldsymbol{R}} \quad (15)$$

The main antenna's gain in interference band is G_m, and the auxiliary antennas' gain is G_a. The system has M auxiliary channels. Here the spatial filtering output response of ST-ASLC system is

$$H(e^{j\varphi}) = \begin{bmatrix} 1 & \frac{G_a}{G_m}\boldsymbol{W}_{opt}^H \end{bmatrix} \begin{bmatrix} 1 \\ \boldsymbol{S}_a' \end{bmatrix} = 1 + \frac{G_a}{G_m}\boldsymbol{W}_{opt}^H \boldsymbol{S}_a' \quad (16)$$

Where

$$S'_a = [e^{j\varphi}(1, e^{j2\pi\frac{f}{f_0}r}, \cdots e^{j2\pi\frac{f}{f_0}(K-1)r}), \ldots, e^{jM\varphi}(1, e^{j2\pi\frac{f}{f_0}r}, \cdots e^{j2\pi\frac{f}{f_0}(K-1)r})]^T, \varphi = 2\pi\frac{d}{\lambda_0}\sin\theta\frac{f}{f_0} \quad (17)$$

3 Simulations

Suppose that an ASLC system has three auxiliary channels, the interference comes from the direction of 8^0, JNR = 30 dB, SNR = 30 dB, the equivalent distance between main and auxiliary antennas' phase center is 40 times of the wavelength (the actual number is larger), and the main and auxiliary channels are isometric and linear arrays. a_i and b_i are normal distribution random numbers whose means are both 0 and variances are 0.12 and 10^0 respectively. The simulation runs 100 times and then calculates the average. Figure 3 shows the curves of interference cancellation ratio with interference bandwidth for single-auxiliary-channel and three-auxiliary-channels ASLC systems when $r = 1$. Apparently, the wider the interference bandwidth is, the smaller the interference cancellation ratio will be.

Figure 4 shows the relation between delay lines and interference cancellation ratio when $r = 1$ and $B = 0.05$. The more the delay lines are, the higher the interference cancellation ratio will be. But the more the delay lines are, the more complex the system will be, and the more the amount of calculation is. The amount of weights has a direction ratio with $(KM)^3$ [9]. So we must choose the right amount of delay lines for the balance between the improved interference cancellation ratio and the amount of calculation.

Figure 5 shows the relation between delay time and interference cancellation ratio when $K = 10, B = 0.05$. We can see that the changes of the delay time have

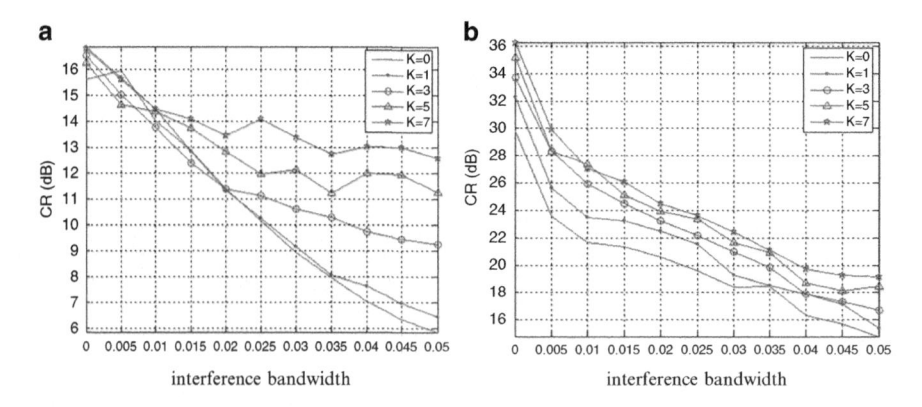

Fig. 3 ST-ASLC system CR with interference bandwidth (**a**) ($r = 1, M = 1$) (**b**) ($r = 1, M = 3$)

Fig. 4 CR with delay lines ($r = 1, B = 0.05$)

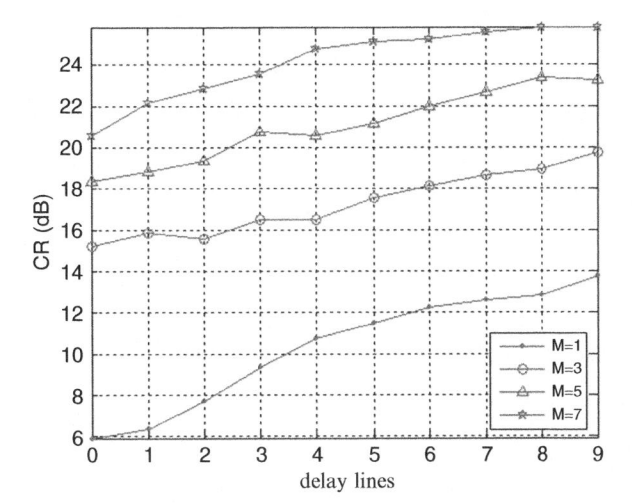

Fig. 5 CR with r ($K = 10$, $B = 0.05$)

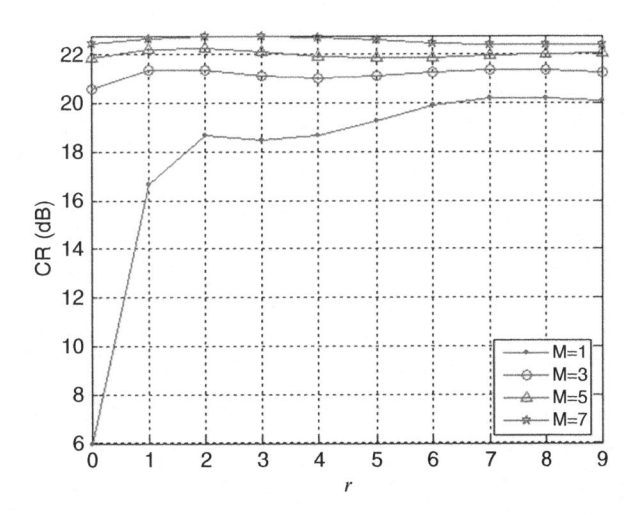

little effect on interference cancellation ratio. But there exists an optimal delay time ratio. Actually; we should choose the optimal r according to different ASLC structures, different delay lines and different interference bandwidth. Figure 6 shows the ST-ASLC system's spatial filtering amplitude response. We can see that the notch of single-auxiliary-channel spatial filtering amplitude response is significantly widened. This is the essential reason for improving the performance of ASLC under wide-band interference by the delay lines.

Under no amplitude and phase error conditions, Fig. 7 shows the changing curves of ASLC system's CR with auxiliary channels. Figure 8 gives the changing curves of single –auxiliary-channel ST-ASLC system's CR with delay lines. We can see that adding auxiliary channels and increasing delay lines have the same effect as for the single-auxiliary-channel ASLC system.

Fig. 6 ST-ASLC spatial filtering response under wide-band interference ($K=4, r=1$)

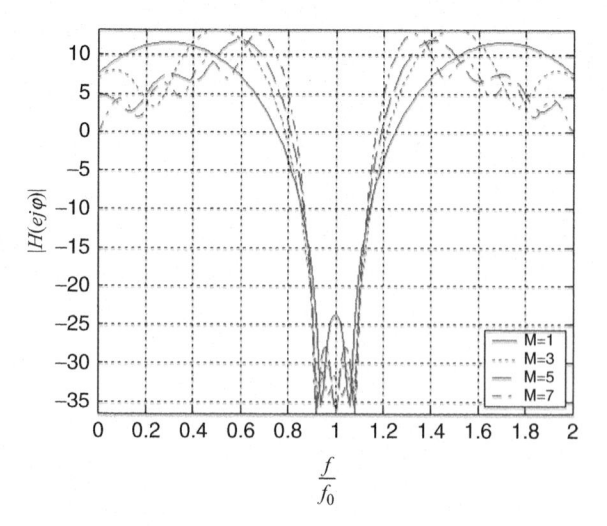

Fig. 7 ASLC CR with auxiliary channels

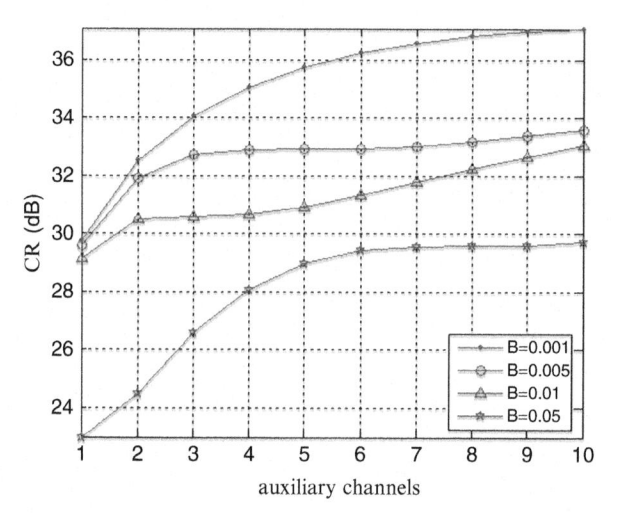

Fig. 8 ST-ASLC CR with delay lines ($r=1, M=1$)

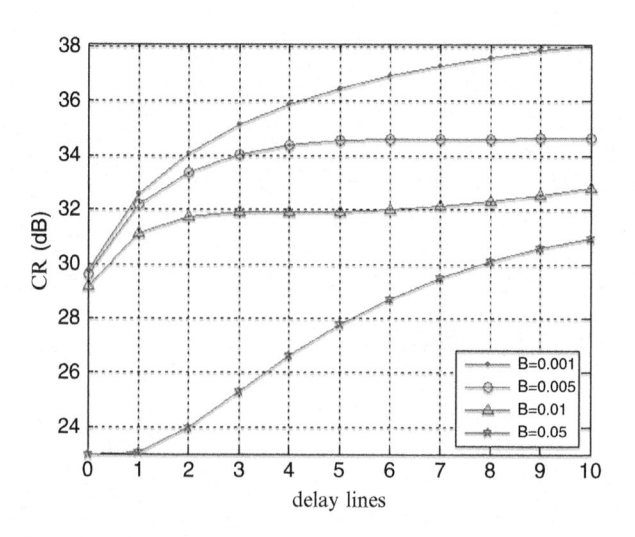

4 Conclusion

Through the theory analysis and simulations above, conclusions can be drawn as follows:

Wide-band interference can reduce the single-channel ASLC system's interference cancellation performance. The wider the interference bandwidth is, the lower the single-channel ASLC cancellation ratio is. Or in other words, nearly the single-auxiliary-channel ASLC system doesn't have the anti-jamming ability under wide-band interference. For example, single-auxiliary-channel ASLC interference cancellation ratio is only 6 dB when B = 0.05.

Auxiliary channels and delay lines increasing can both improve the system's interference cancellation ratio. When there are no amplitude and phase errors, increasing the delay lines and the auxiliary channels have the same effect as for the single-auxiliary-channel system. Multi-auxiliary-channel system will be better than the same delay lines single channel system when the channel numbers are equal to the delay lines'.

The advantages of increasing the auxiliary channels lies in improving the system's degrees of freedom and eliminating multiple interferences while it has the disadvantages of high costs and expensive expenses. The biggest advantage of increasing the delay lines is low cost. But it is unable to work with multiple interferences.

References

1. Xingcheng Li, Yong-shun Zhang (2008) Analysis of ASLC performance under wide-band jamming [J]. Mod Radar 3:34–36 (In Chinese)
2. Yaohuan Gong (2003) Adaptive filtering [M], 2nd edn. Electronic Industry Press, Beijing, p 7 (In Chinese)
3. Steinhardt AO et al (2000) Sub-band STAP processing [A]. 1th IEEE sensor array and multi-channel signal processing workshop [C], IEEE Press, New York, pp 1–6
4. Zhang Y et al (2001) Adaptive array processing for multi-path fading mitigation via exploitation of filter banks [J]. IEEE Trans Antenna Propag 49(4):505–516
5. Compton RT Jr (1988) The relationship between tapped delay line and FFT processing in adaptive array [J]. IEEE Trans Antenna Propag 36(1):15–26
6. Wisler DJ et al (1988) The development of a base-band cross-pol canceller [A]. IEEE ICC 88 [C]. 3, 1349–1354
7. Xingcheng Ling (2008) Research on electronic interference techniques of ASLC system [D]. Doctoral Dissertation of Air Force Engineering University (In Chinese)
8. Jun Jiang, Xingcheng Li, Wei-Hua Ren (2009) Performance analysis of ASLC with amplitude and phase error [J]. Electron Opt Control 3:34–36 (In Chinese)
9. Sien-chang, Charies Liu et al (1997) Wideband and wide angle side-lobe cancellation technique [P]. US Patent 3202990, 29 Jul 1997

Velocity Error Analysis of INS-Aided Satellite Receiver Third-Order Loop Based on Discrete Model

Dong-feng Song, Bing Luo, Xiao-ping Hu, An-cheng Wang, Pu-hua Wang, and Kang-hua Tang

Abstract Traditionally, continuous model is usually applied to analyze the velocity error of INS-aided GNSS receiver. However, the actual implemented system is discrete. In consequence, the analysis results based on continuous model are iffy. Aimed at this problem, this paper establishes a discrete model of INS-aided third-order carrier loop according to the actual system. Then, the velocity error caused by discretization is investigated and the error mechanism is analyzed. Analysis and simulation results show that the delay of aiding information is the main error source of INS aided GNSS receiver in high dynamic scene. To improve the dynamic performance of INS-aided carrier loop efficiently, the delay of aiding information should be shortened as much as possible.

Keywords INS-aided • High dynamic • Discrete model

1 Introduction

Third-order PLL (Phase Locked Loop) is often used in a GNSS (Global Navigation Satellite System) receiver's tracking loop. The effects of noise increase with increasing loop bandwidth, while dynamic tracking errors increase with decreasing loop bandwidth [1]. So in high dynamic scene, the tracking loop is often out of lock due to a great Doppler frequency shift. But the tracking loop aided by INS velocity information can adapt to the high dynamic scene and has a better dynamic performance [2, 3]. The continuous model and transfer function of INS-aided tracking loop in complex frequency domain are given, and the simulation results based on continuous model show a remarkable improvement in dynamic performance [4].

D.-f. Song • B. Luo (✉) • X.-p. Hu • A.-c. Wang • P.-h. Wang • K.-h. Tang
College of Mechatronics and Automation, National University
of Defense Technology, Changsha, China
e-mail: ruobing@nudt.edu.cn

Most references have built continuous model in complex frequency domain and proved the advantages of INS-aided loop using continuous model. But the continuous model cannot reflect the actual quantitative velocity error in a specific dynamic scene. In this paper, the discrete simulation model of INS-aided third-order carrier loop with the reference of actual system in z domain is gradually established and theoretical analysis of the velocity error is given. The whole paper is organized as follow: (1) Introduction; (2) The establishment of the continuous third-order carrier PLL model and the INS-aided carrier PLL model; (3) Building up the third-order carrier PLL discrete model and INS-aided PLL model in z domain based on the actual GNSS receiver's loop; (4) Calculating the velocity errors of aforementioned four models in a specific high dynamic scene and comparing corresponding results; (5) The mechanism analysis on discrete INS-aided carrier loop velocity error; (6) Conclusion.

2 Continuous Third-Order Carrier PLL Model and INS-Aided Model

Third-order PLL is a basic part of GNSS receiver's carrier loop and determines the dynamic performance of the loop. The research of the third-order carrier PLL model in complex frequency domain is very important.

2.1 Model 1: Continuous Model of Third-Order PLL in Complex Frequency Domain

Third-order PLL is sensitive to jerk signal and stays stable while noise bandwidth is equal to or below 18 Hz. The third-order PLL model is established using Matlab/Simulink tool as shown in Fig. 1.

In Fig. 1, NCO means Numerical Controlled Oscillator and NCO is usually modeled as an integrator. With the help of formula (1), the third-order loop parameters $k_3 = 12079.21, k_2 = 579.10, k_1 = 55.07$ are calculated when $B_n = 18Hz$ [5].

$$k_3 = w_0^3, k_2 = 1.1w_0^2, k_2 = 2.4w_0, B_n = 0.7845w_0 \qquad (1)$$

The closed-loop transfer function is:

$$\frac{\hat{\varphi}}{\varphi} = \frac{k_1 s^2 + k_2 s + k_3}{s^3 + k_1 s^2 + k_2 s + k_3} \qquad (2)$$

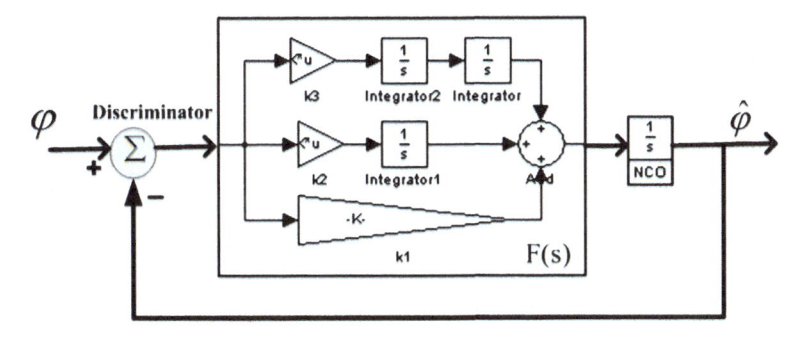

Fig. 1 The third-order PLL model

Fig. 2 INS-aided PLL model

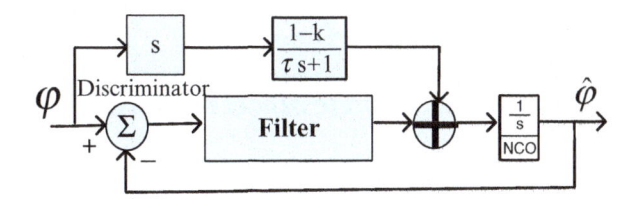

2.2 Model 2: Continuous Model of INS-Aided Third-Order PLL in Complex Frequency Domain

INS's velocity information is used to aid PLL. We can establish the INS-aided model shown in Fig. 2, where $\frac{1-k}{\tau s+1}$ represents the aided velocity error, k represents velocity scaling error and τ represents constant delay time.

The transfer function of INS-aided carrier loop is inferred:

$$\frac{\hat{\varphi}}{\varphi} = \frac{(1-k+k_1\tau)s^3 + (k_1+k_2\tau)s^2 + (k_2+k_3\tau)s + k_3}{\tau s^4 + (1+k_1\tau)s^3 + (k_1+k_2\tau)s^2 + (k_2+k_3\tau)s + k_3} \quad (3)$$

In most cases, k has little effect on the performance of the loop, so it is set to 0; the output rate of inertial device is usually 200 Hz, and the output rate of inertial navigation solution information is often 100 Hz while using two-sample method navigation-solution. INS velocity information changes every 10 ms, so τ is set to 0.01 s.

3 Discrete Third-Order Carrier PLL Model and INS-Aided Model

GNSS receiver is a discrete control system, so establishing the discrete model of carrier loop is very necessary. And only the discrete model can reflect the correct velocity information of receiver's carrier loop.

Fig. 3 Discrete third-order PLL model

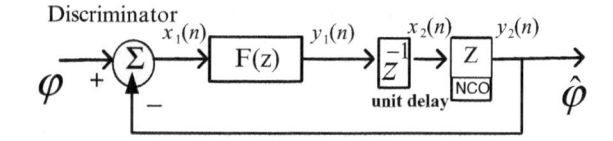

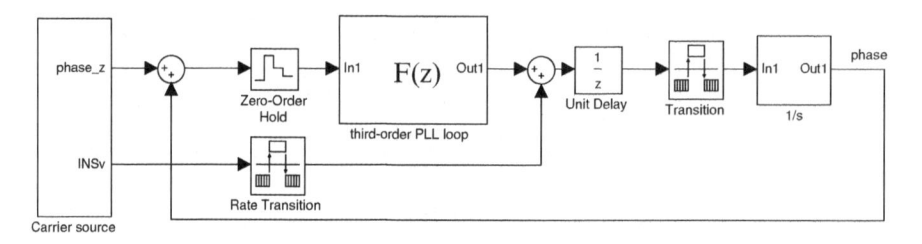

Fig. 4 INS-aided third-order PLL discrete model

3.1 Model 3: Discrete Model of Third-Order PLL in z Domain

In a discrete model in z domain, the analog integrator is often replaced by a rectangular digital integrator. Thus the third-order PLL is modeled as $F(z)$ which is shown in Fig. 3. Noting that third-order PLL's input sequence is $x_1(n)$, and the output sequence is $y_1(n)$. We can work out the third-order loop transfer function as follow ($T_1 = 1$ms):

$$\frac{Y_1(z)}{X_1(z)} = \frac{k_1 z^{-2} - (2k_1 + k_2 T_1)z^{-1} + (k_1 + k_2 T_1 + k_3 T_1^2)}{z^{-2} - 2z^{-1} + 1} \quad (4)$$

In the design of actual receiver, the control quantity of the current period cannot immediately act on the plant. There exists a one-control-period delay as z^{-1} shown in Fig. 3. The receiver's IF sampling frequency is in the order of 10^6, so we set NCO Sequence $x_2(n)$, $y_2(n)$ update rate to 0.1 MHz. This two-period model is shown in Fig. 3.

3.2 Model 4: Discrete Model of INS-Aided Third-Order PLL in z Domain

For the actual INS information rate being 100 Hz, we must establish a three-period model in Matlab/Simulink environment. And the INS information update period is equal to τ in Fig. 2. We use the correct INS velocity to aid third-order PLL directly, which means setting k in Fig. 2 to zero. Thus this discrete model's parameters are the same with those in Fig. 2 (Fig. 4).

4 Simulations Results in a Specific High Dynamic Scene

Third-order PLL is sensitive to jerk signal and the steady-state error is 0 when acceleration signal exists. So we should test the models in a high dynamic scene with both jerk and acceleration signals.

4.1 Simulation Scene

We set a high dynamic scene which includes jerk signal and acceleration signal: the track stays 10 g/s from 0 to 10 s, then the track stays 100 g from 10 to 20 s. At last, the track stays -10 g/s during 20 ~ 30 s. This scene is shown in Fig. 5:

4.2 The Simulation Results of the Different Models

Figures 6, 7, 8, and 9 show velocity error in the dynamic scene in Fig. 5.

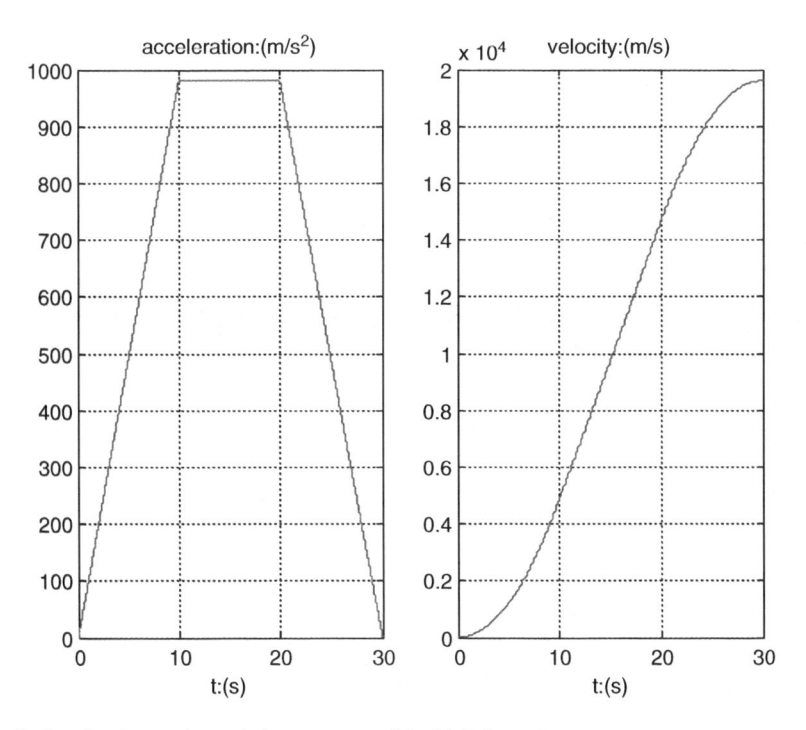

Fig. 5 Acceleration and speed change curve of the high dynamic scene

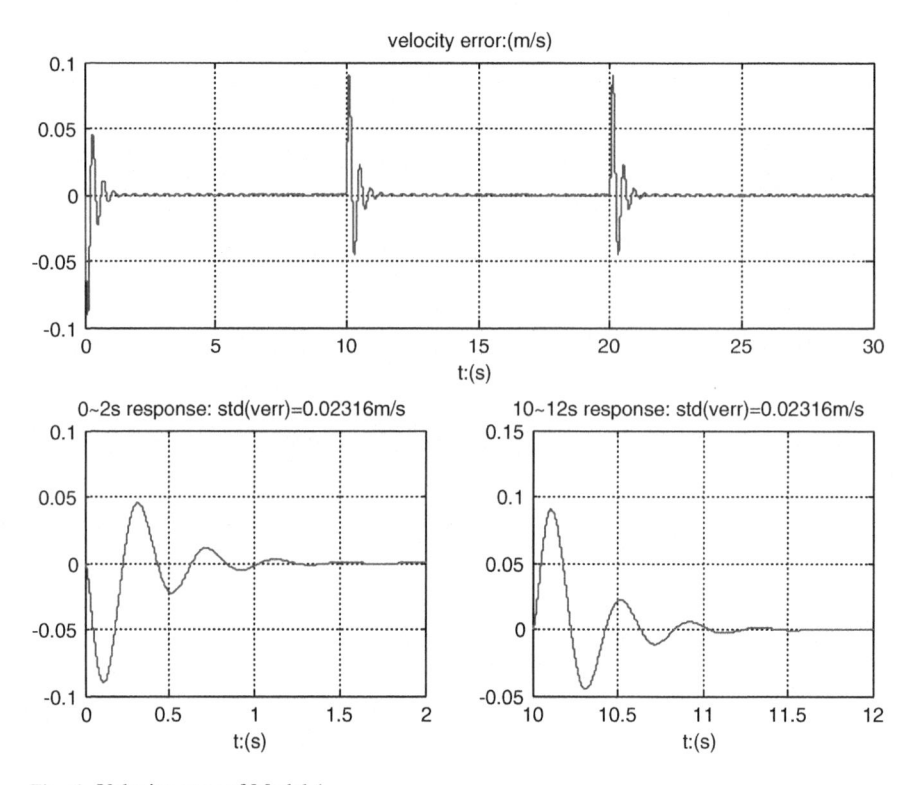

Fig. 6 Velocity error of Model 1

The features of these four figures are concluded in Table 1. The simulation steady-state errors in jerk and acceleration scene of Model 1 and Model 2 are the same with theoretic error. And the shock amplitudes of Model 2 are much smaller than that of Model 1, which proves the theoretic advantage of INS-aided loop. But the features of Model 3 and Model 4 are much worse than that of Model 1 and Model 2. As Model 3 and Model 4 are more similar to actual system, the continuous model of carrier loop and INS-aided carrier loop cannot reflect the velocity error of actual system. At the same time, Figs. 8 and 9 give more details of velocity error.

5 The Mechanism Analysis on Velocity Error of Model 4

Model 3 is the discrete system of Model 1 and while Model 4 is the discrete system of Model 2. We should draw some comparison figures to show the difference between them so as to get the reason of the bigger velocity error. In actual GNSS receiver, the period of carrier loop is usually 1 ms, which cannot be shortened because of the limitation of CPU speed. Thus the research of how to use the INS

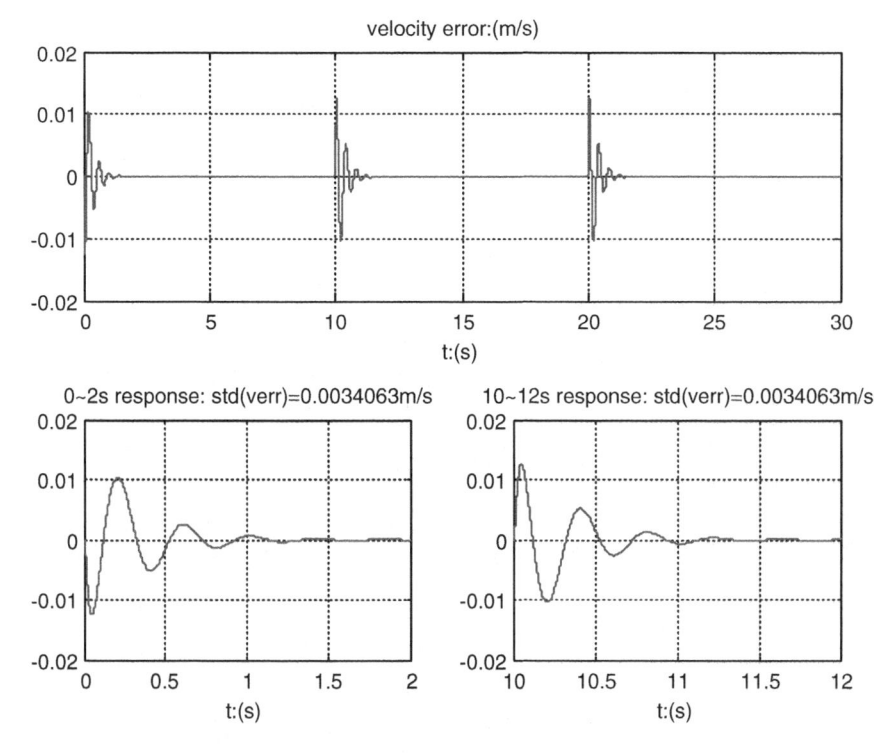

Fig. 7 Velocity error of Model 2

information properly to decrease velocity error is very important. Normally, the aided velocity error at k + 1 time is consists of three parts shown in Eq. 5.

$$\delta \vec{v}_{rx,k+1} = \delta \vec{v}_{rx,k} + \vec{a}_{rx,k}\tau + \vec{j}_{rx,k}\tau^2/2 \qquad (5)$$

Where, $\delta v_{rx,k}$ is velocity error at k time, $a_{rx,k}$ is acceleration at k time, $j_{rx,k}$ is jerk at k time. At time of 15 s, $\delta v_{rx,k} = 0, j_{rx,k} = 0$. From Fig. 10, at time of 15 s there exists a sudden change with the aided velocity shown in Eq. 6.

$$\Delta V = \vec{a}_{rx,k}\tau \qquad (6)$$

Thus, $\Delta V = \vec{a}_{rx,k}\tau$ is the source of the shocking velocity error of INS-aided carrier loop. And the shocking amplitude depends on the features of NCO and third-order PLL. Figure 11 shows the state trajectory of the velocity error during the whole scene. We can see the velocity error's shocking period is the same as aided velocity's changing period.

To suppress velocity error, the sudden change of ΔV must be decreased, which means that τ should be decreased. If $\tau = 0$, it means the INS information update rate is 1 kHz as carrier loop, and simulation results shows that the velocity error will be

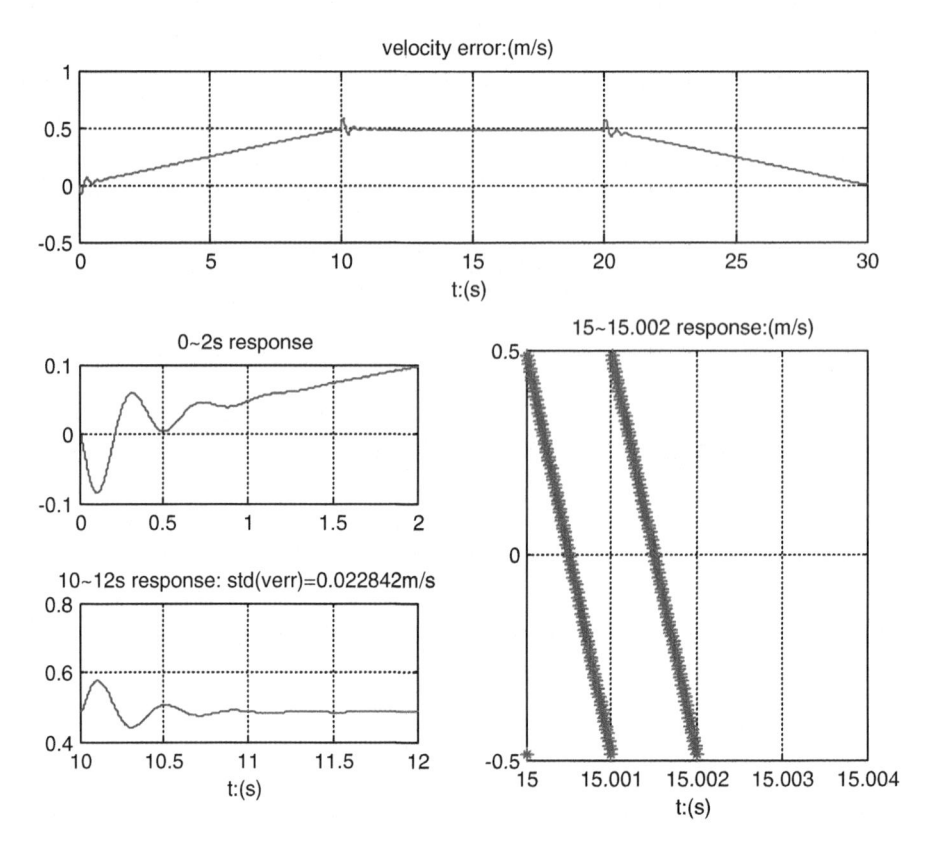

Fig. 8 Velocity error of Model 3

0 just like Model 2. But INS's update rate cannot be as fast as carrier loop because of the limitation of INS device. To decrease τ, Yong Luo [6] uses interpolation and extrapolation method.

6 Conclusion

This paper introduces how to establish the discrete model of INS-aided third-order carrier loop and puts forward the analysis of velocity error of discrete model in a specific high dynamic scene. To decrease velocity error, τ should be decreased. This conclusion guides us to decrease the INS delay time as the most efficient method to improve the dynamic performance of carrier loop. Only when τ is short enough, could, INS information help to improve the loop.

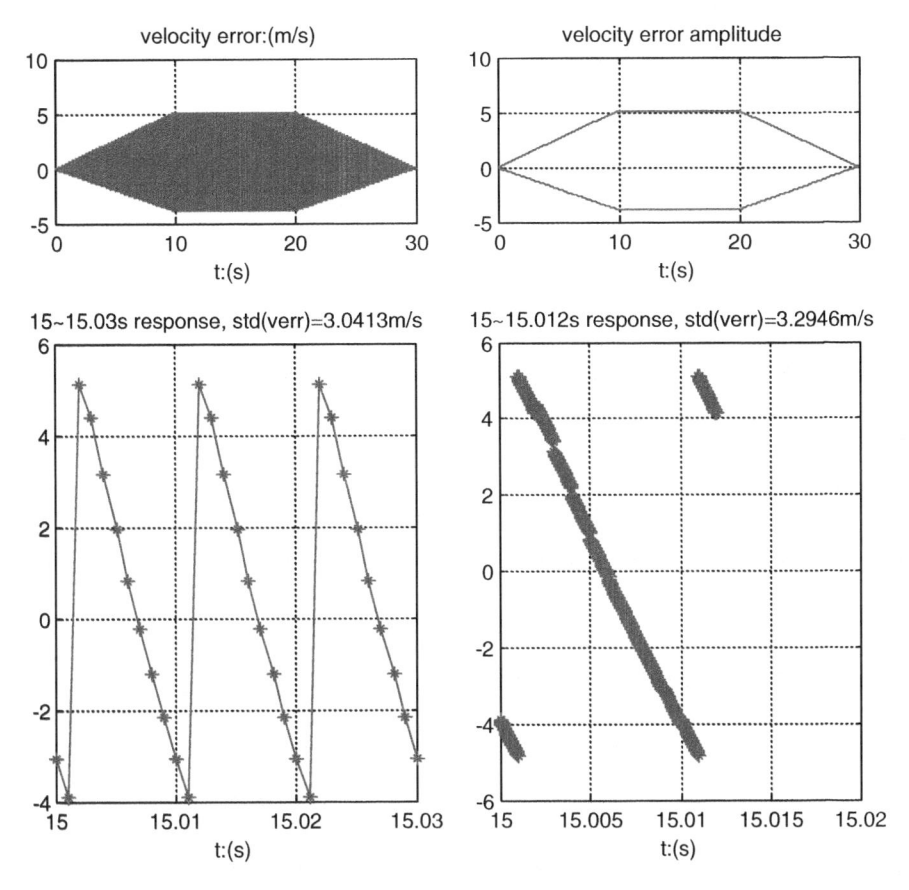

Velocity Error Analysis of INS-Aided Satellite Receiver Third-Order Loop 1037

Fig. 9 Velocity error of Model 4

Table 1 Simulation results comparison

Time	0 ~ 10 s (Jerk)		10 ~ 20 s (Acceleration)	
Item	Shocking amplitude(m)	Steady-state error(m)	Shocking amplitude(m)	Steady-state error(m)
Model 1	−0.09 ~ 0.045	0	−0.045 ~ +0.09	0
Model 2	−0.012 ~ +0.01	0	−0.01 ~ +0.012	0
Model 3	−0.08 ~ +0.06	Increase	−0.44 ~ +0.57	0.48
Model 4	Increase	Shocking increase	−3.9 ~ +5.1	Shocking

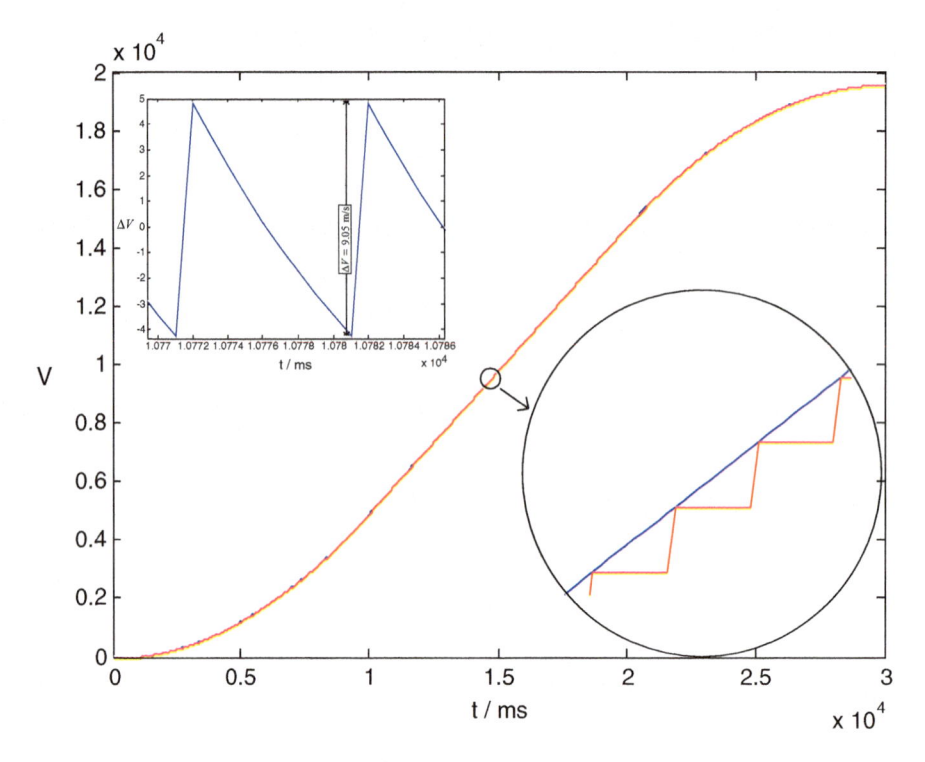

Fig. 10 The error between aided velocity and real velocity

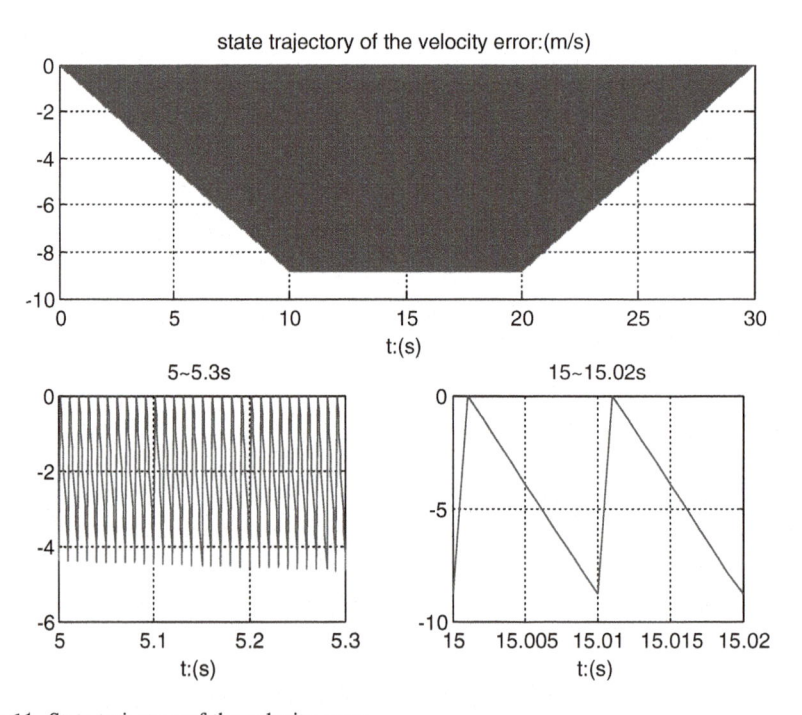

Fig. 11 State trajectory of the velocity error

References

1. Lian-yang Han (2009) Research on high dynamic INS aided GPS carrier tracking algorithm. Aerosp Control 27(3):65–71 (In Chinese)
2. Schmidt GL, Phillips RE (2003) INS/GPS integration architecture performance comparisons [R]. Navigation sensors and integration technology, RTO-EN-SET-064. NATO, London, pp 127–132
3. Xiu-feng He, Jian-ye Liu, Xin Yuan (1996) Performance analysis of GPS receiver aided by velocity signal provide by inertial system. Aerosp Control 4:40–46 (In Chinese)
4. Zhong-qing Yu, Jie Zhang, Gong Cheng (2008) Research on GNSS receiver tracking loop using inertial velocity. Wirel Commun Technol 1:34–36, 41 (In Chinese)
5. Kaplan ED (2002) Understanding GPS principles and applications[M]. Artech House, Boston, pp 185–187
6. Yong Luo, Wen-qi Wu, Xiao-feng He, Yao Guo (2011) Doppler interpolation method based on extrapolation and CIC filter. J Chin Inert Technol 19(2):64–68 (In Chinese)

Data-Processing for Ultrasonic Phased Array of Austenitic Stainless Steel Based on Wavelet Transform

Xiaoling Liao, Qiang Wang, and Tianhong Yan

Abstract In order to suppress the large number scattering signal echoes caused by the coarse grain structure which is tend to cover up the defect signal, resulting in the defect misjudgment in the detection of austenitic stainless steel material by ultrasonic phased array, the wavelet transform was applied to deal with the experimental data. By introducing the theory of wavelet transform and based on the study of ultrasonic testing characteristics of austenitic stainless steel welds, ultrasonic phased array testing were performed in an austenitic stainless steel welds test block, a real-time S-scan detection image was chosen for quantitative analysis from an UT acquisition system. The experimental results show that the wavelet transform can enhance the signal-to-noise ratio of the austenitic stainless steel ultrasonic phased array inspection effectively and help identify the location of defects. The researches demonstrate that this method achieves the defect measurement for the 90 mm thick of austenitic stainless steels.

Keywords Ultrasonic phased array • Ultrasonic testing • Austenitic stainless steel welds • Wavelet transform

1 Introduction

Austenitic stainless steel is a face-centered cubic crystal structure of the iron-based alloys, it has been widely applied in harsh environments, especially in the field of aerospace and nuclear industry for its non-magnetic, high toughness and excellent plasticity performance.

X. Liao • Q. Wang (✉)
College of Quality and Safety Engineering, China Jiliang University, Hangzhou, China
e-mail: qiangwang@cjlu.edu.cn

T. Yan
College of Mechanical and Electrical Engineering, China Jiliang University, Hangzhou, China

Ultrasonic phased array system for non-destructive testing (NDT) has been used increasingly in recent years. Compared with conventional single element transducers, the ultrasonic phased array can perform multiple inspections without the need of reconfiguration.

Due to the imperfect welding technology, grain noise usually appears in NDT ultrasonic phased array testing, which caused by unresolvable reflectors and the defects tend to locate in austenitic stainless steel welds and lead security issues. This noise can be suppressed by certain de-noising algorithms, such as band pass filter, low pass filter and wavelet de-noising algorithm.

In order to satisfy the security requirements in these fields, non-destructive tested for these material is necessary. In addition, more and more researches were done to take effective measurements to suppress the noise in the ultrasonic. Several acoustic and ultrasonic testing techniques are trying to be used to work for the inspection such as acoustic emission, guided waves and phased array [1]. Among these techniques, phased array is good at checking for degradation in thick wall coarse-grain materials like austenitic stainless steel [2–4].

Because of some key advantages over Fourier analysis, wavelet analysis has been widely utilized in signal estimation, classification and compression. Wavelet transform cut up data into different frequency component, and reconstruct the useful components by thresholding policy. Also with its multi-resolution characteristics, it is widely used in ultrasonic signal de-noising processing to improve the signal-to-noise ratio [5].

In this paper, conventional ultrasonic testing problems are analyzed combined with the principle of phased array [6, 7], and wavelet transform is made through the ultrasonic phased array testing experiment on defects in the depths of 40 mm in the austenitic stainless steel block. The choice of the mother wavelet is also discussed in this paper [8]. Choosing the appropriate wavelet based on the correlation coefficient and denoising has been applied.

2 Experimental Studies

2.1 Experimental Instruments and Method

Pulse-echo method was chosen and Omniscan MX2 detector by Olympus was employed in this experiment. Detection parameters are shown in Table 1. The probe has 64 array elements but only 16 of them simultaneous excitation, its frequency is 5 MHz and oil was used as the coupling agent in this experiment. Besides, the S-scan angle is 40–70° and the defect in the depth of 40 mm is the main defect we detect.

In order to simulate the defects, we made a test block. The material of the test block is 304 and it is 90 mm thick. Five side–drilled holes with a diameter of 2 mm were machined in the block (Fig. 1). Three of them are in the austenitic stainless

Table 1 Detection parameters

Probe model	Wedge model	Scan style
5L64-A12	SA12-N55S	S Scan 40–70°

Fig. 1 The structure and size parameters of the test block

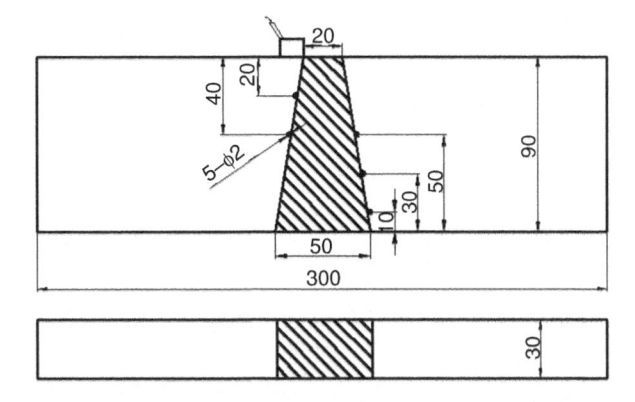

steel and their depths are 10, 30 and 50 mm. Two of them are in the austenitic stainless steel welds and their depths are 20 and 40 mm. The probe radiate by refraction through a Plexiglas wedge and produces transverse waves.

2.2 Wavelet Transform

In many applications in signal and image processing the observed data are influenced by noise. An important question in this text is how to estimate the underlying "clear" signal from the noisy observations.

Wavelet transform (WT) based methods is useful for analyzing signals. In overcoming the problems associated with Fourier transform methods, they analyze the signals in time – frequency domain. So the information regarding both the time-domain and frequency-domain are available. Choosing a wavelet member from a wavelet family is at best a trial-and error method.

The Wavelet transform Ws(a, b) of a signal x(t) is given by [9].

$$Ws(a,b) = \int_{-\infty}^{\infty} x(t)\psi\left(\frac{t-b}{a}\right)dt \qquad (1)$$

Where $\psi(.)$ is the mother wavelet and a, b are the dilatation and translation coefficients respectively.

Based on the principle of suppressing the noise signal, Wavelet transform procedures can be summarized as (i) wavelet transform of the noisy register; (ii) pruning and/or thresholding of the coefficients in the trans-formed domain; (iii) reconstruction of the de-noised signal by inverse transform [10].

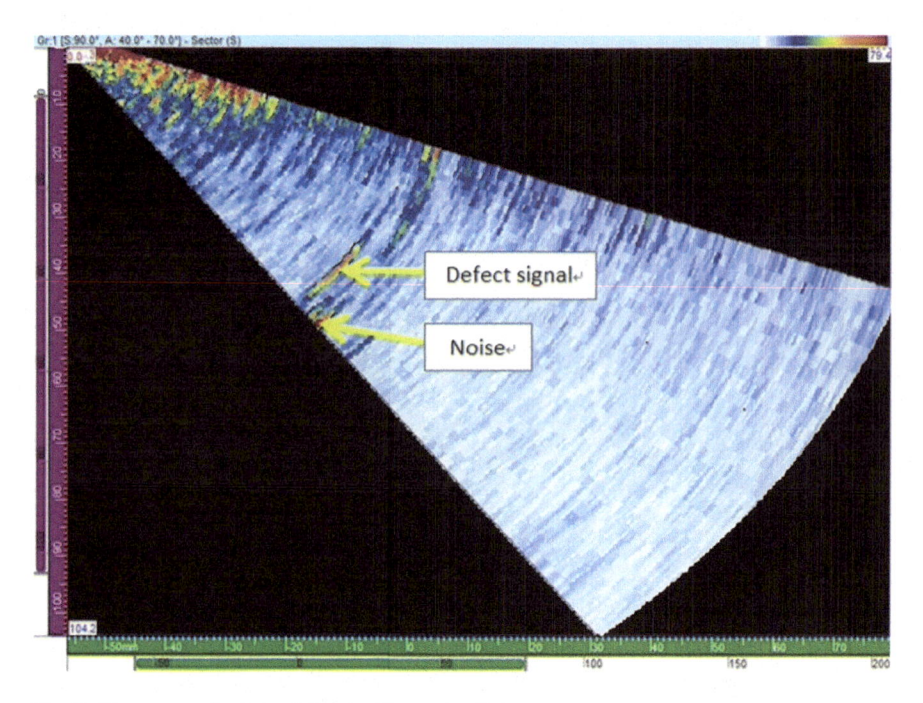

Fig. 2 The test result of the defect in 40 mm depth

According to recently researches, in the detection of ultrasonic echo waves on both time and frequency domains, different kinds of mother wavelet in the Symlet family displays different features. So the mother wavelet of Sym8 was selected in this paper.

2.3 Experimental Data Analysis

To conduct a quantitative analysis, the data of the ultrasonic S-scan image in angle 45° was extracted. In addition, the signal amplitude is expressed by percentage of the total image height in this paper.

As shown in the S-scan images obtained from UT acquisition system (Fig. 2), due to the effects of noise and attenuation caused by coarse grain structure, the signal-to-noise ratio of defect signal has become quite low, defect in 45 mm was caused by instrument(Marked in Fig. 2).

In Fig. 3, it presents the analysis results of ultrasonic plot data after wavelet transform. The mother wavelet in this paper was sym8, and the 3rd-order wavelet, the 5th-order wavelet and 8th-order wavelet was used to process the plot. Comparing the Fig. 3a–d, it's clear that the wavelet transform can suppress the grain

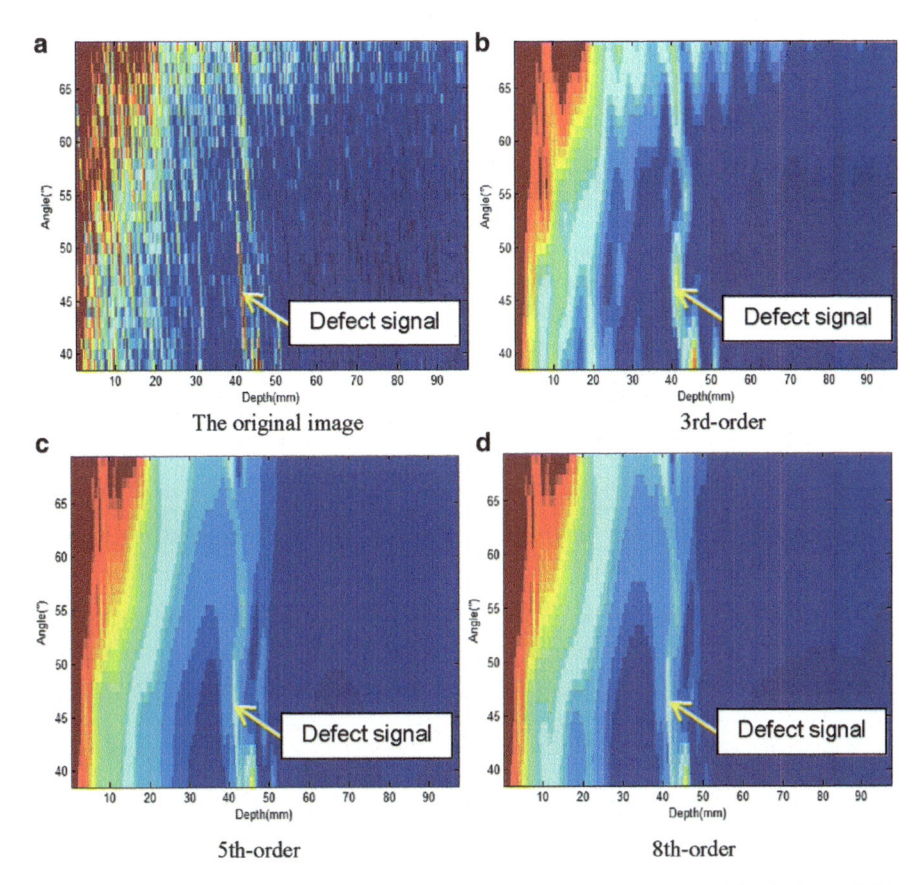

Fig. 3 The effect of transform with mother wavelet sym8 for (**a**) the original plot and (**b**) 3rd-order wavelet transform and (**c**) 5th-order wavelet transform and (**d**) 8th-order wavelet transform

noise well. Particularly, the 3rd-order wavelet transform displays the satisfied performance in suppressing grain noise.

Almost all the depth of noise has been attenuated effectively, and the noise signal curve is very smooth. But the defect in 45 mm caused by instrument is still retain. However, to some extent, it also reduce the useful signal.

3 Conclusion

In this paper we discuss the noise models and the different approaches of processing the noise in the domain of various wavelet transforms. In particular, we focus on how the wavelet transforms with different order affect the processing results. Furthermore, we state three different orders wavelet processing images to discuss the impact.

Wavelet transform approach is presented for analyzing noisy signals. The adopted method shows that the wavelet transform can be effectively applied to austenitic stainless steel measurement.

1. The wavelet de-noising analysis can suppress the grain noise of ultrasonic phased array systems with 64 linear piezoelectric elements for NDT effectively. It can suppress the noise amplitude, smooth the signal curve, and retain the backwall echo signal, though it also can reduce the helpful signal.
2. Ultrasonic phased array technology can effectively detect the defects of austenitic stainless steel. But to the welds' detection, the SNR is low, so in order to increase the defect detection rate, mathematical methods is needed to improve the SNR of deep defects.
3. To coarse grained material and anisotropic structure like austenitic and austenitic welds, the path of the ultrasonic wave would change but the change is irregular.

Acknowledgements This work is supported by Key Program of Science and Technology Planning Project of Zhejiang Province, China under Grant 2011C11079.

References

1. Song S-J, Shin HJ, Jang YH (2002) Development of an ultra sonic phased array system for non destructive tests of nuclear power plant components. Nucl Eng Des 214:151–161
2. Steve M, Jean-Louis G, Olivier R et al (2004) Application of phased array techniques to coarse grain components inspection. Ultrasonics 42:791–796
3. Chassignole B, El Guerjouma R, Ploix M-A et al (2010) Ultrasonic and structural characterization of anisotropic austenitic stainless steel welds: towards a higher reliability in ultrasonic non-destructive testing. NDT&E Int 43:273–282
4. Dong H, Qiang W, Kun X et al (2012) Ultrasonic phased array for the circumferential welds safety inspection of urea reactor. Procedia Eng 43:459–463
5. Grosse CU, Finck F, Kurz JH et al (2004) Improvements of AE technique using wavelet algorithms, coherence functions and automatic data analysis. Constr Build Mater 18:203–213
6. Ruiju H, Lester W, Schmerr J (2009) Characterization of the system functions of ultrasonic linear phased array inspection systems. Ultrasonics 49:219–225
7. Mendelsohn Y, Wiener-Avnear E (2002) Simulations of circular 2D phase-array ultrasonic imaging transducers. Ultrasonics 39:657–666
8. Kun X, Qiang W, Dong H (2012) Post signal processing of ultrasonic phased array inspection data for non-destructive testing. Procedia Eng 43:419–424
9. Rodríguez MA, San JL, Lázaro JC et al (2004) Ultrasonic flaw detection in NDE of highly scattering materials using wavelet and Wigner–Ville transform processing. Ultrasonics 42:847–851
10. Sgarbi M, Colla V, Cateni S et al (2012) Pre-processing of data coming from a laser-EMAT system for non-destructive testing of steel slabs. ISA Trans 51:181–188

The Non-stationary Signal of Time-Frequency Analysis Based on Fractional Fourier Transform and Wigner-Hough Transform

Jun Han, Qian Wang, and Kaiyu Qin

Abstract In order to solve the problem of non-stationary signal analysis, a time-frequency analysis method is introduced in this paper. By comparing with several common time-frequency analysis techniques, this paper proposes a non-stationary signal time-frequency analysis based on the fractional Fourier transform (FRFT) and the Wigner-Hough transform. The simulation results indicate that the FRFT combined with Wigner-Hough transform is a useful method to perform high-resolution non-stationary signal analysis.

Keywords Time-frequency analysis • Fractional Fourier transform • Wigner-Hough transform

1 Introduction

The study of non-stationary signal in both the time and frequency domains simultaneously is very important. It is encountered commonly in many research areas such as radar and sonar analysis, signal detection and parameter estimation, image processing, and fault diagnosis etc. The practical motivation for time–frequency analysis is that classical Fourier analysis assumes that the signal is infinite in time or periodic, while the signal in practice is short duration, and has substantially changed over their duration. This is poorly represented by traditional methods, which motivates time–frequency analysis. One of the most basic forms of time-frequency analysis is the short-time Fourier transform (STFT). In the continuous-time case, the function required to be transformed is multiplied by a window function which is

J. Han (✉) • Q. Wang • K. Qin
Astronautics & Aeronautics institute, University of Electronic Science and Technology of China, Chengdu, China
e-mail: hanjun192835@126.com

nonzero for only a short period of time. The Fourier transform (a one-dimensional function) of the resulting signal is taken as the window is slid along the time axis, resulting in a two-dimensional representation of the signal [1, 2]. The main advantage of this method is its easiness in implementation by using the fast Fourier transform. Since it is a linear time-frequency representation, the crucial drawback of this method is that windowing the signal leads to a tradeoff in time resolution and frequency resolution. At the same time, the rectangular window is often used in the short time Fourier transform, this approach has more serious boundary effects. So we often need to use other more complex window function which adds the complexity of the calculation [3]. Compared with the short time Fourier transform, the Wigner-Ville (WVD) distribution is one of commonly used bilinearity time-frequency distribution. It has many superior properties, especially to the single component of the non-stationary signal analysis. However, due to the bilinearity time-frequency distribution of WVD, the WVD for multi-component of the non-stationary signals will lead to serious cross terms which degrade the energy accumulation [4].

The fractional Fourier transform (FRFT) is one family of linear transformations generalizing the Fourier transform. It can be thought of as the Fourier transform to the n-th power, where n need not be an integer — thus, it can transform a function to any intermediate domain between time and frequency. Its applications range from filter design and signal analysis to phase retrieval and pattern recognition. Moreover, it has been applied to different applications such as high resolution SAR imaging; sonar signal processing, blind source separation [5]. The FRFT can make the signal of time-frequency analysis have more choices. Wigner-Hough transform is a shape-matching technique proposed by Hough in 1962, it can test the detected image of the parametric curves in the parameter space and coalesce the formation with the corresponding curve parameters about the peak point, and get the image of each curve parameter [6]. For discrete limited image, the idea of the Hough transform is quantified the parameter space which made by all possible line parameters into a finite number of parameters table. In application, we first transform signal with the Wigner-Ville, then transform signal with the Hough, and now we get the Wigner-Hough transform.

2 The Relationship Between the FRFT and Other Time-Frequency Analysis

The FRFT can be understood as rotation of the signal for arbitrary α-angle in the time-frequency plain [7]. The FRFT can be defined by Eq. 1:

$$X_p(u) \stackrel{def}{=} \{F^\alpha[x(t)]\}(u) = \int_{-\infty}^{+\infty} x(t) K_p(t,u) \mathrm{dt} \qquad (1)$$

The transform kernel $K_p(t,u)$ is:

$$K_p(t,u) \stackrel{def}{=} \begin{cases} \sqrt{\dfrac{1-j\cos\alpha}{2\pi}} \exp\left[j\dfrac{t^2+u^2}{2}\cot\alpha - jut\csc\alpha\right], \alpha \neq n\pi \\ \delta(t-u), \alpha = 2n\pi \\ \delta(t+u), \alpha = (2n+1)\pi \end{cases}$$

Where p and α are the order of FRFT and the rotation angle, $\alpha = p\frac{\pi}{2}$. The energy of the white noise was uniformly distributed in the whole time-frequency plane. The probability of energy accumulation is very small in the fractional Fourier domain. The energy of linear frequency modulation signal (LFM) is accumulated in the fractional Fourier domain. We can estimate the initial frequency and chirp rate of the LFM signals through the location of the energy accumulation in the two-dimensional fractional plane. So that, we can complete the parameters estimate. The FRFT has very close contact with other time-frequency analysis methods.

2.1 Relationship Between the FRFT and WVD

The WVD of signal $x(t)$ can be defined as follows: Set $x(t)$ for a continuous time signal, The WVD can be expressed by Eq. 2:

$$X(t,\omega) = \int_{-\infty}^{\infty} x\left(t+\frac{\tau}{2}\right) x^*\left(t-\frac{\tau}{2}\right) e^{-j\omega\tau} d\tau \tag{2}$$

Substitute $\tau = \frac{\tau}{2} + t$ into the above Eq. 2, we can get Eq. 3:

$$X(t,\omega) = 2e^{2j\omega t} x^*(2t-\tau) e^{-2j\omega \tau} d\tau \tag{3}$$

Use the FRFT motion properties, we can express $x^*\left(t-\frac{\tau}{2}\right)$ as follows:

$$X^*\left(t-\frac{\tau}{2}\right) = \int_{-\infty}^{\infty} X_{\alpha-z+2t\cos\alpha} e^{-j2t^2\sin\alpha + j2zt\sin\alpha} K_\alpha(\tau,z) dz \tag{4}$$

So, we get Eq. 5:

$$X(t,\omega) = 2e^{2j\omega t} \int_{-\infty}^{\infty} X_\alpha^*(-z+2t\cos\alpha) *$$

$$e^{-j2t^2 \sin\alpha \cos\alpha + j2z \sin\alpha} \int_{-\infty}^{\infty} x(\tau) e^{-2j\omega\tau} K_\alpha(\tau,z) d\tau \tag{5}$$

We can get Eq. 6 from use the frequency shift properties of the FRFT [8].

$$X(t,\omega) = 2e^{2j\omega t} \int_{-\infty}^{\infty} X_p(z + 2\omega \sin \alpha) \tag{6}$$

$$X_p^*(-z + 2t\cos\alpha)e^{-j2(t^2+\omega^2)\sin\alpha + 2jz\sin\alpha - 2jzt\omega\cos\alpha}dz$$

Substitute $\varepsilon = z + 2\omega \sin \alpha$ into the above Eq. 6, we can get Eq. 7:

$$X(t,\tau) = 2e^{2j\omega t} \int_{-\infty}^{\infty} X_p(\varepsilon)X_p^*(-\varepsilon + 2t\cos\alpha + 2\omega\sin\alpha)* \\ \exp\begin{pmatrix} 2j(\omega^2 - t^2)\sin\alpha\cos\alpha + \\ 2j\varepsilon(t\sin\alpha - \omega\cos\alpha) - 4j\omega t\sin^2\alpha \end{pmatrix} d\varepsilon \tag{7}$$

This expression is the WVD in the (t, ω) coordinate system. We would like to change to a new coordinate system (u, v) According to the following formula coordinates $\begin{cases} u = t\cos\alpha + \omega\sin\alpha \\ v = -t\sin\alpha + \omega\cos\alpha \end{cases}$ we transform (t, ω) into (u, v), and after simplification, then Eq. 7 can be written as Eq. 8:

$$X(t,\omega) = 2e^{2juv} \int_{-\infty}^{\infty} X_\alpha(\varepsilon)X_\alpha * (2u - \varepsilon)e^{-2jv\varepsilon} d\varepsilon \tag{8}$$

From the above discussion, the right side of equation is the WVD of $X(u)$. So we can see that, in considering the significance of the axis of rotation, the WVD of $X(u)$ is same with the WVD of $x(t)$.

2.2 Relationship Between FRFT and Wigner-Hough Transform

The Wigner-Hough transform is a straight line integral projective transformation. S. Kay and G.F. Boudreax-Bartels firstly applied Hough transform to the WVD [6]. The transform, often called Radon-Wigner transform (RWT) or Wigner-Hough transform (WHT), have been used to facilitate the analysis and synthesis of the non-stationary signals[8]. To the given finite energy signal $x(t)$, we define the Wigner-Hough transform domain to the mapping of the parameter field of (f, g), then The Wigner-Hough transform can be defined as Eq. 9:

$$WHT_x(f,g) = \int_{-\infty}^{\infty} \int x\left(t+\frac{\tau}{2}\right) x^*\left(t-\frac{\tau}{2}\right) * e^{-j2\pi(f+gt)\tau} d\tau dt \qquad (9)$$

Compared with the WVD, the Wigner-Hough transform has the advantage of noise suppression and suppress cross terms. The modulus square of the signal $x(t)$, P order FRFT transform is just the Wigner-Hough transform on the α direction [9]. Use of this relationship, many properties of the Wigner-Hough can be applieddirectly in the FRFT transform [10]. Meanwhile, the Wigner-Hough combined with the FRFT will achieve better results in the time-frequency analysis.

3 Simulation Results

In the simulation, the non-stationary signal which consists of the two transient signals of different positions is analyzed. The simulation conditions were as follows: Signal 1 is the product of a Gaussian amplitude modulated signal whose signal points are 128, time center is 45 and a constant frequency modulation signal whose normalized frequency f_0 is 0.25. Signal 2 is the product of a Gaussian amplitude modulated signal whose signal points are 128, time center is 85 and a constant frequency modulation signal whose normalized frequency f_0 is 0.25. The simulation signal is sum of signal 1 and signal 2.

Figure 1 shows the time-domain waveform of non-stationary signals, which consists of two components. Figure 2 shows the STFT with the window function of the 65-point hamming window. This time the frequency resolution is better, but the two components of the signal can't be distinguished on the timeline. The signal was analyzed with the WVD and the time-frequency diagrams are shown in Fig. 3, we can clearly see that signal exists cross-connection interference item. We first transform signal with the Wigner-Ville, then transform signal with the Hough, and now we get the Wigner-Hough transform, Fig. 4 shows the signal's Wigner-Hough

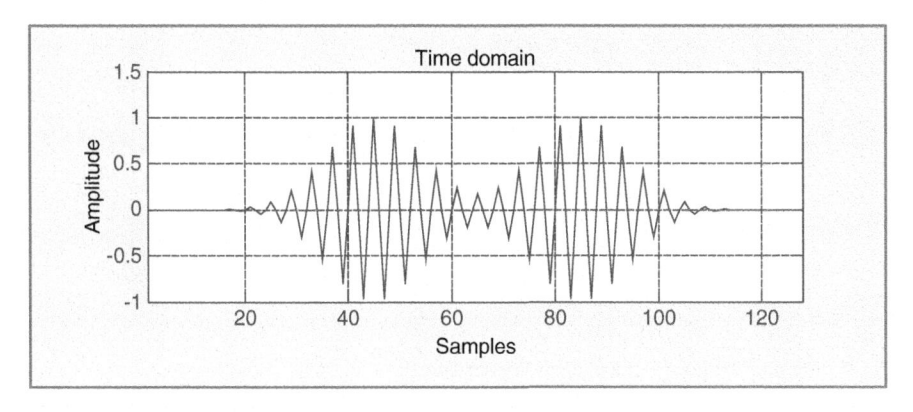

Fig. 1 Time domain waveform of the simulation signal

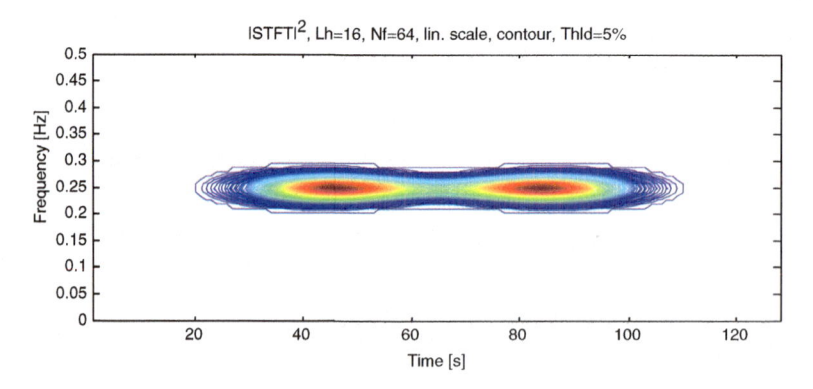

Fig. 2 The simulation signal's time-frequency graph about STFT

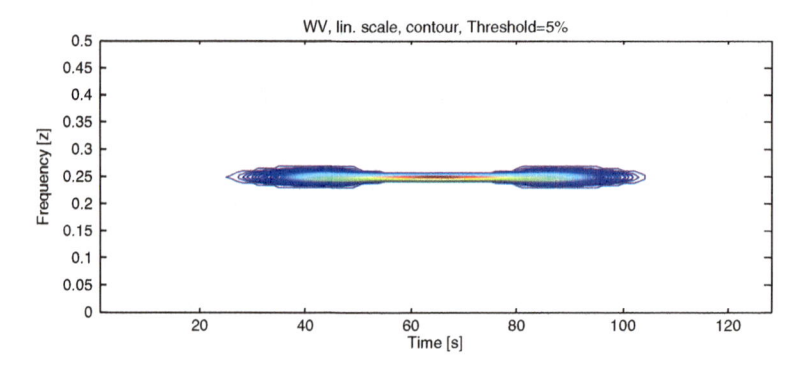

Fig. 3 The time-frequency graph about Wigner-Ville

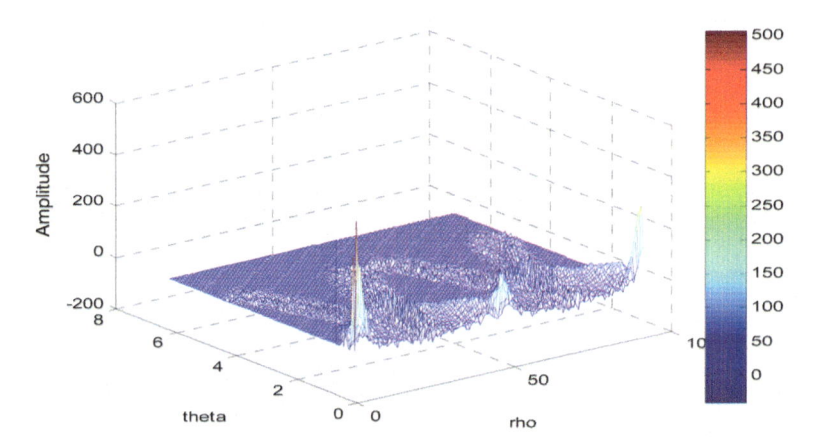

Fig. 4 The time-frequency graph about Wigner-Hough

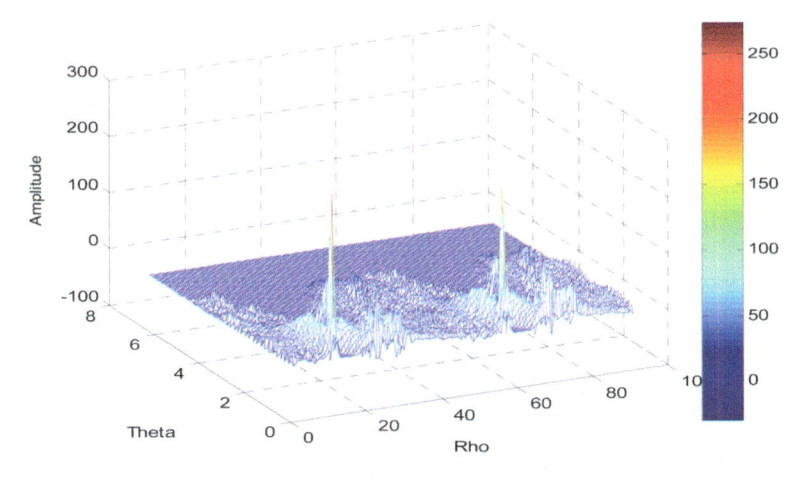

Fig. 5 The time-frequency graph about the FRFT and WVD

transform. We can see there are two peaks in the (ρ, θ) plane, which corresponds to the simulation signal. Although the Wigner-Hough transform do good for suppress the cross terms, but still not enough obviously.

In this paper, firstly, I take the FRFT to the simulation signal, the simulation signal in the original coordinate system (t, ω) counter-clockwise α angles, and then transform it with Wigner-Hough. The results showed in Fig. 5. We can see more clearly that signal in (ρ, θ) plane has two peaks, and the peaks in the middle of the three-dimensional time-frequency graph.

4 Conclusion

In conclusion, the combination of FRFT and the Wigner-Hough transform provides a method for the time-frequency analysis. This paper firstly took the FRFT to the simulation signal by setting the $K_p(t, u)$, and then transformed it with Wigner-Hough. The results showed that the method not only detected the signal components in the non-stationary signals well, but also easily observed and estimated the instantaneous frequency of the analysis signal, and restored the phase information. This advantage enables the method to be quite useful in time-frequency analysis.

References

1. Qian SE, Chen DP (1996) Joint time-frequency analysis, methods and applications. Prentice Hall PTR, New York, pp 15–67
2. Amein AS, Soraghan JJ (2007) The fractional Fourier transform and its application to high resolution SAR imaging. In: IEEE Proceedings of geoscience and remote sensing. IEEE Computer Society, Barcelona, Spain, pp 5174–5177
3. Barbu M, Kaminsky EJ, Trahan RE (2005) Fractional Fourier transform for sonar signal processing. In: IEEE Proceedings of OCEANS. IEEE Computer Society, Washington DC, USA, pp 1630–1635
4. Deng L (1999) Wavelet denoising of chirp-like signals in the Fourier domain. In: Proceedings of the IEEE international symposium on circuits and system. IEEE Computer Society, Orlando, Florida, USA, pp 540–543
5. Kay S, Boudreax-Bartels GF (1985) On the optimality of the Wigner distribution for detection. In: Proceedings of the IEEE ICASSP-85. IEEE Computer Society, Orlando, Florida, USA, pp 1017–1021
6. Barbarossa S (1995) Analysis of multicomponent LFM signals by a combined Wigner-Hough transform. IEEE Trans Signal Process 43(6):1511–1515
7. Capus C, Brown K (2003) Short-term fractional Fourier methods for the time-frequency representation of chirp signals. J Acoust Soc Am 113(6):3253–3263
8. Boashash B (1992) Estimating and interpreting the instantaneous frequency of a signal-part2: fundamentals. Proc IEEE 80(4):540–568
9. Allen JB (1977) A unified approach to short-time Fourier analysis and synthesis. Proc IEEE 65(11):1558–1564
10. Pei S, Min-Hung Yeh, Tzyy-Liang Luo (1999) Fractional Fourier series expansion for finite signals and dual extension to discrete-time fractional Fourier transform. IEEE Trans Signal Process 47(10):2883–2888

Weakness in a Serverless Authentication Protocol for Radio Frequency Identification

Miaolei Deng, Weidong Yang, and Weijun Zhu

Abstract The design of secure authentication protocols for radio frequency identification (RFID) system is still a quite challenging problem. Many authentication protocols for RFID have been proposed, but most have weaknesses or flaws. We analyze the security of a serverless RFID authentication protocol which recently been presented by Hoque et al. The protocol was expected to safeguard both RFID tag and RFID reader against major attacks, and RFID server is not needed in the protocol. However, our security analysis shows that the authentication protocol is vulnerable to attack of data desynchronization. This attack destroys the availability of the protocol. Furthermore, improvement to overcome the security vulnerability of the protocol was presented.

Keywords RFID • Authentication protocols • Security • Attack

1 Introduction

Radio frequency identification (RFID) has been widely applied, and has the advantages of automatic object recognition. RFID can be used in a great variety of applications such as stock security, supply chain management, retail inventory control, access control or product tracking. In an RFID system, the cost of the tags is low, which implies that the tags have very limited computational capabilities and storage. General -purpose security protocols cannot be applied directly to the RFID system [1].

M. Deng • W. Yang
College of Information Science and Engineering, Henan University of Technology, Zhengzhou, China
e-mail: dmlei2003 @163.com

W. Zhu (✉)
School of Information Engineering, Zhengzhou University, Zhengzhou, China
e-mail: zhuweijun76@163.com

With the wide application of RFID systems, the security of the transmissions between the readers and the tags has received additional consideration [2]. So far, many RFID authentication protocols [3–5] have been put forward to protect RFID communications. These protocols have used the "backend database" (also known as server) model. Three parties are contained in this model, that is, the RFID tag, the RFID reader, and the backend database (or server). The backend database will return the information of the RFID tag to the RFID reader when the backend database verifies the tag and reader. This is possible because the backend database has knowledge of all the tag secrets as well as tag data. However, a prominent weakness of the backend database model is that an always reliable connection between the backend database and the RFID reader is needed. In addition, having a backend database creates a single point of failure, which may result in denial of service attacks [6].

To solve the problems, some authentication protocols have been presented which provide mutual authentication between the tag and the reader without the need for a constant backend database [7–9]. Recently, Hoque et al. suggest a serverless, untraceable authentication and forward secure protocol for RFID tags [10]. Hoque's authentication protocol safeguards both reader and tag against attacks as often as possible without the intervention of server (i.e. backend database). However, our security analysis shows that the authentication protocol is vulnerable to attacks of data desynchronization. This attack destroys the availability of the protocol. Furthermore, the improved serverless RFID authentication protocol is proposed, and it can withstand the attack of data desynchronization.

2 Hoque's Authentication Protocol

Usually, an RFID system comprises the RFID reader, R, the RFID tag, T, and the backend database. Nevertheless, Hoque's RFID system is a serverless system. Therefore, Hoque's serverless RFID system mainly contains two parties, one is a set of RFID tags and the other is the RFID reader R. In addition, a certification authority CA is encompassed in the RFID system to authorize readers to tags.

2.1 Notation and Assumption

In the system, all tags and readers have knowledge of a function $M(.)$ and a pseudorandom number generator $P(.)$. $P(.)$ is a low cost random number generator which applies to the RFID system. $M(.)$ is assumed as an one way hash function. An RFID reader R has a contact list L and a unique identifier r. L and r are obtained from a CA by R. In addition, each tag T includes a unique secret t and a unique identifier id. Subscripts are utilized to describe a particular T or R and their variables. The contact list L_i has the following shape,

Fig. 1 Hoque's RFID authentication protocol

(1) $R_i \to T_j$	:	$request, rand_i$
(2) T_j	:	$n_j = P(seed_{Tj} \oplus (rand_i \| rand_j))$
(3) $R_i \leftarrow T_j$	:	$n_j, rand_j$
(4) R_i	:	$n_i = rand_i$
(5)		for all m from 1 to n //run through list L_i
(6)		Let $n_m = P(seed_m \oplus (rand_i \| rand_j))$
(7)		if $(n_m == n_j)$ then
(8)		Let $s = M(seed_m)$
(9)		$n_i = P(s)$
(10)		$seed_m = M(s)$
(11)		$R_i \to T_j : n_i$
(12) T_j	:	Let $k = M(seed_{Tj})$
(13)		Let $a = P(k)$
(14)		if $(a == n_i)$ then
(15)		$seed_{Tj} = M(k)$
(16)		else
(17)		Reader is not authorized or is an adversary

$$L_i = \begin{cases} seed_1 & : & id_1 \\ \ldots & : & \ldots \\ seed_n & : & id_n \end{cases}$$

where, for $1 \leq j \leq n$, $seed_j$ is initialized by $seed_j = h(r_i \| t_j)$. The initial $seed_j$ is given by CA and stored in R_i's nonvolatile memory. An adversary is denoted as Q.

2.2 RFID Authentication Protocol

Hoque's authentication protocol is shown in Fig. 1.

3 Vulnerability of Hoque's Protocol

Hoque et al. analyzed their proposed RFID authentication protocol and deemed that their protocol is secure. They claimed that the RFID protocol can provide privacy protection and withstand tracking attack, cloning attack, denial of service (DoS) attack, physical attack, and eavesdropping attack.

Fig. 2 Data desynchronization attack

(1) $R_i \to T_j$	:	$request, rand_i$
(2) T_j	:	$n_j = P(seed_{Tj} \oplus (rand_i \parallel rand_j))$
(3) $R_i \leftarrow T_j$	:	$n_j, rand_j$
(4) R_i	:	$n_i = rand_i$
(5)		for all m from 1 to n
(6)		Let $n_m = P(seed_m \oplus (rand_i \parallel rand_j))$
(7)		if $(n_m == n_j)$ then
(8)		Let $s = M(seed_m)$
(9)		$n_i = P(s)$
(10)		$seed_m = M(s)$
(11)	$R_i \to$	: n_i, Q prevent T_j receiving n_i
(12) T_j	:	Do nothing

However, the RFID protocol cannot offer any protection against data desynchronization attack: an adversary Q can easily force an honest tag to fall out of desynchronization with the reader so that it can no longer authenticate itself successfully. The attack can be described as follows (Fig. 2).

In the attack, the adversary Q easily destroys the desynchronization of the $seed_{Tj}$ updating between the reader and the tag. In line (11), Q can interrupt the message n_i from the tag T_j to the reader R_i. Thus, the reader R_i has refreshed the secret $seed_{Tj}$ while the tag T_j will not. Therefore, the shared secret between the tag T_j and the reader T_j may not be identical, which will threw the RFID system into confusion. After a successful data desynchronization attack, because Q makes the reader R_i and the valid tag T_j share the different secrets, R_i will not be authorized by T_j and T_i will not be authorized by R_j yet. The attack destroys the availability of the protocol.

4 The Anti-desynchronization RFID Protocol

We are trying to improve Hoque's RFID authentication protocol and provide an anti-desynchronization authentication protocol in this section.

4.1 Improvement of Hoque's Authentication Protocol

The reader ought to keep the history of the entire seed update in order to settle the data desynchronization attack issue. That is, the reader should keep not only the current records but also the previous records of the seed update process. When the reader cannot authenticate an honest tag because of the data desynchronization attack, it uses the old shared seed again to accomplish the authentication.

In particular, the RFID reader R_i stores contact list L_i and an identifier r_i in its nonvolatile memory. The contact list L_i comprises information about the RFID tags that R_i can access to and each tag contains the current seed $seed_j$, the previous seed $seed_{jp}$ and a unique identifier id. Thus L_i will have the following shape after authenticating itself to CA,

$$L_i = \begin{cases} seed_{1p}, seed_1 & : \ id_1 \\ \ldots & : \ \ldots \\ seed_{np}, seed_n & : \ id_n \end{cases}$$

In a general authentication process, the current seed of T_j, $seed_{Tj}$, will be utilized to accomplish the mutual authentication between the reader and the tag. Nevertheless, if the reader fails to look up the current seed for the desynchronization of the shared secret, it may use the previous seed to complete the authentication. Thus the improved protocol is shown in Fig. 3.

4.2 Security Analysis

The improved RFID protocol is analyzed in this section to estimate whether the protocol satisfies the security requirements or not. Similar to the original Hoque's authentication protocol, the improved protocol can also guarantee the privacy of the RFID tag, and resist the tracking attack, cloning attack, denial of service attack, eavesdropping attack and physical attack. The detailed analysis of security in this respect is overleapt and the similar analysis can be seen in Hoque's paper [10].

The adversary can impede the communication between a tag and a reader, which can threw the RFID system into confusion as before, but the desynchronization resistant mechanism that discussed just now makes the authentication protocol also meet the requirement of desynchronization resistance.

The improved protocol is also forward security. The protocol conceals the information utilized for updating the seed. The seed updating is performed whenever the authentication is successful, therefore future security compromised on an RFID tag will not disclose data previously transmitted.

4.3 Performance Evaluations

A wide adoption of RFID technology requires the RFID protocols not only to be secure, but to be practical and efficient. We analyze the improved RFID protocol by estimating its storage cost and computation cost.

Fig. 3 The improved RFID authentication protocol

(1) $R_i \rightarrow T_j$	:	$request, rand_i$
(2) T_j	:	$n_j = P(seed_{Tj} \oplus (rand_i \parallel rand_j))$
(3) $R_i \leftarrow T_j$	:	$n_j, rand_j$
(4) R_i	:	$n_i = rand_i$
(5)		for all m from 1 to n
(6)		Let $n_m = P(seed_m \oplus (rand_i \parallel rand_j))$
(7)		if $(n_m == n_j)$ then
(8)		Let $s = M(seed_m)$
(9)		$n_i = P(s)$
(10)		$seed_m = M(s)$
(11)		$R_i \rightarrow T_j : n_i$
(12)		else Let $n_m = P(seed_{mp} \oplus (rand_i \parallel rand_j))$
(13)		if $(n_m == n_j)$ then
(14)		Let $s = M(seed_{mp})$
(15)		$n_i = P(s)$
(16)		$seed_m = M(s)$
(17)		$R_i \rightarrow T_j : n_i$
(18) T_j	:	Let $k = M(seed_{Tj})$
(19)		Let $a = P(k)$
(20)		if $(a == n_i)$ then
(21)		$seed_{Tj} = M(k)$
(22)		else
(23)		Reader is not authorized or is an adversary

An RFID tag in improved scheme just stores its one seed for its only one authorized RFID reader. Certainly a tag still requires other memory space for communication and computation. However, computation in the improved protocol only contains hash operation and only one value utilized as authentication needs to be stored, so the required memory space is very limited.

The computation load of RFID tags in the improved protocol is also rather light-weight. The improved authentication protocol contains two hash functions, $M(.)$ and $h(\cdot, \cdot)$. The tag T_j will get $h(r_i \parallel t_j)$ as $seed_{Tj}$ from CA, that is, the hash functions, $h(\cdot, \cdot)$ is computed by CA. Therefore the cost of the protocol may be determined based on the computation of $M(.)$ function. According to the improved RFID authentication protocol described above, it can be seen that $M(.)$ is computed twice, first in line (18) and second in line (21). Therefore, the cost for the protocol is little higher than other protocols [3–5] that require the tag to execute only one hash function. In our scheme, the additional hash functions can avoid disclosing the tag secret to the reader and also permit the protocol to be serverless.

5 Conclusion

Designing of a secure authentication solution for low-cost RFID tags is still an open and challenging problem. In this paper the cryptanalysis of a recent lightweight RFID authentication protocol is proposed. The RFID authentication protocol is vulnerable to attack of data desynchronization. The proposed attack could have been avoided by following a desynchronization resistant mechanism for designing RFID protocols. The improved RFID authentication protocol is forward security, and satisfies the security requirements, such as privacy protection, tracking attack resistance, cloning and physical attack resistance. The improved protocol is lightweight and suitable for the low-cost RFID environment.

Acknowledgements This work is supported by China Postdoctoral Science Foundation (2012M511588), Ph.D. Programs Foundation of Henan University of Technology (2013BS007) and National Natural Science Foundation of China under Grant No. U1204608.

References

1. Chai Q (2012) Design and analysis of security schemes for low-cost RFID systems[D]. Doctor dissertation of Waterloo University, Waterloo
2. Jules A (2006) RFID security and privacy: a research survey. IEEE J Sel Areas Commun 24(2):381–394
3. Lee YK, Batina L, Singele D (2010) Low-cost untraceable authentication protocols for RFID systems[C]. In: Proceedings of the 3rd ACM conference on wireless network security, ACM Press, New Jersey, pp 55–64
4. Deng Miaolei, Ma Jianfeng, Zhou Lihua (2009) Design of anonymous authentication protocol for RFID. J Commun. 30(6):24–31 (in Chinese)
5. Moessner M, Khan GN (2012) Secure authentication scheme for passive C1G2 RFID tags. Comput Netw 56(1):272–286
6. Duc DN, Kim K (2011) Defending RFID authentication protocols against DoS attacks. Comput Commun 34(3):1196–1211
7. Tan CC, Sheng B, Li Q (2007) Serverless search and authentication protocols for RFID[C]. In: Proceedings of the 5th annual IEEE international conference on pervasive computing and communications, IEEE Press, New York, pp 34–41
8. Ahamed SI, Rahman F, Hoque E (2008) S3PR: secure serverless search protocols for RFID [C]. In: International conference on information security and assurance, IEEE Press, Busan, pp 187–192
9. Deng Miaolei, Wang Yulei, Qiu Gang et al (2009) Authentication protocol for RFID without back-end database. J Beijing Univ Posts Telecommun 32(2):68–71 (in Chinese)
10. Hoque ME, Rahman F, Ahamed SI et al (2010) Enhancing privacy and security of RFID system with serverless authentication and search protocols in pervasive environments. Wirel Pers Commun 55:65–79

The Radar Images Smooth Rolling

Peng Gao

Abstract The main problem of achieving the radar images rolling demo is how to achieve the goal of smooth rolling. The problem was mainly caused by the high computational load for image processing and low computing efficiency. In order to solve this problem, a complete solution is brought up in this paper. In this solution, the computational load is reduced by compressing the raw format radar images. A new compression method for RAW format radar image is raised. The image processing time is reduced by choosing the right gray equilibrium and Image Mosaic method. In this paper, a new gray equilibrium method is raised based on the histogram equalization method. And the gray correlation matching algorithm with iterative search method based image pyramid is used to splice images. Moreover, the calculation efficiency is increased by using two threads for control data reading and processing. The parallelization between CPU and I/O can be achieved. Meanwhile, the parallel computing ability of Multi-core CPU can also be used. The solution has proved the efficiency by detailed description and examples. The high resolution radar images can be rolled smoothly with less calculation and memory consumption and have satisfactory image quality. So this solution can effectively achieve the goal of smooth rolling.

Keywords Radar image · Roll · Compression · Gray equilibrium · Multi-threaded

1 Introduction

Usually the radar images showed to uses and audiences are static [1], such as paper image and electronic image. But if showed the continuous images in same strip, the method of static display is difficult to promises continuous experience.

P. Gao (✉)
Master, Center for Earth Observation and Digital Earth Chinese Academy of Sciences, Institute of Remote Sensing and Digital Earth Chinese Academy of Sciences, Beijing, China
e-mail: foree@foxmail.com

By comparison, showing these images in rolling way is much better. For achieving these images rolling demo, there are many problems need to be solved in the process of achieving the images smooth rolling, one of the most critical is how to achieve the goal of smooth rolling. By now, there is no complete solution for solving radar images smooth rolling and the related issues. So In this paper, a complete solution is brought up and proved the efficiency by detailed description and examples.

The radar image for this solution is the most common RAW format. The 16-bit integer RAW high resolution radar image will be used in this paper. This type of images always has huge volume. This will lead to a huge amount of calculation and Further influence the computing efficiency. So compressing the radar image is very necessary. At present there have been many research results for radar image compression [2–4], but no specific to RAW format radar image. In this paper, through the analysis of the characteristics of this kind of image, a new data progressing algorithm is raised. Using this algorithm, the 16-bit image can be compressed to 8-bit. This will reduce the computation time effectively.

Grey equilibrium is also the implementation steps for RAW format image rolling demo. Because of most of the gray details in RAW image are in low gray level, the image without gray equilibrium show black or with odd light spot. Even after image compression processing, the image are still dull. So the grey equilibrium processing are essential for raising the image visual effect. Because of the process need a number of calculations, the choice of the equilibrium method concerns the image quality and smooth roll. In this paper, a new method come is raised based on the histogram equalization method [5]. This method takes the same calculation as histogram equalization but better processing effects. So using this method will put less pressure on smooth rolling.

The RAW images do not just need gray equilibrium, but also need image mosaic processing. It's also time-consuming. Because the images in the same strip are overlapped in a small area near and along the vertical direction near the boundary, we can choose the gray correlation matching algorithms [6]. Through the pyramid iterative search strategy [7], the gray correlation matching algorithms will spend much less calculated. The images are in rolling, so the stitching precision can only be required no visual differences for users. By verification, this method can reach the stitching precision.

Besides, a novel double thread scheduling method is used to control the image processing, read, and display in this solution. Combining multi-threading parallel computing [8] thoughts, this method realizes the parallel operation between the CPU and I/O, also realizes the multi-thread parallel computing. It's useful for increasing the calculation efficiency and the image rolling smoothly.

To sum up, in order to achieve the goal of the radar images smooth rolling, meet the image quality requirements and user experience, much research is developed aimed at data compression, gray equilibrium, image mosaic and multi-threading control. As a result, many effective methods are brought up.

2 Solution

The solution stated above, mainly includes three aspects: image processing, image mosaic and multithreading control. The image processing includes image compression and gray equilibrium. Images are calculated and displayed by the two threads. Their functions are alternative during the image rolling. Figure 1 is the flow chart for this solution. From now on, we will introduce the three mainly aspects by combining the flow chart above.

2.1 Multithreaded Control

As shown in Fig. 1, the whole process of calculation and display are completed in two threads. But in Fig. 1 the flow chart shows the flow when the program has run a period of time and the two threads are stability alternative. It's subtle differences with the flow at the beginning of the program running. Combined with the flow chart, I will introduce the processing from initial rolling to steady rolling.

At the beginning of the program, the thread-1 read two images once. This is different with the Fig. 1. After reading, It will be checked whether the thread-1

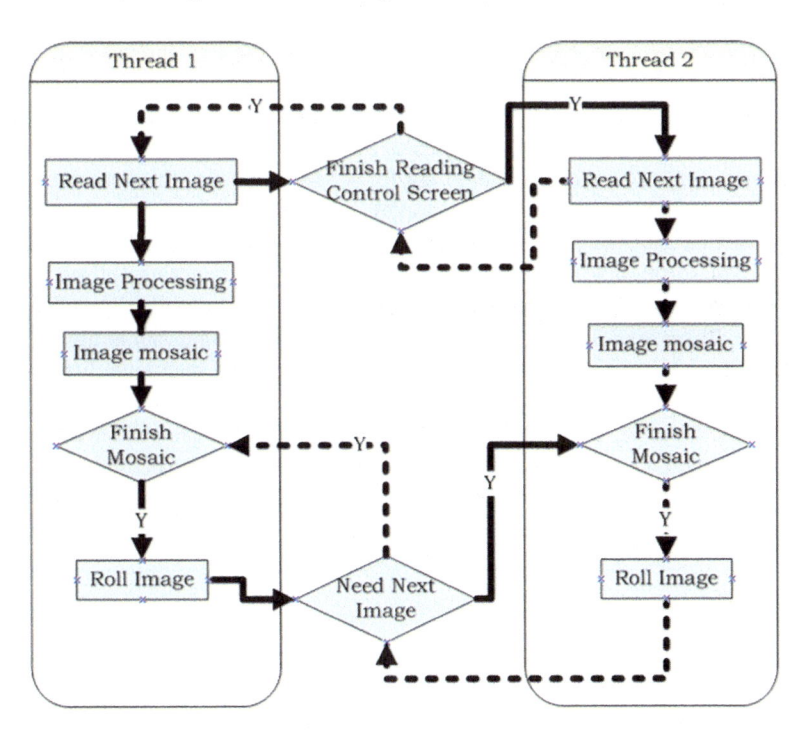

Fig. 1 Solution flow chart

controls the screen display. If so, continue; if not, wait. Because the program is just in the initial state, no one control the screen display. So thread-1 continues. Then the thread-1 notifies the thread-2 to read next image. At the same time, the thread-1 begins to compress image and equilibrate the gray. After these, the two images will be spliced. Then, it will be checked whether the screen need display new image. Usually, it will happen when the old image has showed completely.

After thread-1 has finished the image reading, the thread-2 will be noticed to read next image. After the thread-2 finished reading, the same notice will be sent to thread-1 from thread-2. If thread-1 is controlling the screen display, and the image is still rolling, the notice will be suspended. When the new image is needed, thread-1 will release the old image memory, and begins to deal with the notice which has be suspended.

When thread-2 receives the request for show next image from thread-1, it will be checked whether the image processing and image mosaic has finished. If so, control the screen and show new image; if not, wait until they finish. Now the thread-1 has worked as the thread-2 before, reading and process image in background. In proportion, thread-2 becomes the controller instead of thread-1. From now on, the images will roll continuous as the Fig. 1.

Above is the basic flow that two threads control the screen display alternately in order to achieve the goal of the image rolling demo. When finish themselves rolling task, each thread release the right of controlling screen and image memory, then begins to read next image. In this way, there are always no more than two images in memory. Moreover, in the process of image rolling, the two threads will not be destroyed and new thread or variables will not be created, so the total used memory will not increase.

Besides, when a thread is reading data, the other is always processing data or displaying images. So they will never ask for the I/O control at the same time, As a result, the I/O and CPU will work in parallel. If the time for reading is short, it will be appear that the two threads are using CPU in the same time. On a dual core processor platform, it will take full advantage of parallel computing ability of CPU. Fortunately, the multi-core processor has become the mainstream. These designs above are all for avoiding the stumble rolling when the images roll at high speed.

Unfortunately, because of the runtime environment and data size differs from each other, the maximum speed is not unfixed with smooth rolling. Through the analysis of the flow, we can find out the most likely scenario which causes the rolling suspend is that the time for reading and processing image is too long, so that image mosaic has not finished when the screen require a new image. So, we will introduce the appropriate method for processing and splice image.

2.2 Image Processing

In this paper, the 16-bit integer radar image is used. It's $16,384 \times 16,384$ pixels and 512 MB. If such large data is immediately used to gray equilibrium or image mosaic

Fig. 2 The image after compressing

or display on screen, the efficiency will be very low. The time can't meet the requirements that the smooth rolling need short processing time. So before using the image data, it must be compressed firstly.

A light and shade moderate RAW image will be a black image browsed directly. So the image doesn't list here. The reason is that the main nonzero grays are mostly in the low gray level. For most RAW image, the higher 6–7 bit of gray value are mostly 0, the nonzero gray bit are almost in front 12 bits, the last 4 bits are always 0. Moreover, the first bit are always 0. So all these bits which are always 0, can be compressed and omitted.

Due to the limitation of the human eye, the small gray variance of too much light or dark pixels is difficult to distinguish, but is easier distinguished for middle tone. So the middle tone should be kept while compressing image data. The low 1-bit or 2-bit should be compressed. For the reason that the most screen can only show the 8-bit gray image, the 16-bit RAW image can be compressed to 8-bit without influencing the effect for screen display. Through the above analysis, the compression method in this paper is that the high six bits and low two bits will be ignored, only the middle eight bits will be reserved. Using this method, the image data will be compressed to half size.

The image compressed is showed in Fig. 2. As shown, the image isn't wholly dark, and already shows some gray details. So the data compression can be regarded as gray equilibrium.

Although the visual effect in Fig. 2 has been improved greatly than raw image, it's still dull and difficult to differ ground information. The reason for this is that most gray are still in the low level. So farther gray equilibrium need to be introduced to image processing.

At present there have been many method for image equilibrium [9], such as time domain enhancement technology. For radar image, many methods can be used,

Fig. 3 The image after image equilibrium

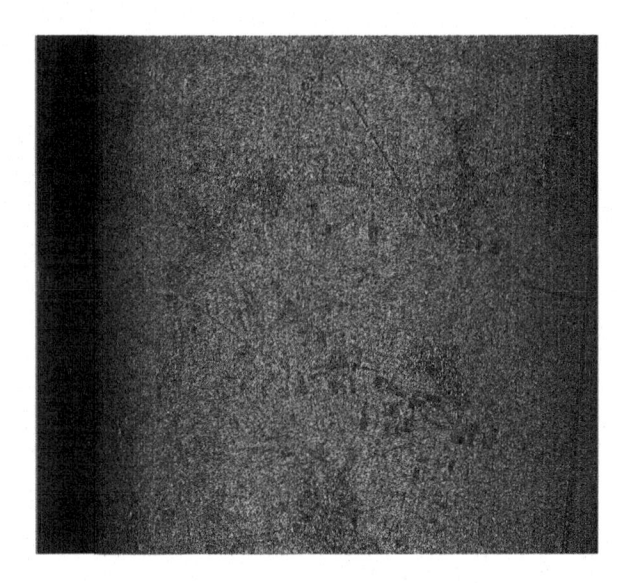

such as wavelet transform, histogram equalization, histogram specification and so on. The image smooth rolling required the time for image compression should not vary long, so the method which is high-computational is not suitable. The histogram equalization is the fastest method, but its treatment effect is unsatisfactory. This method remarks the middle gray and weakens the boundary gray, so the image shows too much white. Unfortunately, the most important ground characteristics in radar image are showed by high and low gray. So use the original histogram equalization method will lose many ground characteristics. In order to avoid such, the original histogram equalization should be changed to weaken middle tone little, and enhance the boundary tone.

A new simple but effective gray equilibrium has been raised in this paper. Calculating image histogram is the first step, how many pixels in each gray level (0–255) will be known. Than find the minimum gray level 'Min' in which the first pixel appears from level 0 to level 255. In the same way, find the maximum gray level 'Max' in which the first pixel appears from level 255 to 0. Than choose the gray level, to which the total pixels are first exceed 5.5 % of the whole image from 'Min' level, as the 'NewMin'. Also in the same way, choose the gray level, to which the total pixels are fist exceed 5.5 % of the whole image from 'Max' level, as the 'NewMax'. In the end, scan all pixel. If its grey level is less than or equal to 'NewMin', set its grey level as 0; if its grey level is more than or equal to 'NewMax', set its grey level as 255; the rest will be set as '(original level − 'NewMin') * 255/('NewMax'−'NewMin')'.

The image processed in this way will be shown as Fig. 3. Compare with Fig. 2, the image after gray equilibrium is much better. The difference between the new method and original histogram equalization is that the new method save the pixel gray level which belongs to 5.5 % of the whole pixels in the highest level and 5.5 % of the whole pixels in lowest level, at the same time, weaken the middle gray level.

Fig. 4 The image after mean processing

Though enlarged the image, we can see many speckle noise in the image, in this paper, the easiest mean processing is used to remove the noise. Compare Fig. 4 with Fig. 3, although the mean processing is sample but works well. In view of the requirement that smooth rolling needs speed data processing, other more complex method will not be considered, even they may have better treatment effect.

2.3 *Image Mosaic*

At present image mosaic algorithm [10] has already been very mature. The three main types are the mosaic algorithm based on region gray level correlation calculation, the mosaic algorithm based on attribute correlation, and the mosaic algorithm based on similar interpretation.

Because of the images are always rolling, the requirement for image mosaic precision is not very high, but for the image processing speed is high. The image mosaic for the image in same trip is always along the vertical direction, and the overlapping regions are small in the boundary region of image. So the mosaic complexity is low and the calculation is little. For these feature, the mosaic algorithm based on region gray level correlation calculation is very suitable. The characteristics of this method are simple and well registration performance while there are not large deformations, and also good adaptability. The image data used for this method should keep the original gray, so the data after compressing is suitable. The iterative search method based image pyramid has been used for this mosaic algorithm in this paper. In this method, the rectification begins from the lowest level, and set the level as the center for next level rectification, until up to original image level (the pyramid bottom). In this way, the calculation will be reduced.

Fig. 5 The image after mosaic

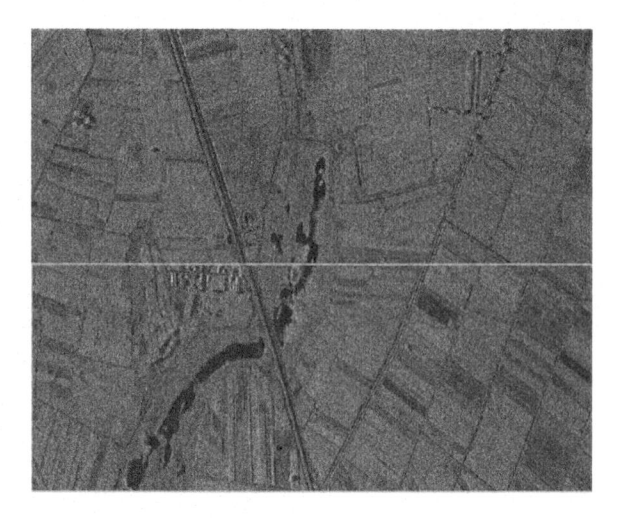

Figure 5 is the partial close-up of mosaic image of the adjacent image. The mosaic effect is satisfactory. Even there are small offset, but it's difficult to find during the image rolling. So the mosaic algorithm based on region gray level correlation calculation can meet the requirements of smooth rolling and mosaic effect. Though description all respect of this solution combining with the demonstrations, the solution has been proved to be effective for solving the problem of radar images smooth rolling.

3 Conclusion

A new solution has been raised for radar image rolling demo in this paper. The solution has been proved to be effective by detailed description and demonstrations. It not only has applicative value, but also is instructive and meaning full to the related project. Moreover, a new method for image compression and an improved method for gray equilibrium, have been raised aiming at the features of radar image and the combining requirement of image smooth rolling. Although these methods have strong pertinence, they will work well in other projects with some modifications.

References

1. Zhang Xin (2011) Radar image display and processing[D]. Dalian Maritime University, Dalian
2. Dong Pengshu, Zhange Chaowei, Jin Jiagen, Xie Youcai (2009) Compression and implementation of high resolution radar signals[J]. Syst Eng Electron 31(1):54–56
3. Pan Shoudao, Zhu Jie (2012) The research of a new radar image compression method based on BAQ algorithm[J]. Electron Meas Technol 35(2):65–68

4. Wang Renlong (2009) The research on radar image compression based on wavelet transform [D]. Harbin Engineering University, Harbin
5. Gonzalez RC, Woods RE (2003) Digital image processing[M], 2nd edn. Prentice Hall, London, p 793
6. Zhang Hongbing (2009) Research on algorithm of remote sensing image mosaics[J]. Xidian University, Xi'an
7. Jenkinson M, Smith S (2001) A global optimization method for robust affine registration of brain images[J]. Med Image Anal 5(2):143–156
8. Andrews GR (2002) Foundations of multithreaded, parallel, and distributed programming [M]. Addison Wesley, New Jersey, p 664
9. Guo Hongyan (2010) Research on synthetic aperture radar image enhancement and detection technology[D]. University of Electronic Science and Technology of China, Sichuan
10. Zhou Juntai (2011) Studies of technologies and methods on image mosaic[D]. Hunan University of Technology, Hunan

Index of Volume 1

A
An-cheng Wang, 59–66
Anqi Shangguan, 195–200
Antang Zhang, 185–193

B
Baoliang Guo, 3–12
Bing Chen, 383–391
Bing Liu, 85–91
Bing Luo, 59–66
Bing Xu, 457–465
Bo Fan, 185–193

C
Chang-liang Liu, 541–547
Changchun Tang, 219–229
Chao Xu, 335–340
Cheng Lin, 153–160
Cheng'an Wan, 427–437
Chongru Liu, 239–247
Chuantao Li, 509–515
Chunjie Chen, 75–82
Cunbo Jiang, 67–73

D
Danjing Li, 457–465
Danzhen Gu, 75–82
Dawei Meng, 93–103
Deying Yi, 427–437
Ding Wang, 47–57
Dong-feng Song, 59–66
Dun-wen Song, 211–218

E
Eng Gee Lim, 319–323

F
Fan Yang, 67–73
Fang Du, 363–369
Fei Guo, 439–446
Fei Yang, 67–73
Feifei Zhang, 419–426
Feng Hong, 563–570
Feng Sun, 259–266
Feng Wang, 555–562
Feng Yan, 211–218
Fengqing Qin, 293–301
Funaki Tsuyoshi, 541–547

G
Gang Deng, 23–33
Guang Tian, 127–134
Gujing Han, 491–500
Guo-jun Zhao, 373–381
Guoliang Zhan, 449–455
Guoquan Ren, 127–134

H
Haifeng Li, 239–247
Haijian Shao, 115–125
Haijian Zhuo, 177–184
Haijiang Wang, 169–174
Haijun Wang, 525–532
Hang Xu, 373–381
Haodong Ma, 383–391

He Zhu, 419–426
Heguo Hu, 325–332
Heng Wan, 457–465
Heyou Cheng, 267–273
Hongbo Dong, 161–167
Hongbo Li, 363–369
Honglin Gao, 533–538
Hongluan Zhao, 231–236
Hongxia Zhan, 275–282
Hongzheng Fang, 383–391
Hua Qiu, 23–33
Huali Sun, 563–570
Hui Guo, 85–91
Huifan Xie, 525–532
Huiqun Zhao, 449–455

J
Jialiang Li, 533–538
Jialiang Wu, 185–193
Jiamei Liu, 363–369
Jian Zhang, 419–426
Jianan Lou, 509–515
Jianbing Meng, 249–257
Jiangchuan Niu, 185–193
Jianhua Wang, 457–465
Jianhua Yu, 509–515
Jianlei Shi, 311–317
Jianshe Liu, 185–193
Jiayan Zhang, 177–184
Jiazhi Yang, 67–73
Jie Chu, 509–515
Jin Ma, 283–291
Jinfei Tang, 533–538
Jing Li, 343–350, 439–446
Jinhua Liu, 127–134
Jinyong Yin, 353–361
Jinze He, 93–103
Jinzhi Sun, 85–91
Jinzhuang Lv, 525–532
Junfeng Cui, 115–125

K
Ka Lok Man, 319–323
Kai Ding, 467–473
Kaiyu Wan, 319–323
Kangyi Zhang, 219–229

L
Lei Guo, 145–152
Lei Yu, 427–437, 467–473
Li Ma, 501–508

Li-ming Tu, 343–350
Li Yuan, 353–361
Liang Zhang, 107–113
Liang Zhao, 419–426
Lichen Shi, 15–22
Lifang Pei, 335–340
Lijing Tong, 449–455
Lili Wu, 411–417
Lingling Zhang, 153–160
Long Wang, 15–22

M
Meng Cheng, 135–143
Meng Zhang, 555–562
Mengdi Wang, 311–317
Ming Lei, 467–473
Ming-ming Jiang, 59–66
Mingyue Ma, 47–57

N
Nan Zhang, 319–323
Ningping Yao, 161–167

P
Pengfei Liu, 517–524
Pengfei Tian, 239–247
Pu-hua Wang, 59–66

Q
Qi-rong Qiu, 311–317
Qian Miao, 211–218
Qing-guo Zhang, 135–143
Qingchao Zhang, 35–45
Qiurong Li, 259–266
Qiushui Yu, 85–91
Quanxi Li, 411–417
Quanyao Peng, 449–455

R
Rui Hou, 153–160
Ruihua Zhang, 267–273
Ruining Yang, 85–91
Ruiqing Ma, 35–45

S
Shanlin Yang, 169–174
Shengli Yi, 67–73
Shi-ying Ma, 211–218

Shimeng Cui, 383–391
Shuqiang Li, 249–257
Suping Wu, 363–369

T

Tao Zheng, 439–449
Ting Cao, 501–508
Tong Wang, 573–579
Tong Zhen, 555–562
Trillion Q. Zheng, 427–437

W

Wangsheng Liu, 201–208
Wei Wang, 35–45, 239–247
Weidong Ma, 401–410
Weini Zeng, 353–361
Wen-hui Zhao, 283–291
Wentao Ruan, 275–282
Wenyan Guo, 501–508
Wuzhi Min, 491–500

X

Xiang Li, 319–323
Xiangli Kang, 35–45
Xiangqiang Liu, 525–532
Xiao Liu, 483–489
Xiaobin Li, 457–465
Xiaohua Li, 201–208
Xiaojuan Bai, 201–208
Xiaoliu Shen, 501–508
Xiaoming Zhang, 555–562
Xiaona Sun, 475–481
Xiaoqian Lu, 517–524
Xiaoxiao Cheng, 419–426
Xin-yu Zhang, 541–547
Xin Zhang, 303–310
Xinbo Yao, 363–369
Xingming Fan, 67–73
Xinhua Wang, 533–538
Xinke Ma, 501–508
Xiu Yang, 75–82
Xuanming Zhao, 3–12
Xuemei Hu, 231–236
Xueyan Bai, 517–524
Xugang Feng, 177–184

Y

Ya-nan Liu, 211–218
Ya'an Li, 201–208
Yafeng Yao, 161–167
Yanpeng Wu, 145–152
Yaofeng Xue, 563–570
Yazhou Zhang, 249–257
Yifan Li, 449–455
Ying Wu, 145–152
Yinghua Yang, 533–538
Yong Cheng, 549–554
Yongli Wang, 393–399
Yongming Xu, 93–103
Youcheng Wang, 219–229
Yu Tian, 533–538
Yue Guo, 467–473
Yuejun Li, 231–236
Yugang Qin, 107–113
Yuguang Liu, 15–22
Yuhong Zhao, 325–332
Yunguang Qi, 127–134
Yunhong Xia, 491–500
Yunhui Zhang, 325–332

Z

Zeng-ping Wang, 343–350
Zengping Wang, 439–446
Zhao Wang, 319–323
Zhaohui Xu, 555–562
Zhaoshuo Wang, 525–532
Zhendong Liu, 231–236
Zheng-tuo Zhang, 343–350
Zheng-zhong Zhang, 283–291
Zhenhua Kang, 249–257
Zhenpeng Xu, 353–361
ZhenYa Wang, 549–554
Zhi-xia Zhang, 541–547
Zhigang Ao, 219–229
Zhipeng Zhang, 231–236
Zhiping Jia, 267–273
Zhiqiang Li, 249–257
Zhishan Duan, 3–12
Zhixia Zhang, 303–310, 483–489
Zhiyong Li, 427–437
Zhongjie Shen, 161–167
Zhonglei Chen, 75–82
Zuowei Chen, 239–247

Index of Volume 2

A
An, X., 711

B
Bai, T., 745
Bai, Y., 985

C
Cao, Y., 657
Chen, J., 783
Chen, L., 995
Chen, M., 665
Chen, X., 683, 985, 1003
Cheng, J., 649
Cheng, Y., 925

D
Deng, M., 1055
Ding, L., 609
Dong, H., 833
Du, X., 955
Duan, Z., 1013

F
Fang, W., 825
Feng, W., 859, 917
Funaki, T., 909

G
Gao, P., 1063
Guo, S., 675

H
Han, J., 1047
Han, Y., 745, 1013
Heng, H., 793
Hu, X.-p., 1029
Hu, Y., 665, 675
Huang, Y., 657, 737

J
Jia, G., 825
Jiang, J.-l., 867
Jiang, W., 843
Jin, L., 851

K
Kang, X., 683

L
Lei, J., 977
Lei, Q., 755
Li, B., 701, 811, 977
Li, G., 649, 737
Li, H., 737
Li, J., 851
Li, M.-z., 891
Li, X., 875, 1019
Li, Y., 693, 701
Li, Z., 891
Liao, Q., 583
Liao, X., 1041
Lin, M., 755
Liu, C.-l., 909
Liu, F.-h., 617

Liu, J., 683
Liu, X., 601, 925
Lou, L., 641
Lu, J., 883, 891
Luo, B., 1029
Luo, F.-Z., 665

M
Ma, H., 727
Ma, X.-p., 617

N
Nie, L., 843

P
Peng, Y., 701

Q
Qi, B., 701
Qi, Z., 977
Qiao, Z., 967
Qin, K., 899, 1047
Qin, Z., 711

S
Shan, L., 825
Shen, J., 941
Shen, Y.-j., 867
Shi, Y., 755
Shi, Z., 825
Si, H., 583
Si, Y., 783
Song, D.-f., 1029
Sun, C., 1013
Sun, M., 765

T
Tan, C., 609
Tang, K.-h., 1029
Tao, X., 801
Tu, G., 883

W
Wang, A.-c., 1029
Wang, C., 629
Wang, H., 601, 941

Wang, K., 891
Wang, M., 985
Wang, P., 641
Wang, P.-h., 1029
Wang, Q., 1041, 1047
Wang, S., 773
Wang, X., 591, 657, 825, 933, 947
Wang, Y., 811
Wang, Z., 925
Wei, H., 917
Wu, H., 883
Wu, X., 719
Wu, Y., 851

X
Xiao, S., 683
Xie, R., 591
Xiong, D., 817
Xiong, H., 793
Xu, H., 773, 967
Xu, M., 711
Xu, Q., 817
Xu, X., 801
Xu, Y., 1013
Xue, H., 773
Xue, X., 833

Y
Yan, T., 801, 1041
Yan, X., 899
Yang, B., 773
Yang, J., 843
Yang, S., 941, 1019
Yang, W., 1055
Yang, Z., 843, 985
Yao, C., 657
Yao, M., 947
Ye, Q., 891
Yin, H., 591

Z
Zeng, D., 955
Zhang, C., 629
Zhang, M., 917
Zhang, S., 629
Zhang, S.-Y., 665
Zhang, W., 583
Zhang, X.-y., 909
Zhang, Y., 955, 1003
Zhang, Z.-j., 617

Zhang, Z.-x., 909
Zhao, D., 833
Zhao, E., 755
Zhao, Q., 719
Zhao, X., 899
Zheng, Y., 591
Zhou, M., 719

Zhou, Q., 875
Zhou, Y., 977
Zhu, C.-q., 891
Zhu, H., 701, 925
Zhu, T., 801
Zhu, W., 1055
Zou, Y., 883

Printed by Printforce, the Netherlands